RoManSy 6

Proceedings of the Sixth CISM-IFToMM Symposium
on Theory and Practice of Robots and Manipulators

Edited by
A. Morecki, G. Bianchi and K. Kędzior

RoManSy 6
Proceedings of the Sixth
CISM-IFToMM Symposium
on Theory and Practice
of Robots and Manipulators

Edited by
A. Morecki, G. Bianchi and K. Kędzior

Sponsored by the CISM-Centre International des Sciences Mécaniques, IFToMM-International Federation for the Theory of Machines and Mechanisms, in association with the IVth Technical Division of the Polish Academy of Sciences

Co-sponsored by the Institute for Aircraft Engineering and Applied Mechanics, Warsaw University of Technology, Technical University of Cracow, Technical University of Mining and Metallurgy in Cracow

The MIT Press
Cambridge
Massachusetts

First MIT Press edition, 1987

© 1987 Editions Hermes

All rights reserved. No part of this book may be reproduced in any form by any electronic or mechanical means (including photocopying, recording, or information storage and retrieval) without permission in writing from the publisher.

This book was set in France
and printed and bound by Imprimerie Laballery in Clamecy (France)

Library of Congress Cataloging-in-Publication Data

RoManSy (6th : 1986 : Udine, Italy)
  Theory and practice of robots and manipulators.

  Includes bibliographies.
  1. Robots — Congresses. 2. Manipulators (Mechanism) — Congresses. I. Morecki, Adam. II. Title.
TJ210.3.R66 1986   629.8'92   87-3839
ISBN 0-262-13226-5

Contents

Organizing and Programme Committee 9
Editorial Note 11

Part 1 Opening Session / Jean Vertut Memorial Session
Eulogy 15

Advanced Teleoperation.Introductory Paper.
The Advanced Teleoperation Project 19
B. Espiau
Advanced Teleoperation (I).
Control and Supervision in Computer Aided Teleoperation 28
P. Gravez, B. Lepers, R. Fournier and G. André
Advanced Teleoperation (II). The Generalized Information Feedback
Concept in Computer-Aided Teleoperation 43
G. André and A. Fournier
Advanced Teleoperation (III). An Integrated Experiment 57
R. Fournier, J.P. Gaillard, N. Fiori and B. Espiau
Analysis of a Robot Wrist Device for Mechanical Decoupling of the End-
Effector Position and Orientation 68
J.C. Guinot and P. Bidaud

Part 2 Mechanics 1 79
Robot Motion: Configuration Analysis of Redundant
and Non Redundant Manipulators 81
M.S. Konstantinov and D.N. Nenchev
Analysis of the Positioning and Orientation Accuracy
in 6R Manipulators (Direct Task) 90
J. Knapczyk and A. Morecki
A Unified Approach to Modelling of Flexible Robot Arms 99
J. Rauh and W. Schiehlen
Solving the Inverse Kinematic Problem for Robotic Manipulators .. 107
L. Sciavicco and B. Siciliano
Determination of the Accuracy of Flexible Automatic
Positioning Module with Clearances 115
V. Natbiladze
Invariant Kinestatic Filtering 120
H. Lipkin and J. Duffy

Part 3 Mechanics 2 129
Redundant Manipulators and Kinematic Singularities
The Operational Space Approach 131
O. Khatib
Modelling and Simulation of Mechanical Process in Hyperstatical
Gripping with n-Contact Points 139
S. Zeghloul, J.P. Lallemand and D. Murguet
Computer Aided Modelling of Pneumo-Hydraulic Robots 148
C. Rzymkowski
A Cartesian Model of Manipulator Kinematics 155
M. Galicki
A Method for Solving the Inverse Problem of Kinematics of Anthropomorphic
Manipulators with Spherical Wrist 164
Peng Shang-xian and Wang Gang
Dynamic Equations of General Robots by Kane's Method 169
Peng Shang-xian and Wang Gang
The Role of Delay in Robot Dynamics 177
G. Stepan

Part 4 Synthesis and Design 1 185
Smart Hand Systems for Robotics and Teleoperation 187
A. K. Bejczy and B. M. Jau
A Mathematical Model of a Flexible Manipulator
of the Elephant's-Trunk-Type 198
G. Malczyk and A. Morecki
Analytical Design of Two-Revolute Open Chains 207
B. Roth
On a Fundamental Study of Micro Mechanical Gripper
Using Shape Memory Alloy (SMA) Actuator 215
K. Matsushima and N. Usui
The Kinematic Design and Mass Redistribution of Manipulator
Arms for Decoupled and Invariant Inertia 221
H. Asada
Graphical-Interactive System for CAD and Simulation
of Manipulation Systems 246
L. Lilov, B. Bekjarov, M. Lorer and V. Atanasov

Part 5 Sensing and Machine Intelligence 1 253
Force Feedback in Telemanipulators 255
R. Bicker and L. Maunder
Theoretical and Experimental Investigations of Optical Fibre
Reflective Sensors for Robotics 267
E. Marszalec, J. Marszalec and A. Morecki
Task Specification and Closed Loop Control of Manipulators
in the Presence of External Sensors 275
R. Palm, A. Moltmann and H.J. Horch
Adaptive Force Control of Grippers Taking into Account
the Dynamics of Objects 283
T. Fukuda, N. Kitamura and K. Tanie
Bilateral Remote Control with Dynamic Reflexion 296
K. Tanie, K. Komoriya, M. Kaneko, T. Ohno and T. Fukuda

Part 6 Control of Motion 1 309

Finger-Arm Coordination Control Method
for Multiple Degrees of Freedom Robot 311
S. Sugano and I. Kato
A Model-Based Expert System for Strategical Control Level
of Manipulation Robots 322
D. Stokić and M. Vukobratović
Robot-Task Adaptability by Semi-Local Correction without Contact
A. Jutard, T. Redarce, J.F. Chabrier and G. Liégeois
Robot Control Synthesis in Conjunction with Moving Workpieces 342
P. Kiriazov and P. Marinov
Dynamic Command Motion Tuning for Robots. A Self Learning Algorithm . 350
G.W. Vernon, J. Rees Jones and G.T. Rooney

Part 7 Sensing and Machine Intelligence 2 359

C-Surface Theory Applied to Force-Feedback Control of Robots 361
J.P. Merlet
Experimental Investigation of Active Force Control
of Robot and Manipulator Arms 369
G. Galatis, A. Hadzistylis and J.R. Hewit
Automatic Grasp Planning. An Operation Space Approach 380
M. T. Mason and R. C. Brost
A Method of Optical Processing in the Robot Vision 388
B. Macukow
Tridimensional Optical Syntaxer 395
A. Oustaloup and P. Melchior

Part 8 Locomotion and Walking Machines 403

Towards Generalized Concepts and Tools for Unconventional Mobile Robots.
General Languages, Mobility Modes 405
J.J. Kessis, J. Penne, J.P. Rambaut and N. Mattar
Mobile Robotic Systems for Use in Unstructured Terrain 412
K.J. Waldron and S. Agrawal
Wall Climbing Vehicle Using Internally Balanced Magnetic Unit 420
S. Hirose
Experimental Development of a Walking Transport Robot 428
B.D. Petriashvili, M.A. Bilashvili and V.O. Margvelashvili
Legs that Deform Elastically 436
H.B. Brown Jr and M.H. Raibert
Features of Mechanisms Synthesis of Walking Robot Propelling Agents . . . 444
V. V. Korenovski and A. Ja. Pogrebnjak
Avoiding Obstacles by a Mobile Robotized Vehicle 455
M. Badida and J. Buda

Part 9 Application and Performance Evaluation 463

The Automation of the Mine Support Erection Technology
with Remotely-Controlled Manipulators 465
Yu. A. Tzeitlin, V. Ya. Potyomkin and L.N. Prokopishin
Experimental Investigations of Robots and Manipulators 472
B. Heimann and S. Tschakarow
Minimization of Vibrations of a Gantry Manipulator During Positioning . . . 480
K. Tomaszewski and A. Golaś

Experimental Evaluation of Feedforward and Computed Torque Control . . 488
C.H. An, C.G. Atkeson, J.D. Griffiths and J.M. Hollerbach
Experimental Research and Development of Methods for Improving
Kinematic and Dynamic Robot Characteristics 496
A.N. Ananjev, E.G. Ananjeva and E.G. Nakhapetjan

Part 10 Synthesis and Design 2 505
Kinematics and Torque Control of Multi-Fingered Articulated Robot Hand . 507
A.E. Samuel and P. Ridley
Progress towards a Robotic Aid for the severely Disabled 517
S. Michalowski
Logical Structures for Collision Avoidance in Assembly with Robots 524
A. Rovetta
Repositioning-Unit for very Fine and Accurate Displacements
Analysis and Design 535
F. Artigue, C. François and J.G. Pontnau

Part 11 Synthesis and Design 3 543
Polyarticulated Mechanical Structure for Decoupling
the Position and Orientation of a Robot 545
M. Fayet and A. Jutard
Application of l-Coordinates in Robotics 555
A. Sh. Koliskor and V.I. Sergeyev
Design of Spring Mechanisms for Balancing the Weight of Robots 564
J.M. Hervé
Structural and Geometrical Systematization of Spatial Positioning
Kinematic Chains Employed in Industrial Robots 568
Fl. Duditza, D. Diaconescu and Gr. Gogu
Tasks and Methods of Constructing Mechanical Facilities
and Control Systems of Industrial Robots Taking into Account their Force
Interaction with the Equipment 576
S.N. Kolpashnikov and I.B. Tchelpanov

Part 12 Control of Motion 2 589
Contribution to Solving Dynamic Robot Control in Machining Process 591
M. Vukobratović and D. Vujić
An Approach to Development of Real-Time Robot Models 603
N. Kirćanski, M. Kirćanski, M. Vukobratović and O. Timénko
Time-Optimal Robotic Manipulator Task Planning 615
S. Dubowsky, M.A. Norris and Z. Shiller
Time-Optimal Motions of Some Robotic Systems 623
L.D. Akulenko, N.N. Bolotnik, F.L. Chernousko and V.G. Gradetsky
Frequency Space Synthesis of a Robust Dynamic Command 633
A. Oustaloup and B. Bergeon
Structure Strategy Problem on a Redundant Manipulator 642
H. Asama, M. Onosato and H. Yoshikawa

Participants 653

Organizing and Programme Committee

Chairman:
Prof. A. Morecki
   Warsaw University of Technology, Al. Niepodległości 222, r.212, 00-663 Warsaw, Poland

Vice Chairman:
Prof. G. Bianchi
   CISM, Piazza Garibaldi 18, 33100 Udine, Italy

Members:
Prof. A. P. Bessonov
   Academy of Sciences of the USSR, Griboedova Street 4, Moscow-Centre, 101000, USSR
Prof. I. Kato
   Waseda University, Faculty of Science and Engineering, Ookudo, Shinjuku-ku, Tokyo 182, Japan
Prof. A. E. Kobrynskij
   Academy of Sciences of the USSR, Griboedova Street 4, Moscow-Centre, 101000, USSR
Prof. M. S. Konstantinov
   Central Laboratory for Manipulators and Robots,
   P.O. Box 97, Darvenitza, Sofia, Bulgaria
Prof. H. Rankers
   Bedrijfsmechanisatie, Landbergstraat 3,
   2628 CE Delft, The Netherlands
Prof. B. Roth
   Stanford University, Dept. of Mechanical Engineering,
   Stanford, California 94305, USA
Dr R. D. Schraft
   Fraunhofer Institute for Production and Automation, University of Stuttgart, P.O. Box 951, Stuttgart, FGR
†Dr J. Vertut
   Commissariat à l'Energie Atomique, B.P. n° 2,
   91190 Gif-sur-Yvette, France
Prof. M. Vukobratović
   Institute "Mihailo Pupin", Volgina 15, P.O. Box 906, Beograd, Yugoslavia

Scientific Secretary:
Dr K. Kędzior
   Warsaw University of Technology, Al. Niepodległości 222, r.216, 00-663 Warsaw, Poland

Secretary:
Dr A. Bertozzi
   CISM, Piazza Garibaldi 18, 33100 Udine, Italy,
   Tel.: (0432) 294989 or 22523

Editorial Note

This volume contains the papers accepted for the Sixth Symposium on Theory and Practice of Robots and Manipulators "RoManSy '86" held in Cracow, Poland, 9-12 September 1986.

"RoManSy '86" was attended by 90 participants from 16 countries (as listed) who were selected experts in the field of robotics.

The Symposium Programme included:
- Opening and Closing Sessions attended by CISM, IFToMM and Polish officials,
- Jean Vertut Memorial Session with five lectures given by the French scientists, the followers and collaborators of Jean Vertut,
- Working Sessions (mechanics, synthesis and design, sensing and machine intelligence, control of motion, locomotion and walking machines, application and performance evaluation),
- Two film and video sessions.

The papers in this book are in the same sequence as the sessions mentioned above. All linguistic and terminology corrections have been kept to a minimum.

The Proceedings of the previous five Symposia are available in the final form. The proceedings of the "RoManSy '73" (Sept. 5-8, 1983, Udine, Italy) may be obtained from Springer-Verlag, Vienna or CISM, Udine, Italy. Those of the "RoManSy '76" (Sept. 14-17, 1976, Jadwisin, Poland) and "RoManSy '78" (Sept. 12-15, 1978, Udine, Italy) may be purchased from Elsevier (Amsterdam, Holland) or PWN-Polish Scientific Publishers (Warsaw, Poland). Proceedings of the "RoManSy '81" (Sept. 8-12, 1981, Zaborov, Poland) may be obtained from PWN (Warsaw, Poland). The Proceedings of the "RoManSy '84" (June 26-29, 1984, Udine, Italy) may be obtained from Kogan Page (London), Hermes Publishing, (Paris).

The next symposium "RoManSy '88" will be held in Udine, Italy, in September 1988.

The Editors express their thanks to Mr S. Menasce for his excellent cooperation in the publishing of the Proceedings.

A. Morecki, G. Bianchi and K. Kędzior

Part 1
Opening Session

Jean Vertut Memorial Session

Eulogy

Jean Vertut was born in Paris in 1929.

He was a brilliant student at Chaptal College and went to the Ecole centrale des arts et manufactures de Paris (Central School for Arts and Factories of Paris) in 1950. He left school as a graduate engineer in 1953, gaining there a reputation for vitality and unconventionality among his colleagues and professors.

He first held a minor position in a firm where he had no change of giving the best of himself. So he quitted and began working for the CEA on 24th September 1956. This gave rise to a tremendous vitality that Jean Vertut fully exploited by bringing up new concepts in mechanics and more generally in electromechanics within "the Radiation Control and Atomic Engineering Service" (Service de contrôle des radiations et de génie radioactif).

He windely contributed to the elaboration of the now internationally accepted standards on protection and transfers in nuclear plants. Jean Vertut's technical achievements are not only due to his ingenious mind and scientific knowledge but also to his human qualities.

As the concept of "protection" moved towards "intervention", he naturally passed from remote handing to teleoperation.

His imagination and ingeniousness kept producing all sorts of devices, mechanisms and prototypes to which numerous patents and industrial applications bear testimony. Some of them, as for instance the manipulator MA 11, are operating in thousands of cells throughout the world and are still being manufacturated.

In the mid sixties, Jean Vertut feels it is essential to step into a new technological phase: he wants to break free from the idea of physically linking of mechanical manipulators without losing their dexterity.

Automation and sensors impose, however, limitations on the machine's performances. With appearance of the MA-23 manipulator these problems were solved: the first prototype developed in 1975 is still considered as one of the best.

Jean Vertut was then able to understand that this technological leap forward offered vast possibilities of application in various environments, may they be hostile or not. Jean Vertut was then ahead of his time when he designed the Virgule machine (1970-72), and thus tackled the range of problems associated with the conception of today's mobile robots (locomotion, piloting, transmission, control, etc.). The basic concepts of robot automation operating in hostile environments were developed as early as that time, thus paving the way for the first MA 23 manipulator called OCEANO (1976) for the exploration of deep seas. He also helped to work out the national programme for handicapped people, the so-called SPARTACUS programme for which he designed the MAT 1 (1979) characterized by a new mechanical structure and amazing dynamic performances.

Yet these multiple areas of interest never damped his efforts in the field of nuclear research as is shown by his numerous inventions: MERITE (1977), MAM (1978), MIR (1981).

As the French programme ARA is launched in 1980, Jean Vertut is given the charge of Advanced Teleoperation. With this programme he demonstrates that computer aided teleoperation makes it possible for men and machines to perform tasks together and give good results. In this type of research concepts are almost more important than the purely technological considerations. And they put French robot automation in a leading position. Professor T.B. Sheridan writes in the introduction to the book by Vertut and Coiffet, *Remote operation and technological evaluation*, published by Hermes: "I personally think that Jean Vertut's laboratory and the associated teams of the CEA have taken the lead and that over the past few years the "French School" has showed the way, closely followed by our Japanese competitors, and perhaps more recently, by the US".

He was on his way along this road to success — a success he too scarcely shared the joys of — when he departed from this life on Saturday May 25th 1985 at age of 56.

Jean Vertut made a lasting impression on the international community of robotics. This eager and tireless traveller bubbling over with new ideas wove and enlivened a vast network of human relations throughout the world.

He was a member of various national and international organisations and he readily held conferences in congresses, associations and universities of all countries. This remarkable speaker made his point perfectly plain when talking about technology. The audience listened admiringly to his brilliant lectures, embellished by picturesque comparisons and a language of gesture. This man with a comprehensive mind was warm and enriching.

Moreover, he was interested in miscellaneous topics such as gesture anthropology (he was one of the directors of Marcel Jousse, an institution he ran until he died), pottery (he was a founding member and the treasurer of the Francine Dulpierre association), speleology, and of course archeology: his passion for prehistory and his profound knowledge of prehistoric iconography made him an undisputed expert in this field.

This enthusiastic humanist explained that his interest in prehistoric art involved both the beauty of the art and an opportunity through the art to understand man's unique place in nature.

We will keep in mind the lessons he taught us until his last day: he wanted us to be open-minded and mindful of others so that we could understand the world and the human beings beyond the scope of words.

R. FOURNIER

Jean Vertut joined the IFToMM Family in 1973. During the first CISM-IFToMM RoManSy '73 Symposium he presented an important paper on Teleoperators.

From that time Jean was very involved in IFToMM activity. As a member of our Technical Committee for Robots and Manipulators he was one of the most enthusiastic collegues, who helped us in preparation of the RoManSy events in 1976, 1978, 1981, 1984 and 1986. Step by step the French scientists followed him in this activity. RoManSy '86 may be the last example of French activity in robotics; over 16 papers were accepted for presentation and the symposium was opened by a special Jean Vertut Memorial Session. Jean's friends and collaborators are continuing and developing of his ideas.

Jean was the first President of the IFToMM Committee in France. He collected a large group of scientists interested in Theory of Machines and Mechanisms. During the last IFToMM World Congress, which was held in New Delhi on Dec. 1983, Jean Vertut was elected as a member of the Executive Council of IFToMM.

The memory of Jean as an excellent engineer, designer and outstanding personality, will be forever with us.

A. MORECKI

Advanced Teleoperation
Introductory Paper
The Advanced Teleoperation Project

B. Espiau

IRISA, Rennes, France

ABSTRACT

   This paper is an introduction to three further technical papers on advanced teleoperation. We first give a brief history of this research project, recording the part played by Jean Vertut. We present the basic concepts of advanced teleoperation, starting with the necessity for coexistence of autonomy and transparency, emphasizing the virtual level in relation to the real level. In the following section we introduce the technical papers, the first of which is devoted to generalized master slave control and supervisory systems, the second to generalized information feedback, and the third to a description of an integrated experimental test bed. The paper ends with some thoughts on the future of advanced teleoperation.

1. INTRODUCTION: A BRIEF HISTORY

   Teleoperation and robotics have as a common basis the fact that in both cases man acts on an environment through a machine. The main difference arises during the

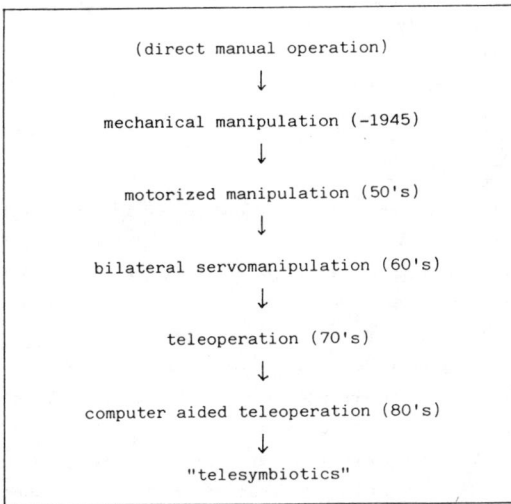

Figure 1

execution of a task: in teleoperation, man remains a significant part of the system, while in robotics man is removed from the system when it is working.

Historically, studies in teleoperation began much earlier than those in robotics, for two main reasons:
  (i) the necessity, just after the second world war, of devising remote handling devices for nuclear applications
  (ii) the unavailability before the sixties of the theoretical and technical tools which would enable the first robots to be constructed

The development of remote and teleoperator systems (figure 1) has always been closely related to the evolution of the available technology, and it may be expected that, in the future, teleoperation will rely more and more on the progress made in robotics, mainly in the development of decision and sensing capabilities. At the same time, the fields of application for teleoperation have extended from nuclear technology to underwater work, and, more recently, to spatial manipulation, including aids for disabled or handicapped people, and manufacturing applications. Anticipated extensions of the field might be concerned with rescue, service and intervention systems, thus entering the domains of so-called "third generation robotics" (see (8) for a complete history of teleoperation).

As is well known, in the field of teleoperation (research, development and applications), the earliest major contributions were made by a few individuals, including R. Goetz, A. Bejczy ((1)), E. G. Johnsen ((3)), T. Sheridan ((4)), some others, and, of course, Jean Vertut ((5) to (9)). A member of CEA (French Council for Nuclear Energy), and famous designer of remote vehicles and manipulators, Jean Vertut started in 1981 a research project in advanced teleoperation, within the framework of the national CNRS project: "Advanced Robotics and Automation."

Eight French research laboratories worked in an "Advanced Teleoperation Group," with the assistance of industrial observers ((5)). At its inception the aim was to study basic components of a computer-aided teleoperation system, specifically
  - task description and efforts display
  - sensor-based control
  - man/machine control and task sharing
  - performance measurement and evaluation
  - synthetic vision
  - system integration

It was decided to focus the above studies onto common test facilities, in order to gather all the contributions into a single system, and also to study the interaction between the subsystems. Over the last two years it has become clear that more complex levels of systems analysis could be achieved, such as supervisory control or generalized information feedback.

The project ended this year with the achievement of so-called "canonical experiments," performed on the teleoperation test site, which included all the functions of advanced teleoperation studied in the laboratories.

Unfortunately, Jean Vertut died on May 26th, 1985, so he will never see the final results of the research work he had inspired and led during these years. In spite of this sad event, the Advanced Teleoperation Group has carried on its work, and we dedicate the results here presented to the memory of Jean Vertut, in homage to a friend and an outstanding scientist.

The remainder of this introductory paper is organized as follows: in section 2, we briefly present the basic concepts of what we guess "advanced teleoperation" would be; in section 3, we introduce the three technical papers which describe the results of the group's work. Section 4 concludes the paper by examining possible perspectives in advanced teleoperation.

Note

References listed at the end of this paper concern only general aspects of teleoperation. Specialized references are provided in each of the following three papers.

2. SOME FACTS ABOUT "ADVANCED TELEOPERATION"

2.1 Background

A teleoperation system has to interact with an environment in a complex way. The worksite is generally located in a "hostile" world; "hostile" means that this world is not the world in which the human operator daily lives, or might easily perform the desired tasks. With regard to robotic tasks, the applications requiring teleoperators involve complex, non-repeatable and non-deterministic tasks. Moreover, a task of teleoperation is generally performed only once, or at most a few times.

The aim of "advanced teleoperation" is to increase the ability of the system to become more adaptable with regard to disturbances or unknown factors in the world, in order to become more efficient in performing the part of the tasks which may not be fully described in advance.

Let us start the analysis from figure 2, stated by Jean Vertut in (7), in which robots lie symmetrically on both sides of the automaticity axis. Over this axis

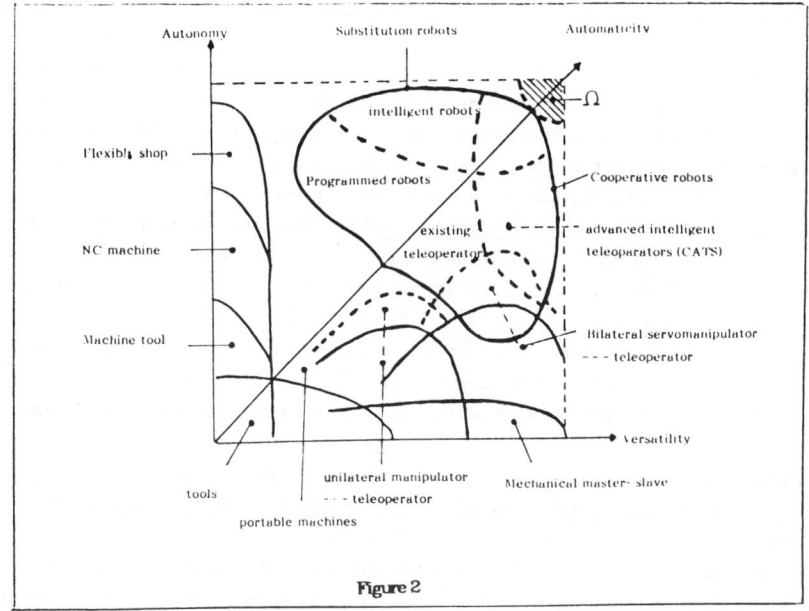

Figure 2

are substitution robots, and under it cooperative robots. The Ω zone is the ideal, where "intelligent" teleoperators and "intelligent" robots are closer to becoming the same thing.

More precisely, as in any robotics systems, three main "metafunctions" are included in a teleoperation system: perception, decision and action. The main difference comes from the fact that these functions have to be shared between man (or men) and machine(s); further, in advanced teleoperation, this function-sharing may vary

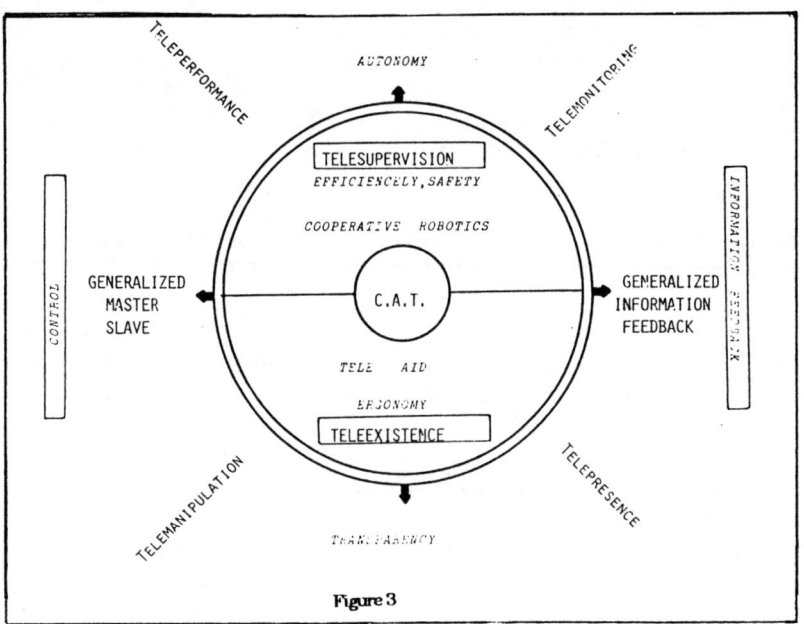

Figure 3

in time. Figure 3 ((10)) illustrates this point in more detail. North and South extremes in this figure respectively represent the aims of full-machine and full-man control. Notice that, in the first case, we need autonomy, which means: research of fully automatic perceptive and decision making techniques, whose efficiency would be sufficient at each control level to ensure the required adaptivity; while in the second, the objective is transparency of the system, assuming that the best result would be obtained by a man directly performing the task, if it were possible.

Advanced teleoperation, and its subset "computer aided teleoperation" (CAT) lies between the two, given that

Proposition 1: man without assistance may be unable to perform certain tasks for many obvious reasons

Proposition 2: full autonomy is unattainable at present (and within the near future)

The aim of studies in advanced teleoperation is to examine each stage of the process and to obtain an optimal exploitation of human and machine resources at all times.

The consequences of this for the executive and perceptive aspects of the system are shown on West and East parts of the scheme 3: the concept of generalized master slave control extends the telemanipulative techniques with various functions of assistance, but physical control by man remains present. Generalized information feedback includes the notion of telepresence, while extending it to non-anthropomorphic sensing.

2.2 Structure of a CAT System ((10))

Figure 4 shows a general description of an ideal advanced teleoperation system.

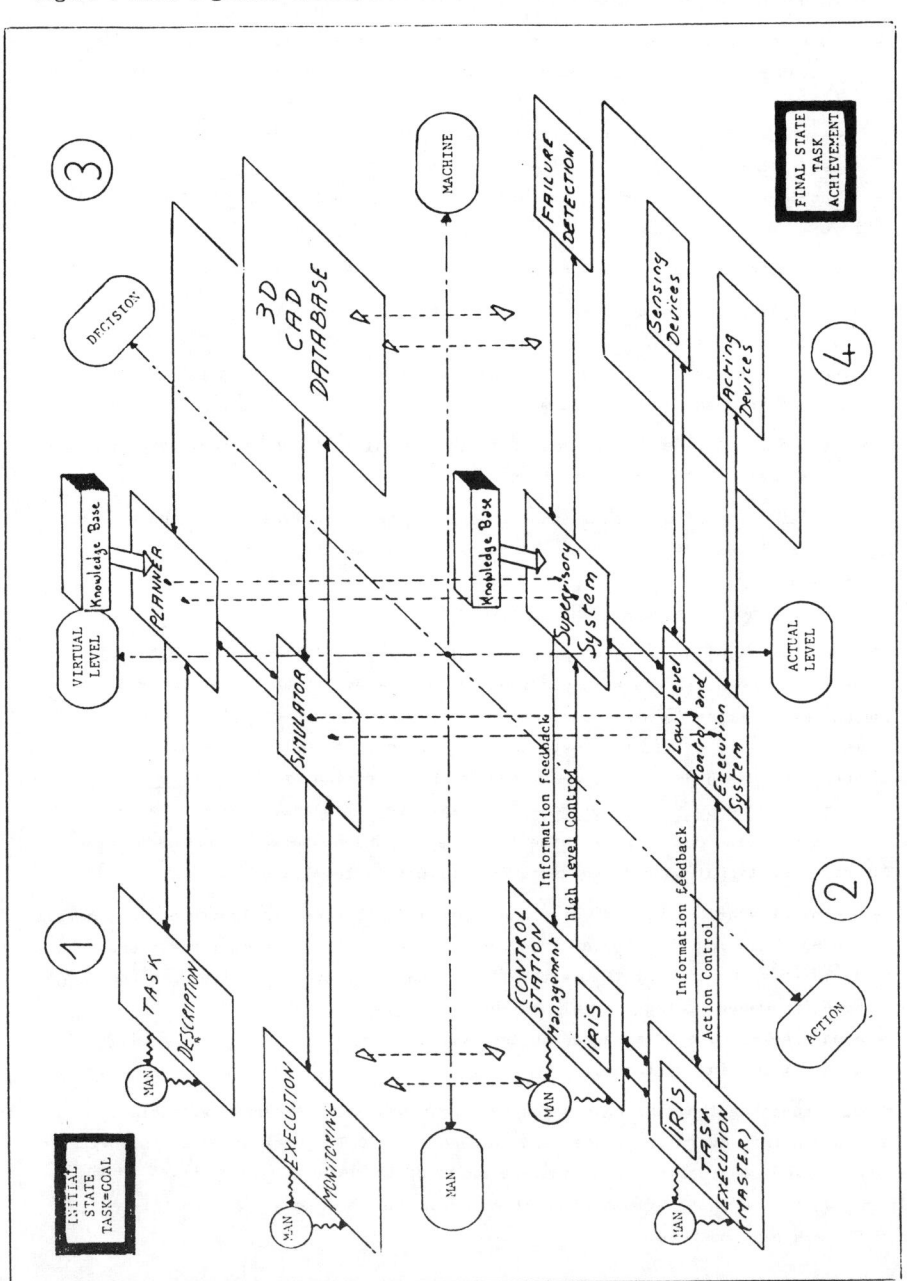

FIGURE 4

It is immediately apparent that a supplementary dimension is added by introducing the notion of a "virtual system," in parallel to the actual one, with its own machine representation, and all the associated functions of off-line simulation, (task preparation, and knowledge integration); and on-line intervention in the actual system, (for the purposes of planning, controlling and displaying, while at the same time possibly updating the knowledge and data bases).

The main functional subsystems associated with this virtual level generally are:

- a 3D environmental database with the corresponding CAD facilities
- a system representation allowing the simulation of interactions with the environment
- a knowledge base associated with a task descriptor and planner
- a man/machine interface for off-line communication
- a machine/machine interface for bidirectional communication with the "actual" level

Let us outline two of the essential features of this set of subsystems:

1. The ability to perform realistic task simulation for the purposes of validation, evaluation or training
2. The possibility of updating all knowledge and data bases from information derived from the actual system

The second level thus is concerned with the actual world and actual task achievement. Here, CAT involves two main things:

1. The existence of a complex (and variable) "filter" between human operator(s) and final tasks (i.e. between master and slave worlds in the classical terminology).
2. The presence of several control levels

In fact, we need, not only to perform real-time manipulation tasks, but also to manage all the other elements provided by the system, which is not limited to classical master/slave control and simple force/vision feedback. It is obvious that the complexity of all the instantaneous working possibilities requires a specific management and decision system, dedicated to the selection of both pertinent information and adequate control modes (for example bilateral, sensor-referenced, autonomous modes). But this decision level may be shared between man and machine.

The main functional subsystems associated with this level are:

- a control station, with display and control facilities, including action, motion and manipulation oriented devices, and with a facility for multiple-operators
- a real-time control system, with all the required levels, including a knowledge base for managing the control station
- a machine/machine interface with the "virtual" level
- a set of sensing systems

without forgetting the working devices (manipulators, mobile carriers, tools).

Let us end this overview of the main features of a CAT system with the following remark: if we consider the four extreme parts of the scheme 3 as numbered in the figure, we notice that, by combining them in various ways they will constitute specific systems, such as:

1,2,3,4:   Complete CAT system
1,(2),3:   Simulation or training system
1,3,4,:   Robotics system
2,4:   Teleoperator (from classical bilateral M/S to teleexistence system)

This shows the generality of the proposed so-called CAT scheme. Unfortunately, such a system does not fully exist at present; in the advanced teleoperation project, we have studied some basic components of the ideal structure, focusing on control aspects and information feedback problems, through a common experimental background. These studies are reported in the following three papers, which we now briefly introduce.

3. INTRODUCTION TO THE RESULTS OF THE "ADVANCED TELEOPERATION PROJECT" ((6)),(9))

3.1 Control and Supervision

This paper is devoted to the problem of sharing control between man and machine. The aim of the realized system was to improve both reliability and accuracy in task execution while reducing the stress and physical worktension of the human operator. Two aspects constitute the proposed approach:

The first ("man-oriented") aspect is the "generalized master/slave control," a bidirectional process based on both aided-effector control and generalized force feedback. Various classes of local working modes may be provided.

The second ("task-oriented") aspect is an interactive supervisory system which allows us to select dynamically the control modes which are, at each instant, the most efficient for a given task. This supervisory system, which reduces the human overhead involved in managing the complex set of modes may be understood as a first step towards a general "operating system for advanced teleoperation."

4. CONCLUDING REMARKS

This helps clarify some basic concepts in advanced teleoperation. The implementation on a common test site, through canonical experiments, performed in real time, has allowed us to start experimental evaluation of the "systems" aspects.

However, limits of the present version have also been pointed out; several components or subsystems are not yet well defined. This is why further studies will have to integrate supplementary tools, implicitly present in the general scheme of figure 4, but not yet ready to be implemented in the actual test plant:

4.1 Tools from Computer Science:

It may be shown that increasing the capabilities of a CAT system and thus its complexity, quickly leads to a dramatic overload for human operators. Further, in the same way as elementary tasks can be more or less controlled by a machine, high level decision tasks have to be integrated into an advanced teleoperator, with two main aspects: first, task description tools, decision systems and planners are required for robotics-like tasks (navigation, automatic sub-tasks). Further, some specific features of teleoperation also require tools of knowledge engineering, to select at each time the control modes, the I/O devices, the relevant information feedback; to help conduct highly complex tasks; to manage and update complex knowledge and data bases, etc. Tools of artificial intelligence, like expert systems, are thus expected to play an important role in future advanced teleoperators.

At the hardware level, it may be noticed that among the various required communication channels, some demand high performance in data flow transfer; further, as man remains somewhere in the loop, and also as some applications may require a fast response time, it is not possible in some of the system to tolerate significant delay. For these reasons, the low level of advanced teleoperators needs high performance networks and communication protocols taking into account major real-time constraints.

4.2 <u>Tools from Man and System Science</u>:

An important part of the activity of a future advanced teleoperator will involve simulation and training operations. This requires, on one hand, accurate models of the elements of the system and of their connections, and on the other, tools for creating, updating, and using such models. An example of an application which needs accurate process models is found in earth-to-orbit teleoperation, where transmission delays are significant.

We may divide the main useful models into three classes:
- dynamic models of man/machine loops; for example, the dynamic properties of the generalized master/slave control modes
- decision-making models; this, for example, is necessary for building expert systems
- system models: 3D databases, models of communication channels, of failures and disturbances, etc.

Moreover, it is necessary not to forget to use the concepts and the methods of ergonomy at two levels: in an a-priori way, to contribute to the design of man-machine communication systems, and a-posteriori, for the evaluation of advanced teleoperator systems, the improvement of their performances, and the adjustment of some parameters of the previously mentioned models.

From a practical point of view, this requires the implementation of updated technologies; real-time 3D database management and image synthesis, efficient communication networks, high performance AI machines.... For those reasons it is obvious that we are at the first stage of advanced teleoperation. However, as the need for these is increasing, in particular with regard to space applications ((2)) and for several projects on remote intervention systems, we hope that the realization of truly intelligent advanced teleoperators will be effective in the next ten years.

Let us end this paper by recalling that this research project was launched by Jean Vertut, who was its leader until his death. We hope that the last part of the project has been carried out with fidelity to his ideas, and we all thank him for having led us to share both his passion and his knowledge of robotics and teleoperation.

REFERENCES

(1) BEJCZY,A.K.: Robotics as Man Extension Systems in Space, 9th Symposium IFAC, Budapest, July 1984
(2) JENKINS, L.M.: Telerobotic Work System Concepts, AIAA/NASA Symposium on Automation, Robotics and Advanced Computing for the National Space Program, Washington, September 1985
(3) JOHNSEN, E.G., CORLISS, W.R.: Human Factors Applications in Teleoperator Design and Operation, Wiley Interscience Editor, 1971
(4) SHERIDAN, T.B., JOHANNSEN, G.: Monitoring Behaviours and Supervisory Control, Plenum Press, New York, 1976
(5) VERTUT, J., FOURNIER, R., ESPIAU, B., ANDRE, G.: Advances in a Computer Aided Bilateral Manipulator System, Symposium: Robotics and Remote Handling in Hostile Environments, Gatlinburg, USA, April 1984
(6) VERTUT, J., FOURNIER, R., ESPIAU, B., ANDRE, G.: Sensor-Aided and/or Computer-aided Bilateral Teleoperator System (SCATS), 5th RoManSy, Udine, Italy, June 1984, Kogan Page and Hermes editors
(7) VERTUT, J.: Advances in Computer Aided Teleoperation Systems in the frame of the French Advanced Robotics and Automation (ARA) Project, Conference ICAR '83, Tokyo, 1983
(8) VERTUT, J., COIFFET, P.: Teleoperation, (I) and (II), Hermes Publishing, France, 1984 and 1985, Kogan Page and Prentice Hall in English
(9) VERTUT, J., ESPIAU, B., and al.: Advanced Teleoperation Group/ARA Project; annual reports: September 1982 (Poitiers), November 1983 (Besancon), September 1984 (Toulouse); CNRS, FRANCE
(10) ANDRE, G.: Systeme de Teleoperation Assistee par Ordinateur, ARA Report, to appear, IRISA 1986

Advanced Teleoperation (I)
Control and Supervision in Computer Aided Teleoperation

P. Gravez, B. Lepers*, R. Fournier** and G. André***

*CAL, Lille, France, **CEA Rennes, France, ***IRISA, Rennes, France

SUMMARY

In this paper is presented the architecture of a complete control system to be used in advanced teleoperation, based on two main hierarchical levels:

The first part describes the "generalized master/slave control" as a bidirectional process based both on aided effector control and on generalized force feedback. The coupling algorithms between master and slave take into account various functions: scaling of shifting in force or position, weight compensation, generation of artificial geometric constraints, sensor referenced loops, coordination with the motions of a carrier.

The second part presents an interactive "supervisory system" allowing dynamic selection of control modes which are at all times the most efficient for a given task. The previously defined elementary modes are classified as: manual, automatic coupling/recoupling, semi-automatic; these include many functions of parallel control sharing between the operator and the computer. The mode allocation is realized upon the basis of criteria developed from a general description of the task, from man-machine communications, and from external data.

The problem of off-line reconfiguration and on-line mode activation are also considered.

1. INTRODUCTION

Generally speaking, a teleoperation machine consist of two workstations:
- a master station where control and restitution means are sited;
- a remote station comprising operative and sensory systems.

In classical teleoperators, there is a dual relationship between the relative components of the two stations. For example, both the master and the slave manipulators are geometrically servoed by the same bilateral position or rate control loops. In a computer aided teleoperation (CAT) machine, this relationship is managed by a computer-mediator. In this way the symmetry is broken, giving rise to numerous "coupling patterns" between master and slave resources; these patterns are called <u>control modes</u>. Each of them might be very complex and we analyse the mode control problem in terms of the behaviour of the man-machine system. At this

Advanced Teleoperation (I) 29

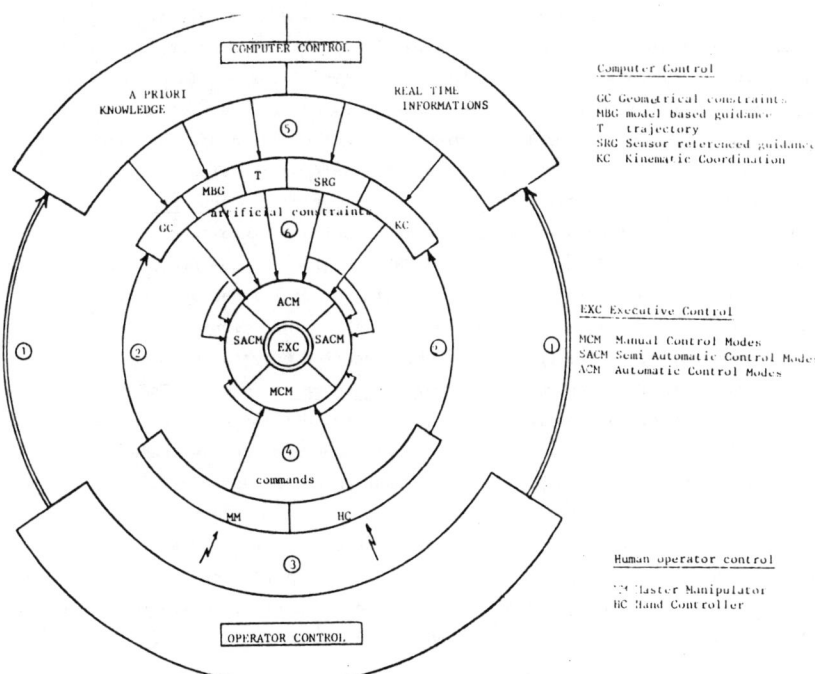

Figure 1 : Diagram of the Cooperative Control Concept

higher comand level, the narrow gap between CAT and robotics justifies the use of sophisticated control systems, implemented by a distributed processing architecture.

A structural hierarchy similar to that considered by Kohout (9) for the representation of movement in the brain could be evoked. A given task corresponds to that of a nerve bundle orientation; the individual patterns manage information flow and organize themselves into structural hierarchies. In our representation, this will correspond to the definition of local, global and metaglobal modes in a natural multilayered hierarchy (12). This paper lays emphasis on the first two control layers. The lower level, called the generalized bilateral control system (GBCS), was developed in the ARA project of the CEA and results in an extended control mode catalogue. A short description of the GBCS is given in the second paragraph. In order to optimize man-computer cooperation, the control modes are managed at the supervisory level as will be described. These considerations lead us to a CAT machine controlled by two operators: the executing operator or task operator (T.O.) and the supervising operator or management operator (M.O.). The study of the coordination level was carried out by the CAL and originates from research concerned with optimal allocation of tasks between man and computer based on the relative capacities of the two partners, as measured by spy systems (3). In this field, a multicriteria methodology, which takes into account the reliability constraint, has been developed (5). The main problem that will be evoked here is the integration of the algorithms that assume a dynamical approach to the hard/soft adaptive computing structure needed to achieve real time efficiency. We therefore describe the Sarah system, which corresponds to the supervisory level.

2. GENERALIZED BILATERAL CONTROL SYSTEM (2)

As described earlier (13) (2) the term "generalized bilateral control" (GBC) is used to define a bilateral control with force-reflecting feedback where the master and the slave arms are geometrically and, consequently, kinematically and dynamically dissimilar. Ideally, the system may be considered as a bidirectional process between two main frames: one linked to the grip, where the operator's hand acts, and the other linked to the end effector of the slave manipulator.

The first step is to try to obtain a bilateral system as transparent as possible. The second step is to modify the basic process by generating additional constraints or compensations in order to obtain some adaptivity with respect to the task requirements and therefore to define computer aided control modes. From this point of view three key features of GBC are:
- generalized kinematic coordination: computer control allows use of master and slave manipulators with different configuration.
- task-referenced assistance: task oriented control of the effector can be developed

in particular by defining functional coordinates while improving the operator
'feeling' through force feedback.
- the degree of autonomy (versus degree of assistance): the definition of powerful
control functions increases the slave autonomy (i.e. by generating automatic subtasks).

2.1 Coordination system

Given the models of the master and the slave manipulators, direct and inverse
mathematical transformations in position, velocity and forces are implemented:
(i) typically from the master to the slave, priority being given to the task with
a position control loop. The slave's control takes into account: the command
from the master, some functions under computer control and additional terms including
various external information coming from the sensory system, the environment simulator,
and the mobile transporter...
(ii) on the other side, from the slave to the master; the control may generate
a kinaesthetic feedback, i.e. force acting on the operator's hand taking into account
different factors like: constant force bias for gravity supression, force processed
by the computer, information from force sensors. Furthermore the notion of kinematic
coordination between master and slave manipulators may be extended to other requirements such as: task effective coordination, manipulator-mobile carrier coordination,
manipulator-manipulator coordination on the work site.

2.2 Task-referenced control

Coordination functions and task-referenced control functions are defined in
terms of functional coordinates and coupling algorithms.

As shown on the diagram GBC allows description of numerous "hybrid" control
modes in particular by partitioning d.o.f. of the end effector's motions. Various
schemes have been investigated, taking into account:
- operator's commands
- artificial constraints
- sensory feedback

As mentioned in reference (2) efficient modes have been experimented with and
evaluated: position indexing and scaling, force indexing and scaling, generation
of artificial geometrical constraints, compliance control with force sensor referenced
control, environmental model referenced control, motion coordination of a telemanipulator mounted on a mobile transporter.

2.3 Control Mode Classes

The following classes are classically defined according to the degree of autonomy:
- manual modes: where the operator performs the task with natural dexterity.
This leads to the **teleexistence concept**.
- semi-automatic modes: where the operator can act in parallel or in series with

the computer. An important part of our research activity has been devoted to this
topic. This also leads to the <u>teleassistance concept</u>.
- automatic modes: where autonomous control, including adaptive control loops,
is achieved. This leads to the <u>telerobotic concept</u>.

The experimental system which has been developed at CEA in the context of the
ARA project is described in paper 3. Real time computer control is implemented
on a minicomputer connected to several microcomputers. In the near future we will
investigate real time networks including distributed real time kernels and high
speed communication through a serial bus.

3. PRINCIPLES OF TASK DESCRIPTION IN C.A. TELEOPERATION

In robotics, conventional programming methods are based on explicit programming
languages. On the other hand, "interactive control" in C.A. teleoperation requires
a structure of interconnected modular control modes in a specific organizational
hierarchy taking into account the sharing of the knowledge base between the human
operator and the machine.

3.1 Interactive control hierarchy

The classification has three levels (12):
(i) At the lowest level, the machine executes control modes. This is MOLOC, a
local mode which defines a "generalized state" of the executive control system
i.e. parallel configuration of the GBCS (generalized bilateral control system)
and the GIFS (generalized information feedback system). A catalogue includes numerous
examples of MOLOC as elementary instructions at a tactical level.
(ii) At intermediate level, we define GLOMODE (or global mode) as a structure of
interconnected MOLOCS (in a similar way as a molecule is composed of atoms) including
elementary strategic functions such as selection of the degree of autonomy (between
manual, semi-automatic, automatic MOLOCS) or taking into account reconfiguration
procedures. The control sharing at GLOMODE level emphasizes human decision-making
capabilities.
(iii) At the highest level (METAGLOMODE level), the specification of a complex
operation (inspection, maintenance, dismantling) should make possible the generation
of macroglomode as an orderly list of the GLOMODES necessary for performing the
task. Ideally each level of control should correspond to a knowledge base comprising:
- a set of rules for devolving a current level description of a given task into
  a lower level description;
- a dynamic model of the work environment.

The rules for devolution are specific to the control level, whereas the environment
model may be shared. In practice the problems of applying artificial intelligence
systems in a real-time context are such that to generate low level commands based

on a task description and a knowledge base, is not yet on-line realizable. Therefore a different approach is adopted which consists of an off-line preparatory phase (on one hand) and the calling of operators on-line to take into account the specific aspects of the workspace (on the other hand).

3.2 Task preparation

Generally speaking, the preparation phase provides the command system comprising:
- a <u>generic description</u> of the task which may be quite partial;
- a <u>law of adaptation</u> to the environment which specifies the feedback to be taken into account and is passed on to the system or to the operator, as well as the rules governing man-machine cooperation.

In this way manual, assisted and automatic functions are to be found at each control level.

At the MOLOC level, a set list of command modes will be sufficient. It is however more advantageous to have an extendable set of GLOMODES thereby integrating a programming system into the CAT machine. This system enables the operator to adapt existing GLOMODES or establish new ones in order to carry out any frequently encountered sub-tasks. As regards the highest level, an artificial intelligence system is justified by the non-repetitive aspect of teleoperation tasks which make difficult classical programming. Working off-line, it provides a script, i.e. essentially a rough decomposition of the overall task into sub-tasks (1) (8). Later this script could be interactively completed on-line at the METAGLOMODE level. These considerations are summarized in figure 2.

4. THE SARAH SYSTEM. PRINCIPLES

The <u>Sarah System</u> has been integrated into the C.A. Teleoperation system developed at CEA Saclay to take the command at the GLOMODE level into account. It consists basically of a supervisor supported by a micro-computer Micral 90-50 and a MacIntosh in the role of the operator interface. Initially designed to work autonomously, Sarah was eventually foreseen in the frame of a complete control system including the METAGLOMODE level. The fundamental characteristics of Sarah are:
(1) to have the capacity of generating linked series of control modes;
(2) to offer the operators the greatest possible interactivity by allowing them to act at any level as long as machine reliability is not affected;
(3) to help the operators in their work by selecting pertinent information and building up a constructive dialogue;
(4) to make possible the inclusion of extra new sensors, additional functions as well as an interface with an artificial intelligence system.

Note that points (2) and (3) may be conflicting and that the success of a system such as Sarah lies in compromise.

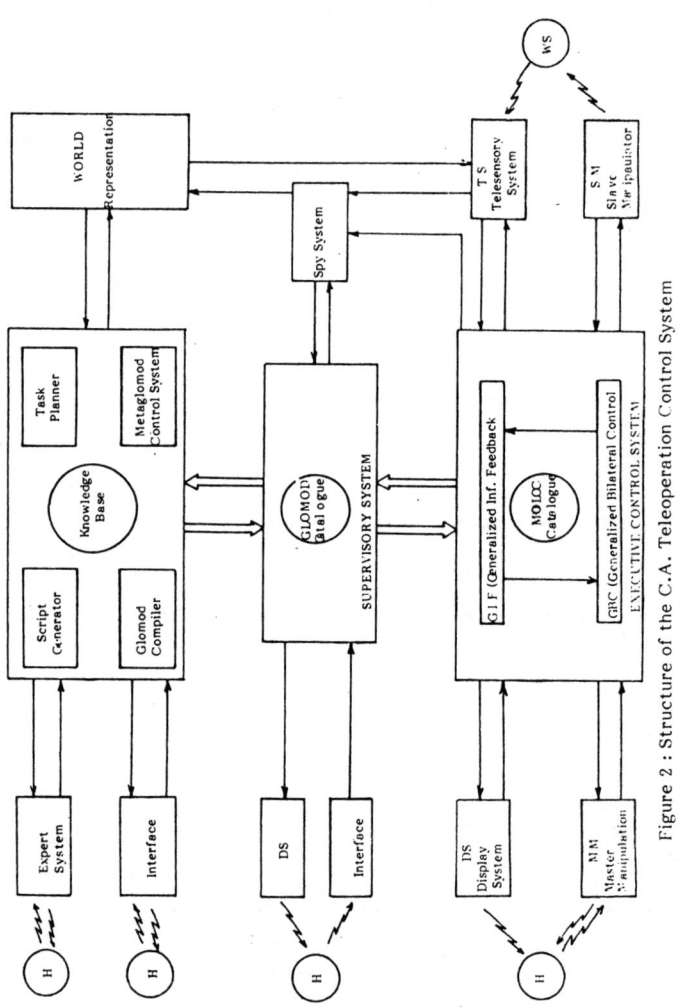

Figure 2 : Structure of the C.A. Teleoperation Control System

4.1 Modular Principle - Network Structure

As regards point 1, the carrying out of an efficiently linked series of control modes is only possible when certain conditions are respected.
- first it is necessary that the material set up of the teleoperator be adapted to the mode or the GLOMODE required.
- in some cases, before control is effected, there should be a transition phase to modify the teleoperator state. This could be a realignment of the master to the slave, with predefined motion of the carrier.
- furthermore a GLOMODE calls for factors which need to be acquired possibly through on-task learning.

Generally speaking, using a control mode requires the availability of some resources whether material or not. These may be made available by going through intermediate stages which themselves call for linked series of modes. Nevertheless the initial GLOMODE will sometimes have to be de-actived. Given this, one of the roles of the supervising system will be to manage as much as possible the intermediate phases, so simplifying the task of the operator.

Concerning the adaptibility of compiled GLOMODES to a real environment, the solution retained for Sarah is based on a network structure. The linked series created in this way offers several possibilities for carrying out the same teleoperation work. Two examples of GLOMODE are given - figures 3a and 4a. On-line, by the choice of the track to be followed in the network, the human operator has at his disposal a flexible command tool capable of acting in unforseeable circumstances. However, if carrying out a complex linked series fails to realize the corresponding task, the natural answer is to have available simpler GLOMODES. It is therefore advantageous to offer the operator a set of modular GLOMODES at different levels, going from control modes to highly sophisticated sequences.
A further advantage of the modular requirement is to use existing linked series to compile more complex GLOMODES, in the frame of an enlargeable dictionary. These considerations have lead us to develop Sarah in a Forth environment (1) (10).

4.2 The FORTH System

Compiling and interpreting threaded codes, Forth makes possible the creation of highly structured programmes built on procedures which may continually be carried out in an autonomous manner. Moreover, it features a remarkable capacity of fast computation memory use as well as speed of developing software. Using all these advantages, Sarah is built around a kernel which basically consists of a Forth machine adapted to working on GLOMODES and their network structure.

4.3 GLOMODE Compilation

At the GLOMODE level, the off-line preparation phase consists in compiling series of control modes which will be carried out on-line. The first stage is to define

the network relative to the existing GLOMODES. Some instructions in the control
field make possible the implementation of loops and moreover the parameter names
can be indexed to loop counters. We are therefore able to define a GLOMODE doing
n times the same linked series using different parameter names at each loop and
to input on-line the value of n. On the other hand, the modularity requirement
prescribes the networks to have only one entry mode and one exit mode, making it
necessary sometimes to use the atomic GLOMODE stop.

During the second stage, programming consists in associating with each node
of the network:
- a lower level GLOMODE
- a parameter module
- a parameter field
- a control field
- a label field

The memory structure for each node is given in figure 5.

5. THE SARAH KERNEL

5.1 MOLOC Scheduling

On the ARA teleoperation machine, a MOLOC is activated by a signal from the
supervisor towards the GBCs sending a message and a list of parameters. The human
operator can interrupt the mode at any time, thereby showing that some abnormal
situation has arisen. If the task is being performed correctly, the GBCs or the
executive operator, depending on whether the mode is automatic or cooperative,
stops the current MOLOC when the task is completed. In this way, the human operator
can intervene through:
- an operator interruption
- a "natural" ending of a cooperative control mode

When a MOLOC is stopped, a report containing the origin of the end of mode is
sent to the higher level, sometimes followed by a list of parameters to be memorized.
The control mode series are supplied to the supervisor system in the form of <u>GLOMODE
structures</u> comprising:
- elements making it possible to evaluate on-line the degree of realizability of
the GLOMODE according to the state of the machine and of the operator as well as
the quality of information supplied by the sensor systems
- a program corresponding to the activation of an isolated MOLOC (atomic GLOMODE) or
a network structure where each node is associated with a previously defined GLOMODE
(atomic or not).

5.2 The Evolution Menu

Whilst in operation Sarah follows the implementation of a control mode with a process of selection of the next one to be activated. The MOLOC having been finished, the system provides the operator with an <u>evolution menu</u> which displays all of the linked series or of the control modes possible in the network, taking into account available information on the state of the teleoperation machine. Besides this, the options "return to previous mode" and "end of GLOMODE" are systematically offered. Depending on the operator's choice, Sarah decides on the mode and sets it off. If the different proposals do not provide a reply to a given situation, then the GLOMODE can be abandoned in order to run a more appropriate one. At this stage, the system assistance to the operator consists of providing him with a limited selection of possible immediate evolutions and secondly setting the problem of order in terms of task elements instead of only referring to the involved control modes. During an inspection task, a typical situation would have the evolution menu propose a sensor-assisted surface following motion, either automatic or manual. Then a robot or teleoperated move towards a starting point would be offered, and in the case of automatic teleoperated motion, the menu would only include the MOLOCS involved in a collision avoidance move.

5.3 Configuration modes and learning modes

As we have seen, to carry out a linked series it is necessary to go through intermediary stages which frequently are themselves linked series called configuration GLOMODES. These comprise parameter acquisition, carrier movement or master slave realignment modes and may be set off in several ways.

(1) First, it is possible to explicitly include <u>configuration GLOMODES</u> in the off-line compiled linked series. The drilling shown in figure 4 thus contains a learning phase of useful positions followed by automatic carrying out of the sub-task.
(2) Then certain intermediary stages such as realignment are automatically governed by Sarah.
(3) Lastly, in an on-line operation a menu enables the operator to select a configuration GLOMODE; thus when an automatic motion sends back the message "end of mode" activated by "unattainable position," the operator sets off carrier shifting by means of the <u>configuration menu</u>, then with "return to the previous mode" in the evolution menu he again starts off the motion which this time may be satisfactorily carried out (figure 3b).

Possibilities (2) and (3) are particularly interesting in that they make it possible to avoid going through intermediary stages during off-line preparation and thereby improve on-line flexibility.

6. PARAMETER AND CONTROL FIELDS

Additionally, the Sarah system includes mechanisms for parameter transfer and control of GLOMODE evolutions.

6.1 Mechanism associated with the parameters

The linked series comprising co-operative as well as automatic modes calls for numerous parameters whose on-line management would be tedious. From this point of view, Sarah is based on a three category classification.

(i) the MOLOC level parameters whose values are automatically memorised by the system after each modification, then restituted in the same manner when necessary. For instance, the weight of the object grasped by the end effector and used for the weight suppression function will be found in this class.

(ii) the GLOMODE level parameters managed in the pre-compiled linked series (motion rate, distance instruction for proximity sensored control loops, etc. The values of most of these parameters may be fixed off-line given the involved sub-task; nevertheless the flexibility requirements make it necessary to enable on-line modifications.

(iii) the METAGLOMODE level parameters (position, distance, etc) whose management is based on a knowledge of the work environment and therefore relies on the operator or on the script generator. With the Sarah system, an essential feature of the system is to allow parameter transfer between the different network structures hierarchically nested in the GLOMODES.

At the compilation stage, some parameter values being undefined, their management relies on symbolic names. On-line acquisition and learning procedures enable the human operator to load or modify the value tables. On the other hand, some parameters require computation and, during the series compilation, we therefore associate with each node a lower level GLOMODE as well as a parameter field containing a list of the parameter names to be used:
- either to pass them to the GBCs to carry out a control mode (occasionally after computation),
- or to record them following acquisition, learning or computation.

In this way, Sarah makes it possible to define a task such as the drilling of n parallel holes between positions A and B following and axis C. The on-line acquisition of a value may occur at any time through the configuration menu. If this value is not known at the beginning of the GLOMODE where it will be used, the system will ask for it while leaving to the operator the possibility of putting off its acquisition or its learning until **absolutely necessary**.

6.2 Definition of Transitions

Regarding the control of the evolution, **experience shows that** it is tedious to display the evolution menu at each end of mode. In most cases, it is possible from a node in the GLOMODE structure to define a privileged successor corresponding to the normal evolution of a task. We have therefore associated with each node a control field specifying automatic transitions (without the operator intervening)

Advanced Teleoperation (I) 39

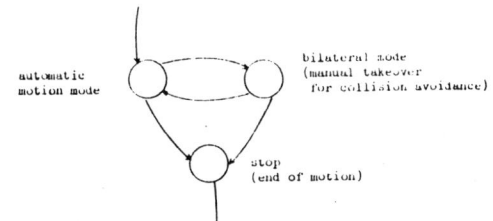

Figure 3A :
Simple Motion Network

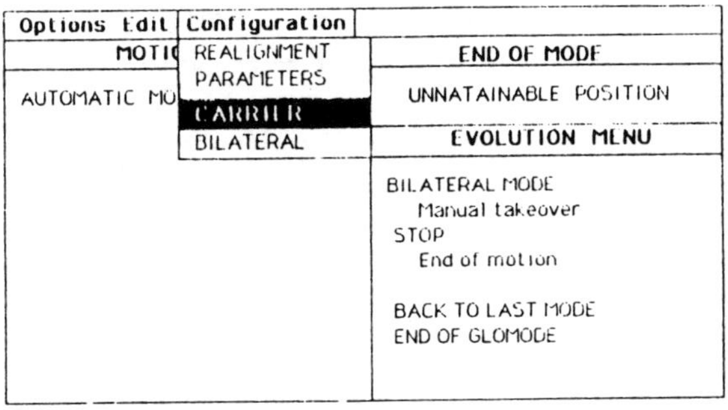

Figure 3B : Evolution and Configuration Menus

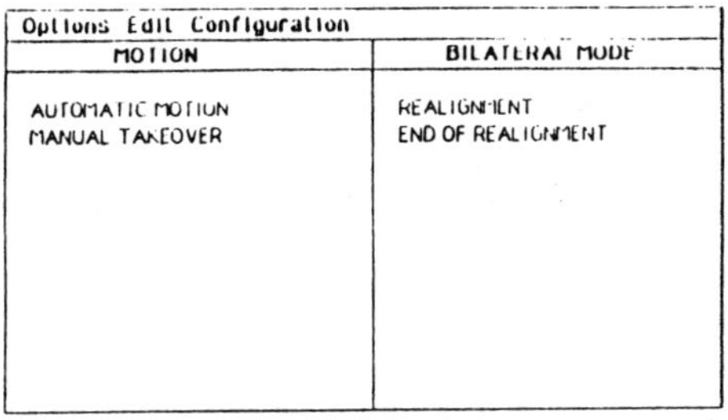

Figure 3C : Manual Takeover

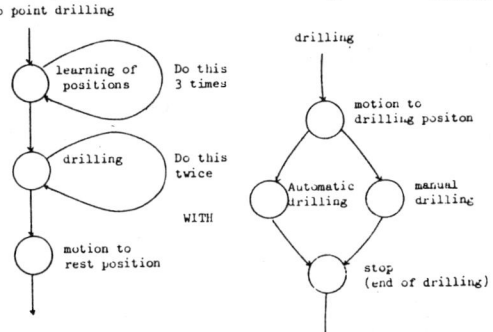

Figure 4A : Network for a Drilling GLOMODE (Simplified)

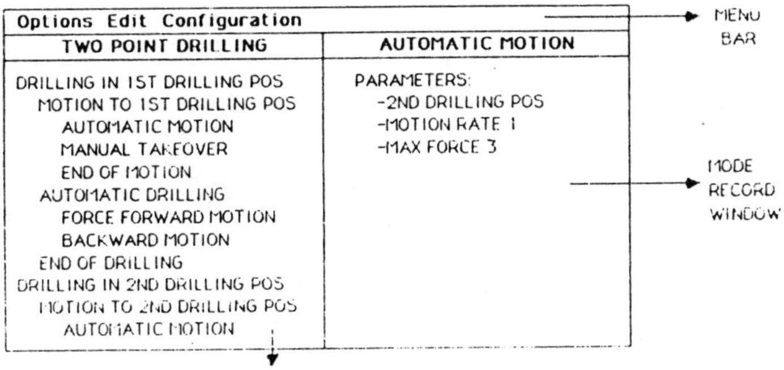

GLOMODE RECORD WINDOW
Figure 4B : Display during Drilling

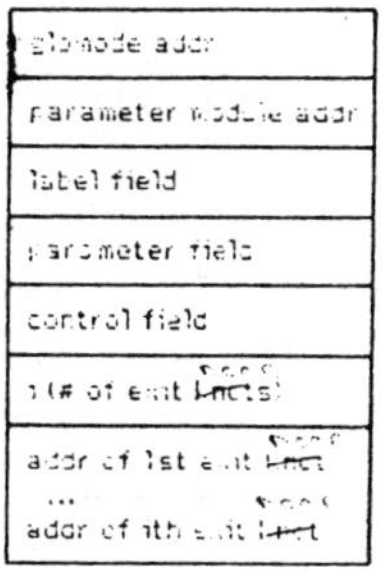

Figure 5 : Memory Structure of a network knot

toward a fixed exit node. These transitions will appear only if the current mode
ends on a normal condition, as for example "end of mode" activated by "end of job"
in robot function or by "operator stopping" (different from operator interruption)
in cooperative function. By extending this approach, some instructions for
<u>conditional automatic transitions</u> have been implemented, linking the choice of
a path to events as the return to zero of a counter, a particular end of mode
condition or an alarm. This mechanism makes possible a Petri net representation
and allows the teleoperation machine to operate automatically according to a
sophisticated behaviour capable of responding to a set of complex situations.

7. MAN MACHINE INTERFACE

7.1 Multiwindow System

Here we present the way in which information relating to the working of the
machine is presented to the management operator, some information being relayed
to the executive operator. The dialogue relies on display in a <u>superimposed window</u>
containing messages released by the system and menus with several items selected
by a mouse or by word recognition. The proposed windows relate to:
- the choice of the linked series to carry out the installation of
- a "control panel" that enables the operator to supervise the evolutions in the
  series
- the selection of intermediary stages
- the on-line management of the parameters

Concerning the control panel, the analysis of the information to be displayed
leads us to distinguish the following classes:

(i) a set of information showing the teleoperator functions in the involved mode
and specifying for instance the role of switches and the type of command applied
to each degree of freedom.

(ii) <u>mode record</u> which gathers the messages sent by the GBCS as "end of realignment,"
"mechanical jig" or "end of mode activated by unattainable position," and enables
the operators to follow the evolution which is taking place according to the
particular aspects of the workspace.

(iii) <u>aim of the mode</u> which specifies to the executive operator the goal of the
MOLOC in the frame of a fixed GLOMODE, for example: "learning a drilling position"
(with a manual mode within an automatic drilling GLOMODE).

(iv) <u>glomode record</u> making it possible to follow the covered path through the
different hierarchical levels of a linked series.

(v) <u>items for choice</u> i.e. the contents of the evolution menu.

7.2 Labels

In order that Sarah might possess necessary information for display purposes,

a <u>label field</u> is associated to the nodes of the network structures constituting the linked series. It contains text items specifying the role of the corresponding GLOMODE (atomic or not) in the frame of the overall series and therefore gives the pieces of information relevant to the aim of modes to be carried out and to the GLOMODE record. The control panel is divided into four zones:
- the menu bar which allows access to the configuration menu
- the <u>GLOMODE record</u> window whose title is the name of the current linked series and which displays the encountered labels showing the hierarchical level
- the <u>mode record</u> window displaying the name of the mode being run and the messages concerning its evolution, except the end of mode conditions which are shown in a particular window.
- the <u>evolution menu</u> window which, with the one of <u>end of mode condition</u>, covers the previous window whenever an end of mode report is sent, and which gives for each possible item the name of the corresponding GLOMODE and the associated label when available; during automatic transitions, only the end of mode condition window is displayed.

An ergonomic study is now being carried out at the Evry's LRE with the aim of condensing the displayed information by the use of colour and graphics simplifying the interpretation of the control panel.

8. CONCLUSION

Compared with other teleoperation systems, the ARA CAT machine enables the human operator to carry out a task using an extended catalogue of control modes. These modes, either manual, assisted or fully automatic are managed by the same control structure, thus giving good flexibility to respond to some unexpected situations. Generally speaking, at a given level, the system can work in cooperative or automatic function independently of the function of the other levels. As in the robotics field, the hierarchical structure leads to "vertical" information exchanges between the different levels involved in the devolution; these being frequently partial or non-existent, a "horizontal" information exchange allows the human operator to compensate for it. At this point, the interaction between the operator and the control system concerned (particularly Sarah) calls for hierarchically non-equivalent orders. At the GLOMODE level, this justifies the extensibility of a linked series catalogue whereas a superior layer is foreseen. In fact, vertical communication emphasizes fixed levels of task description, while the horizontal communication is based on a large set of instruction of different "sub-levels" comprised of two of the previous fixed levels. These considerations lead us to the construction of an interactive extended programming environment (6) which generalizes the notion of operating system. In the near future, the use of an actor oriented methodology and the development of an expert system for task planning will be investigated.

REFERENCES

[1] R.P. ABELSON and R.S. SCHANK, "Scripts, Plans, Goals and Understanding", John Wiley and Sons, 1977.

[2] Guy ANDRE and Raymond FOURNIER, "Generalized end effector control in a Computer Aided Teleoperation System", Proc. ICAR'85, Tokyo, Sept. 85, pp. 337-344.

[3] Patrice AUTECHAUD, Philippe DESODT, Ahmed HABCHI, Daniel JOLLY and Pierre VIDAL, "Hierarchical Control and Man Robot Stucture", 3e Congresso Brasiliero de Automatica, Rio de Janeiro, Sept. 1980.

[4] A.K. BECJZY, G. BEKEY and S. LEE, "Computer Control of Space-borne Teleoperator with Sensory Feedback", IEEE Robotics, St Louis, 1985.

[5] Constantino DIAZ-GONZALEZ et Bernard LEPERS, "Sécurité du Couplage Homme-Machine : l'apport des Méthodes d'Aggrégation de Préférences", Congrès IFAC, Valenciennes, juin 1983.

[6] Richard C. DORF and Richard H. KIRSCHBROWN, "KARMA-A Knowledge-based Robot Manipulation System", Robotic 1, 1985, pp. 3-12.

[7] Tomomasa SATO and Shigeoki HIRAI, "Language-Aided Robotic Teleoperation System (LARTS) for Advanced Teleoperation", Proc. ICAR'85, Tokyo, Sept. 85, pp. 329-336.

[8] Annette KNAEUPER and William B. ROUSE, "A Rule-based Model of Human Problem-solving Behavior in Dynamic Environments", IEEE trans. on S.M.C., Vol. 15, N° 6, Nov/Dec 1985.

[9] KOHOUT, "Representation of Functional Hierarchies of Movement in the Brain", Int. Journ. of M.M.S., 1976.

[10] Peter M. KOGGE, "An Architectural Trail to Threaded-Code Systems", Computer, March 1982, pp. 22-32.

[11] Elisabeth D. RATHER and Charles H. MOORE, "The Forth Approach to Operating Systems", Proc. ACM'76, Oct. 1976, pp. 233-240.

[12] G.N. SARIDIS, "Intelligent Robotic Control", IEEE trans. on A.C., Vol. 28, N° 5, May 1983.

[13] A.K. BEJCZY, S. LEE, "Generalized Control of Robot Arms", Joint Automatic Control Conference, San Diego, 84.

Advanced Teleoperation (II)
The Generalized Information Feedback Concept in Computer-Aided Teleoperation

G. André* and A. Fournier**

*IRISA, Rennes, France, **LAMM, Montpellier, France

SUMMARY

This paper sets the problem of information feedback in computer aided teleoperation from the point of view of supervisory control. Section 1 deals with telepresence versus generalised information feedback (G.I.F.) concepts and the associated perception-action loops in C.A. teleoperation. Section 2 presents the design and implementation of the IRIS experimental system where we underline the contribution of sophisticated technologies: we briefly summarize how force control may be extended to the "remote touch" feeling using proximity sensors. Section 3 focuses on the importance of world model based information for generating realistic or/synthetic, visual or/force feedback. A simulator is firstly used to provide 3D graphic displays. Then it is extended to include a potential field description, which allows the operator to feel in real time artificial magnetic-like forces.

1. INTRODUCTION

Basically teleoperation is a bidirectional system established between a human operator operating at the control station (C.S.) and a remote manipulator performing a task in the worksite (W.S.). It is thus very clear that total efficiency greatly depends on the perception-action loops which are created. Conventional telemanipulators can be viewed as symmetrical systems, where the operator plays roles both of a serial sensory and of a controlling system; this implies natural limitations. The development of the computer aided teleoperation system (CATS) defines a new generation of teleoperators where the evolution induced by telesensing and information feedback synthesis plays a crucial role (from the WS to the CS) as important as the evolution resulting from the computer control system (from CS to the WS). Brief analysis leads to the following conclusions.

On one hand, concerning the control process, the general aim is to improve operational efficiency and safety with regard to the task. Here the major goal is to increase autonomy. This is partially achieved by adding computing capabilities and by generating powerful control functions and several control modes (manual, semi-automatic, automatic). The key problem in the so-called generalized bilateral control and supervisory system is control sharing between man and the computer.

On the other hand, the <u>information feedback process</u> is firstly concerned with the improvement of man-machine interface transparency. Secondly there exists an interface with the control system for generating adaptive control functions. This leads to increased sensory capabilities and develops dynamic selection of information feedback modes. Here the first key problem is to share the information feedback between the man and the computer. The second one is to combine actual and artificial information feedback. It follows that the trends in C.A.T.S. are bi-polar; for the control process as well as the information feedback process:
(i) <u>transparency</u> with the related notions of telemanipulation and telepresence
(ii) <u>autonomy</u> with the related notions of cooperative robotics, telesensing and telesupervision

There is an analogy between the evolution of these two processes.

From this perspective, perception and information feedback in CATS raises several issues:
- the development of sensing devices and acquisition systems: what kind of information is needed?
- the definition of preprocessing systems: what information has to be transmitted to the human and which has to be transmitted to the computer?
- the design of man-machine interfaces: what are the more convenient senses of man for receiving information and what are the procedures to prevent overloading?

2. TELEPRESENCE VERSUS GENERALIZED INFORMATION CONCEPTS

2.1 Telepresence

As shown in Figure 1, telepresence could be defined as a dual relationship between sensory units in the remote worksite and active display units at the control station. Sensory data from the robot manipulator are directly transmitted to the operator's visual, aural, kinaesthetic senses, while robot actuators follow the human operator's movements. So, the operator feels the sensation of presence in the remote site (i.e. as if his hand becomes the end effector and as if his eyes become the stereocameras). Therefore the main characteristic of tele-existence (i.e. combination of telemanipulation and telepresence) which is a bilateral symmetrical process, is exclusively to provide direct man-oriented information feedback, from a typically anthromorphic point of view. Obviously the associated criterion is the improvement of the transparency. In recent years, this approach has been studied by the ETL Laboratory (11).

2.2 G.I.F. Generalized Information Feedback

As essential function of 1.F. (information feedback) is also to provide increasingly efficient assistance during task execution phases. The G.I.F. concept naturally appears with the development of C.A. teleoperation. In fact, the problem is: "given task requirements and human capabilities, to choose, at each time the optimal

configuration of sensors, control modes and feedback (observation modes) to the
operator, to perform the task as well as possible, while preventing the operator
from unnecessary work tension". The important properties follow:
- the process becomes essentially multifunctional and asymmetrical due to the
  following:
   . the original combination of both "real and "artifical" data.
   . the possible transformation of information (interchanging of the nature of
     feedback with regard to the sensory data).
   . the limited number of human operator's input and output channels.
- it becomes interactive: other important features are the adaptive configuration of
  acquisition and restitution devices, and the definition of various coupling modes
  in order to select interfaces on the one hand with the human operator, on the
  other hand with the computer control.

It results in a complex system, where we may emphasize some classifications
of I.F.:
- according to the acquisition principle, there is a combination of:
   . on line (sensor-based data)
   . a priori (model-based) data
- according to the displaying principle, there are:
   . real I.F. (due to direct transmission of sensory data)
   . artificial or synthetic I.F. (coming from a model or processed data)
- according the transformation and interpretation principle
   . man-oriented I.F. (by pointing out the interface with the operator) i.e. realistic
     (real or simulated) displays. The criterion is the maximal transparency with
     a qualitative property of ease of interpretation.
   . task-oriented I.F. (by pointing out the relation with the task): pertinent
     information i.e. providing error notion with respect to the goal of the task.
The criterion here is the guidance capacity with such properties as: rapidity
of interpretation, error measurements,...

Figure 2 illustrates, in a symbolic diagram, the different subsystems and functions
attached to the concept of G.I.F.
- the bottom half space is the actual information domain. Real world (R.W.) is
  observed (function 4) on line through a multisensor system: vision, proximity,
  force sensors which provide preferably real I.F. (function 5) or after processing
  artificial I.F. (function 7).
- the top half space is the artificial information domain. From extraction of
  data from the world model (W.M.) (function 6) artificial information generators
  provide both M.O.I.F. and T.O.I.F. (function 7) mostly in the form of visual
  and kinesthetic feedback. The connections between the two half spaces are a

priori modeling (function 1), building up a model from sensory data (function 2), and identification or parameter updating (function 3).

2.3 Study of Perception-Action loops in C.A. Teleoperation

The perception-action loops result from information sharing between the operator and the computer. Depending on the degree of autonomy, the I.F. function evolves from complete assistance to the operator (manual mode) toward adaptive sensor-referenced control (automatic mode). The last case is widely studied in robotics. On the other hand, from our point of view there are many interesting problems to be investigated in semi-automatic mode. As shown in Figure 3, I.F. are divided into two classes: passive verification information (which proportionally increases with autonomy) and active assistance information. That is the reason why we have mainly studied the combination of both passive and active observation modes particularly when it remains a part of human decision capability.

As mentioned in (4) we define:
- the hybrid (or mixed mode): where the operator is able to act in parallel with the computer. Both visual and force feedback provide partial assistance to the operator while some d.o.f. are under sensor-referenced control.
- the reflex mode: where the operator senses artificial force feedback in order to guide his movements, but remains in series in the loop.

Experimental results show that vision and kinaesthetic senses are the major senses to be involved for ensuring accurate navigation and manipulation performance.

Our objectives were to test original modes. In particular, experiments have been conducted in order to analyse:
- artifical force feedback generated from various sensory data (proximetric, visual, force information)
- visual display of artificial images from a C.A.D. model
- artificial force feedback based on a potential field data base.

3. DESIGN AND IMPLEMENTATION OF THE I.R.I.S. SYSTEM

For providing flexibility the prototype I.R.I.S. (information reflection and information synthetis system) is designed as a modular, distributed acquisition and restitution system, integrated into the C.A. teleoperation system. Figure 4 shows the components under development on the experimental site. In this section, we give an overview of functions attached to sensory systems. The next section will emphasize the information feedback related to C.A.D. models.

3.1 Vision systems

. global 2D vision system: comprised of several cameras providing global vision of the robot and the environment from different viewing positions. Improvements, coming from progress in camera technology (colour, high resolution, high sensitivity infrared technology,...) and increasing mobility, are evaluated.

- "Televise": is an automatic tracking system developed at CEA. The camera is servoed in such a way that it is permanently looking at the work zone, i.e. near the end effector. A microcomputer ensures the autonomous coordination depending on the movements of the manipulator and specific parameters adapted by the supervisor or the operator.

- Stereovision: a telepresence study is carried out by ETCA and IRISA in order to test 3D visual feedback to the operator. Two stereoscopic pairs are developed. The former uses two standard colour TV cameras for global perspective. The latter comprises two CCD micro-cameras MICAM-R (40x40x30mm), 280x200 resolution operating at 100 frames/sec. It is designed in order to be combined with ultrasonic proximity sensors and mounted on the end effector. It thus provides local vision, in particular during inspection tasks. The camera pairs are connected to a multiplexer which generates a specific composite video signal including left and right interlaced images. A viewer controller generates synchronisation signal through wireless infrared transmission to spectacles.

All the previous subsystems provide enhanced visual feedback on a large screen.

- Local vision system: a very small size micro camera ISIGHT 32 (/15x18mm) with low resolution (32x32) has been integrated within the end effector. The dedicated controller provides at high speed (50 times/sec.) geometrical parameters of binary images. This system enables automatic positioning in the same way as proximity sensors, and allows the testing of artificial force feedback generated from visual information.

3.2 The IRISA Multisensor System

This system, developed at IRISA, has been described in detail in reference (1). It comprises an acquisition and preprocessing microcomputer controlling a set of small sensor heads: infrared, ultrasonic, magnetic sensors, arranged on the effector.

Our previous work has demonstrated that proximity sensing greatly contributes to improve the guidance of the effector and enable local adaptivity for relative effector-object positioning.

Through the master arm, the operator can feel artificial sensor-based forces. This kinaesthetic coupling results in a very impressive "remote touch" technique. Useful applications are obstacle avoidance, surface following, automatic searching, centering, grasping. Experimental results, presented in references (1), (4), illustrate typically crosscoupling function, in the frame of G.I.F. concept.

4. USING OF C.A.D. MODELS AND SIMULATION FOR TELEOPERATION

It is well known that C.A.D. systems are more and more used in industrial

robotics for designing robots and work cells, evaluating performance, off-line
programming and optimizing the task. This is achieved from a fully deterministic
point of view. In teleoperation, the problem is really different and more complex.
By taking into account a priori knowledge in the form of a world model, the use
of C.A.D. computer graphics and image synthesis techniques allows definition of
assistance functions which play an important role in the frame of G.I.F. The aim of
a C.A.D. system for teleoperation may be summarized as follows:
- off-line:
  . modelling the manipulator and partially the remote environment
  . programming parts of trajectory, interference checking, defining functional
    regions and artificial constraints.
  . training of the operators as if it was a simulator.
- on-line:
  . providing real-time information feedback to the operator, during task performance
    by:
    - synthetic visual feedback and/or
    - artificial force feedback.

To achieve this goal, we have developed in the frame of the A.R.A. project an
interactive simulation software which includes most previous specifications. The
following sections emphasize the tactical level in C.A. teleoperation.

4.1 Computerized synthetic pictures as visual information (LAMM (8))

The use of realistic artificial images in teleoperation may be required for
several reasons. At first, it could be difficult or very inefficient to put a
camera on the remote site. This could be due to the hostile environment: temperature,
radioactivity, smoke, dust, etc. Furthermore, conventional cameras have limited
mobility. These result in the presence of obstacles between the sensor and the
workspace. Moreover it could produce a transmission delay (like in spatial
applications).

A solution is to build a 3D geometric model of the slave arm and its environment
which is then used to create computerized pictures from artificial cameras. Three
key problems have thus to be solved:

4.1.1. - Modelling the work site

According to the assumptions on world knowledge, geometric modelling can be either
developed directly from a priori data, or reconstructed from sensory data.
Significant results have been obtained by ETL in the latter case (6). However,
most cases are "hybrid" in the sense that the geometric model of the slave environment is partially known, while uncertainties exist on location parameters.

4.1.2. - Updating the Model

When the slave arm is moving, or when the environment is changing, both the model and the resulting picture must be updated. The possible environmental changes include: moving objects, bringing several objects together (grasping), modifying geometrical aspects, etc. The model of the manipulator is updated in real time from the joint positions. At the beginning it is necessary that the model represents exactly the initial conditions. A geometric matcher for recognizing and positioning 3D rigid objects is studied. Algorithms use simple geometric primitives such as point, and line or plane. Interactive procedures allow identification of shapes and positions from direct contact points, or more conveniently, by using proximity sensors.

4.1.3. - Virtual camera positioning and graphical facilities

The synthetic images can be used in order to improve a bad picture or to replace it. The specifications for synthetic pictures are that images are required in "pseudo real time", symbolic and figurative representation should be allowed together in the same picture, extraneous details should be removed from the picture but should remain in the model, computerized pictures and camera pictures should be displayed either at the same time or in turn in the same display.

According to the task requirement, the operator can locate one or several cameras in an optimal manner. A trivial solution is the same location as a real camera in order to get superimposed images to make visual pattern matching. But many other very interesting locations can be chosen, for example a virtual camera mounted on the tool, a moving camera with automatic focusing on a moving frame. Other available adjustments of parameters are: camera orientation, view direction, zooming, etc.

4.1.4. - System description and experiments

At present the prototype system, developed at LAMM, is implemented on a microcomputer INTEL 86380 equipped with powerful MATROX colour graphics boards running under the RMX operating system. This hardware configuration allows the generation of 6000 vectors/sec. Wireframe colour image regeneration takes from 0.3 to 15 s. In the next year it is expected to obtain real time at rate of 1/25 s by using a new synthetic image generator station. The modules of the software are 3D modelling, visualisation and communication software. Real experiments allow evaluation of:
. realistic displays: The use of particular viewpoints depending on the task and the facility of well-chosen 2D projection
. synthetic displays: positioning errors during fine motion and proximity (or force) interaction from graphic indicators.

4.2 Force Reflecting feedback based on the artificial field approach for the guidance of teleoperators (irisa (2) (5))

In addition to the role played by simulation in providing synthetic visual feedback from virtual cameras, an additional potential model can play an important role in generating artificial force feedback. This allows achievement of force reflection without contact in situations such as obstacle avoidance or precise positioning.

Three approaches are currently used to solve the problem of collision free movements: find path in the configuration space, the freeway as an optimal control problem and the potential field approach. Typically the former are global path planning problems, while the latter is a local method. In teleoperation both analyses are required; a C.A.D. system provides easy execution of the following strategy:

(i) - global motion coordination of a redundant manipulator: the manipulator with specific tools mounted on a carrier defines a kinematically redundant system. The IRISA simulator includes not only the manipulator and environment models but also multisensor models (2). Off line task preparation allows description of primary goal (i.e. the end effector trajectory) while a secondary goal (free collision configuration) is automatically generated. Results have been presented in (5). However, in C.A.T., this mode is only regarded as an automatic subtask to reach a predefined work location.

(ii) - local guidance of the end effector under model-based constraints. Assuming now that the 6 d.o.f. manipulator has to perform a given precise local task on a static carrier, our study is focused on the real time guidance capability of the end effector. It is possible that, between no-constraint free movement (in manual mode), and fully-constraint trajectory (in automatic mode), C.A. Tele-operation might include specifically partially-constraint motions in order to increase safety and efficiency. The problem to be solved is the real time generation of attractive or repulsive forces when the effector enters into C.A.D. interactively predefined zones.

The proposed approach is considered in two stages.

4.2.1. - Off-line modelling and preprocessing

. Interactive description of potential zones

Given a geometrical model of the environment, (external representation), the first step consists of interactively describing an additional data base, superposed on the previous one, in the form of other primitives, (internal representation). The required potential fields are subdivided into two functional classes:
- for obstacle avoidance: primitives are closed volumes around given objects of the environment. These are "no-go" regions which will rise to repulsive forces.
- for guidance (assistance for positioning): new primitives are introduced in

the data base; they will give rise to attractive forces or specific constraints.
The operator describes all these regions by using computer graphics facilities.
The potential fields are defined by a geometric model and associated attributes.
Moreover, these models could be adaptive in space (programmable external limit,
stiffness,...) and in time (possibly moving primitives).

. Space subdivision algorithm:

The problem to be solved is, by preprocessing the internal data base, to decrease
the run time computation required to generate artificial forces. The method is
very similar to that employed to speed up image synthesis techniques. The approach
is to subdivide 3D space into a set of elementary convex subregions, for example,
a stack of rectangular cells. A graph is deduced.

4.2.2. - On-line artificial forces generation

. Selection of local potential field primitives:

The first step is a situation analysis which leads to determination of what are
the primitives to be activated. This problem is treated as a search in a graph.
The result is a local subset of the data base as a list of active internal primitives.

. Computation of model interaction:

This step is based on the KHATIB method (7) extended in the frame of C.A. Teleoperation i.e. for generating a force reflecting feedback on the master arm. Forces
are deduced from analytical formulations depending on distance calculation between
point or link and primitives such as: point, line, plane, sphere, cylinder,
parallelepiped, etc. However, since our goal is to obtain real time performance
(50Hz) some simplifications are introduced, in particular only one point or one
link of the manipulator is sensitive to the potential.

4.2.3. - Experimental results

The partition of the work space into repulsive zones, attractive zones and free
space allows the increase of overall safety of motion. Furthermore, we have tested
the precise positioning feasibility. Experiments such as: navigation between
tubes and planes, crossing zone following, obstacle avoidance, guiding along a
straight line, etc., have been achieved. The perception-action loops are analogous
to those which take into account sensory feedback. In addition to artificial
kinaesthetic feedback the operator can observe colour variation on the graphic
scene to denote the presence of the potential field and the force exerted.

5. CONCLUSION

Advanced teleoperation could be viewed as intelligent teleoperation where the
role played by the perception subsystem will become more and more important not
only from an anthropomorphic point of view but also from the perspective of both

supervisory control and autonomous control. We have proposed, in this paper, the
concept of G.I.F. as a multiple information channel system with adaptive configuration capability. Among the wide variety of modes, task-oriented information feedback
enhances the task assistance while telepresence through man-oriented information
improves man-machine communication.

We have pointed out the improvement in the following criteria:
- safety through redundancy of information and the automatic diagnosis ability
- efficiency through pertinent information displays with respect to the task
- transparency of the man-machine interface by the combination of sophisticated technologies

However, at present, the key problems to be solved: the degree of supervision
required from the operator and the interpretation capacity of the computer, the
optimal combination of real and a priori, realistic or symbolic information, the
dynamic selection of I.F. modes according to the task description. This study
is therefore essential to advance the state of the art of computer aided teleoperation
in the direction of cooperative robotics. Just as in executive control it is expected
that the G.I.F. scheduling will benefit from progress in artificial intelligence
techniques.

6. REFERENCES

[1] G.ANDRE:"A Multiproximity Sensor System for the Guidance of Robot End Effectors"Proc. ROVISEC 5, Amsterdam, Oct. 1985.
[2] G.ANDRE, R.BOULIC"C.A.O. de systèmes multicapteurs et simulation de commandes basées sur la perception de l'environnement en Robotique". CAO Robotic Etat de l'Art, Hermès Publishing, 1985.
[3] A.K.BEJCZY:"Kinesthetic and Graphic feedback for intergrated operator control". Proc. 6th Conference Man-Machine Interfaces for Industrial control, 1980.
[4] B.ESPIAU,G.ANDRE:"Sensory-Based Control for Robots and Teleoperators". Proc 5th ROMANSY, Udine, June 1984.
[5] B.ESPIAU,R.BOULIC :"Collision Avoidance for Redundant Robots with Proximity Sensors". Proc 3rd ISRR, Gouvieux, Oct. 1985.
[6] T.HASEGAWA:"A newApproach to teaching object Description for a Manipulator Environment". Proc 12th ISIR, Paris, 1982.
[7] O.KHATIB :"Real Time obstacle avoidance for manipulators and mobile robots". IEEE Conf. on Robotics, St Louis, 1985.
[8] C. QUARO,A.FOURNIER:"Computer Aided bilateral manipulator control using a computerized picture as visual feedback". IASTED 84, Amsterdam, June 1984.
[9] S.SAKANE, M.ISHII, M.KAKIKURA: "Hand-Eye Simulator :A Basic Tool for Off Line Programming of Visual Sensors" ICAR, Tokyo, 1985.
[10] T.B. SHERIDAN:"A review of Recent Research on Supervisory Control of Deep ocean Robotic vehicles and manipulators". Proc 9th IFAC, Budapest, 1984.
[11] S.TACHI, H.ARAI: "Study on Tele-existence (I and II) Design and evaluation of a three dimensional color display with sensation of presence". 5th ROMANSY 84 and ICAR 85.

Advanced Teleoperation (II) 53

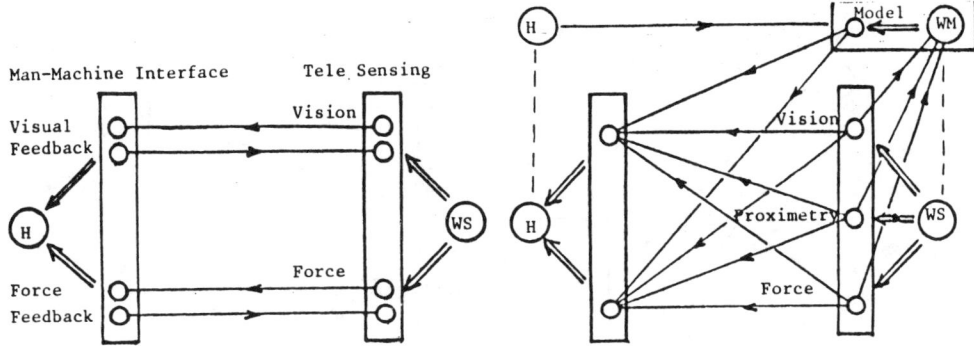

Figure 1: Telepresence and G.I.F. (Generalized Information Feedback) concepts

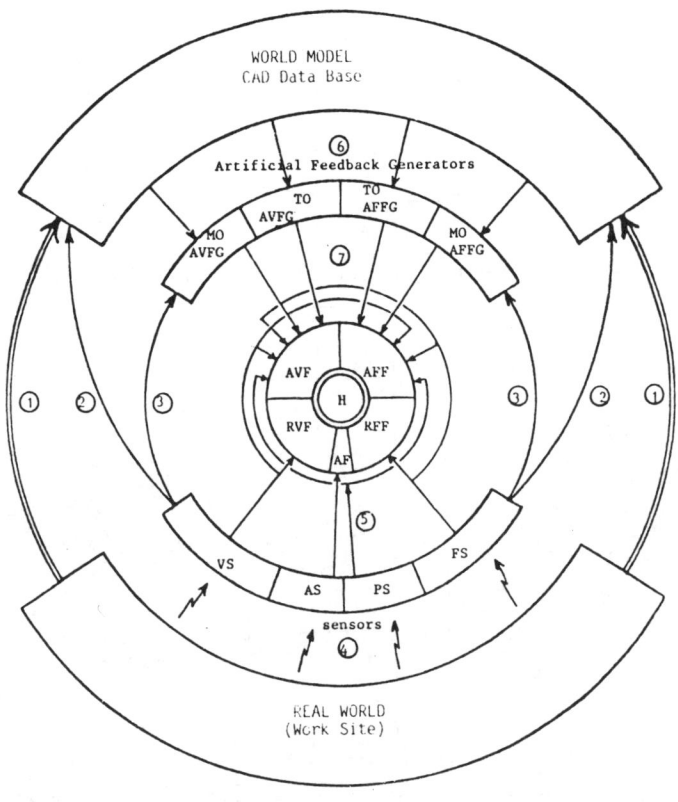

Sensors		Feedbacks		Functions	
VS	Vision Sensor	TO	Task Oriented	①	A priori Modelling
AS	Audio Sensor	MO	Man Oriented	②	Building up a Model
PS	Proximity Sensor	RVF	Real Vision Feedback	③	updating parameters
FS	Force Sensor	AFF	Artificial Force Feedback	④	Sensing
				⑤	Direct Information Feedbacks
				⑥	Data Base extraction
				⑦	Synthetic information displays

Figure 2: The I.R.I.S. (Information Reflection, Information Synthesis) Concept

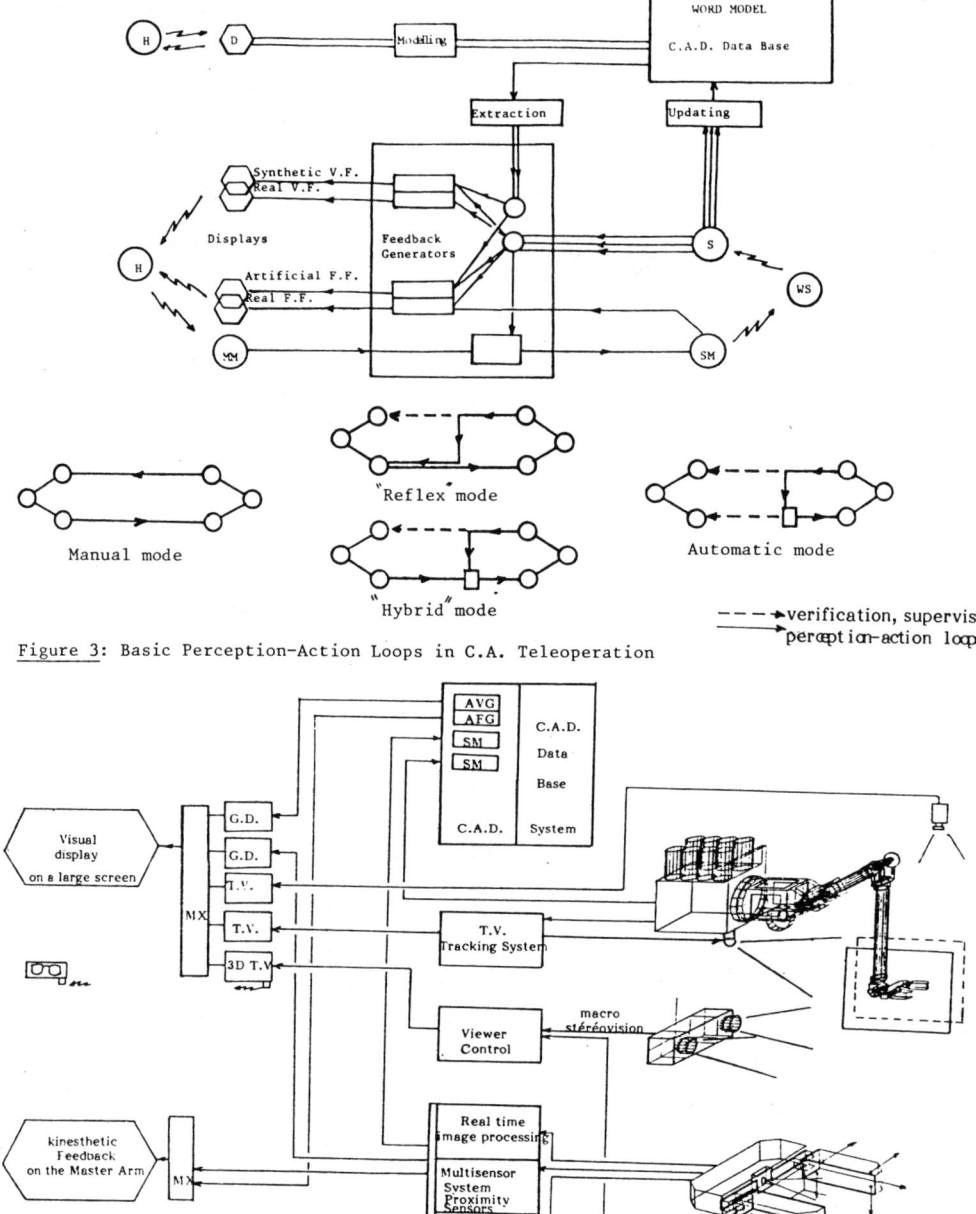

Figure 3: Basic Perception-Action Loops in C.A. Teleoperation

Figure 4 : The I.R.I.S. Experimental System

Advanced Teleoperation (II) 55

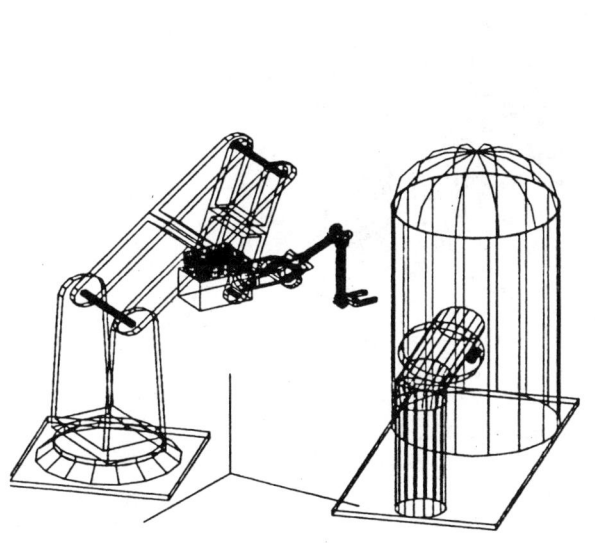
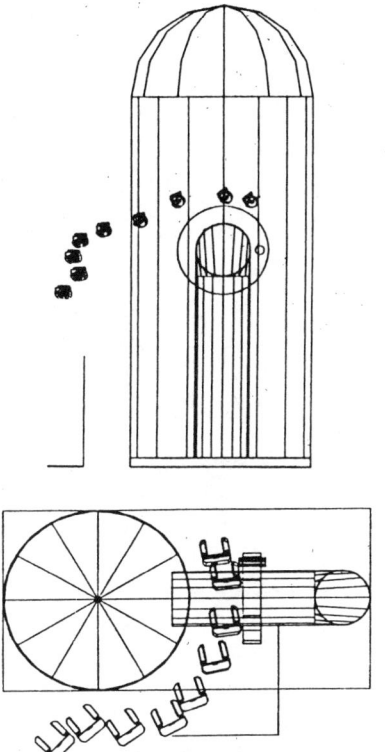

Figure 5: . C.A.D. Model of the Telemanipulator and its environment
. Trajectory of the gripper with collision avoidance
(IRISA)

Figure 6: Examples of Synthetic Pictures (LAMM)

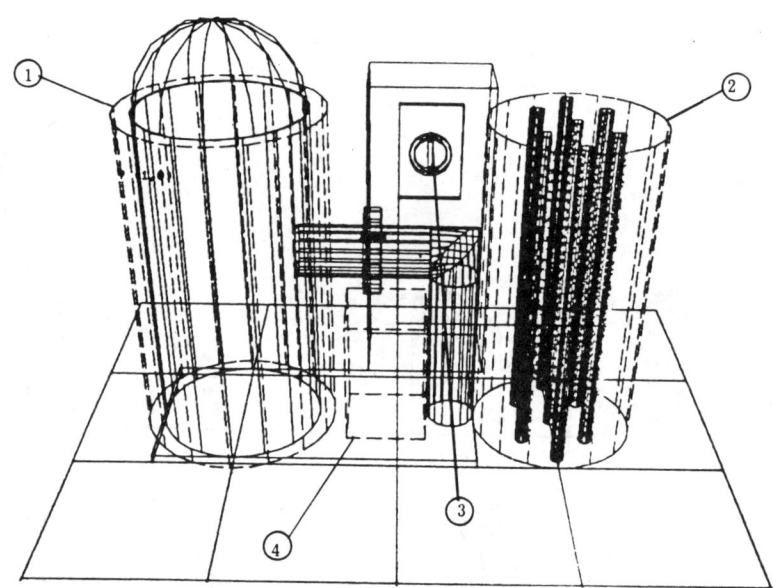

Figure 7 : Examples of interactive description of potential field subregions (IRISA)
①② repulsive regions, ③ attractive region ④ free space

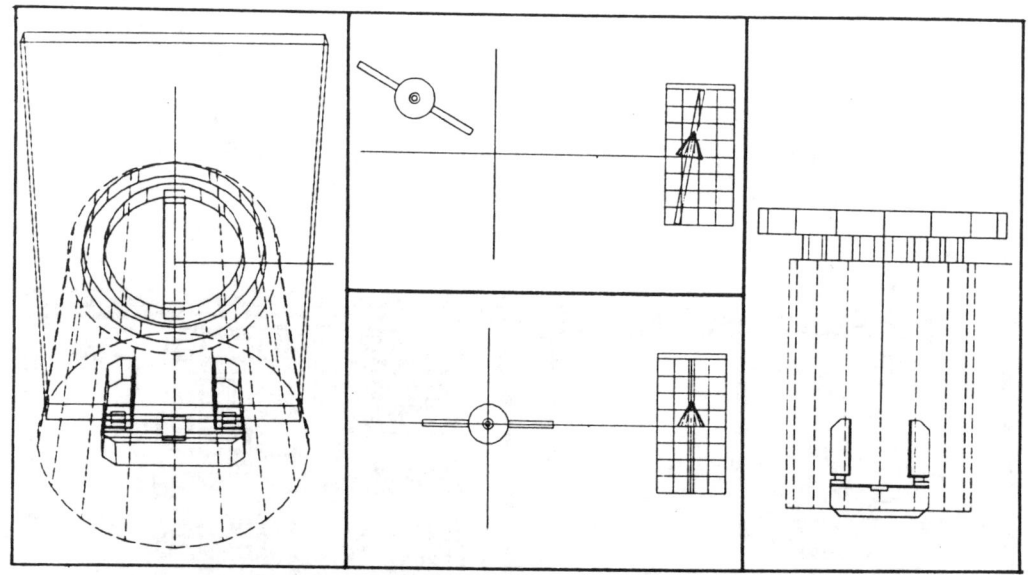

Attractive potential field for the guidance along the axis

Task oriented symbolic indicators

Top view during approaching

Figure 8 : Example of End Effector Guidance functions (IRISA)

Advanced Teleoperation (III)
An Integrated Experiment

R. Fournier (CEA), J.-P. Gaillard (LRE), N. Fiori (LPM),
B. Espiau (IRISA), Rennes, France
and the members of the Advanced Teleoperation Group

ABSTRACT

Cooperation between teleoperation subsystems is considered using an actual example of implementation. The first part of the paper gives a structural description of the site: the remote worksite includes some features of a nuclear plant with a manipulator, carrier, tools, and sensors. Then, a set of so-called "canonical experiments" taken from relevant nuclear applications is presented. Some aspects of ergonomy and evaluation are described in the second part: design of the master control station, with control sharing between an executive operator and an information management operator, and presentation of some evaluation results.

1. INTRODUCTION

The integrated experiments of the Advanced Teleoperation Group have two main motivations: firstly, it is obvious that, in practice, the advanced teleoperation functions may not be studied independently, because a CAT system actually needs to exploit cooperation between them in order to perform highly complex tasks: this is the "system" approach. Further, studying these interactions ensures synergy between all the components of this research project and the design of each functional element thus becomes more efficient.
  Experimenting with CAT features requires both gathering adequate facilities and selection of the test tasks. The nuclear field ((2)) was chosen as a suitable example for the following reasons: the possibility of finding generic tasks, availability of equipment, good theoretical and practical knowledge of the domain in CEA, and relatively easy implementation. The test tasks were selected on the basis of three criteria:
- to be relevant from the point of view of application
- to be complex enough to exemplify CAT functions
- to be simple enough to be performed at a single demonstration site
  This paper describes the main characteristics of the realized experiments. Firstly, we present the experimental system, which includes a remote slave station,

a master control station, and a computer-based management and control system.
Then, we provide details about the performed canonical tasks using an example.
The paper ends with consideration of ergonomical problems ((3), (4)), which are
approached in two ways: the a-priori design of the control station, and the
a-posteriori evaluation of some system performances. In these problems, the two
main human factors taken into account are the physical and cognitive loads during
task execution.

2. SYSTEM ARCHITECTURE AND CANONICAL EXPERIMENTS

2.1 Structure of the System

The main elements which constitute the CAT system, several of which are presented
in figures 1 and 2, are the following ((1)):

2.1.2 Remote Workstation

- **manipulators**: a nuclear MA 23 slave manipulator (1), with its 4-axis PANDA transporter (2); a prototype surgical micromanipulator, geometrically different from
the two master arms, aimed to check various coupling possibilities.
- **visual sensors**: a servoed camera for tracking and effector motions (3); two
automatic and manual colour cameras (4); a stereoscopic black-and-white microcamera.
- **non-visual sensors**: a modular multisensory system, including various proximity
sensors, a laser/microcamera system and a 6-dimensional force/torque wrist sensor.
- **tools**: a set of various pneumatically-driven tools: saw, grinder, screwdriver,
with locally integrated sensing, and an automatic tool-exchanging device.

2.1.2 Master Control Station (MCS)

This is driven by two operators working in collaboration, whose attributes are
described in section 3.1, and who share the input/output devices, as shown in figure 1:
- manipulators: a nuclear MA 23 master arm (5); a universal force-reflecting
mini-master arm.
- visual displays: a video projector using a large screen (6), with a control
and processing system (7) including various video screens for TV cameras; a stereoscopic visual display using polarized spectacles.
- graphic and alphanumeric displays (8): graphic displays dedicated to proximity
and force/torque representation, and to artificial vision; alphanumeric and graphic
terminals for representing local tasks, switching of modes, default and emergency
operations and to support communication with the supervisory system through a
hierarchical menu (see paper 1).
- control desks (9): control of the slave carrier, of the mobile cameras, of the
tracking cameras, and of the audiofeedback from the slave world.

The MCS also includes a speech recognition and synthesis system which is used
for some high level commands, like tool exchange.

2.1.3 Control and Management System (CMS)

We only briefly summarize the principles of CMS, which is more thoroughly described in paper 1. The architecture is distributed to dedicated microcomputers through a serial star network. Some communication channels, which require a high rate of data transfer, use specialized connections (i.e. between the master/slave controller and the multisensory system). When possible, the system architecture is hierarchical, for example in the case of the communication system between the supervisor and the operators which is used during task generation and execution. All the information about state, configuration, alarm, mode switching and planning pass through the network, between the supervisory system and the computer which manages the operator's interface. This computer interprets the commands from the operators, processes the information and generates control instructions for various intelligent terminals listed above.

Further, the slave MA 23 is monitored by a dedicated parallel diagnoser, which checks several electrical and electromechanical components. Its activity is transparent, except in the case of failure detection, where alarms are transmitted to the supervisory system.

2.2 Canonical Experiments

These experiments aim to exploit the new possibilities provided by the CAT system:
- task programming and planning through supervisory control, shared between the computer and a decisional operator
- new control modes sharing the execution of a task between man and machine using generalized master/slave control and generalized information feedback

Details on these two points are provided in papers 1 and 2, and we only recall here that the 3 main groups of control modes are:

Robot and master/slave decoupled modes: These allow performance of automatic subtasks, and definition of decoupling and recoupling procedures;

Unilateral master/slave modes: The slave executes predefined functions, with a passive copy of the master;

Bilateral master/slave modes: They include a large family of elementary modes, from fully manual control to fully automatic control.

In the last case, parallel sharing consists in allocating each variable of the configuration space to one of the control generators: human operator, sensors, computer (programmed motion, generation of geometrical constraints).

All of these functional capabilities have to be checked in the canonical experiments.

2.2.1 The Selected Applications

They have been chosen according to the criteria stated in section 1. They belong to three main classes:

Inspection: The related tasks include some functions of surface following and close stereovision. The principal control mode is the so-called "mixed-mode" ((5)), which combines geometrical constraints and sensor-referenced loops.

Dismantling: The two main tasks of this classical nuclear application are to cut pipes with a grinder, and to cut other kinds of parts using a saw. This typically uses semi-automatic modes, for operator assistance and guidance, like the "reflex mode," where a sensor-based error generates an additional force feedback.

Maintenance: This application is concerned with two kinds of tasks:
- programmed drilling, which needs a learning stage followed by a sequence of fully automatic subtasks;
- mounting/dismantling of a valve, which demonstrates the interest of synthetic vision when some parts are masked or not easily reachable; this example also requires some automatic subtasks like exchanging the tools.

2.2.2 An Example of Task

Figure 3 shows a typical task selected among the canonical experiments: programmed multiple drilling. All the features which are used in this simple task are described in papers 1 and 2.

The whole task is hierarchically described from left to right, with the corresponding states of GMS and GIF in the last part. The time direction starts from the top to the bottom. We may notice that the information feedback includes: vision (synthetic and actual), effort (synthetic and actual, with kinaesthesic or graphic feedback), and visualisation of the proposed menu. The control modes exploit both automatic subtasks which use the points learned and master/slave functions with some assistance like gravity compensation or collision avoidance.

3. ELEMENTS OF ERGONOMY ((5))

3.1 Design of the Master Control Station

As shown in figure 4, the automation process reduces human muscular activity while increasing mental load. It is well known that human efficiency increases with learning and decreases with fatigue or stress. Further, there exists a critical threshold where managing the system produces a heavy mental load which saturates the capabilities of an operator. The MCS has thus been designed from such ergonomic criteria in order to ensure a high degree of safety during the execution of a task.

Activating robot-modes or assistance-modes may be realized through simple on/off signals; but filtering at each time the relevant information for the task requires human intelligence and complex responses. This is why this function is confined to a second operator, named "information management operator" (IMO), the first one being the "task operator" (TO).

The two operators stay in the same room, the TO being in front of the large screen. The IMO stays beside him, behind the video desk, the different functions of which are organized on the basis of ergonomic criteria. The respective functional attributes of these two operators are more precisely described in figure 5.

3.2 Some Elements of Ergonomic Evaluation

To evaluate some aspects of mental and physical loads of an operator, two sets of experiments have been conducted, in two typical tasks with quite different characteristics of dynamics and work tension: drilling and displacement/pointing.

3.2.1 Cognitive Load Evaluation

The studied GMS functions were weight suppression and d.o.f. locking, in a drilling task needing accurate control avoiding damage to the drill. MA 23 master and slave were used, and the operator watched the scene on the large screen with best possible image quality.

To evaluate the cognitive load, a secondary task was imposed, and consisted of deciding "yes" or "no" if an orally presented world included a given letter or not. Both professional and inexperienced operators were selected. The recorded variables were the task time for a drilling operation, the number of broken drills, and the reaction time for the secondary task (RT). Each subject, after a warming-up session, performed 5 trials in each expermental condition, followed by 5 other ones in the reverse order of conditions. Some of the results obtained are presented in table 1.

EXPERIMENTAL CONDITIONS	Weight suppress. d.o.f. Locking	d.o.f. Locking	Weight Suppression	No Assistance
RT (sec)	0.78	0.84	0.84	0.88
Number of broken drills	1	3	3	10

Table 1

The results show that the task time is related to the risk of breaking the drills, and significantly increases after experience, which indicates that the subjects use less pressure to avoid breakage. The assistance factors are weak with regard to the risk factors. The number of broken drills is smaller and the RT decreases with assistance. For such a task, which requires passive muscular activity and fine adjustments, weight suppression and d.o.f. locking reduce the cognitive load.

3.2.2 Muscular Load Evaluation

The simple test-task was to displace and to point an object onto a target. This task was performed by non-professional but trained operators, in order to test various force-feedback rates (R¼, R½) and weight suppression (SCV) or not (N), for two payloads (W5 = 5 kg, W10 = 10 kg). Motions were performed in sagittal/horizontal plane (DS) or in sagittal/vertical plane (DVS). The performance indexes used

were the execution time and positioning errors, while the muscular load was estimated
from the EMG activities of three muscles. Concerning the performance, the experiments
have shown that DS motion was better than DVS motion, that R⅙ gave better results
than R¼, and that weight suppression did not provide improvements for this task.
Some results of the workload evaluation are given in figure 6, and show that DS
motion is better again than DVS motion. Unfortunately, no relevant result is obtained
concerning the interest of suppressing weight for decreasing the muscular load
in this kind of simple task.

3.2.3 <u>Comments on the Ergonomic Studies</u>

These two series of results show the necessity of being careful when speaking
of "operator assistance," and to only define this notion with regard to a task
("aided task"). We also see that, for some dynamic tasks, weight suppression may
sometimes increase the physical load. This is due to the fact that an operator
is physically used to control a motion taking into account gravity, while, when
suppressed, he has some difficulty in using only inertial information. For slow
and precise tasks, an improvement of the quality of work may be observed, with
(for some tasks) a risk of slightly increasing the cognitive load.

4. CONCLUSION

One of the most important next goals in advanced teleoperation research is the
design of a future master control station, with applications extended to cooperative
robots. This research will obviously take advantage of the results of related
work, for example in aerospace applications, and well as the results of the work
described, understood as the starting point of more complete studies. The experimental site provides many functional capabilities on which many new concepts might
be implemented and tested.

Regarding ergonomic aspects, we may expect in the future that the problems of
physical load will become of less importance with regard to the increase in cognitive
load, especially when complex tasks will involve the use of many various control
modes. This is why tools of artificial intelligence are required for progress
in advanced teleoperation.

REFERENCES

[1] VERTUT, J. ; FOURNIER, R. ; ESPIAU, B. ; ANDRE, G. : Advances in a Computer Aided Bilateral Manipulator System, Conference : Robotics and Remote Handling in Hostile Environments, Gatlinburg, USA, april 1984

[2] CLEMENT, G. ; VERTUT, J. ; FOURNIER, R. ; ESPIAU, B. ; ANDRE, G. : An Overview of CAT Control in Nuclear Services, IEEE Conference on Robotics, Saint-Louis, USA, 1985

[3] VERTUT, J. ; ROSSIGNOL, C. : Contribution to Define a Dexterity Factor for Manipulators, 21 th Conference on Remote System Technology, 1973

[4] BOOK, W.J.; HANNEMA, D.J.: Master Slave Manipulator Performances for Various Dynamic Characteristics and Positioning Task Parameters, IEEE Transactions on Systems Systems, Man and Cybernetics, vol SM C-10, pp 764-771

[5] VERTUT, J. and al. : Advanced Teleoperation Group / ARA Project; annual reports : september 1982 (Poitiers), november 1983 (Besancon), september 1984 (Toulouse) ; CNRS, FRANCE

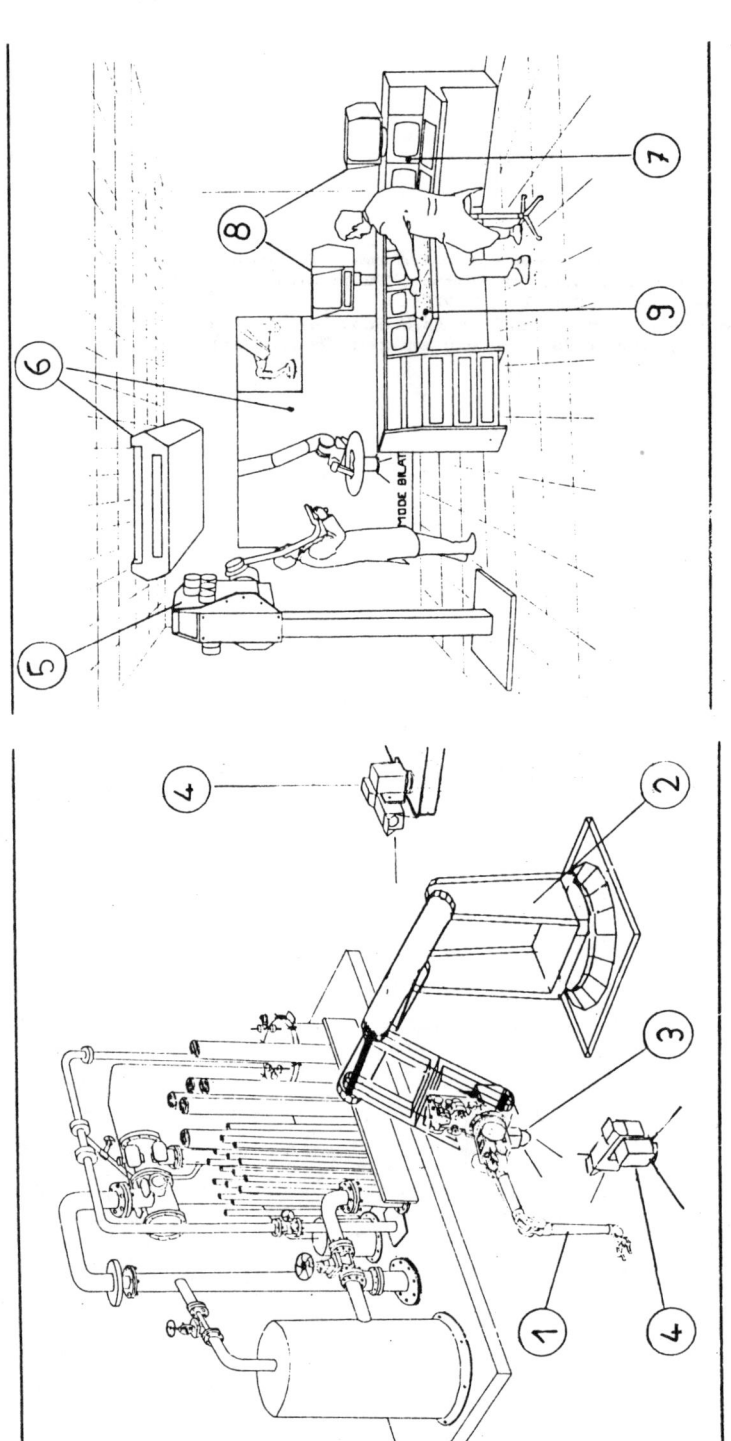

Figure 1A. The Master Control Station

Figure 1B. The Slave Work Site

Advanced Teleoperation (III) 65

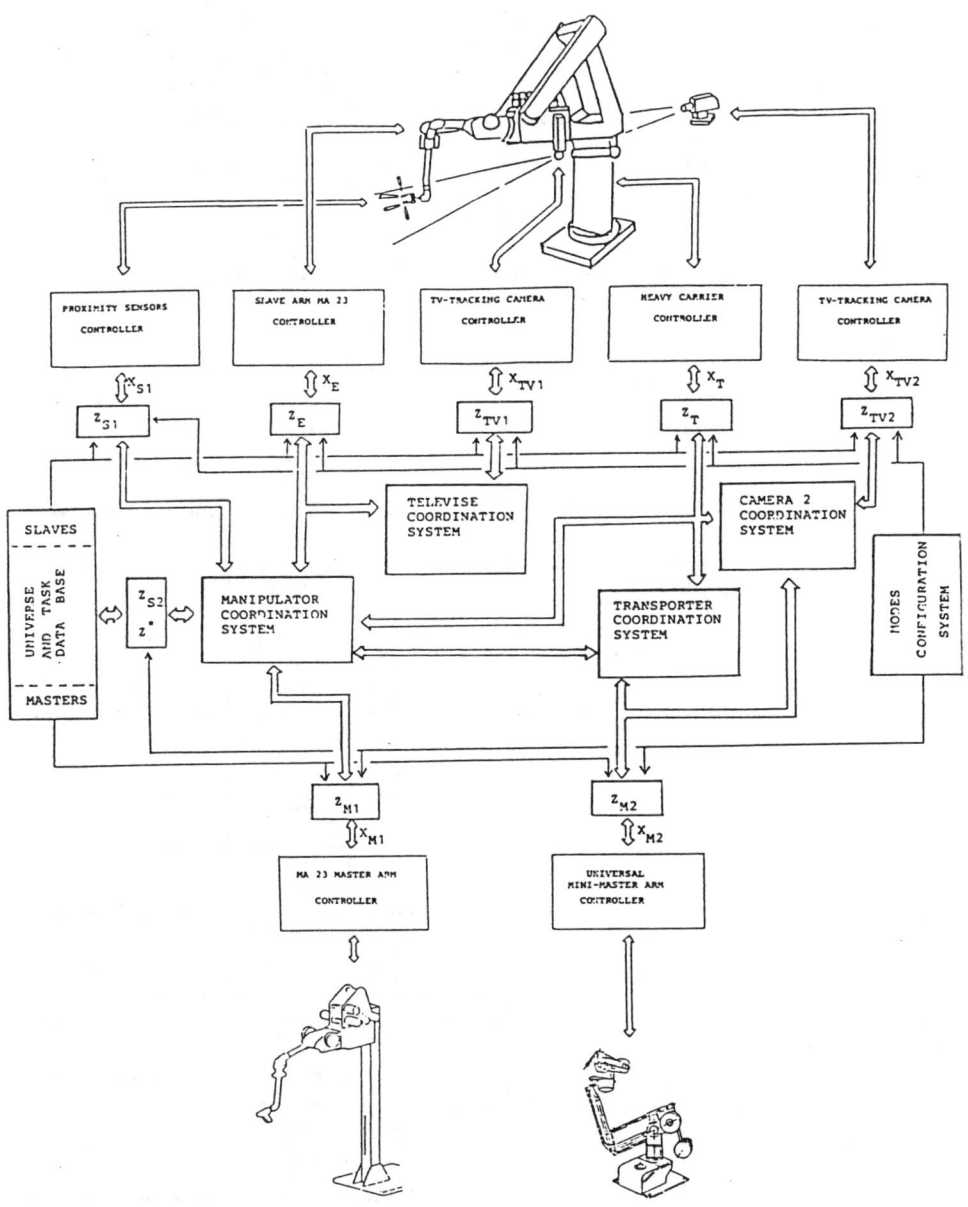

Figure 2. One Configuration of Generalized Master-Slave Control

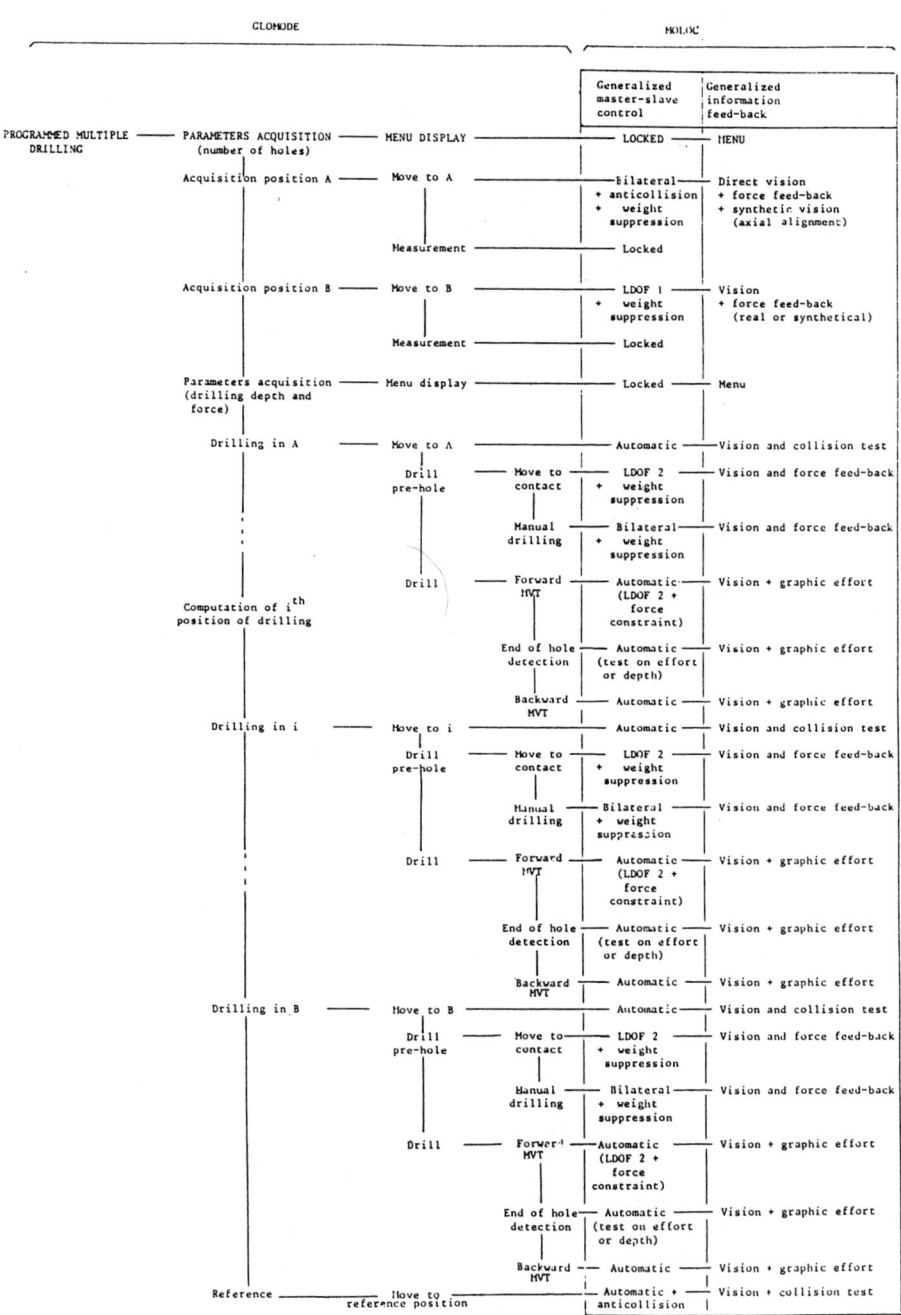

Figure 3. Ara Canonical Experiment Task Decomposition
(Programmed Multiple Drilling)

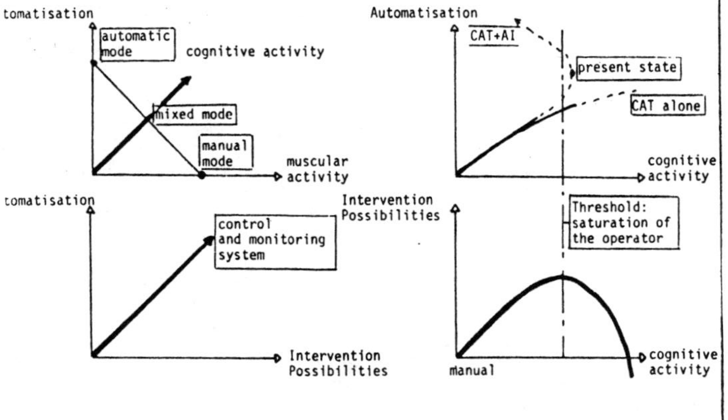

Advanced Teleoperation (III) 67

Level of intervention	TASK OPERATOR (TO)		INFORMATION MANAGEMENT OPERATOR (IMO)
	Tactical	Strategic	
NATURE OF GENERAL FUNCTIONS :			
1- PERCEPTION Available informations :			
- proximity/force feedback	kinesthetic/graphic		graphic
- visual feedback	selected and displayed on the large screen	SYNTHETIC / ACTUAL → All possible	
- audio feedback	speech recognition and voice synthesis		Slave sound feedback
- system state	local, with a simplified menu		all useful levels provided by the supervisory system
2-DECISION			
- task knowledge	local		global
- state memory	short term		long term
- possible decisions	local and limited : selection and validation of transitions through a simple, ergonomic and fast communication system		high level : - selection of the best (the most pertinent) information feedback - global monitoring - management of defaults, alarms, emergency procedures - system reconfiguration
3-ACTION			
- Kind of behaviour	fast and reflex		reflective
- Physical action	Generalized Master/Slave		no action, except the control of camera motions

Figure 5. Functions of the Operators

Figure 4. Influence of Cognitive and Muscular Loads

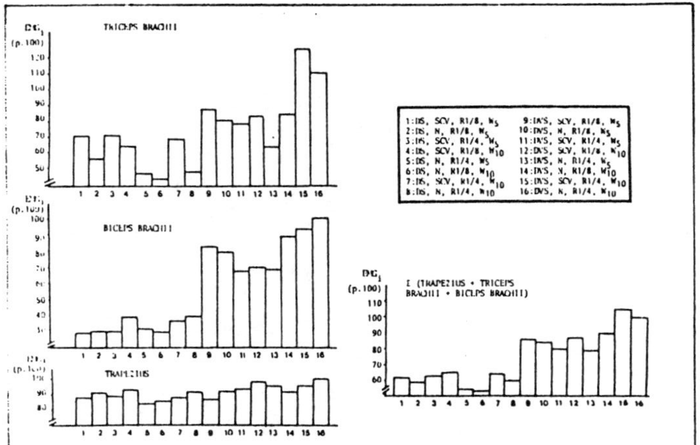

Figure 6. Results of EMG Activities

Analysis of a Robot Wrist Device for Mechanical Decoupling of End-Effector Position and Orientation

J.C. Guinot and P. Bidaud

Laboratoire de Mécanique et Robotique, Paris, France

ABSTRACT

To simplify the control model of the robot (geometric or kinematic) it is useful to uncouple end-effector position and orientation. But for a manipulator with a classical wrist, this is mechanically impossible. In this paper, we propose an orientation mechanism made up of three revolute joints, one of them is obtained by a seven-link mechanism derived from the "Peaucellier inversor". A mechanical system involving a gripper is described.

1. INTRODUCTION

Most industrial robots in use have a series arrangement for the six necessary independent degrees of freedom (d.o.f) in order to position and orient in space the last component of the kinematical chain.

Among the many possible arrangements of the d.o.f. used, in most assembly robots, one chooses the solution which consists of fixing the position of a point using the first three joints and orienting the end-effector using the last three - the "wrist".

The basic design of a robot wrist is a three revolute joint system with two-by-two orthogonal axes converging at a point in the wrist. During typical assembly operations we need to generate pure rotations of the end-effector about one of the points. In this case all the joints must be driven to compensate for the translation of the point under consideration which result from coupling of the different kinematics variables.

In these conditions, the accuracy of such trajectories depends on the coordinated control performance.

To meet these requirements, this paper proposes a mechanical approach.

The original experimental device we describe lets the end-effector have any orientation by keeping a fixed position for one point. Its kinematics is equivalent to a virtual ball joint, of which the centre can be chosen as a point of the manipulated object or the tool-extremity.

2. KINEMATICS OF THE ROBOT WRIST

It is necessary, in order to perform tasks requiring high dexterity, to be able to uncouple the position of the 'p' point of the terminal device and its orientation about this point. Therefore we arrange joint axes of the wrist so that the point of convergency 'O' coincides with point 'p' of the manipulated object, for instance when the joints are arranged as shown in Figure 1.

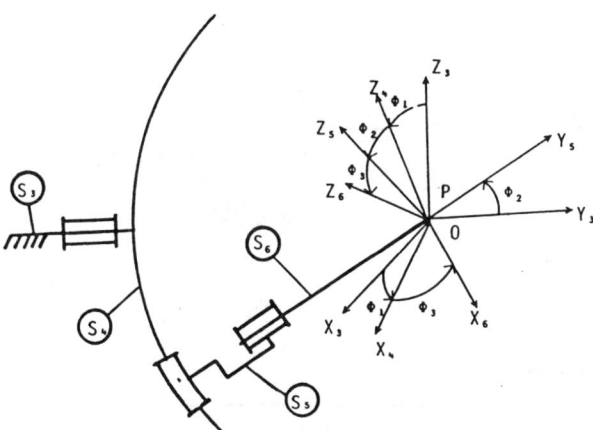

Fig.1. Kinematical arrangement of the wrist.

The index i refers to the local frames on the successive members. The angle $\emptyset_i$ is the angle of joint i. The location of frame R6 fixed to the terminal device can be obtained by means of:

- a rotation $\phi_1 = (X_3, X_4)$ about the axis $'Y_3'$
- a rotation $\phi_2 = (Y_3, Y_5)$ about the axis $'X_4'$
- a rotation $\phi_3 = (X_5, X_6)$ about the axis $'Y_5'$

The global rotation matrix is:

$$M_3^6 = \begin{vmatrix} C_1C_3 - S_1C_2S_3 & S_1S_2 & S_1C_3 + S_1C_2C_3 \\ S_2S_3 & C_2 & -S_2C_3 \\ -S_1C_3 - C_1C_2S_3 & C_1S_2 & -S_1S_3 + C_1C_2C_3 \end{vmatrix}$$

with the usual abbreviations:

$$S_i = \sin \phi_i \quad (i = 1, 2, 3)$$
$$C_i = \cos \phi_i \quad (i = 1, 2, 3)$$

Taking the wrist mounted on the end of the positioning system to be made up of three links and three joints, reproducing a cylindrical, a cartesian... coordinate system (Figure 2).

The frame Ri, with origin O_{i-1} is fixed to the Si body of the kinematical chain.

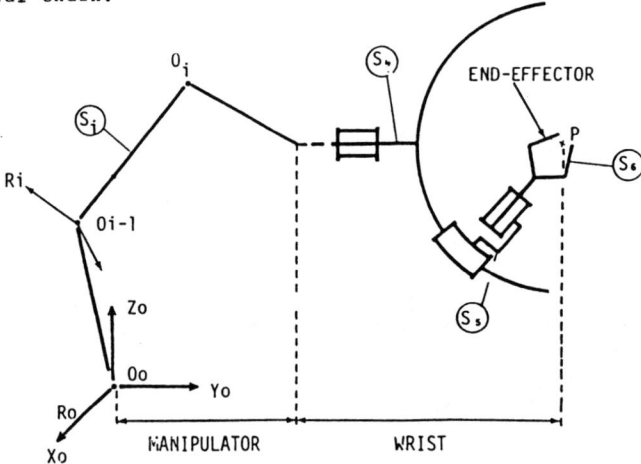

Fig.2. A manipulator with the wrist.

The location of a fixed vector in the end-effector (S6) may be given in the coordinate system R6 by the dual quantity:

$$(\vec{s}_p)_{R6} = (\vec{s} + \epsilon \vec{s}_p)_{R6} \tag{1}$$

$\vec{s}$ is a vector defined by the vectorial relation:

$$\vec{s} = s_{x6}\vec{x6} + s_{y6}\vec{y6} + s_{z6}\vec{z6} \tag{2}$$

where $\vec{x6}$, $\vec{y6}$, $\vec{z6}$ are the unit vectors orienting respectively the X6, Y6, Z6 axes of the frame.

The general relation to obtain the vector representation in the reference frame R6 can be written as:

$$(\vec{s}_o)_{RO} = M_o^6 (\vec{s})_{R6} + \epsilon [\sum_{i=1}^{6} M_o^i (\overrightarrow{O_{i-1}O_i})_{Ri} \wedge M_o^6 (\vec{s})_{R6} + M_o^6 (\vec{s}_p)_{R6}] \tag{3}$$

where M_o^6 is the matrix that transforms the representation of $\vec{s}$ and $\vec{s}_p$ from the coordinate system R6 to the RO coordinate system, M_o^i transforms the representation of vector $\overrightarrow{O_{i-1}O_i}$ from Ri to Ro.

R. Featherstone (1) formed a block matrix $^{R6}_{Ro}X^P_O$. This 6*6 matrix is a representation of the spatial transformation of the quantity $\hat{s}_p$ produced by a translation $\overrightarrow{OP}$ and a rotation expressed in the frame Ro by the product matrix M_o^6.

$$(\vec{s}_o)_{RO} = {}^{R6}_{RO}X^P_0 (\vec{s}_p)_{R6} \tag{4}$$

Any $\vec{s}$ and $\vec{s}_p$ vectors can be expressed by a single vector to give the relationship (5). Its elements represent the orientation components of $\vec{s}$ and the point position ($\vec{s}_o = \overrightarrow{OP} \wedge \vec{s}$) in RO.

$$\begin{vmatrix} s_{x_o} \\ s_{y_o} \\ s_{z_o} \\ s_{o_x} \\ s_{o_y} \\ s_{o_z} \end{vmatrix}_{Ro} = \begin{vmatrix} M_o^6 & | & 0 \\ \text{---} & | & \text{---} \\ M_o^6 \colon (\overrightarrow{OP})_{Ro} \wedge & | & M_o^6 \end{vmatrix} \begin{vmatrix} s_{x_6} \\ s_{y_6} \\ s_{z_6} \\ s_{px} \\ s_{py} \\ s_{pz} \end{vmatrix}_{R6} \tag{5}$$

If we now consider the case Figure 1 with origin in $\vec{P}$, $\vec{s}_p$ is the null vector. Using the relation (1) we get:

$$(\vec{s})_{Ro} = M_o^6 (\vec{s})_{R6} \tag{6}$$

$$\overrightarrow{OP} \wedge M_o^6 (\vec{s})_{R6} = \sum_{i=1}^{3} M_i^o (\overrightarrow{O_{i-1}O_i})_{Ri} \wedge M_o^6 (\vec{s})_{R6} \tag{7}$$

observing that point P is defined by its coordinates in Ro (X_p, Y_p, Z_p) generated by the first three independent variables of the manipulator. The geometrical model inversion is reduced to the study of two systems with three linear equations in three unknowns. In the same manner, it can be shown that the associated kinematical model is equally simple.

Considering the movement of the rigid body S6, in the reference coordinate system 'Ro', described instantaneously axis oriented by the unit vector a_{60}. The velocity of the rigid body S6 is:

$$\sum_{i=1}^{6} (\omega_{i/i-1} + \epsilon V_{i/i-1}) \cdot \hat{a}_{i/i-1} = (\omega_{60} + \epsilon V_{60}) \cdot \hat{a}_{60} \qquad (8)$$

The Jacobian matrix is directly obtained from equation (8) and can be expressed in the form (9), kinematic screw components of the end-effector being expressed in a cartesian coordinate frame.

$$\begin{vmatrix} V_x \\ V_y \\ V_z \\ \Omega_x \\ \Omega_y \\ \Omega_z \end{vmatrix} = \begin{vmatrix} J_v & | & 0 \\ \text{---} & \text{---} & \text{---} \\ & J_w & \end{vmatrix} \begin{vmatrix} \dot{\theta}_1 \\ \dot{\theta}_2 \\ \dot{\theta}_3 \\ \dot{\phi}_1 \\ \dot{\phi}_2 \\ \dot{\phi}_3 \end{vmatrix} \qquad (9)$$

Referring to the 3*3 matrix by Jv related to the first three generalized coordinates and by Jw the 3*6 matrix related to the whole of generalized coordinates, the inverse kinematical model computation inducing the control law comes to the two inversion matrix with a rank 3 and is more simple than for a conventional wrist. That is of an interest when the application needs to coordinate the six d.o.f. of the arm.

Notice that, whatever the down structure of the wrist, it appears a singularity when ϕ_2 becomes equal to zero and consequently a cone of degeneracy (2). The optimal simplicity will be realized when orientation and position will be uncoupled, for instance if the positioning device works in cartesian coordinates.

3. DESCRIPTION OF THE WRIST

The rotating movement of Y_3 and Y_6 axes can simply be obtained with classical revolute joints. The second rotation about X_4 axis needs to bring into the rotation of two others. J. P. Trevelyan studied a similar mechanical arrangement which has been used for a

shearing robot (3). Practically, it's possible to design such an
immaterial joint by combining two planar mechanisms possessing the
following geometric property. In a planar motion of the bodies
constituting the mechanism, a point belonging to one of the bodies
is restrained to move on a circle centered at an immaterial point
out of the mechanism. Fixing on each of these two mechanisms,
situated in parallel planes, the body S6 at the point describing
the circle, theoretically a circular trajectory of the body, can be
obtained, passing through the mechanism's motion center and perpendicular to this plane.

4. GEOMETRICAL CHARACTERISTICS

Mechanical devices realizing this geometric condition may be
obtained from the "Peaucellier inversor" (4) (5).

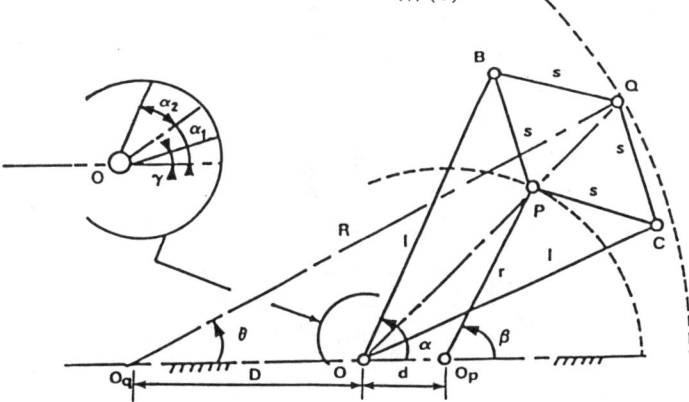

Fig. 3. Peaucellier inversor in which Q draws a circle centered in O_q.

The seven bar linkage considered here is articulated in the fixed
point O and Op. When the point P generates a circle with a radius
r and a center Op, the point Q describes a circle centered in Oq with
a radius R. With the rotations shown in Figure 3 the mechanism's
configuration may be chosen with four variables (l, s, d, r). The
values of R and D must therefore satisfy the conditions:

$$R = r \left| \frac{l^2 - s^2}{r^2 - d^2} \right| \quad (10) \; ; \qquad D = d \left| \frac{l^2 - s^2}{r^2 - d^2} \right| \quad (11)$$

In order to determine the maximal rotation range θ_{max}, physically
limited, we establish the relation.

$$\cos \theta_{max} = \frac{1}{2d} \left[2r + \frac{(d^2 - r^2)(l^2 - s^2 + r^2 - d^2)}{r(l^2 - s^2) + s(d^2 - r^2)} \right] \qquad (12)$$

In the application of two of these mechanism to body guidance, the centers of circular trajectories described by the point Q and respectively Q' (for the second mechanism) must be the same. In addition, to keep out the local mobility of the body S6, the points Oq, Q, Q' must not lie on the same straight line, so that the kinematical parameter of mobility (mc) (6) always stays equal to one. This is very important to the accuracy of the guidance. In these conditions, Figure 4 illustrates the two mechanisms location for a given θ value.

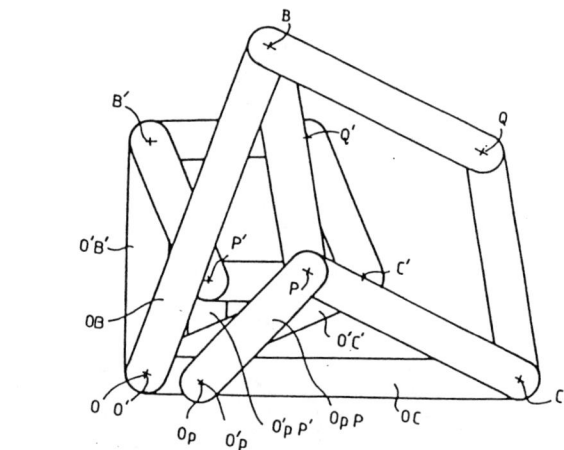

Fig. 4. Two different "Peaucellier inversors", to replace a a OqOq' axis' rotoid joint.

5. ANALYSIS OF STATICS

It has also been found necessary to consider the actuator situation. To control the motion of the body S6 about the X_1 axis, any driving links of one of the two mechanisms can be chosen. Technological problems are similar but the mechanical transmission ratio operated throught the mechanism depends on the choice of the input link.

Referring to Mx, the static screw component of the moment about X_4 axis, in any infinitesimal displacement $\delta\theta$, the couple magnitudes Cα, Cβ, Cγ respectively acting on the fixed joints of the links OB,

OpP, OC, are given by the following relations:

$$Mx \, \delta\theta + C \, \delta\beta = 0 \tag{13}$$

$$Mx \, \delta\theta + C \, \delta\alpha = 0 \tag{14}$$

$$Mx \, \delta\theta + C \, \delta\gamma = 0 \tag{15}$$

where $\alpha = \alpha_1 + \alpha_2$ and $\gamma = \gamma_1 + \gamma_2$ (see notation in Fig. 3)

Thus, the functions $F_{(\beta)} = \delta\theta/\delta\beta$, $G_{(\alpha)} = \delta\theta/\delta\alpha$ and $H_{(\gamma)} = \delta\theta/\delta\gamma$ deduced from the above relations represent the transmission ratio, obtained with different input links.

Their maxima may be written as follows:

$$F_{(\beta)\,max} = \left(\frac{\delta\theta}{\delta\beta}\right)_{max} = \frac{(r^2 - d^2)(r + s)}{r(l^2 - s^2) - s(r^2 - d^2)} \tag{16}$$

$$G_{(\alpha)} = \left(\frac{\delta\alpha}{\delta\theta}\right)_{max} = \infty \tag{17}$$

$$H_{(\gamma)} = \left(\frac{\delta\theta}{\delta\gamma}\right)_{max} = \frac{(r + s)(r^2 - d^2)}{r(l^2 - s^2 + r^2 - d^2)} \tag{18}$$

On comparing these maxima, we conclude $F_{(\beta)} < H_{(\gamma)} < G_{(\alpha)}$ $\forall$ (l, s, d, r). This relation must be taken into account to justify the choice of the driven link O_pP.

6. DESCRIPTION OF THE PROTOTYPE

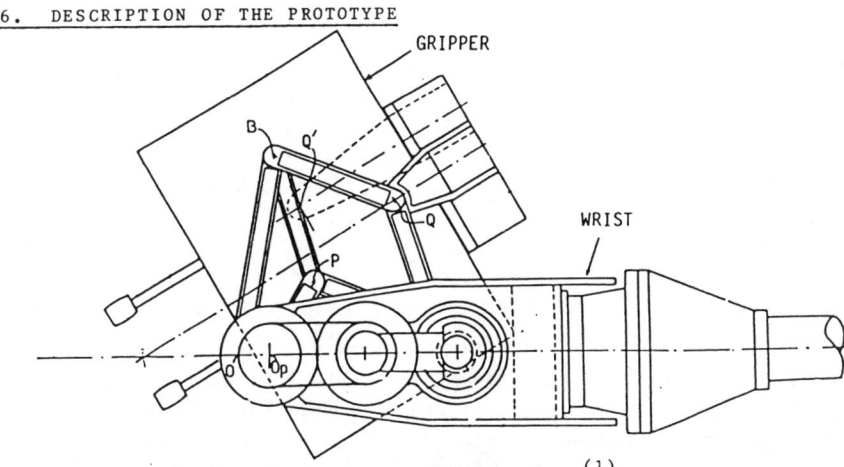

Fig. 5. Front view of prototype of the wrist.[1]
(1) Produced by Service des Prototypes du CNRS.

The lengths of the mechanism's links are determined by attempting to find:
- good access to the objects we have to grasp $D1 > 70$ mm; $D2 > 70$ mm.
- minimal interaction effects of the running clearances: $R1 \gg R2$.
- a better value for the limited rotation $\theta_{max} = \pi/2$.

If k is the ratio R/D equal to r/d, the relation (8) can be written as:

$$F_{(\beta)} = \frac{\delta\theta}{\delta\beta} = \frac{k^2 + 1 - 2k \cos \theta}{k^2 - 1} \tag{19}$$

When two mechanisms of Peaucellier have the same ratio k, their evolution laws are identical. Thus with θ_{max}, R, D and d, we can obtain the other parameters l, s r expressed by:

$$r = kd \tag{20}$$

$$s = kD + \frac{(k^2 - 1)(D + d)}{2(\cos \theta_{max} - k)} \tag{21}$$

$$l = (s^2 + Dd(k^2 - 1))^{\frac{1}{2}} \tag{22}$$

The gripper has been designed to securely grasp objects with different shapes. So as to make sure the object is completely immobilized by the finger, we use a special type of three fingered gripper described in (6).

Previous work has been conducted to analyse the stability of the grasp with a three fingered hand (7). The fingers are independently moved by electrical actuators associated with transmission mechanisms known as the "Balancier d'Olivier Evans". Using isostatic gripping, local sensors have been introduced in each finger to restore the static forces developed on the manipulated object.

7. THE DRIVING UNIT

All the actuators are direct current motors with permanent magnets. These motors are associated with a "harmonic-drive" reducer. The instantaneous joint position is obtained with an incremental optical encoder. We plan driving of mechanisms of Peaucellier simultaneously by two independent actuators for the mass balancing and taking up of play, eliminating backlash.

Fig. 6. Actuator of Mechanisms of Peaucellier.

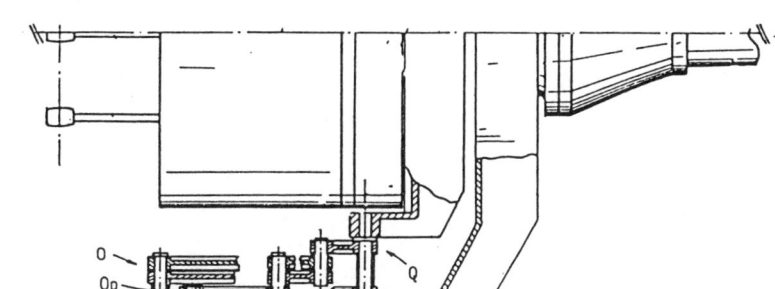

6 - REFERENCES :

|1| R. FEATHERSTONE, "The calculation of Robot Dynamics Using Articulated Body Inertias". Journal of Robotics Research, Vol. 2, n°1 (1985).

|2| R.P. PAUL, C.N. STEVENSON, "Kinematics of Robot Wrists". Journal of Robotics Reasearch, Vol. 2, n°1 (1985).

|3| J.P. TREVELYAN, "Skills for a shearing Robot Dexterity and Sensing", 2nd ISRR, August 1984, Kyoto - Japan.

|4| R. BRICARD, "Leçons de cinématique", Gauthier Villars (1926).

|5| R.S. HARTENBERG, J. DENAVIT, "Kinematic Synthesis of Linkage". Mc Graw-Hill Series in Mechanical Engineering.

|6| R. SIESTRUNCK, "Quelques aspects nouveaux de la théorie des mécanismes et applications". Revue Française de Mécanique, n°45 (1983).

|7| S. ZEGHLOUL, "Analyse de la préhension - Application à un préhenseur tridigital doté de sens tactile", Thèse Doct. 3ème cycle, Poitiers 1985.

|8| J.C. GUINOT, J.P. LALLEMAND, S. ZEGHLOUL, "Etude d'un préhenseur tridigital avec sens tactile pour une opération d'assemblage", Proc. 6th Congress on Theory of Machines and Mechanisms, Dec. 1983, New-Delhi India.

Part 2
Mechanics 1

Robot Motion: Configuration Analysis of Redundant and Nonredundant Manipulators

M.S. Konstantinov and D.N. Nenchev

Robotics Department Higher Institute of Mechanical and Electrical Engineering

ABSTRACT

The subject of this paper is the analysis of singular and boundary configurations of open loop spatial kinematic structures. Redundant as well as nonredundant manipulators are considered. Global and local variables are introduced to classify the current (instantaneous) manipulator configuration (as a regular one,) a a singular, a boundary or a singular-and-boundary configuration. The analysis is based on a linearized inverse kinematic problem solution. The end effector's motion trajectory problem is discussed in connection with the type of a given instantaneous robot configuration. Graphic interpretations are also presented.

1. INTRODUCTION

Recently utilization of robot control in terms of cartesian coordinates has assumed increasing importance. This control method enables adaptive control based on external sensory information (1), force feedback control (2,3), as well as continuous path motion in cartesian coordinates (4) to be used. The robot controller normally has a hierarchical structure, at the following three levels: motion planning, coordinates transformation and servo-control (5).

It is well known that transformation of a gripper's cartesian coordinates (position and orientation) to generalized (joint) coordinates of the manipulator gives rise to difficulties, due to nonlinear analytical terms expressing the relation between these coordinates. In particular, this is the case in so called "singular configurations" (6), where the degree of freedom of the end link decreases. In some other cases, cartesian motion generation is performed correctly, while corresponding joint trajectories, calculated from the coordinates transformation level, lead to physical constraints of joint motion.

Hence, the analysis of controlled motion trajectory feasability requires analysis of every instantaneous arm configuration. Linearized inverse kinematic problem solutions are appropriately employed to check for an irregular configuration; they are well developed and provide a unified approach to nonredundant as well as to redundant manipulators.

2. BASIC NOTATION

The robot task is described by a set of so called <u>task variables</u>, in some cases this is performed redundantly (for example, the gripper orientation in 3 dimensional space can be expressed by a 3x3 matrix or by 2 unit vectors (7)).

We consider only independant task variables, i.e. a nonredundant description of the task. The task variables are represented by a vector
$$\bar{x} = [x_1, x_2, \ldots, x_m]^T, \bar{x} \in R^m.$$
where m denotes the number of independant task variables.

The number of manipulator joints is n and joint coordinates are expressed by a vector $\bar{q} = [q_1, q_2, \ldots, q_n]^T, \bar{q} \in R^n$.
A set of n joint coordinates $q_1, q_2, \ldots, q_n$ defines an arm configuration in R^n. The relation between $\bar{x}$ and $\bar{q}$ is expressed by
$$\bar{x} = F(\bar{q}). \qquad (1)$$
which is nonlinear. Differentiating eq. (1) with respect to time the linear expression
$$\dot{\bar{x}} = J(\bar{q})\dot{\bar{q}}, \qquad (2)$$
is obtained, where $J(J:R^n \to R^m)$ is the Jacoby matrix for $F(q)$. Kinematic control algorithms are based on the equations:
$$\Delta \bar{x}^k = J(\bar{q}^k) \Delta \bar{q}^k, \qquad (3)$$
where $\Delta \bar{x}^k$ denotes a small change of the robot task vector for the k-th arm configuration $\bar{q}^k \in R$, and $\Delta \bar{q}^k = (\Delta q_1, \Delta q_2, \ldots, \Delta q_n)^T$ is a small arm configuration change, which leads to realization of a given $\Delta \bar{x}^k$.

Variables n and m are called <u>global variables</u>, which characterize the robot and the robot task.

It is known that a robot with n joints can perform a task, described by m variables only if a necessary condition $n \geq m$ is fulfilled*. We consider that this assumption holds true for any case. The difference $n-m \geq 0$ is called the <u>degree of global redundancy</u> of the robot with respect to the task.

---
*A sufficient condition for task executability is derived from structural properties of the kinematic chain in (8).

The ability of the robot to perform a given task is defined through its <u>manipulability</u> (7), i.e. the execution of a small change of the task vector $\Delta \bar{x}^k$ by means of a configuration change $\Delta \bar{q}^k$. This depends on w^k - the dimension of R(J) (range of J). It describes quantitatively the degree of freedom of the end link at q^k. The robot task can be realized if w^k = m for every k.

3. BOUNDARY CONFIGURATIONS

Physical constraints, imposed on generalized coordinates define a n-dimensional parallelopiped: $Q=\{\bar{q}: \bar{q}^{min} \leq \bar{q} \leq \bar{q}^{max}\} \in R^n$. Physical constraints are also imposed on the vector change of the arm configuration $\Delta Q=\{\Delta \bar{q}: \Delta \bar{q}^{min} \leq \Delta \bar{q} \leq \Delta \bar{q}^{max}\} \in R^n$. ΔQ is called a <u>set of permitted small configuration changes</u>. Now we can express a set of boundary configurations $Q^b = Q/Q'$, where (see Fig. 1)
$$Q' = \{\bar{q}: \bar{q}^{min} - \Delta \bar{q}^{min} \leq \bar{q} \leq \bar{q}^{max} - \Delta \bar{q}^{max}\} \in Q.$$
A given arm configuration $\bar{q}^b$ is called a boundary one, if $\bar{q}^b \in Q^b$. For any boundary configuration the set of permitted small configuration changes has to be redefined to keep the current configuration within Q (see Fig. 2).

$$\Delta Q^b = \Delta Q \cap \Delta Q''$$

where
$$\Delta Q'' = \{\Delta \bar{q}: \bar{q}^{min} - \bar{q}^b \leq \Delta \bar{q} \leq \bar{q}^{max} - \bar{q}^b\} \in R^n.$$

In the case of a boundary configuration $\bar{q}^b$, the realization of $\Delta \bar{x}^k$ may lead to satisfaction of

$$q_i^{max} \geq q_i > q_i^{max} - q^b,$$
$$q_i^{min} \leq q_i < q_i^{min} - q^b, \quad i=1,2,\ldots n \quad (4)$$

It means that the small change in joint i is permitted, but it cannot be executed due to physical constraints on the i-th generalized coordinate. We assume that $\Delta q_i = 0$, i.e. the i-th manipulator joint is locked. This leads to elimination of the i-th column of J in eq. (3) and to respective reduction of its dimensions. A local variable v^k can be introduced, denoting the degree of mobility (or the number of columns in J after elimination) at q^k. If any of the inequalities in (4) are fulfilled, $v^k < n$, for all other cases $v^k = n$. The difference $n-v^k$ defines the order of a boundary configuration (the number of joints locked).

The executability of the robot task depends on a boundary configuration $q^b \in Q^b$. For example, if the columns v^k of the Jacobian matrix are less than its rows m, than from $w^k = \dim R(J) \leq v^k$ it follows that $w^k < m$ (as it has been shown earlier, this leads to nonexecutability of the robot task).

4. SINGULAR CONFIGURATIONS

The robot is in a singular configuration when

$$w^k < m \qquad (5a)$$

and

$$v^k \geqslant m \qquad (5b)$$

i.e. in a singular configuration rank $J(\bar{q}^k)$ is less than the number of task variables. In accordance to (5b) the reduction of $\dim R(J)$ is not due to a boundary configuration, but only due to linear dependance of the rows of J. Hence, there exist some small changes $\Delta \bar{x}^k$, which are nonexecutable.

An analytic expression of (5) for manipulators with zero degrees of global redundancy (n=m) is:

$$\det J = 0. \qquad (6a)$$

while for manipulators with any degree of global redundancy (n>m):

$$\det JJ^T = 0. \qquad (6b)$$

In a singular configuration small changes of the end link position and orientation lead to significant changes of the arm configuration. The set S of singular configurations for a given ΔQ can be described: $S = \{ \Delta \bar{q} : \Delta \bar{q}_i > \Delta \bar{q}_i^{max}$ or $\Delta \bar{q}_i < \Delta \bar{q}_i^{min}$ for any i, $i = \overline{1, n}\}$.

In some practically important cases explicit solutions can be obtained for eq. (6) in the form:

$$f_j(q_1, q_2, \ldots, q_n) = 0. \quad j = 1.r \qquad (7)$$

These equations define r hyperplanes in R^n. The set S can be represented as an union of r subsets S_j.

From equations (5) it follows that, for a given manipulator, the set of singular configurations depends on the dimension of the robot task. The difference $m - w^k$ is the order of the singular configuration.

The variables w^k and v^k are the local parameters, which characterize any arm configuration q^k. The difference $v^k - w^k$ denotes the degree of local redundancy of the manipulator at q^k. Singular- and-boundary configurations are described by:

$$v^k - w^k \qquad n - v^k . \qquad (8)$$

EXAMPLES

According to the notation, stated above, the dimensions of J at $\bar{q}^k$ are $w^k \times v^k$. If $\bar{q}^k$ is a regular configuration (i.e. a nonsingular and nonboundary one), the equations $w^k = m$ and $v^k = n$ hold true.

For the different values of global variables (m,n) and local ones (w^k, v^k) we distinguish:

1. $n = m$: manipulator without any degree of global redundancy.
1.1. $v^k = n$: a nonboundary configuration.
1.1.1. $w^k = m$: any small change $\Delta \bar{x}^k$ can be executed at $\bar{q}^k$ (Fig. 3a).
1.1.2. $w^k < m$: rank J < rank $(J|\Delta \bar{x}^k)$. There exist some $\Delta \bar{x}^k$, which are nonexecutable due to a singular configuration (Fig. 3b).
1.2. $v^k < n$: a boundary configuration. The degree of global redundancy is zero, hence $w^k < m$. Not every $\Delta \bar{x}^k$ can be performed (Fig. 3c,d). If $v^k - w^k = n - w^k$ it is a singular-and-boundary configuration (Fig. 3e).
2. $n > m$: manipulator with global degree of redundancy.
2.1. $v^k = n$: a nonboundary configuration.
2.1.1. $w^k = m$: any $\Delta \bar{x}^k$ can be executed by means of an infinite number of small configuration changes, due to the global redundancy (Fig. 3f).
2.1.2. $w^k < m$, $v^k > n$: a nonboundary singular configuration (Fig. 3g).
2.2. $v^k < n$: boundary configurations.
2.2.1. $w^k = m$: any $\Delta \bar{x}^k$ can be performed.
2.2.1.1. $v^k > m$: any $\Delta \bar{x}^k$ can be performed by means of an infinite number of small configuration changes (Fig. 3h).
2.2.1.2. $v^k = m$: still any $\Delta \bar{x}^k$ is executable, but there is no local redundancy due to the second order boundary configuration (Fig. 3i).
2.2.2. $w^k < m$, $v^k < n$. $\Delta \bar{x}^k$ cannot be performed, due to the high order boundary configuration (Fig. 3j). If $v^k - w^k > n - v^k$, the configuration is a singular-and-boundary one (Fig. 3k).

In Figures 3 different instantaneous configurations of a planar 3R manipulator are shown. In the case of a manipulator without global redundancy, a 3 dimensional task is defined with x_1 and x_2 for the position of the gripper and x_3 for its orientation.
Equation (1) is:

$$x_1 = \cos q_1 + \cos q_2 + \cos q_3$$
$$x_2 = \sin q_1 + \sin q_2 + \sin q_3$$
$$x_3 = q_3$$

whereby the definition of the manipulator's generalized coordinates can be derived from Fig. 3a, and every link has a unit length. The Jacobian matrix is:

$$J = \begin{bmatrix} \sin q_1 & \sin q_2 & \sin q_3 \\ -\cos q_1 & -\cos q_2 & -\cos q_3 \\ 0 & 0 & 1 \end{bmatrix}$$

The set of singular configurations is defined by means of (5a) and we obtain $\sin(q_2 - q_1) = 0$, or

$$q_1 = q_2, \quad q_1 = q_2 \pm \pi. \tag{9}$$

Let us consider that the manipulator is in singular configuration according to equation 9, but due to a boundary configuration a column from J has to be eliminated ($v^k = 2$). We assume for example that joint 1 reaches its physical constraints. J is rewritten as

$$J' = \begin{bmatrix} \sin q_2 & \sin q_3 \\ -\cos q_2 & \cos q_3 \\ 0 & 1 \end{bmatrix} \tag{10}$$

and the 3 dimensional set Q (Fig. 4a) is transformed into a 2 dimensional one (Fig. 4b). Since rank $J' = w^k \leq 2$, a small change of a task vector with not more than 2 independent task variables can be executed. Hence, a row from J has to be eliminated too and we obtain 3 submatrices J'_j, $j=1,2,3$. Since the order of the boundary configuration is $n - v^k = 1$, it follows from equation 8 that the robot will be in a singular-and-boundary configuration if $w^k \leq 1$, i.e. any of the equations $\det J'_j = 0$, $j=1,2,3$ hold true. The solutions are:

for $j=1 \rightarrow q_2 = 0 \pm \pi$ \hfill (11a)
for $j=2 \rightarrow q_2 = \pm \pi/2$ \hfill (11b)
for $j=3 \rightarrow q_2 = q_3 \pm \pi$, $q_2 = q_3$. \hfill (11c)

From equations 11 it follows that the configuration from Fig. 3d is not a singular-and-boundary one. For the configuration on Fig. 3e equation 11c holds true and this configuration is classified as singular-and-boundary one.

On Figure 4 graphic interpretation of the sets Q, Q^b, ΔQ and S_j (for $-\pi < q_i < \pi$, $i=1,2,3$) is shown.

Different configurations for the case of a manipulator with global degree of redundancy (1 dimensional task - positioning along a vertical line) are shown in Figs. 3f-k. In this case, equations 1 can be written as $x = \cos q_1 + \cos q_2 + \cos q_3$ and the respective Jacobian is $J = (\sin q_1 \quad \sin q_2 \quad \sin q_3)$. The set of singular configurations is derived from $\sin q_1 = 0$, $\sin q_2 = 0$ and $\sin q_3 = 0$. Its graphic interpretation are 27 spheres and parts of spheres, with centre points defined from 3^3 combinations of the values 0, $+\pi$ and $-\pi$ for every joint coordinate.

CONCLUSIONS

In the paper results are obtained concerning the analysis of instantaneous robot configurations. This analysis can be useful in trajectory planning algorithms to avoid or to approach singular- and-boundary configurations.

REFERENCES
1. Paul R. Robot Manipulators: Mathematics, Programming and Control. MIT Press, 1981.
2. Khatib O. The Operational Space Formulation in Robot Manipulator Control. Proc.15th ISIR, Tokyo, Japan, 1985.
3. Konstantinov M.S. et al. Force Feedback Control of Parallel Topology Manipulating Systems. Proc.15th ISIR, Tokyo, Japan, 1985.
4. Paul R. Manipulator Cartesian Path Control. IEEE Trans. Syst., Man & Cyb., Vol.SMC-9, No.11, 1979.
5. Barbara A.J. et al. Hierarchical Control of Robots Using Microcomputers. Proc.9th ISIR, Washington D.C., 1979.
6. Uchiyama M. A Study of a Computer Control of Motion of a Mechanical Arm (1st Rept.) Bulletin of the JSME, Vol.22, No.173 Nov.1979.
7. Hanafusa H. et al. Analysis and Control of Articulated Robot Arms with Redndancy. 8th Triennial W.Congr. IFAC, 1981, Japan.
8. Konstantinov M.S. et al. Motion and structure Systematization of Robots. Izv. VMEI, Sofia 1983 (in Bulgarian).

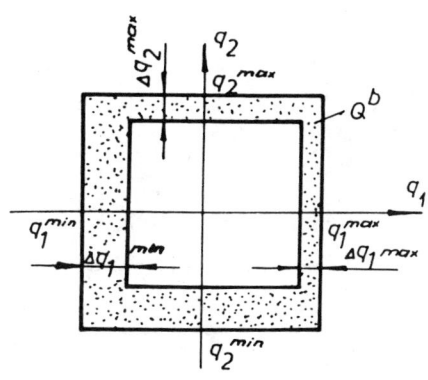

Fig. 1

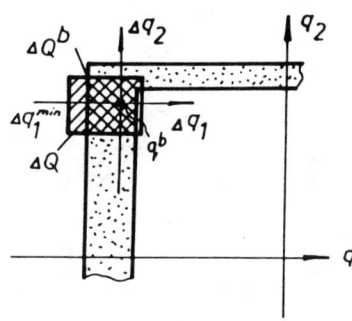

Fig. 2

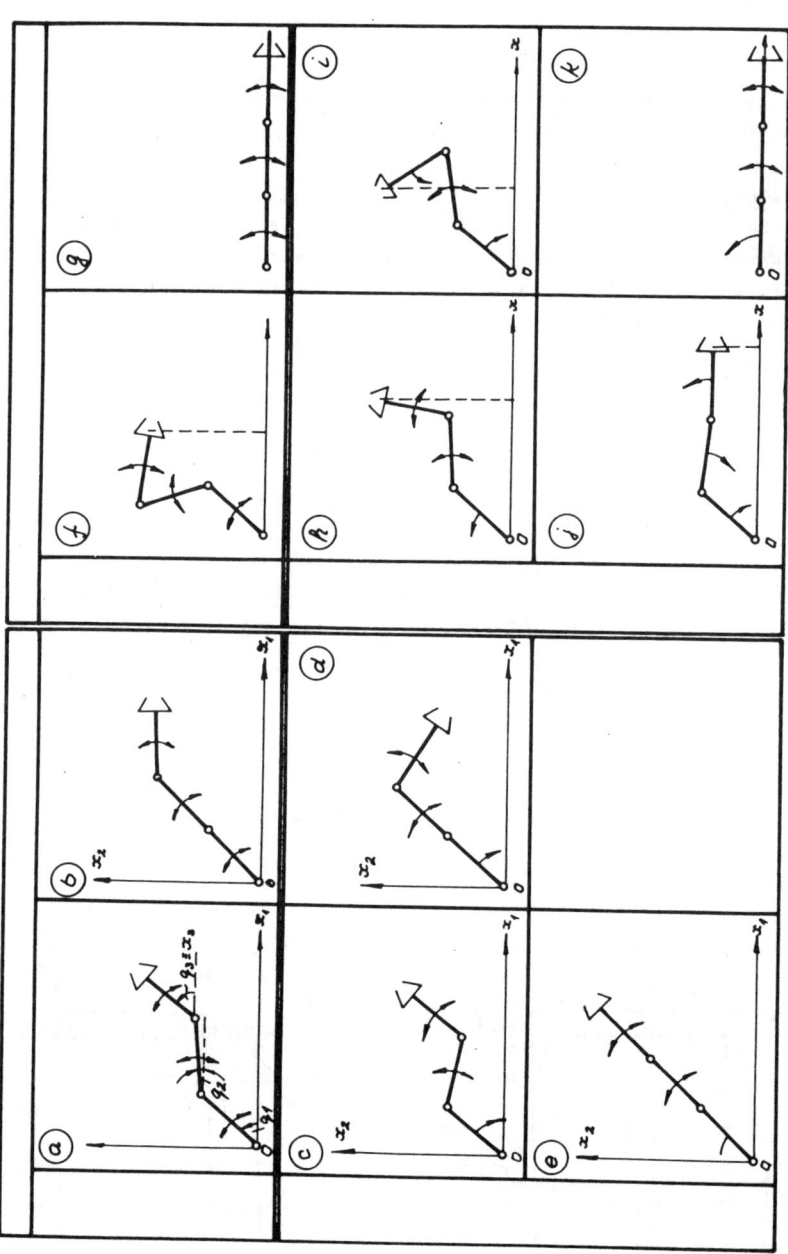

Fig. 3

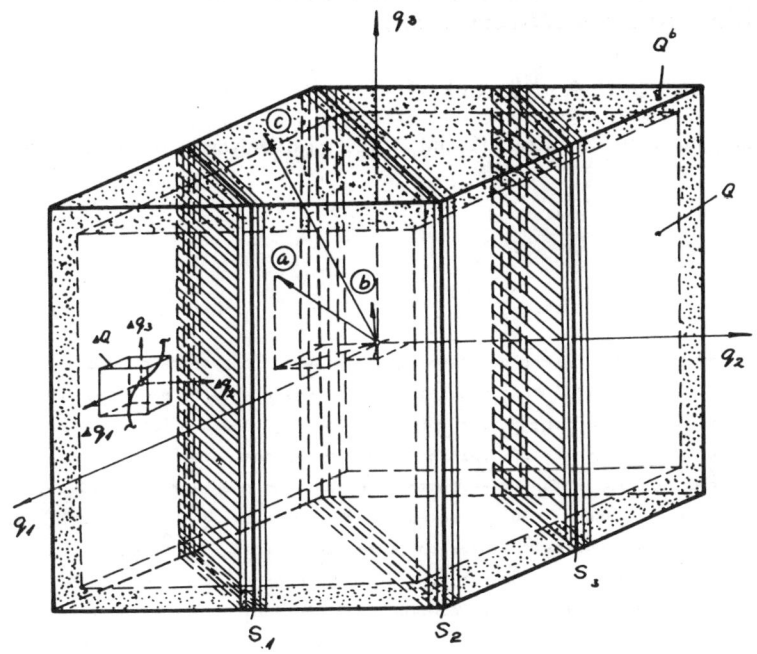

Fig. 4a

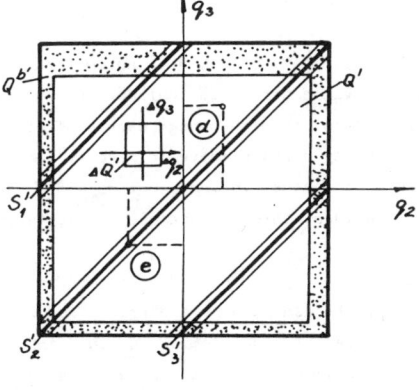

Fig. 4b

Analysis of the Positioning and Orientation Accuracy in 6R Manipulators (Direct Task)

J. Knapczyk* and A. Morecki**

*Technical University of Cracow, Poland, **Technical University of Warsaw, Poland

SUMMARY

A mathematical model of the positioning and orientation errors of mechanical manipulators is developed. This model is used to calculate the expected errors of the end-effector at any point in the workspace for a given error of positioning in the joint. The influence of positioning errors in the joint on the positioning and orientation errors of the end-effector are shown in the form of the Jacobian matrix. The direct task method presented was developed to identify the regions in which the accuracy parameters have the specified values. This region can be used to compare, evaluate and operate the manipulators.

The further step usually aims at providing a method for finding the tolerance of errors of positioning in the joint for the given positioning and orientation tolerance of the end-effector - the so-called inverse task. This task has been solved by means of Paul's method of differentiating the solution of the kinematic equations to obtain the inverse Jacobian, which is easy to evaluate and results in simple expressions.

This method can be used to assist in design, or performance optimization problems by choosing the appropriate performance criteria.

1. INTRODUCTION

A lot of industrial robot applications, such as automatization of assembly of such precise elements as electronic parts, optical devices and semiconductor devices, requires the use of industrial robots with a high degree of positioning accuracy. In order to acheive such accuracy certain criteria have to be met, namely: proper rigidity of the hand, lack of clearance in servo-motors and an adequate servo-system with negligible friction. Besides, this compactness and dexterity of an arm and smoothness of motion are required.

Positioning and orientation errors of the end-effector are functions of the position and configuration of the manipulator and arise from a number of sources, such as dimensional errors and elastic deflections of the links, joint clearances and errors in positioning the joints. These are known as servopositioning errors.

In this paper, a mathematical model has been developed to calculate the expected errors of position and orientation of the end effector at any point in workspace for a given servopositioning error of the individual joint. Joint position error is treated as a small displacement of the joint from its desired position and the method of studying them is usually associated with velocity analysis. The matrix transformation method is used to calculate the position of the hand in terms of its joint positions.

2. ANALYSIS OF THE POSITIONING AND ORIENTATION ERRORS IN MANIPULATORS - A DIRECT TASK

The positioning and orientation errors of the end-effector are interpreted as small variations in the geometric parameters and joint variables. Their cumulative effect on positioning and orientation errors of the end-effector can be found by writing the first order differential of the position and orientation vectors:

$$\Delta \underline{p} = \frac{\partial \underline{p}}{\partial \underline{a}} \Delta \underline{a} + \frac{\partial \underline{p}}{\partial \underline{d}} \Delta \underline{d} + \frac{\partial \underline{p}}{\partial \underline{\alpha}} \Delta \underline{\alpha} + \frac{\partial \underline{p}}{\partial \underline{\theta}} \Delta \underline{\theta} , \quad (1)$$

$$\Delta \underline{\delta} = \frac{\partial \underline{\delta}}{\partial \underline{a}} \Delta \underline{a} + \frac{\partial \underline{\delta}}{\partial \underline{d}} \Delta \underline{d} + \frac{\partial \underline{\delta}}{\partial \underline{\alpha}} \Delta \underline{\alpha} + \frac{\partial \underline{\delta}}{\partial \underline{\theta}} \Delta \underline{\theta} ,$$

where $\underline{p} = \begin{bmatrix} p_x & p_y & p_z \end{bmatrix}^T$ and $\underline{\delta} = \begin{bmatrix} \delta_x & \delta_y & \delta_z \end{bmatrix}^T$ - the position and orientation vectors, $\underline{a} = \begin{bmatrix} a_1 & a_2 \cdots a_n \end{bmatrix}^T$,

$\underline{d} = \begin{bmatrix} d_1 & d_2 & \cdots & d_n \end{bmatrix}^T$, $\underline{\alpha} = \begin{bmatrix} \alpha_1 & \alpha_2 & \cdots \alpha_n \end{bmatrix}^T$,

$\underline{\theta} = \begin{bmatrix} \theta_1 & \theta_2 & \cdots & \theta_n \end{bmatrix}^T$ - the link length, offset distance, twist angle and increment vectors. The angle θ_i is the joint variable and the offset distance d_i is a construction parameter, when i-th joint is a revolute pair. For a prismatic pair, the offset d_i is the joint variable and the angle θ_i is a construction parameter.

$\Delta \theta_i$ is the error of servopositioning the joint or it is a small variation in the geometric parameter θ_i. The sources of this variation can either be elastic deflections or dimensional errors arising due to manufacturing or both.

In this paper only the effects of joints servopositioning errors on the positioning and orientation of the manipulator are considered. The matrix of partial derivative of $\underline{p}$ vector for 6R manipulator is 3 x 6 array:

$$\frac{\partial \underline{P}}{\partial \underline{\theta}} = \begin{bmatrix} \frac{\partial P_x}{\partial \theta_1} & \frac{\partial P_x}{\partial \theta_2} & \cdots & \frac{\partial P_x}{\partial \theta_6} \\ \frac{\partial P_y}{\partial \theta_1} & \frac{\partial P_y}{\partial \theta_2} & \cdots & \frac{\partial P_y}{\partial \theta_6} \\ \frac{\partial P_z}{\partial \theta_1} & \frac{\partial P_z}{\partial \theta_2} & \cdots & \frac{\partial P_z}{\partial \theta_6} \end{bmatrix} \qquad (2)$$

Those partial derivatives can also be expressed as the time derivatives

$$\frac{\partial \underline{P}}{\partial \theta_i} = \frac{\partial \underline{P}}{\partial \theta_i} \qquad (3)$$

These lead to the following notation with the Jacobian matrix $^{(5)}$

$$\begin{bmatrix} \underline{\delta} \\ \underline{P} \end{bmatrix} = J \begin{bmatrix} \Delta \underline{\theta} \end{bmatrix} \qquad (4)$$

where

$$J = \begin{bmatrix} \Gamma_1^x & \Gamma_2^x & \cdots & \Gamma_6^x \\ \Gamma_1^y & \Gamma_2^y & \cdots & \Gamma_6^y \\ \Gamma_1^z & \Gamma_2^z & \cdots & \Gamma_6^z \\ \beta_1^x & \beta_2^x & \cdots & \beta_6^x \\ \beta_1^y & \beta_2^y & \cdots & \beta_6^y \\ \beta_1^z & \beta_2^z & \cdots & \beta_6^z \end{bmatrix} \qquad \underline{\Gamma}_i = \frac{\partial \underline{\delta}}{\partial \theta_i} \quad \underline{\beta}_i = \frac{\partial \underline{P}}{\partial \theta_i} \qquad (5)$$

The components of the Jacobian can be expressed in the end-effector coordinate system. It may be chosen in this way to reduce the computational complexity of determining the Jacobian. The components of the Jacobian with respect to the end-effector coordinate system comes from Paul's work (6). The appropriate equations are:

$$^eT_e = I, \quad {}^{i-1}T_e = {}^{i-1}T_i \; {}^iT_e, \quad (i=6, 5, \ldots 1), \qquad (6)$$

where I is the identity matrix,

for the revolute joint:

$$e_{\underline{r}_i} = {}^{i-1}U_e^T \begin{bmatrix} 0 \\ 0 \\ 1 \end{bmatrix}, \quad (i = 1, 2, \ldots 6) \tag{7}$$

$$e_{\beta_i^j} = \left[{}^{i-1}U_e^j \times {}^{i-1}\underline{r}_{i-1} \right]^T \begin{bmatrix} 0 \\ 0 \\ 1 \end{bmatrix}, \quad (j = 1, 2, 3; \ i = 1, 2, \ldots 6)$$

and for prismatic joint:

$$e_{\underline{r}_i} = 0, \quad (i = 1, 2, \ldots 6), \tag{8}$$

$$e_{\beta_i} = {}^{i-1}U_e^T \begin{bmatrix} 0 \\ 0 \\ 1 \end{bmatrix}, \quad (i = 1, 2, \ldots 6)$$

where

$${}^{i-1}T_i = \begin{bmatrix} c\theta_i & -s\theta_i c\alpha_i & s\theta_i s\alpha_i & a_i c\theta_i \\ s\theta_i & c\theta_i c\alpha_i & -c\theta_i s\alpha_i & a_i s\theta_i \\ 0 & s\alpha_i & c\alpha_i & d_i \\ 0 & 0 & 0 & 1 \end{bmatrix} \tag{9}$$

The top-left 3 × 3 part of the matrix ${}^{i-1}T_i$ gives all of the orientation information and is denoted by ${}^{i-1}U_i$. The first three elements of the right-hand column give all of the relative position information and are denoted by ${}^{i-1}\underline{p}_i$. The position vectors from the end-effector to link i are denoted by $\underline{r}_i$ and from the base to link i by $\underline{p}_i$. Note that ${}^{i-1}U_e^j$ is just the column j of the 3 × 3 orientation part of ${}^{i-1}T_e$. Also note that ${}^{i-1}\underline{r}_{i-1}$ comprises the first three elements of the fourth column of ${}^{i-1}T_e$. Furthermore, $e_{\beta_i^j}$ is just the j-th component of the vector e_{β_i}. The left hand side of the equation (6) can be expressed in the form

$${}^{i-1}T_e = \begin{bmatrix} {}^{i-1}n_x & {}^{i-1}o_x & {}^{i-1}a_x & {}^{i-1}p_x \\ {}^{i-1}n_y & {}^{i-1}o_y & {}^{i-1}a_y & {}^{i-1}p_y \\ {}^{i-1}n_z & {}^{i-1}o_z & {}^{i-1}a_z & {}^{i-1}p_z \\ 0 & 0 & 0 & 1 \end{bmatrix} \tag{10}$$

where $\underline{o} = [o_x \ o_y \ o_z]^T$ and $\underline{a} = [a_x \ a_y \ a_z]^T$ are the orientation and approach unit vectors of the end-effector coordinate system, $\underline{n} = \underline{o} \times \underline{a}$. $\underline{p} = [p_x \ p_y \ p_z]^T$ is the position vector. By using the elements of the matrix ${}^{i-1}T_e$ components of the Jacobian (7) for a revolute joint can be expressed as:

$$e_{\underline{r}_i} = \left[{}^{i-1}n_z \ {}^{i-1}o_z \ {}^{i-1}a_z \right]^T, \tag{11}$$

$$e_{\underline{\beta}_i} = \left[{}^{i-1}(n_y p_x - n_x p_y) \ {}^{i-1}(o_y p_x - o_x p_y) \ {}^{i-1}(a_y p_x - a_x p_y) \right]^T$$

and for a prismatic joint

$$^e\underline{r}_i = \begin{bmatrix} 0 & 0 & 0 \end{bmatrix}^T,$$

$$^e\underline{\beta}_i = \begin{bmatrix} ^{i-1}n_z & ^{i-1}o_z & ^{i-1}a_z \end{bmatrix}^T. \tag{12}$$

The components of the differential translation ($\Delta\underline{p}$) and rotation ($\Delta\underline{\delta}$) vectors, described with respect to the end-effector coordinate frame may be denoted as a 4 x 4 matrix

$$^e\Delta = \begin{bmatrix} 0 & -^e\Delta\delta_z & ^e\Delta\delta_y & ^e\Delta p_x \\ ^e\Delta\delta_z & 0 & -^e\Delta\delta_x & ^e\Delta p_y \\ -^e\Delta\delta_y & ^e\Delta\delta_x & 0 & ^e\Delta p_z \\ 0 & 0 & 0 & 0 \end{bmatrix} \tag{13}$$

where

$$^e\Delta\underline{\delta} = \sum_{i=1}^{6} {}^e\underline{r}_i \Delta\theta_i$$

$$^e\Delta\underline{p} = \sum_{i=1}^{6} {}^e\underline{\beta}_i \Delta\theta_i \tag{14}$$

Premultiplying $^e\Delta$ by ^oT_e we obtain the components of the position and orientation errors described in the base coordinate system

$$\Delta = {}^oT_e \, {}^e\Delta = \begin{bmatrix} 0 & -\Delta\delta_z & \Delta\delta_y & \Delta p_x \\ \Delta\delta_z & 0 & -\Delta\delta_x & \Delta p_y \\ \Delta\delta_y & \Delta\delta_x & 0 & \Delta p_z \\ 0 & 0 & 0 & 0 \end{bmatrix} \tag{15}$$

3. EXAMPLE

We will illustrate the above method by computing the Jacobian and the differential of the position and orientation vectors for the 6R manipulator of the industrial robot RPA-80 (figure 1). This manipulator is defined by the parameters shown in table 1.

Table 1

i	α_i^o	a_i (mm)	d_i (mm)	θ_i^o
1	-90	0	0	-135 ÷ 135
2	0	1000	0	-55 ÷ 55
3	90	0	0	-90 ÷ 90
4	-90	0	900	-90 ÷ 90
5	90	0	0	-90 ÷ 90
6	0	0	450	-90 ÷ 90

The matrices of transformation $^{i-1}T_e$ and the elements of the Jacobian matrix are given in the Appendix.

Software solving the problems of the positioning and orientation error analysis were developed in FORTRAN. All the calculations were done on the SM-4 minicomputer.

The 6R manipulator of the industrial robot RPA-80 is in the following state:

$$\theta_1 = 135^o, \theta_2 = 90^o, \theta_3 = 0^o, \theta_4 = 0^o, \theta_5 = 0^o, \theta_6 = 0^o$$

$$\Delta\theta_i = 0,2.10^{-3} [rd] \quad (i=1,2,\ldots 6)$$

When the link parameters described in Table 1 are substituted into the matrix ^oT_e we obtain

$$^oT_e = \begin{bmatrix} 0 & -0,71 & -0,71 & -1661,68 \\ 0 & -0,71 & 0,71 & 1661,72 \\ -1 & 0 & 0 & 0,01 \\ 0 & 0 & 0 & 1 \end{bmatrix}$$

and the Jacobian matrix

$$J = \begin{bmatrix} -1 & 0 & 0 & 0 & 0 & 0 \\ 0 & 1 & 1 & 0 & 1 & 0 \\ 0 & 0 & 0 & 1 & 0 & 1 \\ 0 & 2350 & 1350 & 0 & 450 & 0 \\ 2350 & -1000 & 0 & 0 & 0 & 0 \\ 0 & -0,01 & 0 & 0 & 0 & 0 \end{bmatrix}$$

The magnitude of the positioning error is

$$\Delta p = \sqrt{\Delta p_x^2 + \Delta p_y^2 + \Delta p_z^2}$$

The contour surfaces, with the magnitude of positioning errors for specified values are plotted in figure 2.

The results obtained show that the tolerances vary from 0.2 to 0.88 mm throughout workspace and are close to those given by the manufacturer of the robot.

4. CONCLUSION

The procedure and computer programs described can be used in the design process of manipulators as well as in these applications.

5. REFERENCES

1. Huang Zhen, Error Analysis of Robot Manipulators and Error Transmission Functions. Proc. of the 15th ISIR, Tokyo 1985, pp.873÷878.
2. Knapczyk J., Morecki A., Hipp R., Analysis of the Positioning and Orientation Accuracy in 6R Manipulator. Prace Nauk. Inst. Cybernetyki Techn.Politechn.Wrocl., Nr 66,1985,pp31÷40.
3. Kumar A., Waldron K.J., Numerical Plotting of Surfaces of Positioning Accuracy of Manipulators. Mechanism and Machine Theory, vol.16,No 4, 1981, pp.361÷368.
4. Morecki A., Kaczmarczyk A., Knapczyk J., Strzelecki S., Is Compactness and Simplicity Needed in Robotics? Proc. of 14th ISIR , Goeteborg 1984, pp.
5. Orin D.E. and Schrader W.W., Efficient Computation of the Jacobian for Robot Manipulators. The Inter. Journal of Robotic Research, vol.3, No 4, 1984.

APPENDIX

The ^1T_e and ^0T_e matrices are simplified by the introduction of variables $\theta_{23} = \theta_2 + \theta_3$. This should be done whenever manipulator joint axes are parallel ($\alpha_2 = 0$).
Substituting the expressions for $^{i-1}\underline{n}$, $^{i-1}\underline{o}$ and $^{i-1}\underline{a}$ in eqs(11) we obtain

$$^e_{\Gamma_1}x = -s_{23}(c_4c_5c_6-s_4s_6) - c_{23}s_5c_6 ,$$

$$^e_{\Gamma_1}y = s_{23}(c_4c_5s_6+s_4c_6) + c_{23}s_5s_6 ,$$

$$^e_{\Gamma_1}z = -s_{23}c_4s_5+c_{23}c_5 ,$$

$$^e_{\beta_1}x = \left[s_{23}(c_4c_5s_6+s_4c_6)+c_{23}s_5s_6\right]d_6 + s_{23}(s_4c_5c_6+c_4s_6)d_4 -$$
$$- \left[c_{23}(c_4c_5c_6-s_4s_6)-s_{23}s_5c_6\right]d_2 + c_2(s_4c_5c_6+c_4s_6)a_2 ,$$

$$e_{\underline{\beta}_1}^y = \left[s_{23}(c_4c_5c_6-s_4s_6) + c_{23}s_5c_6\right]d_6 + s_{23}(-s_4c_5s_6+c_4c_6)d_4 -$$
$$-\left[-c_{23}(c_4c_5s_6+s_4c_6)+s_{23}s_5s_6\right]d_2 + c_2(-s_4c_5s_6+c_4c_6)a_2 \, ,$$

$$e_{\underline{\beta}_1}^z = s_{23}s_4s_5d_4 + c_2s_4s_5a_2 - (c_{23}c_4s_5+s_{23}c_5)d_2 \, ,$$

The equations become much simpler as we move closer to link 6.

$$e_{\underline{r}_2} = \begin{bmatrix} s_4c_5c_6+c_4s_6 \\ -s_4c_5s_6+c_4c_6 \\ s_4s_5 \end{bmatrix} \, ,$$

$$e_{\underline{\beta}_2} = \begin{bmatrix} (c_4c_6-s_4c_5s_6)d_6 + (c_4c_5c_6-s_4s_6)d_4 + s_3(c_4c_5c_6-s_4s_6)+c_3s_5c_6a_2 \\ -(c_4s_6+s_4c_5c_6)d_6 - (c_4c_5s_6+s_4c_6)d_4 - s_3(c_4c_5s_6-s_4c_6)+c_3s_5s_6\,a_2 \\ c_4s_5d_4 + (s_3c_4s_5-c_3c_5)a_2 \end{bmatrix} \, .$$

$$e_{\underline{r}_3} = \begin{bmatrix} s_4c_5c_6+c_4s_6 \\ -s_4c_5s_6+c_4c_6 \\ s_4s_5 \end{bmatrix} \, ,$$

$$e_{\underline{\beta}_3} = \begin{bmatrix} (c_4c_6-s_4s_6c_5)d_6 + (c_4c_5c_6 - s_4s_6)d_4 \\ -(c_4s_6-s_4c_5c_6)d_6 - (c_4c_5s_6 + s_4s_6)d_4 \\ c_4s_5d_4 \end{bmatrix} \, ,$$

$$e_{\underline{r}_4} = \begin{bmatrix} -s_5c_6 \\ s_5s_6 \\ c_5 \end{bmatrix} \, , \quad e_{\underline{\beta}_4} = \begin{bmatrix} s_5s_6d_6 \\ c_6d_6 \\ 0 \end{bmatrix} \, ,$$

$$e_{\underline{r}_5} = \begin{bmatrix} s_6 \\ c_6 \\ 0 \end{bmatrix} \, , \quad e_{\underline{\beta}_5} = \begin{bmatrix} c_6d_6 \\ -s_6d_6 \\ 0 \end{bmatrix} \, ,$$

$$e_{\underline{r}_6} = \begin{bmatrix} 0 \\ 0 \\ 1 \end{bmatrix} \, , \quad e_{\underline{\beta}_6} = \begin{bmatrix} 0 \\ 0 \\ 0 \end{bmatrix} \, .$$

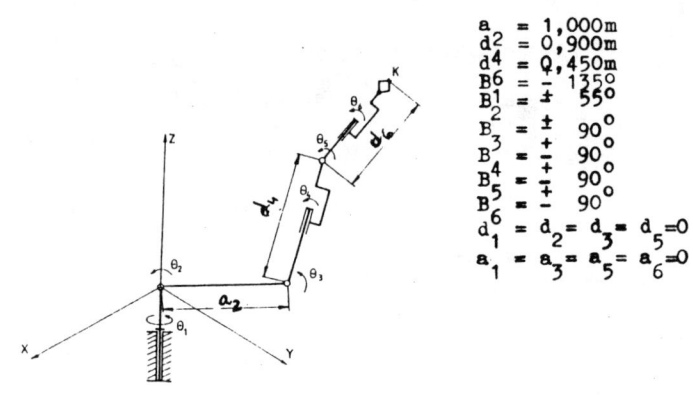

$a_2 = 1,000\text{m}$
$d_2 = 0,900\text{m}$
$d_4 = 0,450\text{m}$
$B_6 = \pm 135°$
$B_1 = \pm 55°$
$B_2 = \pm 90°$
$B_3 = \pm 90°$
$B_4 = \pm 90°$
$B_5 = \pm 90°$
$d_1 = d_2 = d_3 = d_5 = 0$
$a_1 = a_3 = a_5 = a_6 = 0$

Fig. 1

$\theta_1 = 0°$
$\theta_2 = -55°, -45°, \ldots, 55°$
$\theta_3 = -90°, -80°, \ldots, 90°$
$\theta_4 = 90°$
$\theta_5 = -90°, -80°, \ldots, 90°$
$\theta_6 = 90°$
$\Delta\theta_i = 0, 2 \cdot 10^{-3}$

1 $\Delta p < 0,55$
2 $\Delta p < 0,70$
3 $\Delta p < 2,00$

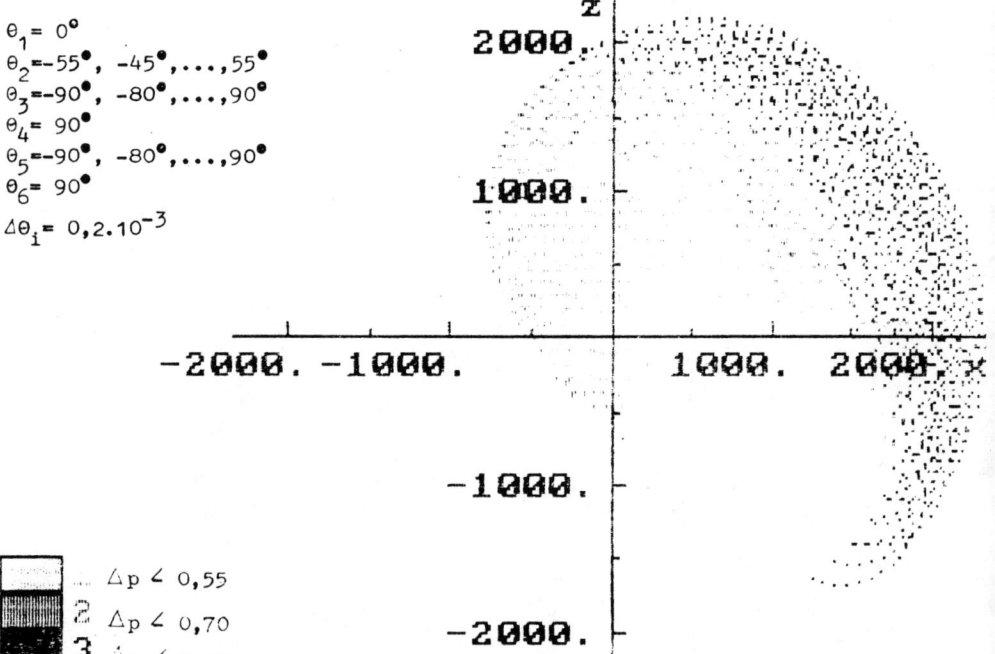

Fig. 2

A Unified Approach to Modelling of Flexible Robot Arms

J. Rauh and W. Schiehlen

Institute B of Mechanics, University of Stuttgart, F.R.G.

Summary

Increasing demands on velocity and accuracy of robot motions require the consideration of the elasticity of robot arms in the controller and structural design. The first and most essential step is the mathematical modeling. The transmission flexibility and the torquer dynamics can easily be modeled by proven rigid body formalisms. The structural elasticity of robot arms, however, has to be regarded with more detail.

There have been numerous papers concerning the modelling approach which can be summarized essentially in three categories:

  i) A series of rigid bodies and elastic springs.
  ii) A series of finite elements.
  iii) One continuous element subject to modal analysis.

For an optimal design it is hard to find out which model is the best one. A comparison is usually omitted in literature.

This paper presents a unified approach including all the mentioned models. For this purpose the method of multibody systems is extended summarizing the rigid body motions as well as the elastic motions. The three different approaches are compared using a simple example. It turns out that the different methods are strongly related to each other. Moreover, for a sufficient number of degrees of freedom even the numerical results are equivalent.

1. Introduction

For the dynamical analysis of robots and for the design of appropriate control laws, a mathematical model of the mechanical system has to be found. In the past, it was often sufficient to model a robot as a multibody system consisting of rigid bodies. The equations of motion of such a multibody system can be generated with computer aid using numerical or symbolical formalism as in ADAMS |1| or NEWEUL |2|, respectively.

With increasing demands on velocity and accuracy of robot motions, the elasticity can not be neglected. There are two different reasons for discussing a robot as a flexible structure. The first one is the elasticity in the joints due to drive coupling and transmissions. Such flexible joints can easily be modelled with the above metioned multibody formalism. The second reason is the elasticity of the link itself resulting from the long and slender form of leighweight robot arms. For a dynamical analysis links can be modelled by: a series of rigid bodies and elastic springs, by a number of finite beam elements, or by one continuous element subject to modal analysis.

Examples of each of the above mentioned models are found in the literature. While a series of rigid bodies and elastic springs can be treated again by the known multibody formalism, the finite element and the modal analysis approach require special consideration. The well known program packages for finite element analysis are usualnny restricted to small motions with respect to the inertial frame. Extensions have to be made for the large nonlinear motion of the finite elements, see e.g. Sunada and Dubowsky |3|. The same questions arise with the application of one continuous beam incuding modal analysis, see e.g. Singh, VanderVoort and Likins |4|.

In this paper, a unified approach including all the mentioned models is presented. The local equations of motion generated for one flexible link are the fundamental basis of the robot global equations. A comparison of the different methods is presented for a simple example.

2. Local Equations of Motion of Flexible Links

For an optimal design of flexible robot arms it is hard to find out which model is the best one. Usually, different approaches are used by different authors and comparisons are rare in literature. As a unified approach the principle of D'Alembert can be used. This has been shown in [5] in great detail. D'Alembert's principle will now also be applied to the three models of flexible links.

2.1 Method of Multibody Systems

The method of multibody systems has to be extended for modeling of flexible robot arms. The procedure is : partition rigid links and include massless beams for consideration of the structural elasticity. This results in an increased number of degrees of freedom of the robot arm. Consequently, better accuracy can only be achieved by higher system order.

A flexible link B_j is modeled using p_{ej} elastic elements B_{jk}, consisting of two rigid bodies B_{jk1} and B_{jk2}, connected by a massless beam of length ΔL, see Fig. 1. This approach is an extension of the so-called lumped mass representation found in the finite element terminology.

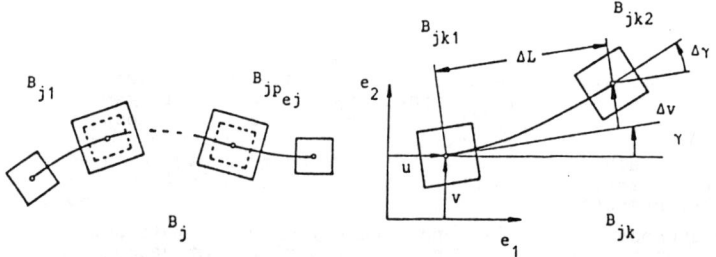

Fig. 1: Multibody System

Formulating Newton's and Euler's equations for the bodies B_{jk1} and B_{jk2} and applying D'Alembert's principle leads to the local equations of motion for the elastic element B_{jk}

$$M_{jk}(x_{jk})\ddot{x}_{jk} + K_{jk}x_{jk} + k_{jk}(x_{jk},\dot{x}_{jk}) = q_{jk}(x_{jk},t) , \qquad (1)$$

where M_{jk} is the inertia matrix, K_{jk} is the stiffness matrix due to the massless beam, k_{jk} is the vector of generalized gyroscopic forces and q_{jk} is the vector of generalized applied forces due to gravity and drive motors. The vector x_{jk} summarizes the generalized coordinates of the free elastic element given the rigid body motions u, v, γ and the elastic deformations Δv, $\Delta \gamma$. It has to be pointed out that the result is only correct if the quadratic terms of Δv, $\Delta \gamma$ are considered.

Connecting the elastic elements, see Fig. 1, and applying D'Alembert's principle once more, the local equations of motion for the flexible link B_j are obtained as

$$M_j(x_j)\ddot{x}_j + K_j x_j + k_j(x_j, \dot{x}_j) = q_j(x_j, t) \quad , \qquad (2)$$

where the matrices have the same meaning as in (1). The position vector x_j includes the rigid body motions u, v, γ and all elastic deformations.

2.2 Method of Finite Element Systems

In this case, the flexible link B_j is modeled by p_{ej} finite beam elements B_{jk} of length ΔL, Fig. 2.

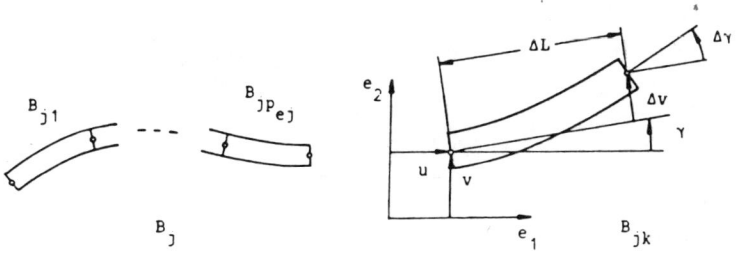

Fig. 2: Finite Element System

The equations of motion for a single finite beam element are obtained from Cauchy's equation of motion by applying D'Alembert's principle with respect to the finite element

$$\int_{B_{jk}} \delta w^T a \, dm + \int_{B_{jk}} \delta e^T H\, e \, dV = \int_{B_{jk}} \delta w^T g \, dm + \sum_m (\delta w_m^T f_m + \delta d_m^T l_m) \quad . \quad (3)$$

The acceleration vector a, the strain vector e and the virtual displacement δw, the virtual rotation δd and the virtual strain δe are found from the position vector of a material point

$$r = r(\varrho_{jk}, x_{jk}(t)) \qquad (4)$$

using the kinematical shape function. Here, the vector ϱ_{jk} represents the material coordinates of the beam element, the vector x_{jk} represents the generalized coordinates of the free elastic element as shown in Fig. 2. The gravitational load is given by g, the forces and moments f_m and l_m are acting on the left and right node, $m=1,2$. The resulting equations of motion are

$$M_{jk}(x_{jk})\ddot{x}_{jk} + K_{jk} x_{jk} + k_{jk}(x_{jk}, \dot{x}_{jk}) = q_{jk}(x_{jk}, t) \quad , \qquad (5)$$

and they can be condensed to the equations of motion for the flexible link regarding the corresponding constraints, see Fig. 2, and applying D'Alembert's principle once more,

$$M_j(x_j)\ddot{x}_j + K_j x_j + k_j(x_j, \dot{x}_j) = q_j(x_j, t) \quad . \qquad (6)$$

Even if these equations look similar (as (1) and (2),) at least the inertia matrices and the vectors of gyroscopic forces are quite different.

2.3 Method of Continuous Systems

The method of continuous systems uses the infinitesimal elastic element for modeling the flexible beam, Fig. 3.

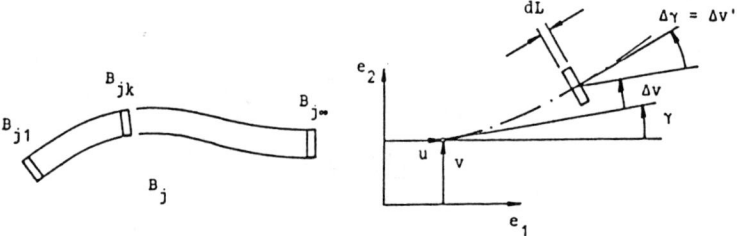

Fig. 3: Continuous System

The flexible link B_j is composed of p_{ej} elements B_{jk}, $p_{ej} \to \infty$, with infinitesimal length dL. Then, D'Alembert's principle (3) has to be applied with respect to the cross section A_{jk} of the elastic element B_{jk}. Now, the position vector (4) depends not only on the generalized coordinates but also on generalized functions

$$r = r(\rho_{jk}, u(t), v(t), \gamma(t), \Delta v(\rho_1, t), \Delta v'(\rho_1, t)) \qquad (7)$$

The stated problem can be solved by separating the generalized functions in space and time

$$\Delta v(\rho_1, t) = J(\rho_1)\, y_e(t) \quad , \qquad (8)$$

where $J(\rho_1)$ is an infinite-dimensional vector of shape functions and the vector $y_e(t)$ summarizes the corresponding generalized coordinates. Then, D'Alembert's principle has to be applied once more with respect to the whole link, a procedure which is also known as modal analysis. The shape function vector can be found from the corresponding eigenvalue problem using the boundary conditions of the whole beam.

The vector of the generalized coordinates for the whole is now

$$x_j = [\; u(t) \quad v(t) \quad \gamma(t) \quad y_e(t) \;] \qquad (9)$$

and the equations of motion are obtained as

$$M_j(x_j)\ddot{x}_j + \Omega_j^2 x_j + k_j(x_j, \dot{x}_j) = q_j(x_j, t) \qquad (10)$$

Due to the modal analysis, the stiffness matrix Ω_j^2 is a diagonal matrix. Further, it has to be mentioned that in applied modal analysis only a finite number of shape functions is used. Thus, the modal analysis is also an approximation method.

3. The global equations of motion

For a robot arm of p_r rigid links and p_f flexible links, the local equations of motion can be summarized by

$$M_i \ddot{x}_i + k_i = q_i \quad , \quad i=1(1)p_r \quad , \tag{11}$$
$$M_j \ddot{x}_j + K_j x_j + k_j = q_j \quad , \quad j=1(1)p_e \quad ,$$

Regarding the corresponding constraints and applying D'Alembert's principle again, the global equations of motion remain as

$$M(y)\ddot{y} + Ky + k(y,\dot{y}) = q(y,t) \tag{12}$$

where M is a symmetric positive definite inertia matrix, k is a vector of centrifugal forces, q is a vector of generalized applied forces and y means the vector of the remaining generalized coordinates.

If the vector y is partitioned into vector y_r of large rigid body motions and vector y_e of small elastic motions, and the applied generalized forces are restricted to the gravity forces q_g and drive forces q_d, the equations of motion show the following structure:

$$\begin{bmatrix} M_r(y_r) & M_{re}(y_r) \\ \hline M_{er}(y_r) & M_e \end{bmatrix} \begin{bmatrix} \ddot{y}_r(t) \\ \hline \ddot{y}_e(t) \end{bmatrix} + \begin{bmatrix} 0 & 0 \\ \hline 0 & K_e \end{bmatrix} \begin{bmatrix} y_r(t) \\ \hline y_e(t) \end{bmatrix}$$
$$+ k(y,\dot{y}) - q_g(y) = q_d(t) \quad . \tag{13}$$

The structure of the global equations of motion is strongly nonlinear, in the case of rigid and flexible robot arms, respectively. However, the number of degrees of freedom for a robot with a flexible arms is larger independently of the chosen model. It remains therefore the question on the accuracy of the results, which will be investigated for a simple example.

4. Simple Beam Pendulum

For comparison of different modeling approaches, a simple beam pendulum with a discrete mass M is analysed with respect to an equilibrium position, see Fig. 4. In this case of small linear vibrations without gravity and drive, the eigenvalue problem appearing for the continuous system model can be completely solved. The corresponding characteristic equation is as follows

$$\cosh \lambda_j \sin \lambda_j - \sinh \lambda_j \cos \lambda_j + 2 \frac{M}{m} \lambda_j \sinh \lambda_j \sin \lambda_j = 0 \quad . \tag{14}$$

For the eigenfrequencies it follows

$$\omega_j = \sqrt{\frac{EI}{m\,l^3}} \, \lambda_j^2 \quad , \tag{15}$$

where m is the mass of the beam and l is its length. The numerical computed eigenfrequencies will serve as a reference solution.

4.1 Modeling by Rigid Bodies and Beams

The equations of motion are given as

$$M \ddot{y} + K y = 0 \quad . \tag{16}$$

The vector of generalized coordinates is for p elastic elements with $f = 2p+1$ degrees of freedom

$$y = [\gamma_0, v_1, \gamma_1, \ldots, v_p, \gamma_p] \quad . \tag{17}$$

For p=3, e.g., the inertia matrix is

$$M = \begin{bmatrix} (19m_e+9M)l_e^2+6I_e & & & & & & \\ 2m_e l_e & 2m_e & & & \text{symm.} & & \\ 2I_e & 0 & 2I_e & & & & \\ 4m_e l_e & 0 & 0 & 2m_e & & & \\ 2I_e & 0 & 0 & 0 & 2I_e & & \\ 3(m_e+M)l_e & 0 & 0 & 0 & 0 & m_e+M & \\ I_e & 0 & 0 & 0 & 0 & 0 & I_e \end{bmatrix} \tag{18}$$

with the equivalent inertia parameters

$$m_e = \frac{1}{2}\frac{m}{p} \; , \quad I_e = \frac{1}{12} m_e l_e^2 \; , \quad l_e = \frac{l}{p} \tag{19}$$

and the stiffness matrix is

$$K = \frac{EI}{l_e^3} \begin{bmatrix} 0 & & & & & & \\ 0 & 24 & & & \text{symm.} & & \\ 0 & 0 & 8l_e^2 & & & & \\ 0 & -12 & -6l_e & 24 & & & \\ 0 & 6l_e & 2l_e^2 & 0 & 8l_e^2 & & \\ 0 & 0 & 0 & -12 & -6l_e & 12 & \\ 0 & 0 & 0 & 6l_e & 2l_e^2 & -6l_e & 4l_e^2 \end{bmatrix} \tag{20}$$

The eigenfrequencies of this model are compared with the exact solutions given above. The relative errors for the first two eigenfrequencies are plotted depending on the number of degrees of freedom used, see Fig. 5.

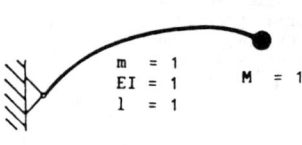

Fig. 4: Beam Pendulum

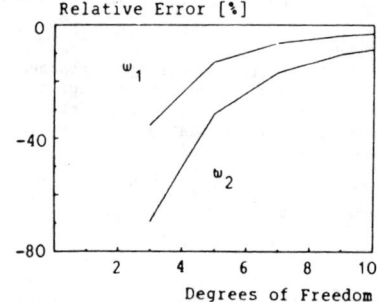

Fig. 5: Multibody System

4.2 Modeling by Finite Elements

The equations of motion look like (16). However, the inertia matrix (18) is now given with $m_e = \frac{m}{p}$ as

$$M = \begin{bmatrix} 9(m_e + M)l_e^2 & & & & & & \\ m_e l_e & \frac{26}{35} m_e & & & \text{symm.} & & \\ \frac{1}{15} m_e l_e^2 & 0 & \frac{2}{105} m_e l_e^2 & & & & \\ 2 m_e l_e & \frac{9}{70} m_e & \frac{13}{420} m_e l_e & \frac{26}{35} m_e & & & \\ \frac{1}{15} m_e l_e^2 & -\frac{13}{420} m_e l_e & -\frac{1}{140} m_e l_e^2 & 0 & \frac{2}{105} m_e l_e^2 & & \\ (\frac{27}{20} m_e + 3M) l_e & 0 & 0 & \frac{9}{70} m_e & \frac{13}{420} m_e l_e & \frac{13}{35} m_e + M & \\ -\frac{13}{60} m_e l_e^2 & 0 & 0 & -\frac{13}{420} m_e l_e & -\frac{1}{140} m_e l_e^2 & -\frac{11}{210} m_e l_e & \frac{1}{105} m_e l_e^2 \end{bmatrix} , \quad (21)$$

while the stiffness matrix is identical with (20). The relative errors for the eigenfrequencies are plotted in Fig. 6 depending on the number of degrees of freedom used here.

4.3 Modeling by a Continuous Beam

Firstly, the eigenfrequencies of the beam without discrete mass, $M=0$, are calculated from (14). The corresponding eigenfunctions are

$$v_1(x) = k_1 x ,$$
$$v_i(x) = k_i (\sin \lambda_i x + \frac{\sin \lambda_i}{\sinh \lambda_i} \sinh \lambda_i x) , \quad i = 2(1)f , \quad (22)$$

where the constants k_i follow by normalization as

$$k_1 = \sqrt{\frac{3}{m}} , \quad k_i = 1 / \sqrt{\frac{m}{2}(1 - \frac{\sin^2 \lambda_i}{\sinh^2 \lambda_i})} , \quad i = 2(1)f . \quad (23)$$

The dynamic boundary condition due to the discrete mass M is

$$F(t) = -M \ddot{v}(1,t) = -M \sum_{i=1}^{f} v_i(1) y_i(t) . \quad (24)$$

The equations of motion can be rewritten as

$$M \ddot{y} + \Omega^2 y = 0 , \quad \Omega^2 = \{ 0, \omega_2^2, \omega_3^2, \ldots, \omega_f^2 \} , \quad (25)$$

where the inertia matrix M is given as

$$M = \begin{bmatrix} 1 + M c_1^2 & & & \\ M c_1 c_2 & 1 + M c_2^2 & \text{symm.} & \\ M c_1 c_3 & M c_2 c_3 & 1 + M c_3^2 & \\ \vdots & \vdots & \vdots & \ddots \end{bmatrix} \quad (26)$$

with

$$c_1 = k_1 , \quad c_i = 2 k_i \sin \lambda_i , \quad i = 2(1)f . \quad (27)$$

Further, the generalized elastic coordinates used in (25) are different from (17). The relative errors for the eigenfrequencies of this model are plotted in Fig. 7.

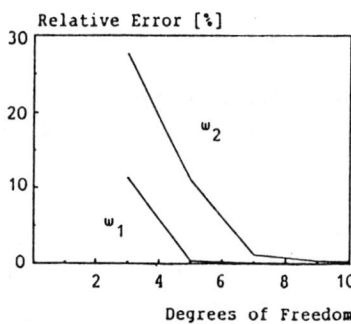

Fig. 6: Finite Element System   Fig. 7: Continuous System

5. Conclusion

In this paper, three different approaches to the modelling of the flexible robot arms has discussed. Modeling the elasticity by rigid bodies and beams is a straight-forward extension of the method of multibody systems. The available formalisms are well suited to handle this kind of approach. The resulting eigenfrequencies are found to be too low, since the inertia is overemphasized.

The finite element approach requires the formulation of additional inertia matrices and vectors to include also the large motions of the links. The shape functions may be the same as used in standard finite element methods; this means unchanged stiffness matrices. The resulting local equations of motion can easily be added to other local equations of motion generated by multibody formalisms. The improvement of the modelling reliability is paid by more effort. The resulting eigenfrequencies are found to be too high.

The approach using a continuous system and modal analysis yields local equations of motion where the generalized coordinates do not have a geometrical meaning. For obtaining a sufficient modelling accuracy, it is important to use shape functions of a simplified homogeneous problem. While this was very easy in the given example, this will be more difficult in the general case.

A comparison of the three approaches shows that the motions are equivalent, if a sufficient number of degrees of freedom is chosen. Therefore, the choice of the modeling approach in robot dynamics is to some extent a matter of taste.

6. References

[1] M.A. Chace: Methods and Experience in Computer Aided Design of Large-Displacement Mechanical Systems. In:. Computer Aided Analysis and Optimization of Mechanical System Dynamics, ed. E.J. Haug, Springer, 1984.

[2] W.O. Schiehlen: Dynamics of Complex Multibody Systems. SM Archives, 9/1984.

[3] W. Sunada, S. Dubowsky: The Application of Finite Element Methods to the Dynamic Analysis of Flexible Spatial and Co-Planar Linkage Systems. Journal of Mechanical Design, July 1981, Vol. 103/643.

[4] R.P. Singh, R.J. VanderVoort, P.W. Likins: Dynamics of Flexible Bodies in Tree Topology - A Computer-Oriented Approach. Journal of Guidance, Vol. 8, No. 5.

[5] W.O. Schiehlen: Technische Dynamik, Teubner, 1986.

Solving the Inverse Kinematic Problem for Robotic Manipulators

L. Sciavicco and B. Siciliano

Universita' di Napoli, Dipartimento di Informatica e Sistemistica, Napoli, Italy

ABSTRACT

Based on a dynamic approach, a general solution algorithm for the inverse kinematic problem for robotic manipulators is presented. It requires only the computation of direct kinematics. Two-stage algorithms are then derived for three basic kinematical structures in order to comply with the mechanical constraints of each structure. Applicability of the algorithm to redundant manipulators with obstacle collision avoidance and limited joint range availability is finally shown.

INTRODUCTION

The crucial point in advanced control of robotic manipulators is the capability of transforming the task space coordinates into the configuration space coordinates, that is solving the Inverse Kinematic Problem. As a matter of fact control is implemented in the joint space, identified by the n-dimensional joint vector $\underline{q}$, whereas motion is specified in the Cartesian space, by means of an m-dimensional Cartesian vector $\underline{x}$. The direct kinematic problem allows to specify in a straightforward manner [1] the relationship between the joint variables and the Cartesian variables as

$$\underline{x} = \underline{f}(\underline{q}), \qquad (1)$$

where $\underline{f}$ is a continuous nonlinear function, whose structure and parameters are known. It is unique, while the inverse transformation

$$\underline{q} = \underline{f}^{-1}(\underline{x}) \qquad (2)$$

is not unique and, because of complexity of (1), is hard to be expressed analitically.

The most common approach for solving the inverse kinematic problem is certainly to obtain a closed-form solution to (2), [2]. Only manipulators with a spherical wrist, however, allow closed-form solutions. The problem becomes more critical for kinematically redundant manipulators, for which the number of degrees of freedom exceeds the six coordinates which are usually required to specify the position and the orientation of the end effector in the Cartesian space.

An iterative technique based on a nonlinear optimization algorithm for solving (2) has been proposed in [3]. Though it seems to be quite general, being applicable to any kinematical structure, it involves a great amount of computation to converge to the desired solution, which makes it impractical for tracking control purposes.

The approach illustrated in this paper is based on a dynamic formulation of the problem first proposed in [4], and later also in [5]. The main advantage of the method is that it only makes use of direct kinematics (1). Besides uniquenesss of the solution is assured, computation time is drastically reduced, joint velocities are automatically generated and occurrence of kinematic singularities may not represent a drawback. Even if

the resultant algorithm is   general and manipulator-independent, an even
smaller number of computations are required and better results are achieved
if the algorithm is customized to the particular kinematical structure.
Furthermore, the same dynamic concept which is at the basis of the method
allows the solution of the inverse kinematic problem for redundant
manipulators in presence of obstacles in the workspace and/or with mechanical
constraints imposed of the joint variables.

THE GENERAL SOLUTION ALGORITHM

The inverse kinematic problem is conceived as a dynamical one in order to get
a general solution algorithm which requires only the computation of direct
kinematics (1). Let $\hat{q}(t)$ be a solution of (1) relative to a given Cartesian
trajectory $\hat{\underline{x}}(t)$. The following error vector $\underline{e}(t)$ can be defined between the
Cartesian trajectory and the corresponding one obtained from the algorithm
state variables $\underline{q}(t)$,

$$\underline{e}(t) = \hat{\underline{x}}(t) - \underline{x}(t). \tag{3}$$

Recall that Cartesian velocities are related to joint velocities through the
Jacobian matrix $J(\underline{q})$ associated with the relationship

$$\dot{\underline{x}}(t) = J(\underline{q})\dot{\underline{q}}(t). \tag{4}$$

In order to assure the convergence of $\underline{q}(t)$ to $\hat{\underline{q}}(t)$, error dynamics is
involved, i.e. via (4) (dropping the time dependence),

$$\dot{\underline{e}} = \dot{\hat{\underline{x}}} - J(\underline{q})\dot{\underline{q}}. \tag{5}$$

With the choice

$$\dot{\underline{q}} = \gamma J^T(\underline{q})\underline{e}, \quad \gamma = \alpha + (\underline{e}^T\dot{\hat{\underline{x}}})(\underline{e}^T JJ^T\underline{e})^{-1}, \quad \alpha > 0 \tag{6}$$

the dynamic system of fig. 1 assures that $\underline{e} \to \underline{0}$, and then $\underline{q}(t) \to \hat{\underline{q}}(t)$. This
issue can be recognized by considering the error Lyapunov function $V = .5\underline{e}^T\underline{e}$
and verifying that its derivative is negative definite in virtue of (6), [4],
[5]. The choice of α in (6) determines the convergence rate of the closed loop
system of fig. 1. Starting with the same initial conditions $\underline{q}(0) = \hat{\underline{q}}(0)$ will
always guarantee good tracking accuracy, avoiding thus the solution
nonuniqueness drawback concerned with those techniques which provide joint
trajectories as result of an interpolation between a finite set of via points
[6].

As it had been anticipated, only direct kinematics is to be computed,
which drastically reduces the computational burden. In   fact only one
iteration per cycle is to be done   as far as the digital implementation of
the algorithm for trajectory tracking is concerned. The implementation of the
algorihtm on a single dedicated microprocessor system has been realized in lab
and is described in [7].

It should be emphasized also that the dynamical system of fig. 1 will
produce joint velocities $\dot{\underline{q}}(t)$ at no additional cost, which is very useful for
advanced control, see model reference adaptive control [8] for instance.

Finally it might be noted that, since no inversion is required ($\underline{r}^{-1}, J^{-1}$),
the algorithm may provide solutions even in case of kinematic singularities,
on condition that the trajectory in the task space is properly planned. To be
more specific, no problem will arise if the singularity belongs to the
trajectory, whereas in case of a trajectory passing in the proximity of a
singularity, indeed, high velocities are mechanically involved; in this case

the convergence of the algorithm is likely to be derated.

NONREDUNDANT MANIPULATORS

A robot task is naturally specified in terms of the end effector Cartesian vector $\underline{x}^T = (p_x, p_y, p_z, \alpha, \beta, \gamma)$ with respect to the base frame; p_i's are the components of the end effector position vector $\underline{p}$, and α, β, γ are the Euler angles (or roll, pitch and yaw angles) which define its orientation. In order to obtain a unique specification of the orientation, however, a unit approach vector $\underline{a}$ and a unit sliding vector $\underline{s}$ can be adopted; the unit normal vector $\underline{n}$ which usually completes the frame at the end effector is redundant since $\underline{n} = \underline{s} \times \underline{a}$. The above vectors can be easily determined from the Euler angles, [9]. In view of the preceeding, for any robot kinematical structure, (1) becomes

$$\underline{p} = \underline{f}_p(\underline{q}) \quad \underline{s} = \underline{f}_s(\underline{q}) \quad \underline{a} = \underline{f}_a(\underline{q}) \tag{7}$$

subjected to the constraints

$$\underline{s}^T\underline{s} = \underline{a}^T\underline{a} = 1 \quad \underline{s}^T\underline{a} = 0. \tag{8}$$

Similarly (4) gives

$$\underline{\dot{p}} = J_p(\underline{q})\underline{\dot{q}} \quad \underline{\dot{s}} = J_s(\underline{q})\underline{\dot{q}} \quad \underline{\dot{a}} = J_a(\underline{q})\underline{\dot{q}} \tag{9}$$

subjected to

$$\underline{s}^T\underline{\dot{s}} = \underline{a}^T\underline{\dot{a}} = 0 \quad \underline{s}^T\underline{\dot{a}} + \underline{a}^T\underline{\dot{s}} = 0. \tag{10}$$

Nonredundant manipulators require six degrees of freedom to identify uniquely the position and the orientation of the end effector. Typical kinematical structures have three revolute joints $(\theta_4, \theta_5, \theta_6)$ at the end effector, whereas the first three joints (q_1, q_2, q_3) are either all revolute, such as the PUMA arm [10], or two revolute and one prismatic, such as the JPL arm [11]. As far as the last three joints, three basic configurations are illustrated in fig. 2. The case a) is of particular interest since it is possible to decouple the position of the end effector from its orientation (spherical wrist). The cases b) and c) may also occur in practical robot design. The three structures can be conveniently characterized through the following constraints on the geometric parameters of the last three joints. More specifically, the lengths a_n and the distances d_n, [1], are respectively in the three cases:

a) parallel axes: $a_4 = a_5 = d_5 = 0$ (fig. 2a),
b) two-by-two intersecting axes: $a_4 = a_5 = 0$, $d_5 \neq 0$ (fig. 2b),
c) nonconverging axes: $a_4 \neq 0$, $a_5 \neq 0$ (fig. 2c).

With reference to fig. 2, the position vector $\underline{p}$ and the approach unit vector $\underline{a}$ are always independent of the last rotation. Hence

$$\underline{p}' = \underline{p} - d_6\underline{a} \tag{11}$$

can be assumed as position vector, and still indicated by $\underline{p}$ without loss of generality. Furthermore the position vector $\underline{p}$ depends on

a) the first three joint variables (q_1, q_2, q_3),
b) the first four joint variables $(q_1, q_2, q_3, \theta_4)$,
c) the first five joint variables $(q_1, q_2, q_3, \theta_4, \theta_5)$,

respectively in the three cases of fig. 2. As a consequence the vector of joint variables $\underline{q}$ in (1) can be partitioned as

$$\underline{q}^T = (\underline{q}_p^T \mid \underline{q}_h^T), \tag{12}$$

where $\underline{q}_p$ are the joint variables which determine the position $\underline{p}$, and $\underline{q}_h$ are the remaining joint variables which, together with $\underline{q}_p$, determine the orientation $\underline{s}$, $\underline{a}$. Accordingly the general algorithm can be partitioned into two stages (first determine $\underline{q}_p$, then $\underline{q}_h$) as follows in the three cases of fig. 2, [12].

Spherical wrist

For this structure (fig. 2a) the vector $\underline{q}$ in (12) is partitioned into

$$\underline{q}_p^T = (q_1 \; q_2 \; q_3) \qquad \underline{q}_h^T = (\theta_4 \; \theta_5 \; \theta_6). \tag{13}$$

The resulting two-stage inverse kinematic algorithm is

$$\dot{\underline{q}}_p = \gamma_p J_p^T \underline{e}_p, \quad \gamma_p = \alpha_p + (\underline{e}_p^T \dot{\hat{\underline{p}}})(\underline{e}_p^T J_p J_p^T \underline{e}_p)^{-1}, \quad \alpha_p > 0 \tag{14a}$$

$$\dot{\underline{q}}_h = \gamma_h (J_s^T \hat{\underline{s}} + J_a^T \hat{\underline{a}}), \tag{14b}$$

$$\gamma_h > (\|\dot{\hat{\underline{s}}}\|_{max} + \|\dot{\hat{\underline{a}}}\|_{max} + \|\dot{\underline{q}}_p\|_{max})(|\Lambda(J_{sp})| + |\Lambda(J_{ap})|)(|\lambda(J_s)| + |\lambda(J_a)|)^{-1}$$

where $J_p = \partial \underline{f}_p / \partial \underline{q}_p$, $J_{sp} = \partial \underline{f}_s / \partial \underline{q}_p$, $J_{ap} = \partial \underline{f}_a / \partial \underline{q}_p$, $J_s = \partial \underline{f}_s / \partial \underline{q}_h$, $J_a = \partial \underline{f}_a / \partial \underline{q}_h$; $\Lambda(A)$ and $\lambda(A)$ denote the maximum and the minimum eigenvalue of matrix A respectively. Further details on the derivation of (14) can be found in [13].

Two-by-two intersecting axes

In this case also θ_4 concurs to determine the position vector $\underline{p}$ (fig. 2b). The partition of $\underline{q}$ in (12) gives

$$\underline{q}_p^T = (q_1 \; q_2 \; q_3 \; \theta_4) \qquad \underline{q}_h = (\theta_5 \; \theta_6). \tag{15}$$

In order to obtain a unique solution for $\underline{q}_p$, the following mechanical constraint is to be incorporated in the first stage:

$$\hat{\underline{a}}^T \underline{z}_4 = \cos\alpha_5, \tag{16}$$

where α_5 is the constant twist angle between the fifth and the sixth link, $\hat{\underline{a}}$ is given in the Cartesian space and $\underline{z}_4$ is the unit vector along the fifth link axis, which is determined through $\underline{q}_p$. The inverse kinematic algorithm results then, [14], [15],

$$\dot{\underline{q}}_p = \gamma_p (J_p^T \underline{e}_p + J_{z4}^T \hat{\underline{a}} e_{z4}), \quad \gamma_p > (\|\dot{\hat{\underline{p}}}\|_{max} + \|\dot{\hat{\underline{a}}}\|_{max})|\lambda \begin{pmatrix} J_p \\ \hat{\underline{a}}^T J_{z4} \end{pmatrix}|^{-1} \tag{17a}$$

$$\dot{\underline{q}}_h = \gamma_h (J_s^T \hat{\underline{s}} + J_a^T \hat{\underline{a}}), \tag{17b}$$

$$\gamma_h > (\|\dot{\hat{\underline{s}}}\|_{max} + \|\dot{\hat{\underline{a}}}\|_{max} + \|\dot{\underline{q}}_p\|_{max})|\Lambda(J_{sp}^T J_{sp} + J_{ap}^T J_{ap})|^{.5}|\lambda(J_s^T J_s + J_a^T J_a)|^{-.5}$$

where $e_{z4} = \cos\alpha_5 - \hat{\underline{a}}^T \underline{f}_{z4}(\underline{q}_p)$, and $J_{z4} = \partial \underline{f}_{z4}/\partial \underline{q}_p$.

Nonconverging axes

In this last case five degrees of freedom determine the position of $\underline{p}$ (fig. 2c). The partition obviously results

$$\underline{q}_p^T = (q_1\ q_2\ q_3\ \theta_4\ \theta_5) \qquad q_h = \theta_6. \tag{18}$$

For the same reason as above, the two following constraints must be added in order to get a unique solution for $\underline{q}_p$:

$$\underline{\hat{a}}^T \underline{z}_4 = \cos\alpha_5 \tag{19a}$$

$$\underline{\hat{a}}^T \underline{a} = 1. \tag{19b}$$

The resulting algorithm is, [12],

$$\underline{\dot{q}}_p = \gamma_p (J_p^T \underline{e}_p + J_{z4}^T \underline{\hat{a}} e_{z4} + J_a^T \underline{\hat{a}}), \quad \gamma_p > (\|\underline{\dot{p}}\|_{max} + \|\underline{\dot{\hat{a}}}\|_{max}) |\lambda \begin{pmatrix} J_p \\ \underline{\hat{a}}^T J_{z4} \\ \underline{\hat{a}}^T J_a \end{pmatrix}|^{-1} \tag{20a}$$

$$\dot{q}_h = \gamma_s (J_s^T \underline{\hat{s}}), \quad \gamma_s > (\|\underline{\dot{\hat{s}}}\|_{max} + \|\underline{\dot{q}}_p\|_{max}) |\lambda (J_{sp}^T J_{sp})|^{.5} |\lambda (J_s^T J_s)|^{-.5} \tag{20b}$$

where it might be noted that, since $\underline{a}$ is determined in the first stage, the second stage is only required to align $\underline{s}$ with $\underline{\hat{s}}$ by means of θ_6.

REDUNDANT MANIPULATORS

A manipulator is termed kinematically redundant if the number of degrees of freedom exceeds the number of task space coordinates. Redundancy can be conveniently exploited to solve the inverse kinematic problem with obstacle collision avoidance and/or limited joint range availability. In these cases solutions have been proposed, based on the use of the generalized inverse, [15]-[19]. It seems, however, that the amount of computation involved is still too large for real time control.

With reference to the general scheme of fig. 1, the two kinds of constraints can be successfully incorporated in the dynamic approach if the task space state vector in (1) is enlarged

Suppose first that, while the manipulator is tracking a desired trajectory in the task space, one or more links along its kinematical structure happen to be much too close to an obstacle in the workspace. Since the inverse kinematic algorithm (6) provides joint configurations which are adjacent to each other as the manipulator proceeds, one or more constraints can be introduced in order to avoid the collision with the obstacle.

More precisely, let $\underline{c}_i$ (i=1,...,n) indicate the position vectors of those points of the obstacle which are closest to each link l_i of the manipulator; a point at minimum distance from the obstacle on each link is automatically individuated and let $\underline{p}_i$ indicate the corresponding position vector. Both vectors are defined with respect to the same base frame, see fig. 3 for a planar example. If the distance between the two points $\|\underline{d}_i\|$, where $\underline{d}_i = \underline{p}_i - \underline{c}_i$, is less than a threshold distance $\hat{d}$, there is a danger of a collision, and the joint velocities which represent the control inputs to the system of fig. 1 should be modified accordingly to the new situation. This can be accomplished as follows. Define the errors

$$e_{dj} = .5(\hat{d}^2 - \underline{d}_i^T \underline{d}_i), \quad j = 1,...,k, \tag{21}$$

where k is the number of active constraints. Differentiating (21) with respect to time gives

$$\dot{e}_{dj} = - J_{di} \underline{\dot{q}} \tag{22}$$

where

$$J_{di} = \underline{d}_i^T J_{pi}, \qquad (23)$$

being J_{pi} the Jacobian matrix associated with vector $\underline{p}_i$. Then the control (6) can be modified into

$$\underline{\dot{q}} = G_d(J^T \underline{e} + J_d^T \underline{e}_d) \qquad (24)$$

which still guarantees that $\underline{x} \to \underline{\hat{x}}$, but also assures that $\underline{d}_i^T \underline{d}_i \to \hat{d}^2$; G_d is a positive definite diagonal matrix.

In this way the motion of those degrees of freedom which influence the motion of the point $\underline{p}_i$ is braked preventing the link l_i from approaching the obstacle. As a matter of fact, a link which is candidate to a collision, in virtue of (24), is forced to move tangentially around the imaginary sphere of center in $\underline{c}_i$ and radius $\hat{d}$.

Although (24) is the basis of the obstacle avoidance scheme proposed here, proper decision making in charge of a higher control level is equally important to successful operation of the algorithm. The threshold distance $\hat{d}$, which should include the thickness of manipulator links, must be programmed accordingly to the sample rate at which $\underline{p}_i$ and $\underline{c}_i$ are updated so as to get a security gap. In addition, in order not to introduce a discontinuity, the feedback gains of G_d should be tapered as a function of distance.

It must be remarked, however, that the computational burden of the pure inverse kinematic algorithm remains contained as it may be checked in (24).

Conceptually similar is the inclusion of mechanical constraints on joint variables into the inverse kinematic scheme (6). If the joint variables q_i are constrained between two extremal values q_{imin} and q_{imax}, i.e.

$$q_{imin} \leq q_i \leq q_{imax}, \qquad i = 1,\ldots,n, \qquad (25)$$

it is possible to define again a threshold $\hat{d}'$ and the errors

$$e_{qj} = \hat{d}' - d_{qi}, \qquad j = 1,\ldots,r, \qquad (26)$$

where either $d_{qi} = q_i - q_{imin}$ or $d_{qi} = q_{imax} - q_i$, depending on which limit is involved. Progressing as above yields the modified control of type (6)

$$\underline{\dot{q}} = G_q(J^T \underline{e} \pm \underline{e}_q) \qquad (27)$$

which assures that $\underline{x} \to \underline{\hat{x}}$ and $d_{qi} \to \hat{d}'$, the sign - applying for q_{imin} and the sign + for q_{imax}; G_q is a positive definite diagonal matrix. In this way the joint variables are prevented from approaching either of the two limits. Remarks on the adequate choice of $\hat{d}'$ and G_q as for the case of obstacle avoidance are in order also in this case.

Last but not least, it must be emphasized that in order to comply with all the constraints given by (21) and (26) it must be checked that

$$k + r \leq n - m, \qquad (28)$$

in other words the enlargement of the error space can be made up to cover the degrees of redundancy available which are in number of $n - m$.

CONCLUDING REMARKS

This paper presented a general solution algorithm for the inverse kinematic problem for robotic manipulators. For each kinematically nonredundant structure the algorithm is conveniently partitioned into two stages so as to better cope with the kinematical structure.

Simulation studies dedicated to several robotic manipulators can be found in [7], [12]-[15] and have not been fully reported here due to lack of space. Tracking errors have always resulted to be contained, on the average of 1 mm (position) and .1°(orientation) for typical velocities of 1m/sec and 90°/sec. Steady-state errors are practically null due to the closed loop structure of the dynamic system of fig. 1; this issue proves that a precise solution can be provided by the algorithm in all those cases when computation time is not the main concern and the purpose is just to get the set of joint variables corresponding to a given configuration of the end effector. It must be emphasized also that uniqueness of the solution is always assured, and a kinematic singularity along the given trajectory does not involve any large errors, as proved for instance by simulation results derived in [14].

Finally it has been shown how the same dynamic approach can be successfully adopted for solving the inverse kinematic problem for redundant manipulators in presence of obstacles in the workspace and/or with mechanical constraints imposed on the joint variables

REFERENCES

[1] J. Denavit and R.S. Hartenberg, "A kinematic notation for lower-pair mechanisms based on matrices," J. Appl. Mech.,June 1955.
[2] R.P. Paul, B. Shimano and G.E. Meyer, "Kinematic control equations for simple manipulators." IEEE Trans. Syst., Man, Cybern., vol. 11, no. 6, June 1981.
[3] A.A. Goldenberg, B. Benhabib and R.G. Fenton, "A complete generalized solution to the inverse kinematics of robots," IEEE J. Rob. Autom., vol. 1, no. 1, March 1985.
[4] A. Balestrino, G. De Maria and L. Sciavicco, "Robust control of robotic manipulators," Proc. 9th IFAC World Congress, Budapest, Hungary, July 1984.
[5] W.A. Wolovich and H. Elliott, "A computational technique for inverse kinematics," Proc. 23rd CDC, Las Vegas, NV, Dec. 1984.
[6] R.P. Paul, Robot Manipulators: Mathematics, Programming, and Control. Cambridge, MA, MIT Press, 1981.
[7] G. De Maria and R. Marino, "A discrete algorithm for solving the inverse kinematic problem for robotic manipulators," Proc. '85 ICAR, Tokyo, Japan, Sept. 1985.
[8] A. Balestrino, G. De Maria and L. Sciavicco, "An adaptive model following control for robotic manipulators," ASME J. Dynamic Syst., Meas., Contr., vol. 105, Sept. 1983.
[9] J.Y.S. Luh, "An anatomy of industrial robots and their controls," IEEE Trans. Autom. Contr., vol. 28, no. 2, Feb. 1983.
[10] C.S.G. Lee, "Robot arm kinematics, dynamics, and control," IEEE Computer, vol. 15, no.12, Oct. 1982.
[11] V.D. Scheinman, "Design of a computer manipulator," Stanford A.I. Lab, Memo. AIM-92, 1969.
[12] G. De Maria, L. Sciavicco and B. Siciliano, "A general solution algorithm to coordinate transformation for robotic manipulators," Proc. '85 ICAR, Tokyo, Japan, Sept. 1985.
[13] A. Balestrino, G. De Maria, L. Sciavicco and B. Siciliano, "A novel approach to coordinate transformation for robotic manipulators," IEEE Trans. Syst., Man, Cybern., submitted for publication.
[14] L. Sciavicco and B. Siciliano, "Coordinate transformation: a solution algorithm for one class of robots," IEEE Trans. Syst., Man, Cybern., submitted for publication.

[15] L. Sciavicco and B. Siciliano, "An inverse kinematic solution algorithm for robots with two-by-two intersecting axes at the end effector," Proc. IEEE Int. Conf. Rob. Autom., San Francisco, CA, April 1986.
[16] A. Liegeois, "Automatic supervisory control of the configuration and behavior of multibody mechanisms," IEEE Trans. Syst., Man, Cybern., vol. 7, no. 12, Dec. 1977.
[17] C.A. Klein and C.H. Huang, "Review of pseudoinverse control for use with kinematically redundant manipulators," IEEE Trans. Syst., Man, Cybern., vol. 13, no. 2, Feb. 1983.
[18] M. Vukobratovic and M. Kircanski, "A dynamic approach to nominal trajectory synthesis for redundant manipulators," IEEE Trans. Syst., Man, Cybern., Jan./Feb. 1984.
[19] M. Kircanski and M. Vukobratovic, "Trajectory planning for redundant manipulators in the presence of obstacles," Proc. 5th CISM-IFToMM Ro.Man.Sy., Udine, Italy, June 1984.

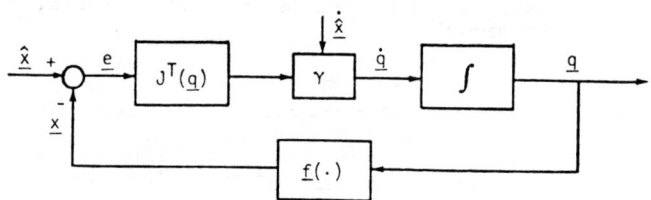

Fig. 1. The general dynamic inverse kinematic algorithm.

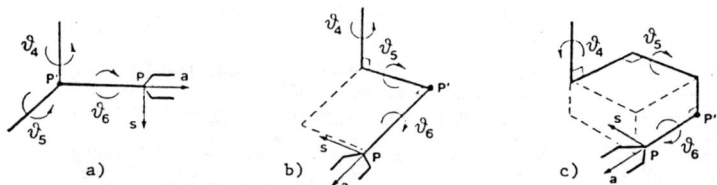

Fig. 2. Three basic kinematical configurations at the end effector: a) spherical wrist, b) 2-by-2 intersecting axes, c) nonconverging axes.

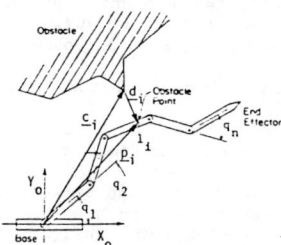

Fig. 3. Geometry of a planar manipulator showing the point nearest to the obstacle.

Determination of the Accuracy of Flexible Automatic Positioning Module with Clearances

V. Natbiladze

Georgian Polytechnic Institute, Tbilisi, USSR

Unmanned production is receiving increasing attention in the machine tool industry. Metal-cutting machine tools using programmed control are often operated by operators with comparatively low qualifications, and industrial robots with controllers are to replace them. In order to match the operation of a metal-cutting machine tool combined with an industrial robot an additional controller is necessary, the design of which presents considerable difficulty. This problem is easily solved by the use of flexible automatic production.

The transition to flexible automatic production calls for radical changes in the equipment used. Industrial robots should not be considered as new devices which can be directly included into the existing range of hardware, but as a new machine requiring the reconsideration of machine equipment, devices, tools and the whole technological process employed. The changes required should be carried out sequentially, step by step.

The modernization of the metal-cutting machine tool is the most immediate and basic problem. This involves the creation of a flexible automatic module, where the industrial robot is combined with the tool providing an indivisible unit with one controller.

The author has designed a flexible automatic module on the basis of the drilling machine, 2P1352 (Figure 1) (1).

The module consists of a frame 1 on which the drilling head 2, the table for fixing the blanks and the robot 4 connected with the controller 5 are installed. The robot is connected with the controller in which the program for placing and removing the pieces is included.

The robot consists of the base 6 fixed to the frame 1; the hollow rod 7 is mounted on the base and can move vertically with the help of a screw and a nut placed inside the rod and interlocked with the displacement mechanism of the tool table 3. The telescopic shaft 8 with its base is fixed on the rod 7 and can make linear and rotating motions, using a screw and a nut placed within the shaft and interlocked with the mechanism of displacement of the table and the rod 7. The gripper 9 is fixed on the end of the telescopic shaft 8.

The construction of the module allows the robot unit to stop when the tool unit is under operation, and when the robot unit is operating, the tool units are stopped, these units being interlocked. The robot unit consists of two motion modules in series connection. Each motion module can rotate and be linearly displaced. The robot has four degrees of freedom, without clamps.

The flexible automatic module operates as follows. In response to the command of the controller 5, the robot 4 grips the blank and places it on the table 3. It then returns to the initial position. The machine tool is switched on with the help of NPC and the table 3 is in the required position, the drilling head is lowered and drilling takes place. The head rises again and turns to the next working position. After the operation is completed the control of table 3 and drilling head 2 is switched off. Then the robot 4 switches on, takes the finished piece from the table 3 and places it on the conveyor without the operator interfering.

Thus, the module provides automation and flexibility with the help of the controller and coordinated operation of its working parts.

Knowing the robot specifications we shall know the tolerance and clearance limits, respectively.

In Figure 2 we describe the robot's schematic diagram. In each of the interconnected joints a clearance and a tension are created. In the case of a moveable joint there is always a clearance.

Introducing the notation of maximum possible clearances in the directions of the axes X, Y and Z:

$|\delta x|, |\delta y|, |\delta z|$ — radial clearance

$|\delta x'|, |\delta z'|$ — length tolerance

$|\delta \varphi_x|, |\delta \varphi_z|$ — angular shift tolerance

Let us consider the location of the points A, B, C and D after linear displacements ΔX and ΔZ. After the displacement of the units, the points have the positions A', B', C' and D' (Figure 3).

Because of the radial deviance the vertical bar deviates by $\delta \alpha$, angle, the value of which is written down as:

$$tg\delta\alpha_1 \approx \delta x_1 = \frac{|\delta z|}{\ell z_2 - \Delta z + |\delta z'|} \quad (1)$$

from $\Delta K'BD$

$$K'D = \sqrt{(\ell x_1 + \ell x + \Delta x + |\delta x'|)^2 + (\ell z - \ell z_2 + \Delta z + |\delta z'|)^2} \quad (2)$$

and also from $\Delta K'DD'$ we can determine the error of the displacement of the point D (of the gripper)

$$\delta_D = K'D \cdot tg\delta\alpha_1 \approx K'D \cdot \delta\alpha_1 \quad (3)$$

In addition, from $\triangle DD'E$ the error of the displacement of the same point D in the direction of the axis X is determined

$$\delta_{D_{X_0}} = \delta_D \cdot \sin\alpha_1 = \frac{\ell z - \ell z_2 + \Delta z + |\delta z'|}{\ell z_2 - \Delta z + |\delta z'|} |\delta z| \tag{4}$$

and it is determined in the direction of the axis Z, too

$$\delta_{D_{Z_0}} = \frac{\ell x_1 + \ell x + \Delta X + |\delta x'|}{\ell z_2 - \Delta z + |\delta z'|} |\delta z| \tag{5}$$

Similarly the angle of deviation $\delta\alpha_2$ is determined at linear displacement ΔX:

$$\delta_{\alpha_2} = \frac{|\delta x|}{\ell x_2 - \Delta X + |\delta x'|} \tag{6}$$

In such a case the point D' reaches point D" and the error of displacement is

$$\delta_{D'} = C''D' \cdot tg\delta\alpha_2 \approx \frac{\ell x - \ell x_2 + \Delta X + |\delta x'|}{\ell x_2 - \Delta X + |\delta x'|} |\delta x| \tag{7}$$

The error of displacement of the point D' in the direction of the axis X is:

$$\delta_{D'_{X_1}} = \frac{\ell x - \ell x_2 + \Delta X + |\delta x'|}{(\ell x_2 - \Delta X + |\delta x'|)^2} |\delta x|^2 \tag{8}$$

and in the direction of the axis Z is:

$$\delta_{D'_{Z_1}} = \frac{\ell x - \ell x_2 + \Delta X + |\delta x'|}{\ell x_2 - \Delta X + |\delta x'|} |\delta x| \tag{9}$$

The tolerated error for the angle of swing $|\delta\varphi_x|$ does not actually affect the error of displacement of the point D, while the tolerance for the angle $|\delta\varphi_z|$ leads to the change of the position of this point in the direction of the axis Y (Figure 4) and consequently the error is equal to

$$\delta_{D''_{Y_0}} = (\ell x_1 + \ell x + \Delta X + |\delta x'|) |\delta\varphi_z| \tag{10}$$

Respectively, the total error in the direction of the axes is:

$$\delta_{D_X} = \delta_{D_{X_0}} + \delta_{D'_{X_1}} = \frac{\ell z - \ell z_2 + \Delta z + |\delta z'|}{\ell z_2 - \Delta z + |\delta z'|} |\delta z| + \frac{\ell x - \ell x_2 + \Delta X + |\delta x'|}{(\ell x_2 - \Delta X + |\delta x'|)^2} |\delta x|^2$$

$$\delta_{D_Y} = \delta_{D''_{Y_0}} = (\ell x_1 + \ell x + \Delta X + |\delta x'|) |\delta\varphi_z| \tag{11}$$

$$\delta_{D_Z} = \delta_{D_{Z_0}} + \delta_{D'_{Z_1}} = \frac{\ell x_1 + \ell x + \Delta X + |\delta x'|}{\ell z_2 - \Delta z + |\delta z'|} |\delta z| + \frac{\ell x - \ell x_2 + \Delta X + |\delta x'|}{\ell x_2 - \Delta X + |\delta x'|} |\delta x|$$

Thus, using formula (11) we can determine in advance the probable error of the displacement of the end point of the gripper of the industrial robot, depending on the mechanism's dimensions and with maximum tolerated clearances taken into account.

The error referred to should be taken into account during the creation of a new system of the flexible automatic module.

REFERENCES

1. A.C. 1138261, Bul. N 5 07.02.85

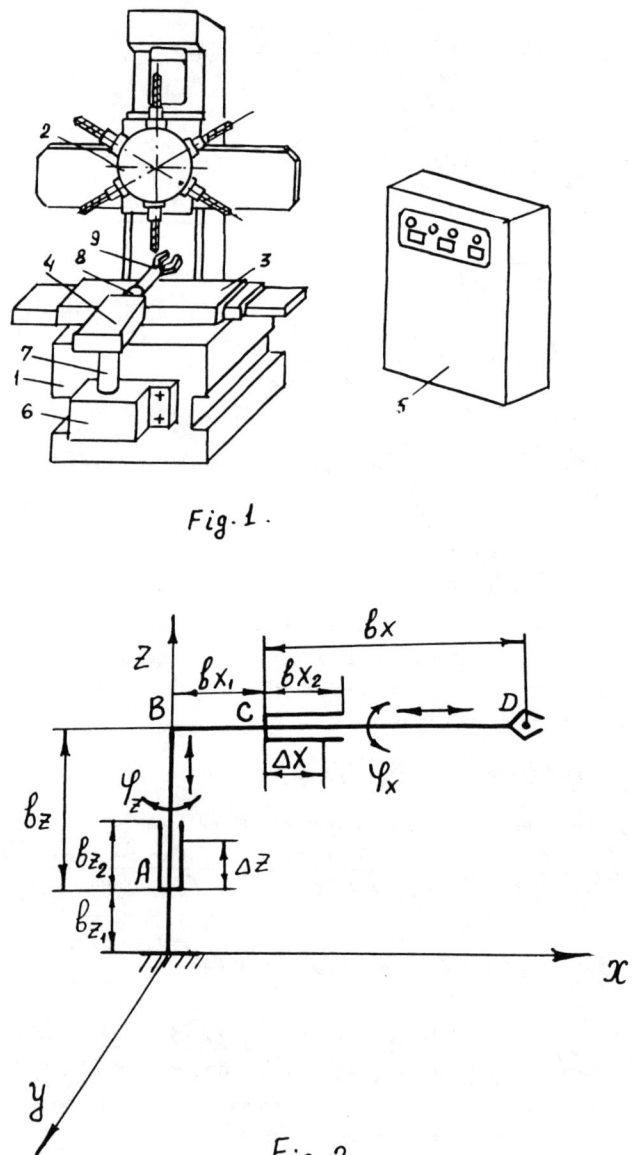

Fig. 1.

Fig. 2.

Accuracy of Flexible Automatic Positioning Module with Clearances 119

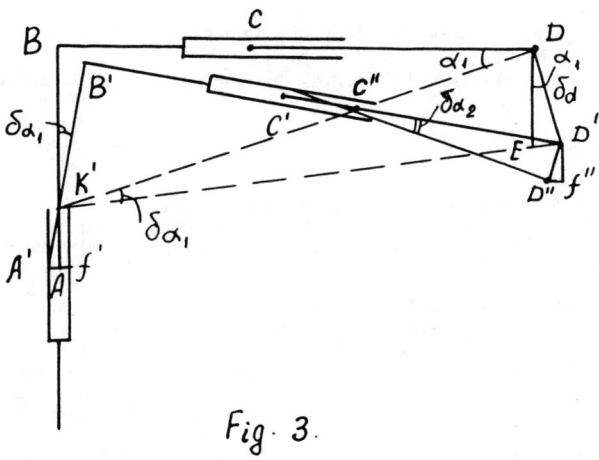

Fig. 3.

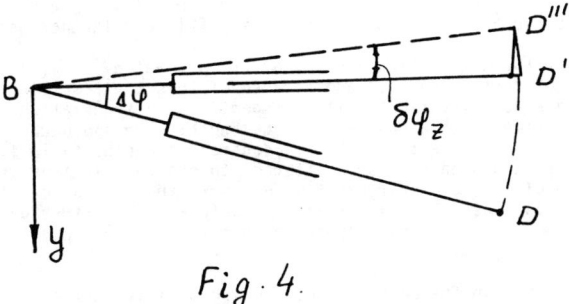

Fig. 4.

Invariant Kinestatic Filtering

H. Lipkin* and J. Duffy**

*The George W. Woodruff School of Mechanical Engineering
Georgia Institute of Technology, Atlanta, U.S.A.
**Center for Intelligent Machines and Robotics University of Florida, Gainesville, U.S.A.

Abstract. Three necessary conditions derived from classical geometry are proposed to evaluate formulations for the simultaneous twist and wrench control of rigid bodies, and for any theory to be meaningful it must be invariant with respect to (1) Euclidean Collineations, (2) Change of Unit Length, and (3) Change of Basis. It is demonstrated in this paper that a previously established theory of hybrid control for robot manipulators (see for example [1-3]) is in fact based on noninvariant principles and is noninvariant with respect to (1) and (2). A new alternative invariant formulation based on screw theory is presented. An example of insertion is included which illustrates both invariant and noninvariant methods.

Introduction. It is useful to introduce the term "kinestatics" to mean the dualistic properties and relations of first order kinematics and statics of a rigid body. The elegant duality afforded by screw theory is used here to develop the kinestatic aspects of the hybrid twist and wrench control of a robotic manipulator which enables the simultaneous control of manipulator motion and its interactions with external constraints. Important applications include insertion tasks, contour tracking and the coordination of multiple robots grasping a single workpiece.
  Hybrid control theory has recently become the subject of a growing research effort which stems from theoretical work proposed in [1]. Apparent experimental verification of this work has been reported in [2], and further generalizations presented in [3].
  It is shown that the theoretical development presented in [1] is based on a mistaken assumption of "orthogonality" and because of this the resulting formulations for hybrid control are not invariant with respect to a translation of origin and neither are they invariant with a change of unit length. The approach developed here (as in [1-3]) is to consider part of hybrid control as a kinestatic filter through which specified twists and wrenches are altered to ensure that they are consistent with a kinestatic model of the environment. However, the kinematic filter development in this paper is based on invariant principles. Further details of this development may be found in [4] and supporting geometrical developments are given in [5,6].

Geometrical Development. In the course of developing an invariant form of kinestatic filtering, it is important to demonstrate that the methodology used is based upon sound fundamental geometric principles. Further, such concepts are useful for simply explaining a mistaken assumption of "orthogonality" which yields a noninvariant formulation.
  As an illustration, let x and U respectively be the homogeneous coordinates of a point and a plane as in [4-6]. The equation,

$$U^T x = U_0 x_0 + U_1 x_1 + U_2 x_2 + U_3 x_3 = 0 \qquad (1)$$

is a projective relation which indicates that the point and the plane are incident. It contains no metrical information relating distance or angle and thus it is incorrect to assert that the point and plane are "orthogonal." Consequently, it is also incorrect to characterize (1) as a "standard inner product" although it may so appear.

Projective collineations, are the most general nonsingular linear transformations and may be expressed as

$$x' = kx, \quad U' = KU. \tag{2}$$

Since the collineation must preserve incidence, then it follows from (1) and (2) that to a nonzero scalar factor,

$$k = K^{-T}. \tag{3}$$

Here and throughout there is no loss in generality by assuming that the scalar factor is unity. Either k or K represents 16 parameters whose ratios represent the 15 independent parameters which are necessary to uniquely specify the transformation. Alternatively, either of the 16 parameters may be constrained respectively by,

$$|k| = 1, \quad |K| = 1 \tag{4}$$

and for simplicity, these conventions are employed in the sequel.

Metrical quantities such as angle and distance may be defined for inner product spaces, i.e. spaces which are endowed with invariant bilinear and quadratic forms (Cayley's Absolute [4-6]). As an example, consider the vanishing bilinear forms,

$$x^T \tilde{\sigma}_\epsilon y = 0, \quad U^T \tilde{\Sigma}_\epsilon V = 0 \quad (5), \qquad \tilde{\sigma}_\epsilon = \begin{bmatrix} I_1 & \cdot \\ \cdot & \epsilon I_3 \end{bmatrix}, \quad \tilde{\Sigma}_\epsilon = \begin{bmatrix} \epsilon I_1 & \cdot \\ \cdot & I_3 \end{bmatrix} \tag{6}$$

where for brevity, (6) and the corresponding quadratic and bilinear forms are referred to as metrics. For $\epsilon = 1, 0, -1$, the metrics (6) are used to define respectively elliptic, Euclidean and hyperbolic spaces. To make the example concrete, consider Euclidean space where $\epsilon = 0$. The second relation in (5) indicates that the planes are $\tilde{\Sigma}_0$-orthogonal, that is, their normals are at right angles. The first relation in (5) indicates that the points are $\tilde{\sigma}_0$-orthogonal. This is an unfamiliar term applied to points in Euclidean space, but it indicates that one or both points lie on the plane at infinity.

Metrical collineations form a subgroup of projective collineations and thus must also preserve the projective properties of incidence. Additionally, they must also leave metrical properties invariant which may be ensured by introducing the constraints,

$$k^T \tilde{\sigma}_\epsilon k = \tilde{\sigma}_\epsilon, \quad K^T \tilde{\Sigma}_\epsilon K = \tilde{\Sigma}_\epsilon. \tag{7}$$

By symmetry, either relation represents 10 distinct equations whose 9 ratios are independent. Thus, as is well-known, it requires 15-9=6 independent parameters to specify a metrical collineation (or isometry). For Euclidean space the admissible collineations form three distinct subgroups, rotations, translations and reflection through the origin (including the identity in each) of which the former two are also rigid body transformations.

The preceding formulations for points and planes may be extended to lines and screws. A line may be described as the join of two points x,y or the meet of two planes U,V. Respectively, the Plücker ray coordinates p, and axis coordinates P, of a line are established as sets of ordered determinants $|xy|$, $|UV|$,

$$p = |xy| = [p_{01} \; p_{02} \; p_{03} \; p_{23} \; p_{31} \; p_{12}]^T, \quad p_{ij} = x_i y_j - y_i x_j \tag{8}$$

$$P = |UV| = [P_{01} \; P_{02} \; P_{03} \; P_{23} \; P_{31} \; P_{12}]^T, \quad P_{ij} = U_i V_j - V_i U_j \tag{9}$$

When p and P represent the same line then to a scalar multiple,

$$p = \tilde{\Delta} P, \quad P = \tilde{\Delta} p \quad (10), \qquad \tilde{\Delta} = \begin{bmatrix} \cdot & I_3 \\ I_3 & \cdot \end{bmatrix}, \quad \tilde{\Delta}\tilde{\Delta} = I_6 \quad (11)$$

where I_n represents the n x n identity matrix and

$$p^T \tilde{\Delta} p = 0, \quad P^T \tilde{\Delta} P = 0, \quad p^T P = 0 \quad (12)$$

vanish identically. Relation (10) is especially important since it establishes the transformation between ray and axis coordinates. When two lines are incident then

$$p^T \tilde{\Delta} q = 0, \quad P^T \tilde{\Delta} Q = 0, \quad p^T Q = 0, \quad P^T q = 0. \quad (13)$$

It should be noted that (8)-(13) are projective relations of which further details on the derivations may be found in [4-6].

Introducing Ball's [7] terminology, a homographic transformation may be represented to a scalar multiple by a 6 x 6 matrix in which the 36 elements have 35 independent ratios, or as preferred here, have a unity determinant. It should be noted that homographic transformations do not necessarily transform lines into lines. However, projective collineations for points and planes also induce projective collineations for lines which form a subgroup of homographic transformations and may be determined from

$$p' = \hat{k} p = |kx \; ky| = |x'y'|, \quad P' = \hat{K} P = |KU \; KV| = |U'V'| \quad (14)$$

where a caret denotes an induced collineation. Since the collineations must preserve incidence properties then it follows from (12) and (14) that

$$\hat{k} = \hat{R}^{-T}, \quad \hat{R} = \hat{k}^{-T} \quad (15), \qquad \hat{k}^T \tilde{\Delta} \hat{k} = \tilde{\Delta}, \quad \hat{R}^T \tilde{\Delta} \hat{R} = \tilde{\Delta}. \quad (16)$$

The relations of (16) are referred to here as the tetrahedron identities since geometrically they are the conditions that the ordered edges (lines) of any tetrahedron are transformed into a similar set. Further, they represent the necessary and sufficient conditions for a (homographic) transformation to be projective. Either of the identities contain 21 distinct relations whose ratios provide 20 independent constraints. Thus induced projective transformations for lines contain 35-20=15 independent parameters which is the same as for points and planes.

Projectively, a screw is defined as a linear combination of lines,

$$p = \mu_1 p_1 + \ldots \mu_n p_n, \quad P = \mu_1 P_1 + \ldots \mu_n P_n. \quad (17)$$

By linearity, it follows that (10) remains valid for screws whereas (12) is valid only when the screw is also a line. Two screws which satisfy (13) are said to be reciprocal. It also follows from linearity that the most general projective collineation of screws is given by $\hat{k}$ and $\hat{R}$.

It is also useful to introduce F. Klein's point-of-view that screws in 3-space may be mapped into points in a 5-space (a 5 dimensional linear variety). Lines in 3-space become points on the hyperquadric (12). Since induced projective collineations must leave this hyperquadric invariant (16), the projective geometry of screws in 3-space is mapped into a pseudo-metrical geometry in 5-space where the pseudo-metric is $\tilde{\Delta}$ and (12) is the pseudo-Absolute. In this sense, when two screws are reciprocal (13), they may also be referred to as (pseudo-) $\tilde{\Delta}$-orthogonal. For brevity, the prefix pseudo- will be omitted in the sequel when the context is clear.

As shown in [4-6], metrical relations for points and planes induce metrical relations for lines and screws,

$$p^T \tilde{\gamma}_\epsilon P = 0, \quad P^T \tilde{\Gamma}_\epsilon P = 0 \quad (18), \quad p^T \tilde{\gamma}_\epsilon q = 0, \quad P^T \tilde{\Gamma}_\epsilon Q = 0 \quad (19), \quad \tilde{\gamma}_\epsilon = \begin{bmatrix} I_3 & \cdot \\ \cdot & \epsilon I_3 \end{bmatrix}, \quad \tilde{\Gamma}_\epsilon = \begin{bmatrix} \epsilon I_3 & \cdot \\ \cdot & I_3 \end{bmatrix}. \quad (20)$$

Equation (18) represents a metrical Absolute which maps to a hyperquadric in 5-space. Two screws which satisfy (9) are said to be $\tilde{\gamma}_\epsilon$- or $\tilde{\Gamma}_\epsilon$- orthogonal and for the Euclidean case where $\epsilon = 0$, the axes of the screws are said to be at right angles or perpendicular. Metrical collineations not only satisfy (16) but must also leave the metrics (20) invariant and therefore,

$$\hat{R}^T \tilde{\gamma}_\epsilon \hat{R} = \tilde{\gamma}_\epsilon, \quad \hat{R}^T \tilde{\Gamma}_\epsilon \hat{R} = \tilde{\Gamma}_\epsilon. \quad (21)$$

Induced metrical collineations for screws contain 6 independent parameters, the same as for points and planes.

Freedom and Constraint. In the following, the underlying geometry is Euclidean although various relations are identified as being projective. It is well-known that the most general instantaneous motion of a rigid body may be described as a twist on a screw and that the most general system of forces and moments is equivalent to a wrench on a screw. Using Plücker's original conventions, a twist T is expressed in axis coordinates and dually a wrench w is expressed in ray coordinates. It is often useful to express T and w in terms of 3 × 1 vectors,

$$T = \begin{bmatrix} V \\ \Omega \end{bmatrix}, \quad w = \begin{bmatrix} f \\ m \end{bmatrix} \quad (22)$$

where Ω is the angular velocity of the body, V is the velocity of a point on the body coincident with the origin, f is the resultant force and m is the resultant moment about the origin.

If a body is free to twist about T while being constrained by w, then T and w are reciprocal and the corresponding rate of virtual work vanishes

$$T^T w = 0 \quad (23)$$

which is a projective relation that is analogous to the last expression in (13). It is also analogous to the incidence of a point and a plane (1) and thus it is incorrect to assert either that T and w are "orthogonal" or that (23) is a "standard inner product." Such misinterpretations of virtual work lie at the very foundation of noninvariant hybrid control theory. (It should be recalled however that reciprocity, as in (23) is equivalent to (pseudo-) Δ-orthogonality.)

When T_B belongs to an n-system of twists B, $0 \le n \le 6$, then w_a belongs to an m=6-n system of reciprocal wrenches a and

$$T_B = B\lambda = B_1\lambda_1 + \ldots B_n\lambda_n, \quad w_a = a\epsilon = a_1\epsilon_1 + \ldots a_m\epsilon_m. \quad (24)$$

These relations are referred to as the freedom equations of T_B and w_a. The constraint equations of T_B and w_a follow from (23), (24)

$$a^T T_B = 0, \quad B^T w_a = 0 \text{ and } a^T B = 0 \quad (25)$$

where the latter equation denotes that the systems are reciprocal.

Consider a manipulator rigidly grasping a workpiece which is constrained by the environment. It is assumed that the body is free to move about any twist in B while being able to (maintain) exert any (reaction wrench) wrench in a. Together, B and a are referred to as a kinestatic model of the environment. Assume that the manipulator is commanded to execute twist T while maintaining wrench w. There are two fundamental cases which must be considered. In the first case assume that T and w are consistent with the model of the environment, that is they satisfy (24) and (25). Under these conditions it is theoretically feasible to execute T and w using a hybrid control scheme with an appropriate control strategy such as damping or compliance control. This problem is referred to as the consistent case and is generally not considered further.

The problem addressed here is the inconsistent case where the specified twist and/or wrench are incompatible with the environment model and may be recognized from

$$T \neq B\lambda, \quad w \neq a\epsilon \quad \text{or} \quad a^T T \neq 0, \quad B^T w \neq 0. \tag{26}$$

It may then be desirable to alter or filter T and w in order to make them consistent with the environment such that the second case is reduced to the first. Specifically, T and w are filtered such that they are respectively in B and a. If an erroneous specification of T and w is not filtered, then the control strategy will attempt to affect an action which is kinestatically unrealizable. The actual outcome is highly dependent upon the details of the control strategy, compensation and physical considerations such as component saturation and friction. It may result in a degradation of accuracy and performance for small errors in T and w to potential instability. Whether it is logical to employ such a filter at all is dependent upon the particular application and range of errors involved and is not considered here.

<u>Kinestatic Filtering</u>. The approach taken is to consider the twist and wrench 6-systems to be each decomposed into complementary subspaces which are related by the vanishing of a bilinear form,

$$g=[a \ \bar{a}], \quad a^T \tilde{\psi} \bar{a}=0 \quad (27), \quad H=[B \ \bar{B}], \quad B^T \tilde{\Psi} \bar{B}=0. \tag{28}$$

It should be noted that this may not always be possible when $\tilde{\psi}$ and $\tilde{\Psi}$ are not positive definite [8]. However, the development here is restricted such that the decomposition is assumed possible. The specified twist and wrench may be decomposed as

$$T=T_B+\bar{T}_B=B\lambda_B+\bar{B}\bar{\lambda}_B \qquad w=w_a+\bar{w}_a=a\epsilon_a+\bar{a}\bar{\epsilon}_a \tag{29}$$

which may be solved for

$$\lambda_B=B^+T, \quad \epsilon_a=a^+w, \quad \bar{\lambda}_B=\bar{B}^+T, \quad \bar{\epsilon}_a=\bar{a}^+w \quad (30), \quad T_B=P_BT, \quad w_a=p_aw, \quad \bar{T}_B=\bar{P}_BT, \quad \bar{w}_a=\bar{p}_aw \tag{31}$$

The superscript + denotes a general left-pseudoinverse which is defined for a general subspace a and symmetrical matrix $\tilde{\Lambda}$ as

$$a^+=(a^T\tilde{\Lambda}a)^{-1}a^T\tilde{\Lambda}, \quad \rho_a=aa^+ \tag{32}$$

where additionally ρ_a represents the corresponding projection matrices in (31). Since g and H in (27), (28) are 6 x 6 and full rank, it may be shown that the inverse in (32) exists. Further, the $\tilde{\Lambda}$-pseudoinverse satisfies the four relations,

$$aa^+a=a, \quad a^+aa^+=a^+, \quad (\tilde{\Lambda}aa^+)^T=\tilde{\Lambda}aa^+, \quad (a^+a\tilde{\Lambda})^T=a^+a\tilde{\Lambda} \tag{33}$$

while the projection matrices satisfy,

$$\rho_a\rho_a=\rho_a, \quad \rho_a+\bar{\rho}_a=I_6, \quad \rho_a\bar{\rho}_a=\bar{\rho}_a\rho_a=0 \qquad (34)$$

Finally, since the filtered twist T_B and wrench w_a are consistent with the environment, the virtual work condition is fulfilled,

$$T_B^T w_a = 0. \qquad (35)$$

<u>Invariance.</u> The preceding formulation of kinestatic filtering is based on an unspecified matrix $\tilde{\Lambda}$. In order that the filtering be well-defined, it must be invariant with respect to:
1. Euclidean rigid body collineations
2. Changes in the unit of length
3. Changes in basis.

A filtering technique which does not satisfy all of these criteria is referred to as noninvariant. It should be noted that the term invariance is used here in its most general sense and includes absolute invariance, relative invariance, contravariance and covariance.

To determine appropriate selections of $\tilde{\Lambda}$, it is useful to express the filtered twist and wrench in expanded form,

$$T_B = B(B^T\tilde{\psi}B)^{-1}B^T\tilde{\psi}T, \quad w_a = a(a^T\tilde{\psi}a)^{-1}a^T\tilde{\psi}w. \qquad (36)$$

After a collineation of space they become,

$$T_B' = \hat{R}T_B = \hat{R}B[B^T(\hat{R}^T\tilde{\psi}\hat{R})B]^{-1}B^T(\hat{R}^T\tilde{\psi}\hat{R})T, \quad w_a' = \hat{k}w_a = \hat{k}a[a^T(\hat{k}^T\tilde{\psi}\hat{k})a]^{-1}a^T(\hat{k}^T\tilde{\psi}\hat{k})w. \qquad (37)$$

In order for these expressions to satisfy the first two invariant criteria it is necessary and sufficient that,

$$\hat{R}^T\tilde{\psi}\hat{R}=\tilde{\psi}, \quad \hat{k}^T\tilde{\psi}\hat{k}=\tilde{\psi}. \qquad (38)$$

However, from (16) and (21) with $\epsilon=0$ the most general form of $\tilde{\psi}$ and $\tilde{\psi}$ must be

$$\tilde{\psi}=\mu_1\tilde{\Delta}+\mu_2\tilde{I}_0, \quad \mu_1\neq 0 \quad (39), \quad \tilde{\psi}=\mu_1\tilde{\Delta}+\mu_2\tilde{\gamma}_0, \quad \mu_1\neq 0 \quad (40)$$

where the scalar inequality is included to preclude the singular case.

Invariant criteria 1 and 2 may be directly verified using (37) and the following collineations in turn,

$$\hat{K}=\begin{bmatrix} R & S \\ \cdot & R \end{bmatrix}, \quad \hat{k}=\begin{bmatrix} R & \cdot \\ S & R \end{bmatrix} \quad (41), \quad \hat{K}=\begin{bmatrix} \delta I_3 & \cdot \\ \cdot & I_3 \end{bmatrix}, \quad \hat{k}=\begin{bmatrix} I_3 & \cdot \\ \cdot & \delta I_3 \end{bmatrix}. \qquad (42)$$

Equation (41) represents the most general Euclidean rigid body collineation where R is a rotation matrix and S is skew symmetric. In (42), δ is a conversion factor such as (100 cm/M). A change in basis may be represented by,

$$B^* = B\phi, \quad a^* = a\varsigma \qquad (43)$$

and substitution of the right sides in (36) yields back the original expressions. It should be noted that the third invariance criterion is actually independent of the choice of $\tilde{\Lambda}$. Further, for the case where T and w are specified as being consistent with the environment, (36) and (37) yield the exact solution independently of the selection of $\tilde{\Lambda}$.

Example. It is useful to illustrate kinestatic filtering with a simple but important example, the insertion of a cylindrical peg in a hole. It is assumed that the peg is partially inserted and that the z coordinate axis is aligned with the peg and hole centerlines. Using $\bar{\psi} = \tilde{\psi} = \tilde{\Delta}$, the environment model and the filtering matrices are respectively,

$$B = \begin{bmatrix} k & \cdot \\ \cdot & k \end{bmatrix}, \quad a = \begin{bmatrix} i & j & \cdot & \cdot \\ \cdot & \cdot & i & j \end{bmatrix} \quad (44), \qquad P_B = \begin{bmatrix} \cdot & \cdot & k & \cdot & \cdot & \cdot \\ \cdot & \cdot & \cdot & \cdot & \cdot & k \end{bmatrix}, \quad p_a = \begin{bmatrix} i & j & \cdot & \cdot & \cdot & \cdot \\ \cdot & \cdot & \cdot & i & j & \cdot \end{bmatrix} \quad (45)$$

where i, j, k represent unit vectors in the coordinate directions. Therefore, P_B filters out all components of the specified twist except for rotations about and translations parallel to the centerlines. Analogously, any specified wrench is filtered by p_a such that it only contains components of force and moment in the i and j directions. Clearly these are the desired actions. The twist filtering (the wrench case is similar), is illustrated by two test twists, unit angular velocities about the z axis (the consistent case) and the y axis (the inconsistent case) respectively,

$$T_1 = \begin{bmatrix} \cdot \\ k \end{bmatrix}, \quad T_{B1} = \begin{bmatrix} \cdot \\ k \end{bmatrix}, \quad T_2 = \begin{bmatrix} \cdot \\ j \end{bmatrix}, \quad T_{B2} = \begin{bmatrix} \cdot \\ \cdot \end{bmatrix} \quad (46)$$

It should be noted that by using both ray and axis coordinates,

$$P_B + P_A = P_B + \tilde{\Delta} p_a \tilde{\Delta} = I_6, \quad p_a + p_b = p_a + \tilde{\Delta} P_B \tilde{\Delta} = I_6, \quad p_a p_b = 0, \quad P_A P_B = 0. \quad (47)$$

Consider now a translation of the origin by -ri without altering the physical problem. The results corresponding to (44)-(46) are,

$$B' = \begin{bmatrix} k & -rj \\ \cdot & k \end{bmatrix}, \quad a' = \begin{bmatrix} i & j & \cdot & \cdot \\ \cdot & rk & i & j \end{bmatrix} \quad (48), \qquad P'_B = \begin{bmatrix} \cdot & \cdot & k & \cdot & -rk & -rj \\ \cdot & \cdot & \cdot & \cdot & \cdot & k \end{bmatrix}, \quad p'_a = \begin{bmatrix} i & j & \cdot & \cdot & \cdot & \cdot \\ \cdot & rk & rj & i & j & \cdot \end{bmatrix} \quad (49)$$

$$T'_1 = \begin{bmatrix} -rj \\ k \end{bmatrix}, \quad T'_{B1} = \begin{bmatrix} -rj \\ k \end{bmatrix}, \quad T'_2 = \begin{bmatrix} rk \\ j \end{bmatrix}, \quad T'_{B2} = \begin{bmatrix} \cdot \\ \cdot \end{bmatrix} \quad (50)$$

yielding the required invariant action.

The form of noninvariant kinestatic filtering which has appeared in the literature [1-3] is based on $\bar{\psi} = \tilde{\psi} = I_6$. When the origin is on the centerlines (44), the noninvariant method yields the same results as (45), (46), i.e. as the invariant method. However when the origin is displaced (48), then (49), (50) are replaced by,

$$P'_B = \begin{bmatrix} \cdot & r^2 dj & k & \cdot & \cdot & -rdj \\ \cdot & -rdk & \cdot & \cdot & \cdot & dk \end{bmatrix}, \quad p'_a = \begin{bmatrix} i & dj & \cdot & \cdot & \cdot & rdj \\ \cdot & rdk & \cdot & i & j & r^2 dk \end{bmatrix}, \quad d = (1+r^2)^{-1} \quad (51)$$

$$T'_1 = \begin{bmatrix} -rj \\ k \end{bmatrix}, \quad T'_{B1} = \begin{bmatrix} -rj \\ k \end{bmatrix}, \quad T'_2 = \begin{bmatrix} rk \\ j \end{bmatrix}, \quad T'_{B2} = \begin{bmatrix} rk \\ \cdot \end{bmatrix}. \quad (52)$$

The first twist, which is consistent with the environment, is left unchanged. However, the second twist is no longer filtered out completely but yields a translation in the z direction. This is a basic problem with the noninvariant method; the result is altered not by changing the physical problem itself but by changing its particular parameterization. The reason for noninvariance is due to the introduction of the identity matrix to construct the filters (51). It is demonstrated in [4-6] that this is actually the metric for elliptic space, i.e. set $\epsilon = 1$ in (20). Since Euclidean rotations form a subgroup of elliptic collineations it is easily verified that rotations do not alter this noninvariant filtering. However, Euclidean translations do not form such a subgroup and this accounts for the noninvariance. It appears that this metric was mistakenly introduced because virtual work (23) was misrepresented as a relation of "orthogonality" rather than reciprocity. Similarly, "norms" such as,

$$T^T T = \Omega^T \Omega + V^T V, \quad w^T w = f^T f + m^T m \qquad (53)$$

and analogous "standard inner products" are without Euclidean meaning and should be avoided.

Conclusion. A systematic geometrical development has led to a new invariant formulation of kinestatic filtering for the hybrid control of manipulators. Further, this development has also shown that an existing theory of hybrid control is actually based upon an elliptic metric and yields noninvariant results in Euclidean space. These findings are not only believed to be significant contributions to the general theory of hybrid control, but the necessary invariance criteria present a rudimentary guideline for evaluating alternative formulations. It should be noted that for the reason of brevity, the only situation considered here is when a general decomposition of the form (27), (28) is possible. Subsequent work will detail the remaining case which involves isotropic subspaces.

Acknowledgments. The authors gratefully acknowledge the financial support of the National Science Foundation (MEA83-04752) and the Westinghouse Research and Development Center, Pittsburgh, PA.

References
[1] Mason, M.T., "Compliance and Force Control for Computer Controlled Manipulators," IEEE Trans. on Systems, Man, and Cybernetics, vol. SMC-11, No.6, June 1981, pp.418-432.
[2] Raibert, M.H., and Craig, J.J., "Hybrid position/force control of manipulators," ASME J. Dyn. Sys., Meas. and Control, vol.105, 1981, pp. 126-133.
[3] West, H., and Asada, H., "A Method for the Design of Hybrid Position/Force Controllers for Manipulators Constrained by Contact with the Environment," Proc. IEEE Intl. Conf. on Robotics and Automation, St. Louis, 1985, pp.251-259.
[4] Lipkin, H., "Geometry and Mappings of Screws with Applications to the Hybrid Control of Robotic Manipulators," Ph.D. Dissertation, Univ. of Florida, 1985.
[5] Lipkin, H. and Duffy, J., "On the Geometry of Orthogonal and Reciprocal Screws," Theory and Practice of Robots and Manipulators, Proc. RoManSy'84: 5th CISM-IFToMM Sym., A. Morecki, et al. Eds., Udine, 1984, pp. 47-56.
[6] Lipkin, H. and Duffy, J., "The Elliptic Polarity of Screws," ASME J. of Mech., Trans. and Auto. in Design, vol.107, Sept. 1985, pp.377-387.
[7] Ball, Sir R.S., Theory of Screws, Cambridge Univ. Press, 1900.
[8] Porteous, I.R., Topological Geometry, 2nd Edition, Cambridge Univ. Press, 1981.

Part 3
Mechanics 2

Redundant Manipulators and Kinematic Singularities
The Operational Space Approach

O. Khatib

Artificial Intelligence Laboratory, Stanford University, Stanford, U.S.A.

Abstract

The operational space formulation has provided a fundamental tool for the description of the dynamic behavior and control of manipulator end-effectors. In this paper, we present the extension of this formulation to redundant manipulator systems. The end-effector equations of motion in operational space of a redundant manipulator are established, and its behavior with respect to generalized joint forces is described. The end-effector is controlled by an operational space control system based on these equations of motion. Asymptotic stabilization of the mechanism is achieved by the use of dissipative joint forces selected from the null space of the Jacobian transpose matrix, consistent with the manipulator dynamics. This allows the elimination of any effects of these additional forces on the end-effector behavior and maintains its dynamic decoupling. We also present a new and systematic approach for dealing with the problems araising at kinematic singularities. The basic philosophy behind this approach is the treatment of the manipulator, at singular configuration, as a mechanism that is redundant with respect to the motion of the end-effector in the subspace of operational space orthogonal to the singular direction.

1. Introduction

Treated within the framework of joint space control systems, redundancy of manipulator mechanisms has been generally viewed as a problem of resolving the end-effector desired motion into joint motions in the sense of some criteria. The manipulator redundancy has been aimed at achieving goals such as the minimization of a quadratic criterion [Whitney 1969, Renaud 1975], the avoidance of joint limits [Liegois 1977, Fournier 1980], the avoidance of obstacles, [Hanafusa, Yoshikawa, and Nakamura 1981, Kircanski and Vukobratovic 1984, Espiau and Boulic 1985], kinematic singularities [Luh and Gu 1985], or the minimization of actuator joint forces [Hollerbach and Suh 1985].

By establishing the end-effector equations of motion and describing its behavior with respect to generalized joint forces, the control of a redundant manipulator is formulated here in terms of finding the operational end-effector forces and generalized joint forces that allow the end-effector to respond to the desired task, while ensuring asymptotic stabilization of the mechansim.

Kinematic singularities is another area that has been also considered within the framework of joint space control and similarly formulated in terms of resolution of the task specifications into joint motions. Generalized inverses and pseudo-inverses have been used, and recently an interesting solution based on the singularity robust inverse has been proposed [Nakamura 1985].

In this paper, we will present a new approach for dealing with the problem of kinematic singularities. In the neighbourhood of a singular configuration the manipulator is treated as a redundant mechanism with respect to the motion of the end-effector in the subspace of operational space orthogonal to its singular direction. The control of the motion in the singular direction is based on the use of the determinant of the Jacobian matrix.

2. Operational Space Formulation

An *operational coordinate system* is a set $\mathbf{x}$ of m *independent* end-effector configuration parameters describing its position and orientation in a frame of reference $\mathcal{R}_0$. For a non-redundant manipulator, the independent parameters $x_1, x_2, \ldots, x_m$ form a complete set of configuration parameters in a domain of the operational space [Khatib 1980] and thus constitute a system of generalized coordinates. The end-effector equations of motion in operational space can be written as [Khatib 1980, Khatib 1983]

$$\Lambda(\mathbf{x})\ddot{\mathbf{x}} + \mu(\mathbf{x},\dot{\mathbf{x}}) + \mathbf{p}(\mathbf{x}) = \mathbf{F}; \qquad (1)$$

where $\Lambda(\mathbf{x})$ designates the kinetic energy matrix, and $\mu(\mathbf{x},\dot{\mathbf{x}})$ represents the vector of end-effector centrifugal and Coriolis forces. $\mathbf{p}(\mathbf{x})$ and $\mathbf{F}$ are respectively the gravity and the generalized operational force vectors. With respect to a system of n joint coordinates $\mathbf{q}$, the manipulator equations of motion in joint space can be written in the form

$$A(\mathbf{q})\ddot{\mathbf{q}} + \mathbf{b}(\mathbf{q},\dot{\mathbf{q}}) + \mathbf{g}(\mathbf{q}) = \Gamma; \qquad (2)$$

where $\mathbf{b}(\mathbf{q},\dot{\mathbf{q}})$, $\mathbf{g}(\mathbf{q})$, and Γ, represent the Coriolis and centrifugal, gravity, and generalized forces in joint space; and $A(\mathbf{q})$ is the $n \times n$ joint space kinetic energy matrix, which is related to $\Lambda(\mathbf{x})$ by

$$A(\mathbf{q}) = J^T(\mathbf{q})\Lambda(\mathbf{x})J(\mathbf{q}); \qquad (3)$$

where $J(\mathbf{q})$ is the Jacobian matrix.

The control of manipulators in operational space is based on the selection of $\mathbf{F}$ as a command vector. In order to produce this command, specific forces Γ must be applied with joint-based actuators. The relationship between the generalized operational forces $\mathbf{F}$ and the joint force vector Γ is given by

$$\Gamma = J^T(\mathbf{q})\,\mathbf{F}. \qquad (4)$$

While in motion, a manipulator end-effector is subject to the inertial coupling, centrifugal, and Coriolis forces. These forces can be compensated for by dynamic decoupling in operational space using the end-effector equations of motion (1). The operational command vector for end-effector dynamic decoupling and motion can be written as

$$\mathbf{F} = \Lambda(\mathbf{x})\mathbf{F}_m^* + \mu(\mathbf{x},\dot{\mathbf{x}}) + \mathbf{p}(\mathbf{x}); \qquad (5)$$

$\mathbf{F}_m^*$ is the command vector of the decoupled end-effector. Using equation (4), the joint forces corresponding to the operational command vector $\mathbf{F}$ in (5) can be written as

$$\Gamma = J^T(\mathbf{q})\Lambda(\mathbf{x})\mathbf{F}_m^* + \tilde{\mathbf{b}}(\mathbf{q},\dot{\mathbf{q}}) + \mathbf{g}(\mathbf{q}). \qquad (6)$$

where $\tilde{\mathbf{b}}(\mathbf{q},\dot{\mathbf{q}})$ is the vector of joint forces under the mapping into joint space of the end-effector Coriolis and centrifugal force vector $\mu(\mathbf{x},\dot{\mathbf{x}})$. $\tilde{\mathbf{b}}(\mathbf{q},\dot{\mathbf{q}})$ can be expressed as,

$$\tilde{\mathbf{b}}(\mathbf{q},\dot{\mathbf{q}}) = \mathbf{b}(\mathbf{q},\dot{\mathbf{q}}) - J^T(\mathbf{q})\Lambda(\mathbf{q})\mathbf{h}(\mathbf{q},\dot{\mathbf{q}}); \qquad (7)$$

where

$$\mathbf{h}(\mathbf{q},\dot{\mathbf{q}}) = \dot{J}(\mathbf{q})\dot{\mathbf{q}}. \qquad (8)$$

A useful form of $\tilde{\mathbf{b}}(\mathbf{q},\dot{\mathbf{q}})$ for real-time control is

$$\tilde{\mathbf{b}}(\mathbf{q},\dot{\mathbf{q}}) = \tilde{B}(\mathbf{q})[\dot{\mathbf{q}}\dot{\mathbf{q}}] + \tilde{C}(\mathbf{q})[\dot{\mathbf{q}}^2]; \qquad (9)$$

where $\bar{B}(q)$ and $\tilde{C}(q)$ are the $n \times n(n-1)/2$ and $n \times n$ matrices of the joint forces under the mapping into joint space of the end-effector Coriolis and centrifugal forces. $[\dot{q}\dot{q}]$ and $[\dot{q}^2]$ are the symbolic notations for the $n(n-1)/2 \times 1$ and $n \times 1$ column matrices

$$[\dot{q}\dot{q}] = [\dot{q}_1\dot{q}_2 \ \dot{q}_1\dot{q}_3 \cdots \dot{q}_{n-1}\dot{q}_n]^T;$$
$$[\dot{q}^2] = [\dot{q}_1^2 \ \dot{q}_2^2 \cdots \dot{q}_n^2]^T. \qquad (10)$$

With the relation (9), dynamic decoupling of the end-effector can be obtained using the configuration dependent dynamic coefficients $\Lambda(q)$, $\bar{B}(q)$, $\tilde{C}(q)$ and $g(q)$. The real-time implementation and the integration of active force control in a unified operational comand vector are presented in [Khatib 1985, Khatib and Burdick 1986].

3. Redundant Manipulators

The configuration of a redundant manipulator cannot be specified by a set of parameters that only describes the end-effector position and orientation. An independent set of end-effector configuration parameters, therefore, does not constitute a generalized coordinate system for a redundant manipulator, and the dynamic behavior of the entire redundant system cannot be represented by a dynamic model in coordinates only of the end-effector configuration. The dynamic behavior of the *end-effector* itself, nevertheless, can still be described, and its equations of motion in operational space can still be established.

Let us first consider the end-effector dynamic response to the application, on the end-effector, of an operational force vector $\mathbf{F}$. The joint forces corresponding to $\mathbf{F}$ are still given by (4). Using the dynamic model (2) and the relation

$$\ddot{\mathbf{x}} = J(q)\ddot{q} + \mathbf{h}(q,\dot{q}); \qquad (11)$$

we established [Khatib 1980] the following equations

$$\Lambda_r(q)\ddot{\mathbf{x}} + \mu_r(q,\dot{q}) + \mathbf{p}_r(q) = \mathbf{F}; \qquad (12)$$

where

$$\Lambda_r(q) = [J(q)A^{-1}(q)J^T(q)]^{-1};$$
$$\mu_r(q,\dot{q}) = \bar{J}^T(q)\mathbf{b}(q,\dot{q}) - \Lambda_r(q)\mathbf{h}(q,\dot{q}); \qquad (13)$$
$$\mathbf{p}_r(q) = \bar{J}^T(q)\mathbf{g}(q);$$

with

$$\bar{J}(q) = A^{-1}(q)J^T(q)\Lambda_r(q). \qquad (14)$$

$\bar{J}(q)$ is actually a generalized inverse of the Jacobian matrix corresponding to the solution that minimizes the manipulator's instantaneous kinetic energy.

Equation (12) describes the end-effector dynamic behavior when the manipulator is submitted to a generalized joint force vector of the form (4). The $m \times m$ matrix $\Lambda_r(q)$ can be interpreted as a *pseudo-kinetic energy matrix* corresponding to the end-effector motion in operational space. $\mu_r(q,\dot{q})$ represents the Centrifugal and Coriolis forces acting on the end-effector, and $\mathbf{p}_r(q)$ the gravity force vector.

Let us now consider the case where an arbitrary joint force vector is applied to the redundant mechanism. Equation (12) can be rewritten as

$$\bar{J}^T(q)[A(q)\ddot{q} + \mathbf{b}(q,\dot{q}) + \mathbf{g}(q)] = \mathbf{F}. \qquad (15)$$

Substituting equation (2) yields
$$F = J^T(\mathbf{q})\Gamma. \tag{16}$$

The matrix $J^T(\mathbf{q})$ describes how the joint space manipulator dynamic forces are reflected at the level of the end-effector.

Lemma

The unconstrained end-effector (12) is subjected to the operational force F if and only if the manipulator (2) is submitted to the generalized joint force vector

$$\Gamma = J^T(\mathbf{q})\mathbf{F} + [I_n - J^T(\mathbf{q})\bar{J}^T(\mathbf{q})]\Gamma_o; \tag{17}$$

where I_n the $n \times n$ identity matrix, $\bar{J}(\mathbf{q})$ is the matrix given in (14), and Γ_o is an arbitrary joint force vector.

When the applied joint forces Γ are of the form (17), it is straightforward from equation (16) to verify that the only forces acting on the end-effector are the operational forces F produced by the first term in the expression of Γ. Joint forces of the form $[I_n - J^T(\mathbf{q})\bar{J}^T(\mathbf{q})]\Gamma_o$ correspond in fact to a null operational force vector.

The uniqueness of (17) is essentially linked to the use in its expression of the generalized inverse $\bar{J}(\mathbf{q})$, which is consistent with the dynamic equations of the manipulator and end-effector. The form of the decomposition (17) itself is general. A joint force vector Γ can always be decomposed in the form (17), and various expressions for Γ associated with various generalized inverses or pseudo-inverses of $J(\mathbf{q})$ can be established.

Let $P(\mathbf{q})$ be a generalized inverse of $J(\mathbf{q})$ and let us submit the manipulator to the joint force vector

$$\Gamma = J^T(\mathbf{q})\mathbf{F} + [I_n - J^T(\mathbf{q})P^T(\mathbf{q})]\Gamma_o. \tag{18}$$

If, for any Γ_o, the end-effector is only subjected to F, equation (18) yields

$$J(\mathbf{q})A^{-1}(\mathbf{q}) = [J(\mathbf{q})A^{-1}(\mathbf{q})J^T(\mathbf{q})]P^T(\mathbf{q}); \tag{21}$$

which implies the identity between $P(\mathbf{q})$ and $\bar{J}(\mathbf{q})$.

4. Control of Redundant Manipulators

Similar to the case of non-redundant manipulators, the dynamic decoupling and control of the end-effector can be achieved by selecting an operational command vector of the form (5). The corresponding joint forces are similar to those of (6),

$$\Gamma = J^T(\mathbf{q})\Lambda_r(\mathbf{q})\mathbf{F}_m^* + \tilde{\mathbf{b}}_r(\mathbf{q},\dot{\mathbf{q}}) + \mathbf{g}(\mathbf{q}); \tag{20}$$

where $\tilde{\mathbf{b}}_r(\mathbf{q},\dot{\mathbf{q}})$ is defined similarly to $\tilde{\mathbf{b}}(\mathbf{q},\dot{\mathbf{q}})$, and can be expressed as

$$\tilde{\mathbf{b}}_r(\mathbf{q},\dot{\mathbf{q}}) = \tilde{B}_r(\mathbf{q})[\dot{\mathbf{q}}\dot{\mathbf{q}}] + \tilde{C}_r(\mathbf{q})[\dot{\mathbf{q}}^2]. \tag{21}$$

Stability Analysis

Under the command vector (20), and with the assumption of a "perfect" compensation (or non-compensation) of the centrifugal and Coriolis forces, the manipulator is subjected to dissipative forces Γ_{dis} due to the velocity damping term $(-k_v \dot{\mathbf{x}})$ in $\mathbf{F}_m^*$, these forces are

$$\Gamma_{dis} = D(\mathbf{q})\dot{\mathbf{q}}; \tag{22}$$

with
$$D(q) = -k_v J^T(q)\Lambda_r(q)J(q). \tag{23}$$

$D(q)$ is an $n \times n$ negative semi-definite matrix of rank m. Although the manipulator is stable, since the condition
$$\dot{q}^T D(q)\dot{q} \leq 0; \tag{24}$$

is satisfied [Mingori 1970], this redundant mechanism can still describe movements that are solutions [Rumiantsev 1970] of the equation
$$\dot{q}^T D(q)\dot{q} = 0. \tag{25}$$

An example of such a behavior is shown in Figure 1a. The end-effector of a simulated three-degree-of-freedom planar manipulator is controlled under (20). The end-effector goal position coincides with its current position, while the three joints are assumed to have initially non-zero velocities (0.5rad/s has been used).

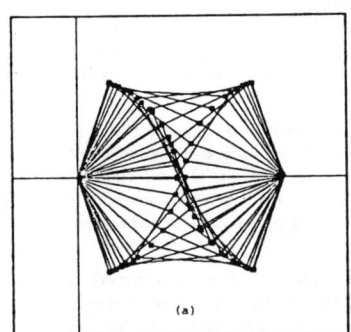
(a)

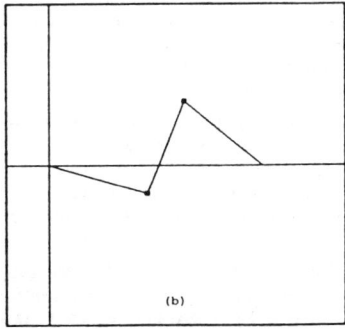
(b)

Figure 1: Stabilization of a Redundant Manipulator.

Asymptotic stabilization of the system can be achieved by the addition of dissipative joint forces [Khatib 1980]. These forces can be selected to act in the null space of the Jacobian matrix [Khatib 1985]. This precludes any effect of the additional forces on the end-effector and maintains its dynamic decoupling. Using (17) these additional stabilizing joint forces are of the form

$$\Gamma_{ns} = [I_n - J^T(q)\overline{J}^T(q)]\Gamma_s. \tag{26}$$

By selecting
$$\Gamma_s = -k_{vq} A(q)\dot{q}; \tag{27}$$

the vector Γ_{ns} becomes
$$\Gamma_{ns} = \Gamma_s + J^T(q)\Lambda_r(q)F_s; \tag{28}$$

with
$$F_s = k_{vq}\dot{x}. \tag{29}$$

Finally, the joint force command vector can be written as

$$\Gamma = J^T(q)\Lambda_r(q)(F_m^* + F_s) + \Gamma_s + \tilde{b}_r(q,\dot{q}) + g(q). \tag{30}$$

Under this form, the evaluation of the explicit expression of the generalized inverse of the Jacobian matrix is avoided. The matrix $D(q)$ corresponding to the new expression for the dissipative joint forces Γ_{di} in the command vector (30) becomes

$$D(q) = -[(k_v - k_{vq})J^T(q)\Lambda_r(q)J(q) + k_{vq}A(q)]. \tag{31}$$

Now, the matrix $D(q)$ is negative definite and the system is asymptotically stable. Figure 1b shows the effects of this stabilization on the previous example of a simulated three-degree-of-freedom manipulator.

5. Singular Configurations

A *singular configuration* is a configuration q at which some column vectors of the Jacobian matrix become linearly dependent. The mobility of the end-effector can be defined as the rank of this matrix [Fournier 1980]. In the case of non-redundant manipulators considered here, the end-effector at a singular configuration loses the ability to move along or rotate about some direction of the Cartesian space; its mobility locally decreases. Singularity and mobility can be characterized, in this case, by the determinant of the Jacobian matrix.

Singularities can be further specified by the posture of the mechanism at which they occur. Different types of singularities can be observed for a given mechanical linkage. These can be directly identified from the expression of the determinant of the Jacobian matrix. The expression of this determinant can be, in fact, developed into a product of terms, each of which corresponds to a type of singularity related to the kinematic configuration of the mechanism e.g. alignment of two links or alignement of two joint axes.

To each singular configuration there corresponds a singular "direction". It is in that direction, that the end-effector presents infinit inertial mass for displacements or infinit inertia for rotations. Its movements remain free in the subspace orthogonal to this direction. This behavior extends, in reality, to a neighbourhood of the singular configuration. The extent of this neighbourhood can be characterized by the particular expression $s(q)$ in the determinant of the Jacobian matrix that vanishes at this specific singularity.

The neighbourhood of a given type of singularity $\mathcal{D}_s$ can be defined as

$$\mathcal{D}_s = \{q \mid |s(q)| \leq s_0\}; \tag{33}$$

where s_0 is positive.

The basic concept in our approach to the problem of kinematic singularities can be formulated as follows:

In the neighborhood $\mathcal{D}_s$ of a singular configuration q, the manipulator is treated as a mechanism that is redundant with respect to the motion of the end-effector in the subspace of operational space orthogonal to the singular direction. For the end-effector motion in that subspace, the manipulator is controlled as a redundant mechanism. Joint forces selected from the associated nullspace are used for the control of the end-effector motion along the singular direction. When moving out of the singularity, this is achieved by controlling the rate of change of $s(q)$ according to the value of the desired velocity for this motion at the configuration when $|s(q)| = s_0$. The selection of sign of the desired rate of change of $s(\mathbf{q})$ allows the control of the manipulator posture among the two configurations that it can generally take when moving out a singularity. A position error term on $s(\mathbf{q})$ is used in the control vector for tasks that involve a motion toward goal positions located at, or in the neighbourhood of, the singular configuration.

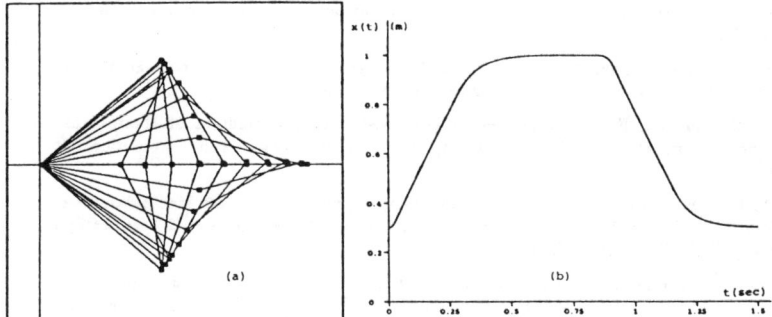

Figure 2: Control at a Singular Configuration.

This approach can be simply extended to configurations where more than one singularity is involved. An example of a simulated two-degree-of-freedom manipulator is shown in Figure 2a. The manipulator has been controlled to move into and out of the singular configuration while displaying two different postures. The time-response of the motion in the singular direction $x(t)$ is shown in Figure 2b.

5. Conclusion

The end-effector equations of motion for a redundant manipulator system have been established, and an operational space control system for end-effector dynamic decoupling and control has been designed. The expression of joint forces of the nullspace of the Jacobian matrix consistent with the end-effector dynamic behavior has been identified and used for the asymptotic stabilization of the redundant mechanism. The resulting control system avoids the explicit evaluation of any generalized inverse or pseudo-inverse of the Jacobian matrix. Joint constraints, collision avoidance, and control of manipulator postures can be naturally integrated in this framework of operational space control systems. Also, a new systematic solution to the problem of kinematic singularities has been presented. This solution constitutes an effective alternative to resolving end-effector motions into joint motions generally used in joint space based control systems.

Acknowledgements

The author would like to thank John Craig, Harlyn Baker, Shashank Shekhar, Joel Burdick, and Bradley Chen for the discussions, comments, and help during the preparation of the manuscript. The support from the National Science Foundation and the Systems Development Foundation are gratefully acknowledged.

References

Espiau, B, and Boulic, R. 1985 (November). Collision Avoidance for Redundant Robots with Proximity Sensors. Preprints of the 3rd International Symposium in Robotics Research, pp. 94-102. Gouvieux, France.

Fournier, A. 1980 (April). Génération de Mouvements en Robotique. Application des Inverses Généralisées et des Pseudo Inverses. *Thèse d'Etat*, Mention Science, Université des Sciences et Techniques du Languedoc, Montpellier, France.

Hanafusa, H., Yoshikawa, T., and Nakamura, Y., 1981. Analysis and Control of Articulated Robot Arms with Redundancy. 8th IFAC World Congress, vol. XIV, pp. 38-83.

Hollerbach, J.M. and Suh, K.C. 1985 (March 25-28). Redundancy Resolution of Manipulators Through Torque Optimization. 1985 International Conference on Robotics and Automation. St. Louis.

Khatib, O. 1980. Commande Dynamique dans l'Espace Opérationnel des Robots Manipulateurs en Présence d'Obstacles. Thèse de Docteur-Ingénieur. École Nationale Supérieure de l'Aéronautique et de l'Espace (ENSAE). Toulouse, France.

Khatib, O. 1983 (December 15-20). Dynamic Control of Manipulators in Operational Space. Proceedings of The Sixth CISM-IFToMM Congress on Theory of Machines and Mechanisms, pp. 1128-1131, New Delhi, India (Wiley, New Delhi).

Khatib, O. 1985 (October). The Operational Space Formulation in the Analysis, Design, and Control of Manipulators. The Third International Symposium of Robotics Research, Paris, France.

Khatib, O. and Burdick, J. 1986 (April). Motion and Force Control of robot Manipulator. The 1986 IEEE International Conference in Robotics and Automation, San Francisco.

Kircanski, M. and Vukobratovic, M. 1984 (June). Trajectory Planning for Redundant Manipulators in Presence of Obstacles. Preprints of the 5th CISM-IFToMM Symposium on the Theory and Practice of Robots and Manipulators, pp. 43-58. Udine, Italy.

Liegois, A. 1977 (December). Automatic Supervisory Control of the Configuration and Behavior of Multibody Mechanisms. IEEE Transactions on Systems, Man and Cybernetics, vol. 7, no. 12.

Luh, J.Y.S. and Gu, Y.L. 1985 (March). Industrial Robots with Seven Joints. Proc. 1985 IEEE International Conference on Robotics and Automation, pp. 1010-1015, St. Louis.

Mingori, D.L. 1970. A Stability Theorem for Mechanical Systems with Constraint Damping. Journal of Applied Mechanics, Trans. of the ASME, pp. 253-258.

Nakamura, Y. 1985 (June). Kinematical Studies on the Trajectory Control of Robot Manipulators. Ph.D. Thesis, Kyoto, Japan.

Renaud, M. 1975 (December). Contribution à l'Etude de la Modélisation et de la Commande des Systèmes Mécaniques Articulés. Thèse de Docteur Ingénieur. Université Paul Sabatier, Toulouse.

Rumiantsev, V.V. 1970. On the Optimal Stabilization of Controlled Systems. PMM, vol. 34, no. 3, pp. 440-456.

Whitney, D.E. 1969 (June). Resolved Motion Rate Control of Manipulators and Human Prostheses IEEE Transactions on Man Machine Systems, no. 10, vol. 2, pp. 47-53.

Modelling and Simulation of Mechanical Process in Hyperstatical Gripping with n-Contact Points

S. Zeghloul, J.-P. Lallemand and D. Murguet

Université de Poitiers - Laboratoire de Mécanique des Solides, Poitiers, France

SUMMARY

In the context of off-line programming of a robot equipped with a nonspecialized gripper, it is interesting to be able to evaluate the object-gripper contact forces in different states of work, such as high speed displacement, assembly operations,... in order to assure the stability of the grasp.

It is also interesting to be able to evaluate the possible movement of an object in a gripper finger under the action of a grasping force. The movement can be harmful to realization of the task, especially assembly.

The work presented here develops methods used to control the influence of gripping in the overall robotic process. The paper consists of three parts. We first present software used to calculate the contact forces in isostatic or hyperstatical configuration of a multi-finger gripper, secondly software for the simulation of movement of n-link robots, and third, software for displaying forces and deformations using animation and construction programs.

INTRODUCTION

For some years, theoretical studies have been directed at the gripping of objects as well as their manipulation using mechanical hands (1 to 5). These studies focus on the kinematical aspect. The analysis of contact forces, possibilities of sliding on contact and potential movement of the object are rare. These aspects are treated in this paper, in which we consider isostatic or hyperstatic gripping using n contact points.

CAD software facilitates the evaluation and the graphical representation of robot function. Several programs for robotic simulation have been developed. Some of these programs operate in planar (2D) environment and others in spatial (3D). We can quote as an example two programs which have robotic functions CATIA(6), EUCLID (7). Other simulators interpret instructions in robot programming language; the simulator developed in Stanford University (8) and the simulator ISR or LM (manipulation language) developed in LIFIA (Grenoble, France) (9). Most of this software needs a large computer and their performance is dependent on the system used.

The second part of the paper describes software which was written in our laboratory for the manipulation and the construction robots. This software is used to simulate an example of the grasping process presented in the third part of this paper. Forces, deformations and sliding are displayed by means of animation and construction programs.

1. HYPERSTATICAL GRIPPING

1.1. Computational Scheme

Under the deforming action of the fingers, the kinematic mobility of the object remains but there appears in O (figure 1) a small movement represented by a six component vector written:

$\vec{d} = (\vec{W}_o, \vec{\theta}_o)^T$ where $\vec{W}_o$ is the displacement of O and $\vec{\theta}_o$ is the absolute rotation

The clamping action introduces a modification of finger parameters expressed by $\vec{p}_i$ (variation of a_i joint coordinates) for each finger i (i = 1 to n). Object clamping is exercised by means of forces acting on some joints of a finger. For the finger i it is represented by the $\vec{S}_i$ vector with a_i coordinates. Sliding or no-sliding is dependent on the coefficient of friction.

a) <u>Equilibrium and behaviour equations</u>

The equations are written as follows:

$$N\vec{F} = \vec{T} \qquad (1)$$

$$J_i^T \vec{F}_i = -\vec{S}_i \qquad i = 1 \text{ to } n \qquad (2)$$

$$\vec{W}_{A_i} = J_i \vec{p}_i + C_i \vec{F}_i \qquad i = 1 \text{ to } n \qquad (3)$$

where
$$\vec{W}'_{A_i} = H_i \vec{d} - \mathcal{C}_i \vec{F}_i \qquad i = 1 \text{ to } n \qquad (4)$$

(1) describes the object's equilibrium under the effect of the contact forces $\vec{F} = (\vec{F}_1, \vec{F}_2, \ldots \vec{F}_n)^T$ where N (6x3n) is the matrix of plucker coordinates of each $\vec{F}_i$ coordinate, and the effect of the external forces and torques $\vec{T} = (T_1, \ldots T_6)^T$.
(2) translates the virtual work principle applied to each finger i using the matrix J_i^T (a_ix3) of the inverse kinematical model.
(3) expresses the displacement $\vec{W}_{A_i}$ of finger's tip A_i; C_i is the compliance matrix (3x3) of finger i.
and finally (4) expresses the finger's displacement $\vec{W}'_{A_i}$ using $\vec{d}$ and the local deformability of the object under the action of the force $\vec{F}_i$; $\mathcal{C}_i$ is the compliance matrix (3x3) of the object as A_i and H_i is the matrix (3x6) obtained by the relation:

$$\vec{W}_o + \vec{A_i O} \wedge \vec{\theta}_o = H_i \vec{d}$$

Earlier relations constitute a linear system with 6n + a + 6 equations and 9n + a + 6 unknowns (a = $\sum_{i=1}^{n} a_i$).

b) Sliding Equations

The introduction of sliding or no-sliding conditions allows us to obtain the 3n missing equations. Supposing that there are n' contacts without sliding and n" contacts with sliding (n' + n" = n), then we have to examine the 2^n possibilities. For the n' contacts without sliding, we have:

$$\vec{W}_{A_i} = \vec{W'}_{A_i} \qquad \text{3n' relations} \quad (5)$$

For the n" contacts with sliding, with $\vec{n}_i$ normal on contact A_i, we obtain:

$$(\vec{W}_{A_i} - \vec{W'}_{A_i}) \cdot \vec{n}_i = 0 \qquad \text{n" relations} \quad (6)$$

$$(\vec{F}_i, \vec{W}_{A_i} - \vec{W'}_{A_i}, \vec{n}_i) = 0 \qquad \text{n" relations} \quad (7)$$

$$\vec{F}_i \cdot \vec{n}_i = \| \vec{F}_i \| \cos\phi_i \qquad \text{n" relations} \quad (8)$$

$$\vec{F}_i \cdot (\vec{W}_{A_i} - \vec{W'}_{A_i}) < 0 \qquad \text{n" relations} \quad (9)$$

where (6) expresses the condition of non-penetration, (7) shows that $\vec{F}_i$ belongs to the sliding plane $(\vec{W}_{A_i} - \vec{W'}_{A_i}, \vec{n}_i)$, (8) shows that $\vec{F}_i$ is on the sliding cone having $\vec{n}_i$ axis and ϕ_i angle, and finally (9) expresses the condition that the friction force is in the direction opposed sliding to the motion. Finally we obtain (9n+a+6) equations in which 2n" (equations (7) and (8)) are nonlinear and n" inequality relations.

1.2. Architecture of software in hyperstatic gripping

The first software developed deals with the gripper developed in the laboratory. Its mechanical program architecture is represented in figure 2. The fingers 2 and 3 and 2 cantilever beams. The grasping force S is applied to the prismatic joint of finger 1 and induces a small displacement p at the base of the finger, a deformation of the 3 fingers and a small displacement u of the object. The program includes different subprograms: 1) data input, 2) choice of contact configuration, 3) calculations.

2. TOOLS FOR CONSTRUCTION AND ANIMATION OF ARTICULATED BODY STRUCTURES

2.1. Geometrical modelling of three-dimensional environment

To simulate a robot's functions the model must reproduce a graphical picture which is quite close to real environment. We distinguish three different entities in the model, the robot and its kinematic structure, the objects to be picked up by the manipulator and the environment in which some elements are considered to be obstacles at the time of trajectory generation.

Geometrical description

The robot model is created by an assembly of rigid bodies. The manipulator is then represented by simple geometrical shapes. We have developed a program for automatic generation of polyhedric shapes (cube, cylinder, cone, sphere, etc.).

Figure 3 represents the manipulator TH8 (RPPRRR) in creation stage. The user chooses sequentially and in an interactive way the primitives representing the robot structure. The data inputs are dimensional parameters of polyhedric shapes. In figure 4 TH8 is presented in its local environment.

Kinematic description

The manipulator is described as a kinematic open chain consisting of n articulated bodies. Two consecutive bodies are connected by a joint: a revolute joint for rotation movement and a prismatic joint for translation. These kinematic characteristics are introduced, together with the model. Other functions are available in the program for geometrical and kinematical modelling of the effector (multi-finger gripper). This software is also able to deal with a multi robot environment. In this case the kinematic characteristics are represented by a tree structure.

2.2. Simulation of movements

The data base created by the software described earlier is used in the animation program. Another program displays the robot and its environment (figure 4). The graphical routine allows several representations: three-dimensional and different plan views.

Control of kinematic structure

For the control, the program generates the goemetrical model automatically. It uses the kinematic characteristics available in the data base representing the robot. The geometrical model is given by the following equation:

$X_i = f(\theta_i)$

where X_i are the cartesian coordinates and orientation angles of the end effector and θ_i joint coordinates.

The control values are transmitted to the program either from keyboard control or are read in data file. In the latter mode, the data file may contain a manipulation task.

Control modes and graphical representation

The display of the trajectory on screen is obtained in several ways.

From the manipulator reference position the user gives the joint coordinates θ_i corresponding to the desired final position; the trajectory is then represented on the screen by the succession of N graphical pictures where N is the number of points of the trajectory.

In the second mode two joint coordinates vectors are specified for two work stations, the display being operated in the same way as the first mode (figure 5).

The third mode allows the trajectory from the current position to the desired position to be displayed; the user then introduces the relative variation of joint coordinates.

Manipulation of environmental objects

As we have mentioned earlier the environmental objects susceptible to being moved by the manipulator are distinguished from fixed objects.

The animation of gripped objects needs the creation of rigid connections between the end effector and the object. The putting down of the object is operated by the destruction of the connection and the update of the data base. Figure 6 shows the manipulator evolution with a gripped object. We can then evaluate the space necessary for object evolution along the task. This allows us to take into account collision problems before the generation of the trajectory.

3. SIMULATION OF GRIPPER'S BEHAVIOUR

3.1 Construction of gripper model

The gripper (figure 2) is built by means of software previously described. In order to take into account finger deformation, the finger is designed with three elements (figure 7). We introduce at some joints two degrees of freedom of rotation around the x axis and y axis. The division into pieces of length h, 2h and 3h allows us to visualize easily and correctly the deflection curve of a cantilever beam. The rotations θ_{ix} and θ_{iy} (i = 1,2) are calculated from the displacement DX and DY of the contact point,

θ_{ix} and θ_{iy} are given by the following relations:

$$\theta_{1x} = \text{arctg} \frac{DY.EY}{8h} \qquad \theta_{2x} = \text{arctg} \frac{DY.EY}{4h} - \theta_{1x}$$

$$\theta_{1y} = \text{artg} \frac{DX.EX}{8h} \qquad \theta_{2y} = \text{artg} \frac{DX.EX}{4h} - \theta_{1y}$$

where EX and EY are scale factors.

3.2 Displaying

As an example, we present the result for the cylinder
(external forces and torques being nil : $\vec{T} = 0$).

The figure 8 shows the results obtained, which bring out the deformability of fingers and the movement of the object due to the grasping force. In order to have a clear figure, the object's displacement has been multiplied by 20 corresponding to EY = 20. EX is connected with EY by the following relation:

$$EX = \frac{1}{DX} [\{R^2 - [Y - EY (u - DY)]^2\}^{\frac{1}{2}} - X]$$

where R is the radius of grasped object, X and Y are the contacts points coordinates with respect to the reference frame XYZ (figure 2).

The forces are represented on the screen by a line whose length is proportional to the intensity of the force. A square represents the point of action. Figure 9 represents the normal and tangential contact forces on the object in the reference position.

On the contour of the object, before grasping, the contact points are represented by empty squares. On the contour of the object, after grasping, the new contact points are represented by full squares. The figure 10 shows clearly object movement and sliding which appears on the two contacts which are on the same side.

CONCLUSION

The paper describes software that are useful tools for the solution of robotic problems by graphical simulation. The programs are implemented on the PDP11 microcomputer with 64 KB memories and written in FORTRAN and assembly language (MACRO-11). The time needed for the computation of a picture such as shown in figure 4 is of the order of 2s.

Utilization of computation of hyperstatic forces and animation programs allows us to display the finger's deformations and the object displacement during the grasping process.

REFERENCES

|1| G. BIANCHI, A. ROVETTA, "On the grasping process for objects of irregular shape", Proc. of the 3rd Ro-Man-Sy, Udine 1978.
|2| A. ROVETTA, "CAD and the Design of Mechanical hands", Proc. of the 5th Ro-Man-Sy, Udine 1984.
|3| J.K. SALISBURY, "Kinematic and force analysis of articulated hand" Ph. D. Standford University, 1982.
|4| S. ZEGHLOUL, "Analyse de la préhension. Application à un préhenseur tridigital doté de sens tactile", Thèse 3è cycle, Poitiers, 1983.
|5| J.C. GUINOT, P. BIDAUD, J.P. LALLEMAND, "Mechanical analysis of a gripper-Manipulator", Proc. of the 2nd I.S.R.R., Kyoto, Japan, 20-23 Aug. 1984.
|6| A. LIEGEOIS, E. DOMBRE, P. BORREL, "Developpement d'un système de CAO et de simulation de Robot manipulateurs" Première Journées annuelles du Programme automatisation et robotique avancées Poitiers, Sept. 1982, pp. 117-128.
|7| A. BARRACO, B. CUNY, G. ISHIOMIN, "Introduction de modèles cinématique et dynamique de robot dans deux logiciels de CAO", CAO Robotique Hermes Publishing, Janvier 1985.
|8| B.I. SOROKA, "Debugging robot program with a simulator", CADCAM-8 conference Anaheim, California November 1980.
|9| C. LAUGIER, "Les apports respectifs des langages symboliques et de la CAO en programmation des Robots". CAO Robotique Hermes Publishing, Janvier 1980.

Modelling and Simulation of Mechanical Process 145

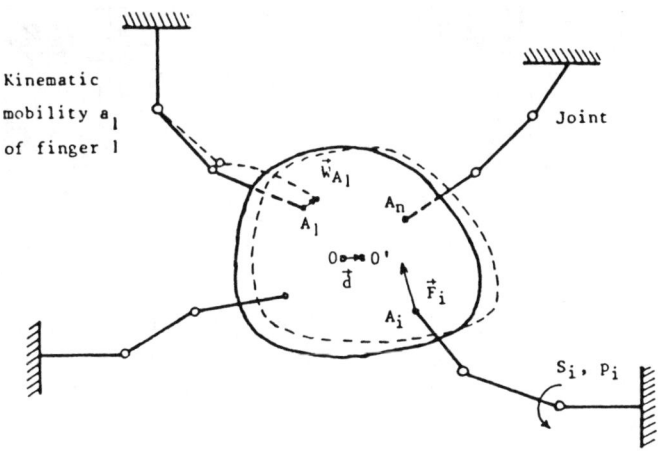

Figure 1 : Schematization of multi-contacts gripper.

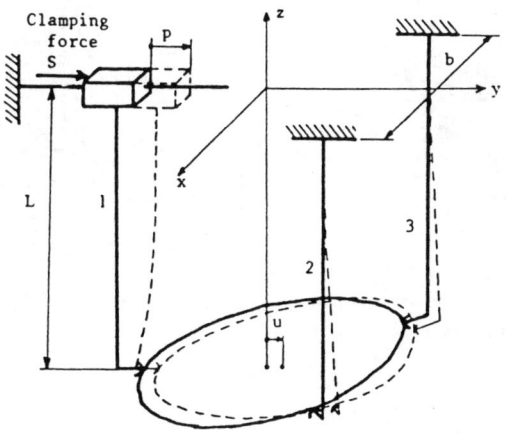

Figure 2 : Mechanical architecture diagram of the gripper
(L = 100 mm ; b = 45 mm).

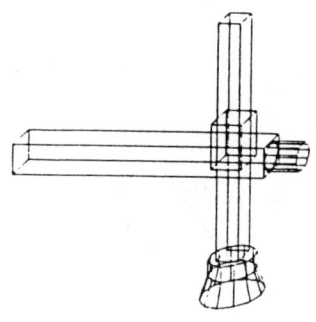

Figure 3 : Manipulator RPPRRR in creation stage.

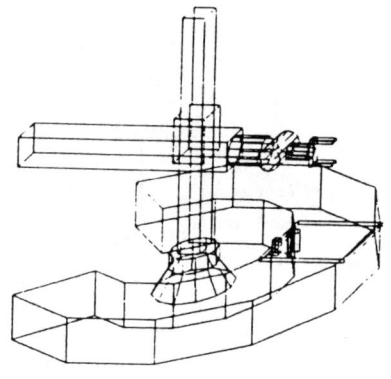

Figure 4 : TH8 in its local environment (reference position).

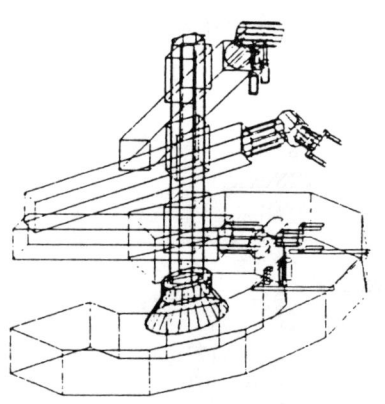

Figure 5 : Evolution of the manipulator between two positions.

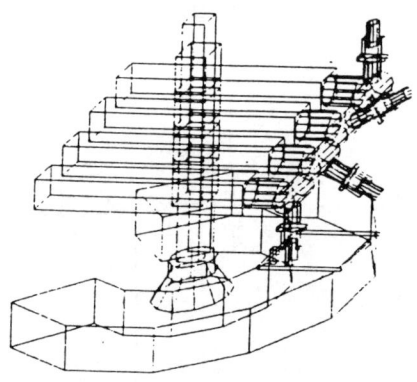

Figure 6 : Animation of a gripped object.

Modelling and Simulation of Mechanical Process 147

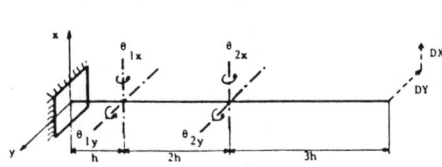

Figure 7 : Degrees of freedom representing the possible deformation of a finger.

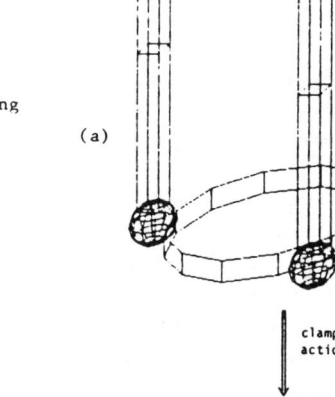

(a)

(b)

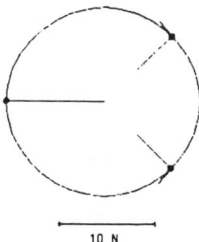

Figure 9 : Visualization of the normal and tangential contact forces under a 10N clamping force.

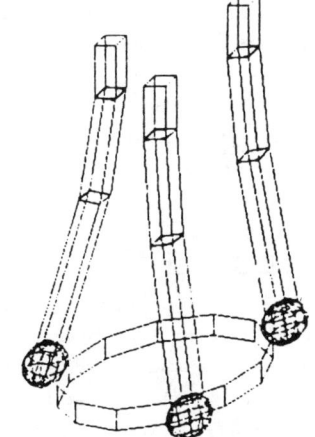

Figure 8 : (a) Gripper in its reference position
(b) Clamping phase. Bending of the fingers is represented with three elements which have 4 degrees of freedom.
- The p joint displacement is 1.715mm
- The u object displacement is 0.715 mm
- The S clamping force is 10 N
- The compliance of finger is 10^{-4} mN^{-1}.

Figure 10 : Visualization of both the displacement of the object and the sliding of contact points.

Computer Aided Modelling of Pneumo-Hydraulic Robots

C. Rzymkowski

Warsaw University of Technology, Warsaw, Poland

The paper describes an algorithm for computer aided modelling of a manipulator with rigid links driven by hydraulic and pneumatic actuators. It is capable of modelling the dynamics of mechanical elements, as well as phenomena taking place in actuators, pneumatic and hydraulic back-up installations. This algorithm automatically generates relations describing generalized forces.

The equations of motion of mechanical elements are derived from Lagrange's equations. Application of Denavit-Hartenberg's formalism permits substitution of differentiation operations by matrix products. Knowledge of pressures in actuator chambers is the basis for determination of driving forces.

The algorithm was transformed into a FORTRAN language program package, which automatically generates, in symbolic form, a complete discrete model of a manipulator of a certain class. Taking into account the special properties of the generated equations provides general procedures, which perform algebraic transformations, and permit utilization of the package implemented on a minicomputer with a main memory smaller than 100 KB. At the same time it avoids frequent use of the external files.

1. INTRODUCTION

The use of computers for simulation is widespread. In robotics computers have been utilized for construction of simulation models since the early seventies. Up until now now a lot of program packages have been written to aid the process of modelling robots as well as to help in synthesis of their control systems. This trend is reflected in the papers presented at successive RoManSy Symposia. (See papers by M. Vukobratovic et al. from "Mihailo Pupin" Institute in Belgrade). The solutions given by different authors are based on several known methods of formulating the laws of mechanics: Newton-Euler's, Gibbs', Appel's, Lagrange's equations or d'Alembert's principle. Most of the existing systems automatically create the equations of motion of mechanical elements leaving the determination of driving forces to the user. The determination of these forces, in the case of pneumatic and hydraulic drives, requires integration of differential equations describing phenomena taking place in the chambers of the actuators. The solution automatically constructs equations describing the drives, so that computer simulation of the mechanical system together with its drives is possible. All the active forces acting on the mechanism, constituting the arm of the robot, are automatically transformed into generalized forces.

The algorithm described here is effective in the case of systems meeting the following demands:
1. The mechanism is an open kinematic chain with rigid links linked together with revolute or prismatic pairs of the V-th class. Pairs of other classes must be substituted by kinematically equivalent assemblies of V-th class pairs.
2. The drive system comprises only pneumatic and hydraulic actuators with reciprocating pistons.
3. The actuators can form closed loops with the elements of the mechanism, so long as their mass is negligible compared to the mass of the elements of the main chain or if their mass can be accurately reduced to the centre of mass of these elements.

2. EQUATIONS OF MOTION OF MECHANICAL ELEMENTS

The motion of a kinematic chain (manipulator) is considered relative to a global, Cartesian frame of reference π_0 fixed to the base. Moveable local coordinate frames are fixed to the elements of the kinematic chain in accordance with Denavit-Hartenberg's formalism, that is: the axis z_i of the frames coincide with the axis of the kinematic pairs.

The general form of the equations describing the manipulator's dynamics is

$$\sum_{l=1}^{N} a_{kl} \ddot{q}_l + \sum_{l=1}^{N} \sum_{j=1}^{N} \Gamma_{k,lj} \dot{q}_l \dot{q}_j = Q_k \; ; \quad k = 1, 2, \ldots, N \tag{1}$$

where N is the number of degrees-of-freedom, and

$$\Gamma_{k,lj} = \Gamma_{k,jl} = \frac{1}{2}\left(\frac{a_{jk}}{q_l} + \frac{a_{lk}}{q_j} + \frac{a_{lj}}{q_k}\right)$$

System of equations (1) has been obtained by transforming Lagrange's equations

$$\frac{d}{dt}\frac{\partial T}{\partial \dot{q}_k} - \frac{\partial T}{\partial q_k} = Q_k \; , \quad k = 1, 2, \ldots, N$$

The kinetic energy of the system of rigid bodies is

$$T = \frac{1}{2}\sum_{i=1}^{N}\left(m_i \underline{v}_i^T \underline{v}_i + \underline{\omega}_i^T I_i \underline{\omega}_i\right) \tag{2}$$

where:
m_i — mass of the i-th element
I_i — matrix of intertia of the i-th element relative to frame π_i fixed to the centre of mass

$\underline{v}_i$ - velocity of the centre of mass of the i-th element relative to π_o
$\underline{\omega}_i$ - angular velocity of the i-th element relative to its centre of mass

This energy should be represented in quadratic form of generalized velocities

$$T = \frac{1}{2}\sum_{k=1}^{N}\sum_{l=1}^{N} a_{kl}\dot{q}_k\dot{q}_l , \qquad (3)$$

where

$$a_{kl} = a_{lk} = a_{kl}(\underline{q}, \underline{\dot{q}})$$

Computing a_{kl} and $\Gamma_{k,lj}$ using Denavit-Hartenberg's formalism enables us to substitute the differentiation operations of the type

$$\frac{\partial}{\partial q_k}/A_{ij} = /A_{k-1,j}\frac{dA_{k,k-1}}{dq_k}/A_{i,k}$$

where $/A_{ij}$ - 4x4 matrix transforming coordinates in π_i into π_j, by multiplication by special matrix $\mathbb{D}_k$, such that

$$\frac{dA_{k,k-1}}{dq_k} = \mathbb{D}_k /A_{k,k-1}$$

Matrix $\mathbb{D}_k$ has two forms:

$$\mathbb{D}_k = \begin{bmatrix} 0 & -1 & 0 & 0 \\ 1 & 0 & 0 & 0 \\ 0 & 0 & 0 & 0 \\ 0 & 0 & 0 & 0 \end{bmatrix} \quad \text{- for a rotary pair,}$$

$$\mathbb{D}_k = \begin{bmatrix} 0 & 0 & 0 & 0 \\ 0 & 0 & 0 & 0 \\ 0 & 0 & 0 & 1 \\ 0 & 0 & 0 & 0 \end{bmatrix} \quad \text{- for a prismatic pair.}$$

This method facilitates transformation of symbolic equations as well as numerical computations.

3. EQUATIONS OF STATE OF THE HYDRAULIC AND PNEUMATIC DRIVES

The equations describing pressures in the chambers of hydraulic and pneumatic actuators may be generated in symbolic form by a computer program, if the following criteria are fulfilled:

1. The installation consists of chambers (the volume of which has uniform distribution of parameters), in which isothermal (in hydraulic or pneumatic chambers) or adiabatic transformations (in pneumatic chambers) take place, with tubing supplying the working medium.
2. The pipes connecting the chambers are short enough, so that the influence of

the inertia of the working medium inside them on the dynamics of the system can be neglected.
3. The non-return valves operate without time-lag.
4. The rate of flow of mass through the tubing and local resistances is described as below:

4a. For pneumatic elements

$G = C\, p_{we}\, \phi(p_{wy}/p_{we})$

where C is the coefficient, which value does not depend on p_{wy}, p_{we}. p_{we}, p_{wy} are input and output pressures of the analised element respectively; ϕ is a function of the quotient of p_{wy}/p_{we}.

4b. For hydraulic elements:

$G = \eta\, \text{sign}\, (\Delta p) \sqrt{|\Delta p|}$

where η is the coefficient, which is not a function of p_{we} and p_{wy}, and

$p = p_{we} - p_{wy}$.

The programs considered may be used even if system does not meet constraints 1 to 4, but then the input data must be modified or otherwise the equations must be corrected. For instance, to analyse the flow in a long tube, the pipe must first be subdivided into discrete elements.

The characteristic $G_p = G_p(p_{we}, p_{wy})$ for several pneumatic elements can be obtained only by numerical computation. For hydraulic installations the only problem needing a numerical solution are tube branchings. The characteristic $G_h = G_h(p_{we}, p_{wy})$ for any combination of parallel and serial connections of elements can be obtained in an analytical form.

The most significant steps of the algorithm, generating "the pressure equations" for the pneumatic (P) and hydraulic (H) subsystems, are listed below.

Algorithm P
1. After a preliminary check of the entered data the program numbers the nodes (the connections of neighbouring elements). This allows a unique assignment of equations expressing the law of conservation of the stream's continuity during the flow through a node.
2. Continuity equations complemented by k relationships of the type

$\delta_i \delta_{i+1} \cdots \delta_{j-1} \delta_j = p_{we}^{ij}/p_{wy}^{ij}$,

where $\delta_i = p_{we}^i/p_{wy}^i$, $p_{we}^i = p_{wy}^{i-1}$, $p_{wy}^i = p_{we}^{i+1}$,

   k - the number of branches in the analysed installation,

   p_{we}^{ij}, p_{wy}^{ij} - input and output pressures of the branch i-j,

permit the determination of pressure quotient at the input and output of the elements of the pneumatic subsystem.
3. The program prints expressions describing the change of pressure against time in

the chambers of pneumatic actuators.

Algorithm H

1. From the data describing the hydraulic installation, a sequence of expressions defining the equivalent coefficients η_z are generated until they have been fully reduced. For serial connections:

$$\eta_z = \left[\sum_{i=1}^{n} \frac{1}{\eta_i^2} \right]^{-\frac{1}{2}}$$

For parallel connections:

$$\eta_z = \sum_{i=1}^{n} \eta_i$$

2. If branches are encountered, adequate continuity equations are developed for these nodes. Pressures existing in these branches can be deduced from the above mentioned equations.

3. Results obtained in step 1 and 2 are sufficient to obtain the mass expenditure in all the branches of the hydraulic installation. This in turn allows the determination of differentials of pressures in the chambers of hydraulic actuators. Equations containing differentials of pressure are integrated together with equations (1) during simulation. Pressure in the chambers of the actuators is the basis for determination of forces exerted by the drives.

4. DETERMINATION OF GENERALIZED FORCES

Two groups of active forces acting on a manipulator can be distinguished. The gravitational forces applied to the centre of mass of the elements form the first group. The directions of these forces relative to a global reference frame π_o are known. The second group consists of forces applied at different points of the mechanism (e.g. forces exerted by the actuators). The directions of these forces relative to a global reference frame are unknown. To determine these directions, knowledge of two points expressed in any coordinate frame, and colinear with the direction of the force vector, is sufficient. In the case of forces exerted by pneumatic or hydraulic drives the points where the actuators are fixed can be taken.

The division of forces into two groups, the expression describing the virtual work of active forces takes the form

$$\delta L = \sum_{i=1}^{N} \underline{F}_i \cdot \delta \underline{R}_i + \sum_{i=N+1}^{n_F} \underline{F}_i \cdot \delta \underline{R}_i + \sum_{i=1}^{n_M} \underline{M}_i \cdot \delta \underline{\phi}_i \qquad (7)$$

where: n_F - total number of active forces (together with gravitational forces),

n_M — the number of active torques,
$\underline{R}_i$ — the position of the point on which the force $\underline{F}_i$ acts expressed in the global reference frame π_o,
ϕ_i — angular displacement in the pairs on which torques M_i act.

The relationship (7) can be transformed into

$$\delta L = \sum_{k=1}^{N} Q_k \cdot \delta q_k \qquad (8)$$

where

$$Q_k = \sum_{i=1}^{N} \underline{F}_i \cdot \frac{\partial \underline{R}_i}{\partial q_k} + \sum_{i=N+1}^{n_F} \underline{F}_i \cdot \frac{\partial \underline{R}_i}{\partial q_k} + \sum_{i=1}^{n_M} M_i \frac{\partial \phi_i}{\partial q_k} \qquad (9)$$

is generalized force.

Because for the first term of the expression (9) — $\underline{F}_i = m_i \underline{g}$, the direction of $\underline{g}$ relative to π_o is known, only the term $\partial \underline{R}_i / \partial q_k$ must be determined. In this case

$$\begin{bmatrix} \underline{R}_i \\ 1 \end{bmatrix} = A_{i,0} \begin{bmatrix} \underline{R}_i^* \\ 1 \end{bmatrix}$$

where $\underline{R}_i^*$ is the position of mass centre of the i-th element relative to π_o.
After simple transformations we obtain

$$\begin{bmatrix} \frac{\partial}{\partial q_k} \underline{R}_i \\ 0 \end{bmatrix} = \begin{cases} A_{k-1,0} D_k A_{i,k-1} \begin{bmatrix} \underline{R}_i^* \\ 1 \end{bmatrix}, & \text{for } k \leq i \\ \underline{0} & \text{for } k > i \end{cases} \qquad (10)$$

Usually in the case of the second term $\underline{R}_i$ does not coincide with mass centre of any element. The way of determination of $\partial \underline{R}_i / \partial q_k$ is similar to one used for (10), but different meaning must be assigned to $\underline{R}_i$. The projections of $\underline{F}_i$ on the axis of the coordinate frame π_o ($\underline{F}_i = \underline{\varepsilon}_i F_i$, where $\underline{\varepsilon}_i$ is the vector of directional cosines) must also be determined.

The algorithm for the determination of the vector of directional cosines, if the direction of force $\underline{F}_i$ is given by specifying the two points P_{i1}, P_{i2}, is as follows:
— first the transformation of coordinates of the points $\underline{P}_{i1}^*$, $\underline{P}_{i2}^*$, expressed in local frames, into coordinates of the global coordinate system π_o, must take place

$$\begin{bmatrix} \underline{P}_{i1} \\ 1 \end{bmatrix} = A_{j1,0} \begin{bmatrix} \underline{P}_{i1}^* \\ 1 \end{bmatrix}, \quad \begin{bmatrix} \underline{P}_{i2} \\ 1 \end{bmatrix} = A_{j2,0} \begin{bmatrix} \underline{P}_{i2}^* \\ 1 \end{bmatrix} \qquad (11)$$

— secondly, their distance must be found

$$s_i = \left[\left(\underline{P}_{i2} - \underline{P}_{i1} \right) \cdot \left(\underline{P}_{i2} - \underline{P}_{i1} \right) \right]^{\frac{1}{2}} \qquad (12)$$

— and in the last step the directional cosines are determined

$$\varepsilon_i = \left(\underline{P}_{i2} - \underline{P}_{i1} \right) / s_i \tag{13}$$

From $\underline{P}_{i1}$ to $\underline{P}_{i2}$ is the positive sense of this vector. If the force $\underline{F}_i$ is exerted by a pneumatic or hydraulic actuator, or if it is necessary because of any other considerations, the following derivative must be determined:

$$\frac{ds_i}{dt} = \ldots = \varepsilon_i \left(\underline{\dot{P}}_{i2} - \underline{\dot{P}}_{i1} \right) \tag{14}$$

where $\underline{\dot{P}}_{i1}$ (similarly $\underline{\dot{P}}_{i2}$) is described by

$$\begin{bmatrix} \underline{\dot{P}}_{i1} \\ 1 \end{bmatrix} = \left[\sum_{k=1}^{j-1} \left(A_{k-1,0} \cdot D_k \cdot A_{j_1,k-1} \right) \cdot q_k \right] \begin{bmatrix} \underline{P}_{i1}^{K} \\ 1 \end{bmatrix} \tag{15}$$

The modulus of the force $\underline{F}_i$ is exerted by a pneumatic or hydraulic drive, as well as the modules of the gravitational force, is determined automatically by the program. The modulus of other forces, as well as of the active torques must be supplied by the user. In the case of actuators with reciprocating movement of the piston the active moments are mainly the effect of friction in the revolute joints.

5. COMPUTER REALIZATION

The algorithm described in the previous sections was transformed into a computer program. It was written in FORTRAN, as it is the most commonly used high level language. No general procedures carrying out symbolic operations were used. This considerably reduced the demand for computer memory and access to external files. The process of creating the equations was sub-divided into two steps. The body of the procedure (continuing the right sides of equations), which is created in the first step can be utilized in the simulation program. The second step reduces the number of instructions used in this procedure.

The package consisted of (in its first version) six programs of the type MAIN and several dozen procedures implemented on SM-4 minicomputer (PDP-11 compatible). Next, the programs were transferred to an IBM PC microcomputer. The existing software solves simple dynamics problems. It can be adapted to solve the inverse problem. The utilization of the constructed simulation models permits an exact determination of controls (e.g. the current in the MOOG type electro-hydraulic transducer). This was not possible when catalogue characteristics or simple linearized models were used.

6. FINAL REMARKS

Because of the limited space an example of application of this package was not given. Such an example will be presented during the symposium.

A Cartesian Model of Manipulator Kinematics

M. Galicki

Department of Applied Mathematics, Higher College of Engineering, Zielona Góra, Poland

SUMMARY

In this paper, a class of multi-unit manipulators, described by using the concept of a "pencil" of vectors with rectangular coordinates, is considered. The internal coordinate representation of the manipulator and the "pencil"-vector description are shown to be synonymous. The concept of a permissible configuration of the manipulator is introduced. Changes in the manipulator's configuration are made using a Cartesian model of the manipulator's kinematics. Finally, an example of planning the manipulator's movement trajectory on the basis of its Cartesian kinematic model is presented.

INTRODUCTION

The typical tasks which robots perform include the shifting of the manipulator's links from a specified initial point to an assigned final point (usually given in the basic system) in the work space. The determination of permissible positions of the manipulator (i.e. those in which the robot's internal coordinates do not exceed the boundary values and the manipulator's links do not intersect the forbidden regions (obstacles) which may crop up on the trajectory planned by the robot) presented in (1), (2), (3), (6), (8) involved kinematic equations and thus the recalculation of internal coordinates into rectangular ones. In paper (4), the above problem was considered using a complicated, so far as calculations were concerned, method of configurational space in which rectangular coordinates were recalculated into the robot's internal coordinates. The recalculation of coordinates was dictated by the fact that the description of the position of the manipulator and of the final point were in different coordinate systems. In this paper, in order to eliminate the time-consuming recalculation of one set of coordinates into another and to make the description uniform, a description of the commonly used class of planar manipulators by means of a "pencil" of vectors determined explicitly by the positions of the centres of two successive kinematic pairs of the manipulator is introduced. The determination of the position of the manipulator's "pencil"-vector representation in the three-dimensional work space is reduced (by resolving this space into two-dimensional subspace) to the determination of rectangular coordinates of some two-dimensional vectors from the assigned table of unit vectors. The change of the manipulator's position (which is responsible for its movement to an assigned final point) is described using a Cartesian model of the manipulator's kinematics which is introduced in the paper. On the basis of the above Cartesian model of the manipulator's kinematics, a computer program has been written in PASCAL for an exemplary task of planning the trajectory if there is a forbidden region.

1. A PENCIL-VECTOR DESCRIPTION OF THE MANIPULATOR

In the paper, a planar polyadic manipulator with $n=0,1,2,..$ Vth class kinematic pairs presented in the figure 1.1 is considered, where:

$0X^1X^2X^3$ is the robot's basic coordinates system

$q=(q^0,..,q^n)$ is the vector of the robot's internal coordinates

$$q_d^i \leqslant q^i \leqslant q_g^i$$

q_d^i, q_g^i is the lower and upper bounds for the i-th coordinate

$1_n=(1_n^1,1_n^2,1_n^3)$ is the vector in basic coordinates connecting the centres of the n-1st kinematic pair with n-th kinematic pair

p^{q^0} is a straight line traced by rotating axis OX^1 by angle q^0

$(Op^{q^0}X^3)$ is the manipulator's working plane at a fixed position of coordinate q^0

$n=1,2,..$

The position of the above manipulator in the basic system can be represented explicitly by successive vectors connecting the centres of the kinematic pairs. Such a representation for the class of manipulators in figure 1.1 is made possible by the following theorem.

Theorem 1. Transformation:

$$T : (q^0,..,q^n) \to (q^0,1_1,..,1_{n+1})$$

where:

$$T(q)=(q^0,P^{(1,2,3)}\{T_0^b(0,0,\|1_1\|,1)^T - T_{-1}^b(0,0,0,1)^T\},..,$$
$$P^{(1,2,3)}\{T_n^b(\|1_{n+1}\|,0,0,1)^T - T_{n-1}^b(\|1_n\|,0,0,1)^T\}) \qquad (1.1)$$

$T_{-1}^b = 1$, 1 - unit four-dimensional matrix,

$T_i^b=T_{-1,0}(q^0)T_{0,1}(q^1)..T_{i-1,i}(q^i)$,

A Cartesian Model of Manipulator Kinematics   157

$$T_{-1,0}(q^0) = \begin{bmatrix} \cos(q^0) & -\sin(q^0) & 0 & 0 \\ \sin(q^0) & \cos(q^0) & 0 & 0 \\ 0 & 0 & 1 & 0 \\ 0 & 0 & 0 & 1 \end{bmatrix}, T_{0,1}(q^1) = \begin{bmatrix} \sin(q^1) & \cos(q^1) & 0 & 0 \\ 0 & 0 & -1 & 0 \\ -\cos(q^1) & \sin(q^1) & 0 & \|l_1\| \\ 0 & 0 & 0 & 1 \end{bmatrix},$$

$$T_{i-1,i}(q^i) = \begin{bmatrix} -\cos(q^i) & \sin(q^i) & 0 & \|l_i\| \\ -\sin(q^i) & -\cos(q^i) & 0 & 0 \\ 0 & 0 & 1 & 0 \\ 0 & 0 & 0 & 1 \end{bmatrix}, P^{(1,2,3)}\{(x,y,z,1)^T\} = (x,y,z),$$
$(x,y,z,1) \in R^4,$

$(0,\ldots 0) \leqslant q < (2\pi,\ldots,2\pi), i=2,3,\ldots,n$, $\|\cdot\|$ - norm of vector,

$P^{(1,2,3)}\{T_{i-1}^b(\|l_i\|,0,0,1)^T\}$ - coordinates of the centre of the i-th kinematic pair in the basic system

is a hetero-valued transformation.

The working space of the manipulator in figure 1.1 can be devolved (taking into account the manipulator's planarity) into a set of the following subspaces (two-dimensional):

$$E^M = \bigcup_{q_d^0 \leqslant q^0 \leqslant q_g^0} /Op^{q^0} x^3/ \qquad (1.2)$$

where

E^M - the manipulator's work space

The devolution of the work space of the manipulator in figure 1.1 implies that its performance can be considered in a two-dimensional work space. In order to introduce a coordinate system in work plane $/Op^{q^0} x^3/$, straight line p^{q^0} should be oriented in such a way that by rotating axis OX^1 by limiting angle q_d^0, it would coincide with straight line $p^{q_d^0}$ (figure 1.1). By giving the direction to straight line p^{q^0}, coordinate system $Op^{q^0} x^3$ is established in work subspace $/Op^{q^0} x^3/$. The translation of vectors l_n to the origin of the above movable coordinate system yields an explicit "pencil"-vector representation of the manipulator in this system. The relationship between coordinates of vectors l_n in basic coordinate system and in movable coordinate system is:

$$(q^0,\ldots,L_n) = (q^0,\ldots,(\text{sgn}(P^{(1,2)}\{1_n\} \circ e_q 0) \|P^{(1,2)}\{1_n\}\|, 1_n^3)) \qquad (1.3)$$

where:

L_n - vector 1_n with coordinates in coordinate system $Op^{q^0} x^3$,

$e_q^0 = (\cos(q^0), \sin(q^0)), P^{(1,2)}\{(x,y,z)\} = (x,y), (x,y,z) \in R^3$

$\odot$ – scalar product of vectors

$\text{sgn}(x) = \begin{cases} 1 & \text{for } x > 0 \\ 0 & \text{for } x = 0 \\ -1 & \text{for } x < 0 \end{cases}$

For further analysis of the manipulator's "pencil"-vector representation a model of digitized work space having the same digitization interval for all the internal coordinates is assumed:

$$E^D = \bigcup_{q_d^0 \leqslant Q^0 \leqslant q_g^0} /Op^{Q^0}x^3/ \qquad (1.4)$$

where

E^D – the digitized work space

$Q^0 = \Delta q [q^0/\Delta q]$, $Q = [Q^0, \ldots, Q^n] = \Delta q [(1/\Delta q)q]$, $\Delta q = 2\pi/M$,

(x) – integer of x, (w) – integer of each component of vector w,

M – an assigned number determining the accuracy of digitization

Without making the considerations less general, it was assumed that the position of the manipulator in space E^M coincided with its position determined in the digitized work space:

$$Q = q$$

In order to speed up the determination of the manipulator's positions in space when changing the values of its internal coordinates, the so-called basic table describing M discrete positions of the unit vector in system $Op^q x^3$ was introduced. The positions are consecutive rotations of the unit vector by angle $2\pi/M$. The basic table is a set of rectangular coordinates corresponding to the positions of the above vector:

Position	0	1	..n..	M-1
$\overline{I}$	$\overline{I}_0$	$\overline{I}_1$	$\overline{I}_n$	$\overline{I}_{M-1}$

(1.5)

where:

$\overline{I}_n$ – a unit vector with rectangular coordinates determined for angle $n(2\pi/M)$, $n = 0, 1, \ldots, M-1$

Taking into account the relations (1.3) and 1.5), the relationships between a "pencil" of real manipulator vectors and the corresponding "pencil" of basic table vectors are as follows:

$$(q^0, \ldots, L_n) = (\Delta q, \ldots, \|L_n\|) \circ (p_0, \overline{I}_{p_1}, \ldots, \overline{I}_{p_n}) \qquad (1.6)$$

A Cartesian Model of Manipulator Kinematics   159

where:
$$\min_{p_n} \|(p_0 \triangle q,..,\tau_{p_n}) - (1,..,1/\|L_n\|) \circ (q^0,..,L_n)\| = 0$$
$$(M_d^0, 0,..,0) \leqslant (p_0,..,p_n) \leqslant (M_g^0, M-1,..,M-1)$$

$(p_0,..,p_n)$ — a vector of indexes (referred to the basic table indexes) corresponding to the positions of the normalized "pencil" of real manipulator vectors in coordinate system $Op^{q^0} x^3$,

$$M_d^0 = \left[q_d^0/\triangle q\right] , \quad M_g^0 = \left[q_g^0/\triangle q\right] ,$$

$$(a_1,..,a_n) \circ (b_1,..,b_n) = (a_1 b_1,..,a_n b_n), (a_1,..,a_n), (b_1,..,b_n) \in R^n$$

The condition which says that the internal coordinates of the real manipulator must not exceed their boundary values can be reduced to an equivalent condition which makes use of the relationships occurring between the indexes of the corresponding basic table vectors:

$$(p_i \oplus M/2) \oplus (M - p_{i+1}) \geqslant p_i^- \qquad p_{i+1} \oplus (M - (p_i \oplus M/2)) \geqslant p_i^+ \qquad (1.7)$$

where:

$$p_i^+ = \left[q_d^i/\triangle q\right] , \; p_i^- = \left[(2\pi - q_g^i)/\triangle q\right], \oplus \text{ —modulo M sum, } i = 1,..,n$$

The condition stating that the real manipulator's links must not intersect was reduced to analytical relationships for the non-intersection of two sections (parameterized by parameters assuming the values in the interval from 0 to 1) which model the above links (that such modelling is possible has been proved in (5)). If:

$$\det(\|L_j\|\tau_{p_j}, -\|L_k\|\tau_{p_k}) = 0 \qquad (1.8)$$

where:

$\det(\cdot)$ — determinant of matrix,
$|j-k| \triangleright 1, \; j,k = 2,3,..,n$

it is assumed that the j-th and k-th links do not intersect. If this is not the case, the following relationships must hold in order that the j-th link did not intersect the k-th link:

$$\frac{\det(\sum_{i=1}^{k-1} \|L_i\|\tau_{p_i} - \sum_{i=1}^{j-1} \|L_i\|\tau_{p_i}, -\|L_k\|\tau_{p_k})}{\det(\|L_j\|\tau_{p_j}, -\|L_k\|\tau_{p_k})} < 0$$

or

$$\frac{\det(\sum_{i=1}^{k-1}\|L_i\|\bar{\tau}_{p_i} - \sum_{i=1}^{j-1}\|L_i\|\bar{\tau}_{p_i}, -\|L_k\|\bar{\tau}_{p_k})}{\det(\|L_j\|\bar{\tau}_{p_j}, -\|L_k\|\bar{\tau}_{p_k})} > 1$$

and (1.9)

$$\frac{\det(\|L_j\|\bar{\tau}_{p_j}, \sum_{i=1}^{k-1}\|L_i\|\bar{\tau}_{p_i} - \sum_{i=1}^{j-1}\|L_i\|\bar{\tau}_{p_i})}{\det(\|L_j\|\bar{\tau}_{p_j}, -\|L_k\|\bar{\tau}_{p_k})} < 0$$

or

$$\frac{\det(\|L_j\|\bar{\tau}_{p_j}, \sum_{i=1}^{k-1}\|L_i\|\bar{\tau}_{p_i} - \sum_{i=1}^{j-1}\|L_i\|\bar{\tau}_{p_i})}{\det(\|L_j\|\bar{\tau}_{p_j}, -\|L_k\|\bar{\tau}_{p_k})} > 1$$

Equations (1.6-1.9) impose bounds on the permissible positions of the manipulator in the work space depending on the values of the indexes corresponding to the positions of the manipulator's links.

2. A CARTESIAN MODEL OF THE REAL MANIPULATOR KINEMATICS

Using the given, in the previous section, relationship (1.6), a definition of the permissible configuration of the real manipulator is introduced.

Definition 1. Permissible configuration K of the real manipulator is a vector described by the relationship:

$$K = (\triangle q, \ldots, \|L_n\|) \circ (p_0, \ldots, \bar{\tau}_{p_n}) \qquad (2.1)$$

the components of which fulfill relationships (1.6 - 1.9)

The concept of the permissible configuration was used to determine the positions of the manipulator in digitized work space E^D. It is assumed that at the initial instant, the vector of the robot's internal coordinates assumes the following values:

$$q^t|_{t=0} = (q_t^0, \ldots, q_t^n)|_{t=0} \qquad (2.2)$$

where:

$q^t|_{t=0}$ — a vector of internal coordinate values at initial instant $t=0$

Taking into account relationships (1.1),(1.3) and (1.6) (assuming that the initial position of the manipulator in work space E^M coincides with its position in digitized space E^D), the permissible configuration of the manipulator at the initial instant is determined:

$$K(t)|_{t=0} = (\Delta q, \ldots, \|L_n\|) \circ (p_0^t, \ldots, T_{p_n}t)|_{t=0} \qquad (2.3)$$

A change in the manipulator's position was obtained through the variation of the indexes' vector coordinates. Taking into account relationship (2.3) and the above, the following definition was introduced:

Definition 2. The Cartesian model of the manipulator's kinematics has the form of the transformation:

$$K(t') = F(K(t), (p_0^t, \ldots, T_{p_n}t)) \qquad (2.4)$$

where:

$$F(K(t), (p_0^t, \ldots, T_{p_n}t)) = (\Delta q, \ldots, \|L_n\|) \circ (p_0^t + k_0, \ldots, T_{p_n}t \oplus \sum_{i=1}^{n} \oplus k_i),$$

$$k_i \in \{-1, 0, 1\}, \quad K(t') = (\Delta q, \ldots, \|L_n\|) \circ (p_0^{t'}, \ldots, T_{p_n}t'),$$

$$p_i \oplus k_1 = \begin{cases} p_i \oplus k_1 & \text{for } k_1 \geq 0 \\ p_i \oplus (M - k_1) & \text{for } k_1 < 0 \end{cases}, \quad i, l = 1, 2, \ldots, n,$$

t' - an instant corresponding to the new position of the manipulator

3. A COMPUTER EXAMPLE

The Cartesian model of the manipulator's kinematics (equation (2.4)) presented in the previous section was applied to the determination of the manipulator's displacements from the initial position to the final position assigned for the situation presented in figure 3.2, where:

S - the assigned position of the manipulator's end at the initial instant

F - the assigned final point

Z - the forbidden region (an exemplary machine tool)

The movement determined by consecutive permissible configurations should proceed in such a way that the manipulator does not enter the forbidden region. In order to by-pass the above region, the method of possible directions presented in (6) was used. To calculate the permissible configurations of the manipulator, the method was modified by means of equation (2.4). Using this method, a computer program was written in PASCAL for the situation presented in figure 3.2 and the following numerical data:

$$M = 126, K(0) = (1, 1, 1, 1) \circ (0, T_{28}, T_0, T_0), p_0^- = p_0^+ = 0, p_i^- = p_i^+ = 10, i = 1, 2, 3$$

The figures 3.2 and 3.3 present several selected permissible configurations of the manipulator determined during the operation of the program.

FINAL REMARKS

The accuracy of the above method depends on the number of basic table vectors. The presented method (of determining permissible configurations) can be generalized to cover the class of manipulators with additional kinematic translation pairs which join the successive rotational pairs as well as the class of manipulators with kinematic pairs of the order lower than V. The Cartesian model of kinematics presented here can also be applied to the determination of the permissible configurations of flexible manipulators of the elephant's trunk type, the model of which is presented in (7).

REFERENCES

[1] - Paul R.P. Robot manipulators:Mathematics,programming and control, The MIT Press,1982
[2] - Brady M. Robot motion:Planning and control,MIT Press,1982
[3] - Brockett R.W. Robotic manipulators and the product of exponentials formula,Harvard University,Cambridge,1983
[4] - Gerke W.Freikollision Bewegungsplannung,Robotersysteme nr1/85
[5] - Udupa S. Collision detection and avoidance in computer controlled manipulators,PhD,California Institute of Technology Pasadena,1977
[6] - Юревич Е.И. Динамика управления роботами,Москва,Наука,1984
[7] - Malczyk G.,Morecki A.Mathematical model of a flexible manipulator of the elephant's trunk type,PREPRINTS ROMANSY,1986
[8] - Кобринский А.А. Манипуляционные системы роботов,Москва,Наука,1985

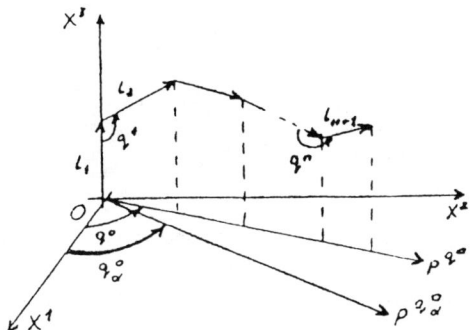

Fig.1.1 Kinematic scheme of manipulator

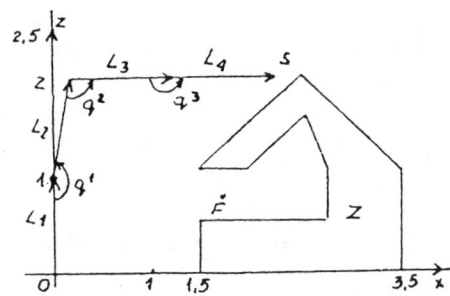

Fig.3.1 An example of the planning of a movement trajectory

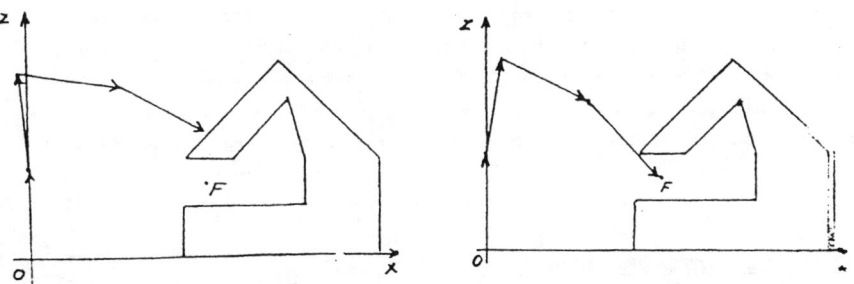

Fig.3.2 The intermediate configuration of the manipulator

Fig.3.3 The final configuration of the manipulator

A Method for Solving the Inverse Problem of Kinematics of Anthropomorphic Manipulators with Spherical Wrist

Peng Shang-xian and Wang Gang

Tianjin University, China

ABSTRACT

The inverse problem of anthropomorphic manipulator kinematics in one that presents difficulties. In this paper is presented the method for solving the inverse problem of kinematics by treating the manipulator as two subsystems - the minimal configuration and the wrist.

In this way, the inverse problem of manipulator kinematics is simplified.

INTRODUCTION

The inverse problem of position is often concerned with the solution of algebraic equations. If we solve these equations using numerical methods, it will occupy a lot of computer time. Pieper (1968) has shown that the inverse problem of position would be very simple if the manipulator had a spherical wrist structure. In this case, the system with six degrees of freedom can be separated into two subsystems with three degrees of freedom.

Generally, the solution of the inverse problem for velocity and acceleration can be written:

$$\dot{\theta} = J^{-1}\dot{x}; \qquad \ddot{\theta} = J^{-1}(\ddot{x} - \dot{J}\dot{\theta})$$

where J is Jacobi's matrix. However, constructing and inversing Jacobi's matrix is very tedious. Featherstone (1983) presented the procedures for solving the inverse problem of velocity for a manipulator with a spherical wrist structure without Jacobi's matrix. Then, Hollerbach (1983) presented procedures for solving the inverse problem of acceleration, based on Pieper's and Featherstone's procedures.

In this paper, we use this partition method to solve the inverse problem of anthropomorphic manipulator kinematics. Besides, a method using rotation matrices and their time differentials is presented for kinematic differential analysis. In this way, they make the procedures even more simple.

2. SOLVING THE INVERSE PROBLEM OF POSITON

The anthropomorphic manipulator with six degrees of freedom is shown in figure 1.

Kinematics of Anthropomorphic Manipulators with Spherical Wrist 165

The axes of the last three joints intersect at a point, so that this manipulator is of a spherical wrist structure.

This system can be divided into two subsystems: one is wrist (figure 1,a); the other is minimal configuration (figure 1,b). Then, figure 1 can be simplified in figure 2.

The procedures for solving the inverse problem of position can be divided into three steps.

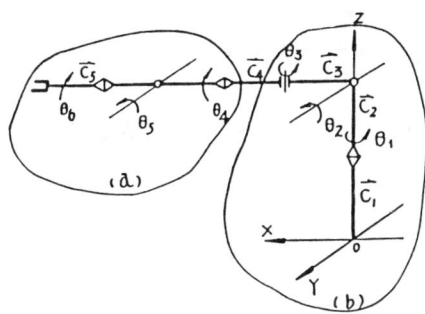 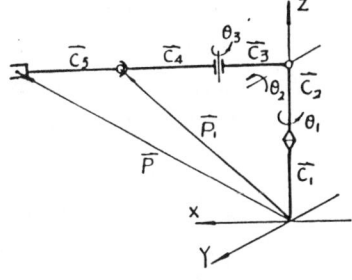

figure 1 figure 2

(1) Finding the vector $\bar{P}_1$

$$[P_1] = [P] - [U][C_5] \qquad (1)$$

where [U] is the orientation matrix of end effector with respect to base frame.

(2) Finding the front three joint angles

From Fig. 2. the position vector P_1 can be expressed as

$$[P_1]=[C_1]+Rot(z^1,\theta_1)[C_2]+Rot(z^1,\theta_1)Rot(y^2,\theta_2)[C_3]+Rot(z^1,\theta_1)Rot(y^2,\theta_2)Rot(z^3,\theta_3)[C_4] \qquad (2)$$

Solving above equations, we can obtain

$$Cs\theta_3 = [(Z_1-C_1-C_2)^2+X_1^2+Y_1^2-C_3^2-C_4^2]/2C_3C_4$$

$$Sn\theta_3 = \pm\sqrt{1 - Cs^2\theta_3}$$

The plus sign in this expression indicates the right hand configuration. The negative gives a left hand configuration

$$\theta_3 = Tan^{-1}(Sn\theta_3/Cs\theta_3) \qquad (3)$$

Then, θ_1 and θ_2 can be formulated easily

$$Sn\theta_2 = -(Z_1-C_1-C_2)/(C_3+C_4Cs\theta_3)$$

$$Cs\theta_2 = \sqrt{1 - Sn^2\theta_2}$$

There is no "-" sign in front of last expression because θ_2 varies from $-\pi/2$ to $\pi/2$.
Hence
$$\theta_2 = \text{Tan}^{-1}(\text{Sn}\theta_2/\text{CS}\theta_2) \tag{4}$$

$$\theta_1 = \text{Tan}^{-1}((Y_1(C_3+C_4CS\theta_4)CS\theta_2 - C_4X_1Sn\theta_3)/(X_1(C_3+C_4CS\theta_3)CS\theta_2 + C_4Y_1Sn\theta_3)) \tag{5}$$

(3) Finding the last three joint angles

The orientation of end effector with respect to base frame can be written

$$[U] = \text{Rot}(Z^1,\theta_1)\text{Rot}(Y^2,\theta_2)\text{Rot}(Z^3,\theta_3)\text{Rot}(X^4,\theta_4)\text{Rot}(Y^5,\theta_5)\text{Rot}(X^6,\theta_6)[I] \tag{6}$$

Then, the orientation of the end effector with respect to forearm is

$$[W] = \text{Rot}^{-1}(Z^3,\theta_3)\text{Rot}^{-1}(Y^2,\theta_2)\text{Rot}^{-1}(Z^1,\theta_1)[U] = \text{Rot}(X^4,\theta_4)\text{Rot}(Y^5,\theta_5)\text{Rot}(X^6,\theta_6) \tag{7}$$

That is

$$\begin{bmatrix} W_{11} & W_{12} & W_{13} \\ W_{21} & W_{22} & W_{23} \\ W_{31} & W_{32} & W_{33} \end{bmatrix} = \begin{bmatrix} CS\theta_5 & Sn\theta_5 Sn\theta_6 & Sn\theta_5 CS\theta_6 \\ Sn\theta_4 Sn\theta_5 & CS\theta_4 CS\theta_6 - Sn\theta_4 CS\theta_5 Sn\theta_6 & -CS\theta_4 Sn\theta_6 - Sn\theta_4 CS\theta_5 \\ -CS\theta_4 Sn\theta_5 & Sn\theta_4 CS\theta_6 + CS\theta_4 CS\theta_5 Sn\theta_6 & -Sn\theta_4 Sn\theta_6 + CS\theta_4 CS\theta_5 \end{bmatrix}$$

From this matrix, the last three joint angles can be formulated

$$\theta_4 = \text{Tan}^{-1}(W_{21}/-W_{31}) \tag{8}$$

$$\theta_5 = \text{Tan}^{-1}((W_{21}Sn\theta_4 - W_{31}CS\theta_4)/W_{11}) \tag{9}$$

$$\theta_6 = \text{Tan}^{-1}(W_{12}/W_{13}) \tag{10}$$

3. Solving the inverse problem of velocity

Suppose the vector $\dot{\vec{P}}$ and $\vec{\omega}_6$ are known. Then the joint angular velocities $\dot\theta_1, \dot\theta_2, \ldots, \dot\theta_6$ can be solved using the following procedures

(1) Finding the linear velocity of wrist

$$\dot{\vec{P}}_1 = \dot{\vec{P}} - \vec{\omega}_6 \times \vec{C}_5' \tag{11}$$

where $[C_5'] = [U][C_5]$

(2) Finding the front three joint angular velocities

Differentiating the equation (2) with respect to time, we obtain

$$[\dot{P}_1] = [B_1]\text{Rot}(Y^2,\theta_2)[C_3]\dot\theta_1 + \text{Rot}(Z^1,\theta_1)[B_2][C_3]\dot\theta_2 + [B_1]\text{Rot}(Y^2,\theta_2)\text{Rot}(Z^3,\theta_3)\cdot [C_4]\dot\theta_1 + \text{Rot}(Z^1,\theta_1)[B_2]\text{Rot}(Z^3,\theta_3)[C_4]\dot\theta_2 + \text{Rot}(Z^1,\theta_1)\text{Rot}(Y^2,\theta_2)[B_2][C_4]\dot\theta_3 \tag{12}$$

where

$$[B_1] = \begin{bmatrix} -Sn\theta_1 & -CS\theta_1 & 0 \\ CS\theta_1 & -Sn\theta_1 & 0 \\ 0 & 0 & 0 \end{bmatrix}; \quad [B_2] = \begin{bmatrix} -Sn\theta_2 & 0 & CS\theta_2 \\ 0 & 0 & 0 \\ -CS\theta_2 & 0 & -Sn\theta_2 \end{bmatrix}; \quad [B_3] = \begin{bmatrix} -Sn\theta_3 & -CS\theta_3 & 0 \\ CS\theta_3 & -Sn\theta_3 & 0 \\ 0 & 0 & 0 \end{bmatrix} \tag{13}$$

Multiplying both sides of equation (12) by $(\text{Rot}(Z^1,\theta_1)\text{Rot}(Y^2,\theta_2))^{-1}$, we can formulate $\dot\theta_1$, $\dot\theta_2$ and $\dot\theta_3$

$$\dot{\theta}_1 = (\dot{X}_{1p}CS\theta_3 + \dot{Y}_{1p}Sn\theta_3)/(C_3CS\theta_2Sn\theta_3) \tag{14}$$

$$\dot{\theta}_2 = -(\dot{Z}_{1p} + C_4Sn\theta_2Sn\theta_3\dot{\theta}_1)/(C_3 + C_4CS\theta_3) \tag{15}$$

$$\dot{\theta}_3 = (\dot{X}_{1p} + C_4CS\theta_2Sn\theta_3\dot{\theta}_1)/C_4Sn\theta_3 \tag{16}$$

(3) Finding the last three joint angular velocities

The angular velocity of the end effector with respect to base frame can be expressed

$$[\omega_6] = \begin{bmatrix} 0 \\ 0 \\ 1 \end{bmatrix}\dot{\theta}_1 + \text{Rot}(Z^1,\theta_1)\begin{bmatrix} 0 \\ 1 \\ 0 \end{bmatrix}\dot{\theta}_2 + \text{Rot}(Z^1,\theta_1)\text{Rot}(Y^2,\theta_2)\begin{bmatrix} 0 \\ 0 \\ 0 \end{bmatrix}\dot{\theta}_3 + \text{Rot}(Z^1,\theta_1)$$

$$\text{Rot}(Y^2,\theta_2)\text{Rot}(Z^3,\theta_3)\begin{bmatrix} 1 \\ 0 \\ 0 \end{bmatrix}\dot{\theta}_4 + \text{Rot}(Z^1,\theta_1)(\text{Rot}(Y^2,\theta_2)\text{Rot}(Z^3,\theta_3)\text{Rot}(X^4,\theta_4))$$

$$\begin{bmatrix} 0 \\ 1 \\ 0 \end{bmatrix}\dot{\theta}_5 + \text{Rot}(Z^1,\theta_1)\text{Rot}(Y^2,\theta_2)\text{Rot}(Z^3,\theta_3)\text{Rot}(X^4,\theta_4)\text{Rot}(Y^5,\theta_5)\begin{bmatrix} 1 \\ 0 \\ 0 \end{bmatrix}\dot{\theta}_6 \tag{17}$$

Collecting the terms containing $\dot{\theta}_1$, $\dot{\theta}_2$, $\dot{\theta}_3$ and (ω_6) in one side of the expression and multiplying both sides by $(\text{Rot}(Z^1,\theta_1)\text{Rot}(Y^2,\theta_2)\text{Rot}(Z^3,\theta_3))^{-1}$.

$$\begin{bmatrix} \omega_{631p} \\ \omega_{632p} \\ \omega_{623p} \end{bmatrix} = \text{Rot}^{-1}(Z^3,\theta_3)\text{Rot}^{-1}(Y^2,\theta_2)\text{Rot}^{-1}(Z^1,\theta_1)$$

we can solve the equations for $\dot{\theta}_4, \dot{\theta}_5$ and $\dot{\theta}_6$

$$\dot{\theta}_4 = \omega_{631p} - \dot{\theta}_6 CS\theta_5 \tag{18}$$

$$\dot{\theta}_5 = \omega_{632p}CS\theta_4 + \omega_{633p}Sn\theta_4 \tag{19}$$

$$\dot{\theta}_6 = (\omega_{632p}Sn\theta_4 - \omega_{633p}CS\theta_4)/Sn\theta_5 \tag{20}$$

4. SOLVING THE INVERSE PROBLEM OF ACCELERATION

Suppose the vectors $\bar{P}$ and $\bar{\ell}_6$ are known. Then the joint angular accelerations $\ddot{\theta}_1, \ddot{\theta}_2, \ldots, \ddot{\theta}_6$ can be solved by using the following procedures:

(1) Finding the linear acceleration of the wrist

$$\ddot{\bar{P}}_1 = \ddot{\bar{P}} - \bar{\epsilon}_6 \times \bar{C}_5' - \bar{\omega}_6 \times (\bar{\omega}_6 \times \bar{C}_5') \tag{21}$$

(2) Finding the front three joint angular accelerations

Differentiating equation (12) and premultiplying both sides by $(\text{Rot}(Z^1,\theta_1)\text{Rot}(Y^2,\theta_2))^{-1}$, we can solve equations for $\ddot{\theta}_1, \ddot{\theta}_2$ and $\ddot{\theta}_3$:

$$\ddot{\theta}_1 = (\ddot{X}_{1pp}CS\theta_3 + \ddot{Y}_{1pp}Sn\theta_3)/C_3CS\theta_2Sn\theta_3 \tag{22}$$

$$\ddot{\theta}_2 = (\ddot{Z}_{1pp} + C_4Sn\theta_2Sn\theta_3\ddot{\theta}_1)/(C_3 + C_4CS\theta_3) \tag{23}$$

$$\ddot{\theta}_3 = -(\ddot{X}_{1pp} + C_4CS\theta_2Sn\theta_3\ddot{\theta}_1)/C_4Sn\theta_3 \tag{24}$$

where

$$\ddot{X}_{1pp} = F_1(\theta_1, \theta_2, \theta_3, \dot\theta_1, \dot\theta_2, \dot\theta_3, \ddot{X}_1, \ddot{Y}_1, \ddot{Z}_1)$$

$$\ddot{Y}_{1pp} = F_2(\theta_1, \theta_2, \theta_3, \dot\theta_1, \dot\theta_2, \dot\theta_3, \ddot{X}_1, \ddot{Y}_1, \ddot{Z}_1)$$

$$\ddot{Z}_{1pp} = F_3(\theta_1, \theta_2, \theta_3, \dot\theta_1, \dot\theta_2, \dot\theta_3, \ddot{X}_1, \ddot{Y}_1, \ddot{Z}_1)$$

(3) Finding the angular accelerations of the last three joints

Differentiating equation (17) and premultiplying both sides by $(\text{Rot}(Z^1, \theta_1)$ $\text{Rot}(Y^2, \theta_2) \text{Rot}(Z^3, \theta_3))^{-1}$, we can solve the equations for $\ddot\theta_4, \ddot\theta_5$ and $\ddot\theta_6$.

$$\ddot\theta_4 = \varepsilon_{631pp} - \dot\theta_6 C S \theta_5 \qquad (25)$$

$$\ddot\theta_5 = \varepsilon_{632pp} C S \theta_4 - \varepsilon_{633pp} Sn\theta_4 \qquad (26)$$

$$\ddot\theta_6 = (\varepsilon_{632pp} Sn\theta_4 - \varepsilon_{633pp} C S \theta_4)/Sn\theta_5 \qquad (27)$$

where

$$\varepsilon_{631pp} = F_4(\theta_1,\theta_2,\theta_3,\theta_4,\theta_5,\dot\theta_1,\dot\theta_2,\dot\theta_3,\dot\theta_4,\dot\theta_5,\dot\theta_6,\ddot\theta_1,\ddot\theta_2,\ddot\theta_3, \varepsilon_{61}, \varepsilon_{62}, \varepsilon_{63},)$$

$$\varepsilon_{632pp} = F_5(\theta_1,\theta_2,\theta_3,\theta_4,\theta_5,\dot\theta_1,\dot\theta_2,\dot\theta_3,\dot\theta_4,\dot\theta_5,\dot\theta_6,\ddot\theta_1,\ddot\theta_2,\ddot\theta_3, \varepsilon_{61}, \varepsilon_{62}, \varepsilon_{63},)$$

$$\varepsilon_{633pp} = F_6(\theta_1,\theta_2,\theta_3,\theta_4,\theta_5,\dot\theta_1,\dot\theta_2,\dot\theta_3,\dot\theta_4,\dot\theta_5,\dot\theta_6,\ddot\theta_1,\ddot\theta_2,\ddot\theta_3, \varepsilon_{61}, \varepsilon_{62}, \varepsilon_{63},)$$

5. CONCLUSION

The spherical wrist structure is often used in modern industrial robots. Therefore when we want to solve the inverse problem of kinematics, it will be very efficient to consider the system as two subsystems.

Rotational matrices and their differentials are used for kinematic analysis when the system is considered. This makes the procedures even more simple. The procedures and formulae for solving accelerations of joints are very similar to those for angular velocities. So if the procedures for joints' angular velocities are carried out, the procedures for joint angular accelerations can be quickly derived.

Reference

1. P. Paul (Robot Manipulators: Mathematics, Programming, and Control, The computer Control of Robot Manipulators). The MIT Press. Second Printing (1982).

2. John M. Hollerbach. (Wrist-Partitioned. Inverse Kinematic Accelerations and Manipulator Dynamics). The International Journal of Robotics Research (1983).

3. R. Featherstone (Position and Velocity Transformations Between Robot end Effector Coordinates and Joint angles). The International Journal of Robotics Research (1983).

Dynamic Equations of General Robots Using Kane's Method

Peng Shang-xian and Wang Gang

Tianjin University, China

ABSTRACT

   This paper presents a method for deriving and solving mathematical models for robot dynamics using Kane's method. The advantages of the procedure that uses Newton's laws are preserved in this kinematic analysis method. These dynamic equations can be used to solve the direct and inverse problems of robot dynamics and can be readily converted into computer codes for numerical calculation.

1. INTRODUCTION

   Newton's laws, and Lagrange's equations are widely used in analysis of robot dynamics. However, these procedures often include unnecessary, or tedious and unwieldy calculations.
   Recently, Kane's dynamic equations have been adopted for obtaining multi-rigid-body system dynamic equations by Huston. These algorithms are not entirely applicable for robots since the industrial robots have only one degree of freedom in each joint. Kane's dynamic equations have also been used successfully for formulating Stanford manipulator's dynamic equations.
   In this paper we present the procedures for obtaining dynamic equations for general robots by using Kane's dynamic equations. In these procedures the advantages of recurrence occuring in Newton's methods are preserved. The algorithms are ideally suited for computer programming.

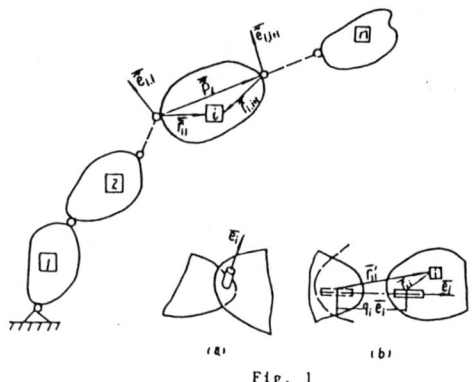

Fig. 1

2. MECHANISM'S CONFIGURATION

The robot with n degree of freedom (d.o.f.) is shown in Figure 1. The joints connecting the mechanism's segments have one d.o.f. each. That d.o.f. may be rotational (Figure 1a) or linear (Figure 1b). S_i is the indicator determining the type of joint. If the joint is rotational, $S_i=0$. If the joint is prismatic, $S_i=1$. $\bar{e}_i$ indicates a unit vector along the axis of joint i.

For each segment a frame for the fixed-body is defined. The origin of such a frame coincides with the center of gravity (c.o.g.) of the segment and the axes are oriented along the inertial principal axes. A vector, expressed with respect to the frame fixed to the body, is designated by notation " ¯ " on the vector.

3. KINEMATICS

(1) <u>Angular velocity and angular acceleration of the i-th segment</u>

The angular velocity of i-th segment is readily obtained as

$$\omega_i = \sum_{j=1}^{n} \vec{w}_{ij} \dot{q}_j \qquad (1)$$

where w_{ij} is the partial angular velocity.

$$\omega_{ij} = \begin{bmatrix} \vec{e}_j(1-S_j) \, ; & j \leq i \\ 0 & j > i \end{bmatrix}$$

The angular acceleration of the i-th segment can be obtained by differentiating equation (1) with respect to time.

$$\vec{\varepsilon}_i = \sum_{j=1}^{i} \vec{w}_{ij} \ddot{q}_j + \vec{n}_i \qquad (2)$$

$\vec{n}_i$ can be obtained by the following expression:

$$\vec{n}_i = \vec{n}_{i-1} + \dot{q}_i(\vec{\omega}_{i-1} \times \vec{e}_i)(1-S_i)$$

(2) **Velocity and acceleration of the i-th segment's c.o.g.**

The velocity of the i-th segment's c.o.g. can be expressed as

$$\vec{V}_i = \sum_{j=1}^{i} \vec{V}_{ij} \dot{q}_j \qquad (3)$$

where vector $\vec{V}_{ij}$ is the partial velocity

$$\vec{V}_{ij} = \begin{cases} \vec{e}_j \times (\sum_{k=j}^{i-1} \vec{P}_k + \vec{r}'_{ii})(1-S_j) + \vec{e}_j S_j & (j<i) \\ (\vec{e}_i \times \vec{r}'_{ii})(1-S_i) + \vec{e}_i S_i & (j=i) \\ 0 & (j>i) \end{cases}$$

where $\vec{r}'_{ii} = \vec{r}_{ii} + q_i \vec{e}_i S_i$.

The acceleration of the i-th segment's c.o.g. can be obtained by differentiating equation (3) with respect to time

$$\vec{a}_i = \sum_{j=1}^{i} \vec{V}_{ij} \ddot{q}_i + \vec{\varepsilon}_i \qquad (4)$$

where vector $\vec{\varepsilon}_i$ can be obtained by solving the recursive expression

$$\vec{\varepsilon}_i = \vec{\varepsilon}_{i-1} + \vec{n}_{i-1} \times (\vec{r}'_{ii} - \vec{r}_{i-1}) + \vec{\omega}_i \times (\vec{\omega}_i \times \vec{r}'_{ii}) + \dot{q}_i (\vec{\omega}_{i-1} \times \vec{e}_i) \times \vec{r}'_{ii}(1-S)$$
$$- \vec{\omega}_{i-1} \times (\vec{\omega}_{i-1} \times \vec{r}_{i-1,i}) + 2\vec{\omega}_{i-1} \times \vec{e}_i S_i \dot{q}_i$$

4. GENERAL FORCE OF INERTIA

(1) **Force of inertia and moment of inertia of the i-th segment**

The force of inertia is

$$\vec{F}^*_i = -m_i \vec{a}_i = -\sum_{j=1}^{i} m_i \vec{V}_{ij} \ddot{q}_j - m_i \vec{\varepsilon}_i$$

From Euler's formula describing the inertial moment, the inertial moment of the i-th segment can be formulated as

$$\vec{M}^*_i = \sum_{j=1}^{i} \vec{b}_{ij} \ddot{q}_j + \vec{b}_i$$

where

$$|\vec{b}_{ij}| = -|T_i||\vec{\omega}_{ij}|; \quad |\vec{b}_i| = |T_i||\vec{n}_i| + |\vec{\Lambda}_i|;$$
$$|T_i| = |A_i||\tilde{J}_i||A_{i-1}|; \quad |\vec{\Lambda}_i| = |A_i|(\tilde{J}_i \vec{\omega}_i) \times \vec{\omega}_i|.$$

A_i is the transition matrix between the frame fixed to the body and the external system. $\widetilde{J}_i$ is the inertial tensor of the i-th segment to the frame fixed to the body.

(2) <u>Contribution of the inertial force and inertial moment of the i-th segment to the general inertial force of the system.</u>
From Kane's dynamic, this contribution to the general inertial force may be expressed

$$(F_k^*)_i = \vec{V}_{ik}\vec{F}_i^* + \vec{\omega}_{ik}\vec{M}_i^* =$$

$$= \sum_{j=1}^{i} (-m_i\vec{V}_{ik}\vec{V}_{ij} + \vec{\omega}_{ik}\vec{b}_{ij})\ddot{q}_j + (-m_i\vec{V}_{ik}\vec{\ell}_i + \vec{\omega}_{ik}\cdot\vec{b}_i)$$

$$(k=1,2,\ldots,n)$$

(3) <u>Contribution of gyroscopic torques to the general inertial force</u>
In some cases the i-th joint is powered by an actuator fitted on i-1-st segment, so that gyroscopic torque exists. The angular velocity of a rotor with respect to i-1-st segment is associated with $\dot{q}_i$. The contribution to general inertial force may be expressed as

$$(F_k^*)_G^{i-1} = (F_k^*)_R^{i-1} + (F_k^*)_I^{i-1}$$

$(F_k^*)_R^{i-1}$ is the contribution of inertial force of a virtual rigid body to the general inertial force. This term can be included in the contribution of the inertial force of the i-1st segment to the general inertial force. To do this, the mass and inertia of the i-1-st segment must include the mass and inertia of gyro. There is only one term to be calculated.

$$(F_k^*)_I^{i-1} = J_{i-1,D}\{\dot{q}_i n_i [-(\sum_{j=1}^{i-1}(\vec{\omega}_{i-1,j}\vec{C}_{i-1,2})\dot{q}_j)\vec{C}_{i-1,1} +$$

$$+ (\sum_{j=1}^{i-1}(\vec{\omega}_{i-1,j}\vec{C}_{i-1,1})\dot{q}_j)\vec{C}_{i-1,2}] - \ddot{q}_i n_i \vec{C}_{i-1,3}\} \cdot$$

$$\cdot \vec{\omega}_{i-1,k} X_{i-1} - J_{i-1,D} \sum_{j=1}^{i-1}(\vec{\omega}_{i-1,j}\vec{C}_{i-1,3})\ddot{q}_j +$$

$$+ \vec{n}_{i-1}\vec{C}_{i-1,3} + \ddot{q}_i n_i] \delta_{ki} X_{i-1}$$

where

$J_{i-1,D}$ is the rotational inertia of gyro; n_i is constant (reduction ratio for rotational joint); $\vec{C}_{i-1,1}$, $\vec{C}_{i-1,2}$, and $\vec{C}_{i-1,3}$ are orthogonal unit vectors set fixed in i-1-st segment. $\vec{C}_{i-1,3}$ is parallel to axe of the rotor.

$$\delta_{ki} = \begin{cases} 1 & (k=i) \\ 0 & (k \neq i) \end{cases}$$

$$X_{i-1} = \begin{cases} 1 & \text{(If gyro exists)} \\ 0 & \text{(If gyro does not exist)} \end{cases}$$

(4) General inertial force of the system

The general inertial force of the system can be derived by adding the above mentioned terms together.

$$F_k^* = \sum_{i=1}^{n} (F_k^*)_i = \sum_{i=1}^{n} (F_k)_I^{i-1} \qquad (k=1,2,\ldots,n)$$

Finally, the F_k^* can be formulated as

$$F_k = \sum_{j=1}^{n} H_{kj} \ddot{q}_j + h_k \qquad (5)$$

where

$$H_{kj} = \sum_{i=1}^{n} (-m_i \vec{V}_{ik} \vec{V}_{ij} + \vec{\omega}_{ik} \vec{b}_{ij} + \alpha_{i-1,j} X_{i-1} \delta_{ki}) -$$
$$- J_{j-1,D} n_j X_{j-1} (\vec{C}_{j-1,3} \vec{\omega}_{j-1,k})$$

$$h_k = \sum_{i=1}^{n} \{-m_i \vec{V}_{ik} \vec{\ell}_i + \vec{\omega}_{ik} \vec{b}_i - J_{i-1,D} (\vec{n}_{i-1} \vec{C}_{i-1,3}) \cdot$$
$$\cdot \delta_{ki} X_{i-1} + J_{i-1,D} \dot{q}_i n_i X_{i-1} [-(\vec{C}_{i-1,1} \vec{\omega}_{i-1,k}) \cdot$$
$$\cdot (\sum_{j=1}^{i-1} (\vec{\omega}_{i-1,j} \vec{C}_{i-1,2}) \dot{q}_j) + (\vec{C}_{i-1,2} \vec{\omega}_{i-1,k}) \cdot$$
$$\cdot (\sum_{j=1}^{i-1} (\vec{\omega}_{i-1,j} \vec{C}_{i-1,1}) \dot{q}_j)]\}$$

5. GENERAL ACTIVE FORCE

(1) External forces contributing to general active force

The external force applied to the i-th segment is equivalent to a single force P_{iF} passing through the c.o.g. together with P_{iM}. The contribution of the external force applied to the i-th segment to the general active force is

$$(F_k)_{pi} = \vec{V}_{ik} \vec{P}_{iF} + \vec{\omega}_{ik} \vec{P}_{iM} \qquad (k=1,2,\ldots,n)$$

(2) **Joint force and moment contribution to the general active force**

In the case of a rotational joint, the force exerted on the i-th segment by the i-1-st segment is equivalent to a force $\vec{F}_{i,i-1}$ together with $\vec{M}_{i,i-1}$. By the law of action-reaction, the force exerted on the i-1-st segment by i-th segment is respectively equivalent to the force $\vec{F}_{i-1,i}$ and $\vec{M}_{i-1,i}$

$$\vec{M}_{i,i-1} = -\vec{M}_{i-1,i} \quad \text{and} \quad \vec{F}_{i,i-1} = -\vec{F}_{i-1,i}$$

The contribution of i-th joint's forces and moments to general active force is

$$(F_k)_{M_i} = (\partial \vec{\omega}_i/\partial \dot{q}_k)\vec{M}_{i,i-1} + (\partial \vec{\omega}_{i-1}/\partial \dot{q}_k)\vec{M}_{i-1,i} +$$
$$+ (\partial \vec{V}_{i-1,s}/\partial \dot{q}_k)\vec{F}_{i-1,i} + (\partial \vec{V}_{i,s}/\partial \dot{q}_k)\vec{F}_{i,i-1}$$
$$(k=1,2,\ldots,n)$$

From $\vec{\omega}_i = \vec{\omega}_{i-1} + \dot{q}_i \vec{e}_i(1-S_i)$; $\vec{V}_{i,s} = \vec{V}_{i-1,s}$; then

$$(F_k)_{M_i} = \begin{cases} \vec{M}_{i,i-1} \vec{e}_i(1-S_i) & (k=i) \\ 0 & (k \neq i) \end{cases}$$

For a prismatic joint, the expression can also be formulated as

$$(F_k)_{F_i} = \begin{cases} \vec{f}_{i,i-1} \vec{e}_i S_i & (k=i) \\ 0 & (k \neq i) \end{cases}$$

Finally, the contribution of the joint's forces and moments to general active force can be written as:

$$(F_k)_{ei} = \begin{cases} M_i(1-S_i) + F_i S_i & (k=i) \\ 0 & (k \neq i) \end{cases}$$

where M_i is the driving moment in the i-th joint, F_i is the driving force in the i-th joint.

(3) General active force of the system
The general active force of the system can be derived as:

$$F_k = \sum_{i=1}^{n} [(F_k)_{pi} + (F_k)_{ei}] = \\ = \sum_{i=1}^{n} (\vec{V}_{ik} \cdot \vec{P}_{iF} + \vec{\omega}_{ik} \cdot \vec{P}_{iM}) + M_k(1-S_k) + F_k S_k \quad (6)$$

6. DYNAMIC EQUATION OF THE SYSTEM

According to Kane's dynamics, the dynamic equation of the system is

$$F_k + F_k^* = 0 \qquad (k=1,2,\ldots,n)$$

Substituting F_k^* and F_k by equations (5) and (6), these formulae may be expressed in a matrix form

$$\mathbb{P} = \mathbb{H} \ddot{\mathbb{Q}} + \mathbb{G} \qquad (7)$$

where

$$P = \begin{bmatrix} \vdots \\ M_k(1-S_k)+f_k S_k \\ \vdots \end{bmatrix} \quad ; \quad \ddot{Q} = \begin{bmatrix} \ddot{q}_1 \\ \vdots \\ \ddot{q}_k \\ \ddot{q}_n \end{bmatrix}$$

$$H = \begin{bmatrix} -H_{11} & -H_{12} & \cdots & -H_{1n} \\ \cdots & \cdots & \cdots & \cdots \\ \cdots & \cdots & \cdots & \cdots \\ -H_{n1} & -H_{n2} & & -H_{nn} \end{bmatrix} \quad ; \quad G = \begin{bmatrix} \vdots \\ \sum_{i=1}^{n} (\overline{V}_{ik} \cdot \overline{P}_{iF} + \overline{\omega}_{ik} \cdot \overline{P}_{iH}) - h_k \\ \vdots \end{bmatrix}$$

7. CONCLUSION

In this paper a method for formulating dynamic equations of robots by using Kane's dynamic equations is presented. This method suits the general configuration of a robot. Partial velocities and partial angular velocities can be readily determined. The governing equations can also be conveniently determined. The algorithms will be even more efficient if they are used for a special robot.

REFERENCES

1. M.Vukobratović and V.Potkonjak, Dynamics of Manipulation Robots (1982)
2. R.Kane, Dynamics, Stanford University (1978)
3. R.L.Huston, Dynamics of Multi-rigid-body System, University of Cincinnati (1978)
4. M.W.Walker, Efficient Dynamic Computer simulation of Robotic Mechanisms, Journal of Dynamic Systems, Measurement and Control (1981)

The Role of Delay in Robot Dynamics

G. Stépán

Department of Mechanical Engineering Technical University of Budapest,
Budapest, Hungary

SUMMARY

Delayed feedback in the control of robots may cause stability problems. Analytical investigations of this effect cannot be found in the literature of robotics. This paper provides a stability criterion which is a basis for the construction of stability charts for parameters of robots with 2÷3 degrees of freedom (DOF).
In the case of more DOF the criterion provides simple estimates for the stable regions in the parameter space and it may also serve as a basis for numerical methods.
The paper gives examples which are the first results of this research.

1. INTRODUCTION

Delay or time lag, or dead time causes instability in dynamic systems. In some cases when unexpected stability problems occur in robotics investigation of the role of delay is required. What are these cases when dead time cannot be left out of consideration?
  Delay may occur in any part of the robot including:
(a) its control system
(b) the information transmission, and
(c) the mechanical part of the robot
  Case (a) may be produced by the delay in the reflexes of a human operator in a master/slave system. However, the situation is very similar in the case of an on-line control through a computer where the sampling period gives rise to time lag. Only the results obtained by simulation or experiment provide the critical value of the delay when stability problems arise and then the strategy of the control is strongly determined by this (1). In the case of a human being, the delay cannot be less that 0.1 second and it is about 0.01 to 0.001 second for computers even in the best case such as when motion equations like those of Appell-Gibbs or Kane (see (2,3)) are used.
  In case (b) transmission delay may cause instability in undersea or space tele-operations. The dead time is equal to the time needed by the ultrasonic or electro-magnetic wave to cover the distance to and fro between the master and the slave.

This value may exceed 0.1 to 1 second (see (4)).

A delayed feedback can be found in the mechanical part of the robot (case c) if it is used in a material forming process like milling, rolling, welding, etc.

During cutting, there occur delays which are proportional to the time period of the revolution of the tool (or workpiece)(5). Moreover, some rolling devices take into consideration distributed delay which is also inversely proportional to the relative velocity of the surfaces of the tool and workpiece (6). However, case (c) is usually not considered a problem of robotics.

This paper gives a necessary and sufficient condition for the asymptotic stability of delayed systems and it provides some conclusions for robots with great delays in case (a) and (b).

2. STABILITY CRITERION

The position of the robot manipulator with n DOF can be described by the n-dimensional vector $\underline{q}$ containing the general coordinates q_k (k=1,n). The differential equation of motion has the form

$$\underline{\underline{D}}(\underline{q})\ddot{\underline{q}} + \underline{F}(\dot{\underline{q}}) + \underline{C}(\underline{q},\dot{\underline{q}}) + \underline{G}(\underline{q}) = \underline{Q} , \qquad (1)$$

where $\underline{\underline{D}}$ represents intertia, $\underline{F}$ is the damping, $\underline{C}$ contains centrifugal and Coriolis forces and $\underline{G}$ represents gravitational forces and elasticity of the system.

Let the end position of the robot be at $\underline{q} = \underline{0}$. The control system determines the general force $\underline{Q}$ as a function of $\underline{q}$ and/or $\dot{\underline{q}}$. If there is a delay in the system, $\underline{Q}$ will depend on $\underline{q}_t$ (and/or $\dot{\underline{q}}_t$) where $\underline{q}_t$ represents the "past" of $\underline{q}$. Mathematically it is defined as follows:

$$\underline{q}_t(\vartheta) = \underline{q}(t+\vartheta), \quad \vartheta \in [-r, 0] ,$$

where $r \geq 0$ is the length of retardation and t stands for time. If we assume $\underline{Q}(\underline{q}_t, \dot{\underline{q}}_t)$ we get a retarded functional differential equation (1) because $\underline{Q}$ depends on functions.

The linearized form of (1) is

$$\underline{\underline{A}}\ddot{\underline{q}} + \underline{b}(\dot{\underline{q}}_t) + \underline{c}(\underline{q}_t) = \underline{0} , \qquad (2)$$

where the constant matrix $\underline{\underline{A}}$ of the general mass is positive definite and

$$\underline{c}(\underline{q}_t) = \int_{-r}^{0} d\underline{\underline{C}}(\vartheta)\underline{q}(t+\vartheta) , \quad \underline{b}(\dot{\underline{q}}_t) = \int_{-r}^{0} d\underline{\underline{B}}(\vartheta)\dot{\underline{q}}(t+\vartheta) . \qquad (3)$$

The matrices $\underline{\underline{B}}$ and $\underline{\underline{C}}$ consist of functions of bounded variation. These Stieltjes integrals mean that there may be one, or more, discrete or "continuous" delays in the system. For example:

$$\int_{-r}^{0} d\underline{\underline{C}}(\vartheta)\underline{q}(t+\vartheta) = \sum_{k=1}^{p} \underline{\underline{C}}_k^d \underline{q}(t-r_k) + \sum_{k=1}^{p} \int_{r_{k-1}}^{r_k} \underline{\underline{C}}_k^c(\vartheta)\underline{q}(t+\vartheta)d\vartheta ,$$

where $r_o = r$ and $r_p = 0$.

If r = 0 then we have an ordinary differential equation and the stability can be examined with the Routh-Hurwitz or Lienard-Chipart criteria. If r > 0 and there is only one discrete time lag in the system, the well-known methods of Nyquist, Bode or Nichols can be used, but there exist some better methods like those in (7,8). If there are more discrete delays than one in the system, the application of the Pontryagin method (9) may be successful in some special cases. To be able to investigate even the most general cases a new criterion has been constructed.

Let the characteristic function D and the functions M and N be defined as follows:

$$D(\lambda) = \det\left(\underline{\underline{A}}\lambda^2 + \int_{-r}^{0} \lambda e^{\lambda\vartheta} d\underline{\underline{B}}(\vartheta) + \int_{-r}^{0} e^{\lambda\vartheta} d\underline{\underline{C}}(\vartheta)\right), \qquad (4)$$

$$M(\omega) = \mathrm{Re}\,D(i\omega), \quad N(\omega) = \mathrm{Im}\,D(i\omega), \quad (i = \sqrt{-1}, \; \omega \in R_+). \quad (5)$$

The positive zeros of M are denoted by $\omega_1 \geq \omega_2 \geq \ldots \geq \omega_m \geq 0$.

Theorem. The end position $\underline{q} = \underline{0}$ of the system is asymptotically stable if and only if

$$N(\omega_k) \neq 0, \; k = \overline{1,m} \quad \text{and} \quad \sum_{k=1}^{m} (-1)^k \mathrm{sign}\, N(\omega_k) = (-1)^n n. \quad (6)$$

The proof can be found in (10).

3. GENERAL STABILITY PROPERTIES

We are going to mention briefly some general results obtained by means of this theorem. First, consider "stiffness" in the system.

The $\underline{q} = \underline{0}$ position is unstable if

$$\det(\underline{\underline{C}}(0) - \underline{\underline{C}}(-r)) < 0. \qquad (7)$$

This expresses the fact that if the influence of the past is too great then the system becomes unstable. In a system

$$\underline{\underline{A}}\ddot{\underline{q}} + \underline{\underline{B}}\dot{\underline{q}} + \underline{\underline{C}}\underline{q} = \underline{\underline{R}}\underline{q}(t-r) + \underline{\underline{P}}\dot{\underline{q}}(t-r)$$

with one discrete delay, (7) means instability if

$$\det(\underline{\underline{C}} - \underline{\underline{R}}) < 0$$

Regarding damping, the statement that: "increasing damping cannot cause instability in retarded systems" is true only for systems with 1 d.o.f. For example, the investigation of human control of a crane has proved that in a quite simple delayed system large dumping may be wrong from the point of view of stability (11). On the other hand, a delayed feedback may stabilise systems with negative damping (i.e. with excitation).

4 EXAMPLES

Figure 1 shows a simple model of the equilibrization of a rod (inverted pendulum).

Similar tricks are often used in robotics to represent the efficiency (the short delay) of control systems and servo-motors (see for example the roller holding trick demonstrated by Yaskawa Co. on a Motoman robot at '85 Int. Ind. Robot Exhibition, Tokyo). Now, equation (1) has the form

$$\begin{bmatrix} m\ell^2/3 & 0.5m\ell\cos q_1 \\ 0.5m\ell\cos q_1 & m \end{bmatrix} \begin{bmatrix} \ddot{q}_1 \\ \ddot{q}_2 \end{bmatrix} + \begin{bmatrix} -0.5mg\ell\sin q_1 \\ -0.5m\ell\dot{q}_1^2\sin q_1 \end{bmatrix} = \begin{bmatrix} Q_1 \\ Q_2 \end{bmatrix}, \quad (8)$$

where

$$Q_1 = 0, \quad Q_2 = F = b_1\dot{q}_1(t-r) + b_0 q_1(t-r).$$

r denotes the delay of reflexes of the operator who tries to equilibrate the rod by means of force F. After the elimination of q_2 and the linearization of (8) at the position $q_1 = 0$, we get the characteristic equation (4):

$$D(\lambda) = \lambda^2 + \frac{6}{m\ell}(b_1 r\lambda e^{-\lambda} + b_0 r^2 e^{-\lambda}) - \frac{6g}{\ell}r^2 = 0.$$

If we investigate this on the basis of the theorem we get the stability conditions for b_1 and b_0 and r:

$$b_0 > g, \quad b_1 = b_0 r\frac{\tan\omega}{\omega}, \quad b_0 < \left(\frac{m\ell\omega^2}{6r^2} + g\right)\cos\omega, \quad \omega \in \left(0, \frac{\pi}{2}\right). \quad (9)$$

(9) does not give any value for b_1, b_0 if

$$\left.\frac{d^2 b_0}{d\omega^2}\right|_{\omega=0} < 0 \quad \Rightarrow \quad r > \sqrt{\frac{\ell}{3g}} = r_{cr}.$$

If the delay of the operator's reflex is greater than r_{cr} then the rod cannot be equilibrated in this way. In the case of $\ell = 0.3$ meters long rod we get $r_{cr} = 0.1$ second.

Figure 2 shows the second example. The $q_1 = 0$ position has to be found with the help of a camera fixed at the end of the elastic arm. There is a kinematical constraint with delay in the control:

$$\dot{q}_2(t) = -k \text{ arth} q_1(t-\tau).$$

The stability investigation of the linearized motion equation

$$m\ddot{q}_1(t) + f\dot{q}_1(t) + sq_1(t) + fkq_1(t-\tau) + sk\int_{-\infty}^{-\tau} q_1(t+\vartheta)d\vartheta = 0$$

gives the stability chart of figure 2 on the plane of the parameters $\tau\sqrt{s/m}$ and $k\sqrt{m/s}$ when the relative damping factor is 0.01. Note that there are considerable stable regions in the case of considerable delays.

Let the last example be the experimental master-slave system based on the force-reflective manipulator MA23 of CNES in ARA project (4). The linearized equations of motion are

$$m_1\ddot{q}_1(t)+f_1\dot{q}_1(t)+k(q_1(t)-q_2(t-\tau)) = Q_1 ,$$
$$m_2\ddot{q}_2(t)+f_2\dot{q}_2(t)+k(q_2(t)-q_1(t-\tau)) = Q_2$$

with subscript 1 for the master and 2 for the slave. τ is the delay in information transmission. Let us suppose that an operator with delay r of reflexes tries to find the $q_1 = q_2 = 0$ position. There is no load on the slave ($Q_2 = 0$) and the operator gets information about the position q_2 of the slave only with delay τ (i.e. by means of a camera):

$$Q_1(t) = -bq_2(t-\tau-r) .$$

If the inertia m_1 and viscous friction f_1 are negligible in the master we get the characteristic function (4):

$$D(\lambda) = m_2\lambda^2 + f_2\lambda + k(1-e^{-2\tau\lambda}) + be^{-(2\tau+r)\lambda} .$$

Thus, there are two different discrete delays in this system: 2τ and $2\tau+r$. The stability condition of the theorem gives a simple estimate for a stable region in the parameter space:

$$\tau(0.44k+b)+rb < f_2 .$$

This result is shown by figure 3.

5. CONCLUSIONS

The method based on the theorem of part 2 can be applied to the stability investigation of end positions of robot manipulators under a control with delayed feedback. The first and second examples prove that necessary and sufficient conditions of asymptotic stability can be determined in the parameter space of simple robots. The second example illustrates a case when a system may get back its lost stability with considerable delays. In the case of the third example the theorem has been used to give estimates (sufficient conditions) for the parameters of the master-slave system and for the delays when the end position is stable.

References

1. Vukobrativič M., Stokič D., "Approximative Models in Dynamic Control of Robotic Systems", Ro-man-sy '84, pp. 104-112.
2. Vukobratovič M., Kircanski N., Real-Time Dynamics of Manipulation Robots, Scientific Fundamentals of Robotics 4, Springer, New York, 1985.
3. **Kane** T.R., Faessler H., "Dynamics of Robots and Manipulators Involving Closed Loops", Ro-man-sy '84, pp. 75-83.

4   Vertut J.,Charles J.,Coiffet Ph.,Petit M.,"Advance of
    the New MA23 Force Reflecting Manipulator System"
    Ro-man-sy '76.
5   Tobias S.A., Machine Tool Vibrations, LTD London-
    Glasgow, 1965.
6   Lasota A.,Rusek P.,"Stability of Self-Induced Vibrations
    in Metal Cutting", Proc. Fifth World Congress on
    Theory of Machines and Mechanisms 1979, pp.1502-1505.
7   Jesipowich J.M.,"On the Stability of the Solutions of
    a Class of Differential Equations with Time Lag"
    /in Russian/, Prikl.Math,Mech., Vol 15. pp. 601-608,
    1951.
8   Hsu C.S.,"Application of the τ-Decomposition Method
    to Dynamical Systems Subjected to Retarded Follower
    Forces", J.Appl.Mech., Vol 37. pp. 258-266, 1970.
9   Hale J.K., Theory of Functional Differential Equations,
    Springer, Berlin, 1977.
10 Stépán G.,"A Stability Criterion for Retarded Dynamical
    Systems", ZAMM, Vol 64. pp. 345-346, 1984.
11 Tran v.D,,Stépán G.,"Stability Investigation of a
    Man-Machine System", under publication

The Role of Delay in Robot Dynamics 183

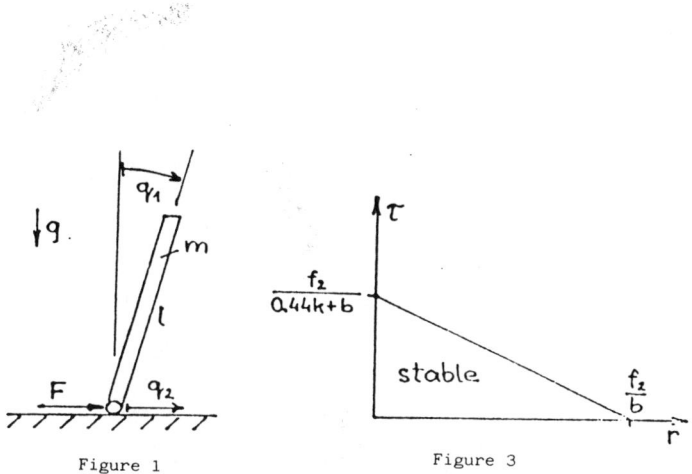

Figure 1

Figure 3

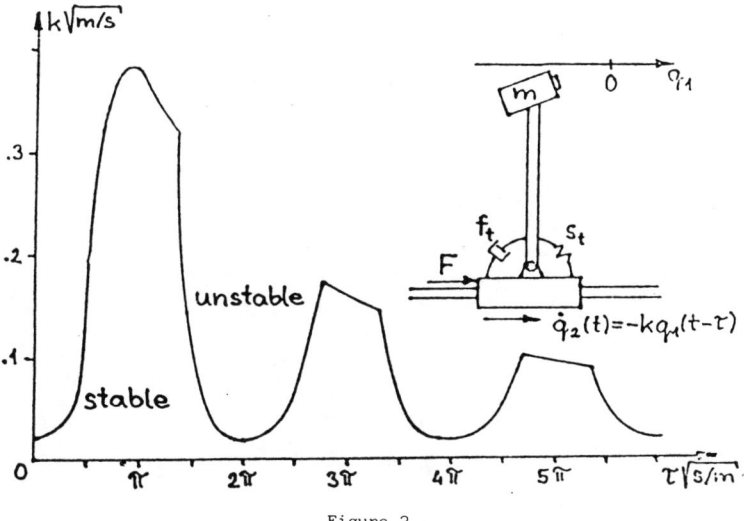

Figure 2

Part 4
Synthesis and Design 1

Smart Hand Systems for Robotics and Teleoperation

A.K. Bejczy and B.M. Jau

Jet Propulsion Laboratory, California Institute of Technology, Pasadena, U.S.A.

SUMMARY

This paper describes two smart hand systems developed at JPL recently for robotics and teleoperator applications. The first unit was designed for potential application on an Orbiting Maneuvering Vehicle (OMV) equipped with an approximately 5 meter size manipulator performing, e.g., satellite servicing and repair tasks. The second unit was designed for experimental use on a human-size industrial robot arm called PUMA 560 to explore issues in versatile object handling and compliance control in grasp actions. The developments followed an integrated design approach by considering mechanism, sensing, electronics, control and displays within an integrated system design architecture. The descriptions in this paper briefly summarize all design aspects of the two smart hand systems.

1. INTRODUCTION

Anticipated space assembly, servicing and repair tasks to be performed by remote manipulators motivated a smart hand design and development effort at the Jet Propulsion Laboratory (JPL) to enhance remote manipulation capabilities in both teleoperator and robotic modes of control. The term smart hand refers to the human analogy which reveals that the hand is both a powerful and delicate mechanism as well as a sensory instrument through which information is received and transmitted. It is also worth noting that, in the proper sense of the word, manipulation is the function of the hand. The function of the arm is to position and orient the hand, act as a mechanical connection and power and sensing transmission link between the hand and the main body of a person. The full functional meaning of the arm rests in the hand.

An initial design study subdivided the mechanical hand (or end effector) design requirements into four principal areas: (i) mechanism and mechanical performance, (ii) sensing and data handling, (iii) control, and (iv) man-machine interface for decision and control. Several conceptual design alternatives were considered and graded [1] according to the end effector categories shown in Figure 1. An interesting conclusion of the design study was that the development of dexterous and anthropomorphic hands in a master-slave control configuration would be desirable and technically feasible. For more on this conclusion, see [1].

Analysis of space assembly, servicing and repair tasks to be performed

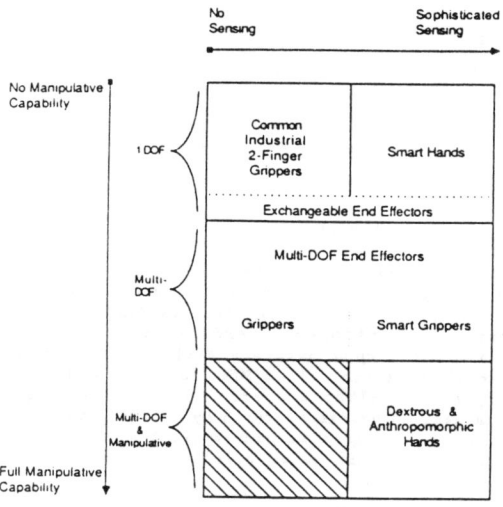

Figure 1. End Effector Categories

by remote manipualtors also lead to the conclusion that an evolutionary approach to the design and development of space end effectors can generate important and needed capability increases. The first step in this evolutionary development approach was the consideration of one degree-of-freedom (dof) parallel claw end effectors equipped with six dof force-torque balance and one dof grasp force sensors and servoable in position, rate and grasp force modes of control. Two smart hands of this category were designed and developed at JPL recently. The basic difference between the two smart hand units is the end effector size and drive mechanism and the local electronics and subsystem interface instrumentation.

In this paper, we briefly describe the two smart hand systems developed at JPL recently. The first unit was designed for potential application on an Orbiting Maneuvering Vehicle (OMV) equipped with an approximately 5 meter size manipulator called Protoflight Manipulator Arm (PFMA) at the Marshall Space Flight Center (MSFC). The second unit was designed for experimental use on a human-size industrial robot arm called PUMA 560. The OMV and PUMA Smart Hand descriptions will be concentrated on the mechanism including the sensors and on the electronics including sensor and control data handling.

2. OMV SMART HAND

The specific design and performance requirements for the OMV smart hand are derived from (i) considerations of typical tasks the hand has to perform, (ii) considerations of the system the hand has to be interfaced with, and (iii) considerations of advanced sensing, control and man-machine interface capabilities which should be demonstrated and tested for performance evaluation.

Typical test tasks are represented by the following ones: (i) Mate and demate a fluid coupling mechanism which has an open area of 10.2 cm by 8.6 cm to reach the handle. (ii) Open and close an access panel by turning a wing nut which is a 0.5 cm thick flat stock with an area of 7.6 cm by 3.8 cm. (iii) Remove and replace a battery module by grasping a square beam handle.

These test tasks dictate the following requirements: a) The hand shall have an outside width no wider than 18 cm. b) The maximum hand opening shall be not less than 6.4 cm. c) The minimum hand closing shall be no more than 0.6 cm. d) The overall construction of the hand shall be so that the hand can reach into the fluid coupling mechanism. e) The hand shall be capable of squeezing with 445 N force. f) The maximum tip force on the hand shall be at least 45 N and the maximum tip torque shall be at least 20 Nm. g) The maximum closing velocity of the hand shall be at least 2.5 cm/sec. h) The gripping action shall have a linear path throughout the travel.

The major system interface requirements are: a) The entire mechanical hand system shall mount to the PFMA wrist. b) The claws shall be intermeshing such that oval, round and square beams as small as 0.6 cm in diameter can be grasped. c) All electrical communication to and from the mechanical hand, including the electrical power, shall be through an existing slip ring subsystem at the last wrist joint. This slip ring subsystem permits the use of altogether seven electrical wires for power and/or signal transmission.

The major sensing, control and man-machine interface requirements are: a) The force-torque sensor mounted to the base of the mechanical hand shall measure forces and torques as applied to the hand in all three orthogonal directions up to 133 N and 68 Nm, respectively, in each direction with a resolution of at least 1 part in 500. b) The grasp force sensor mounted to the base of the claws shall measure grip force up to 535 N with a resolution of at least 1 part in 200. c) The grasp control loop shall be closed locally at the mechanical hand based on commands from the central control computer at the MSFC control station. d) The computer graphics display of sensor data shall permit (i) the use of alternative display formats on the task level and (ii) the fuse of computer graphics with video data on TV monitors. The computer graphics update rate shall be at video update rate on a color display.

2.1 OMV Smart Hand Mechanism

The overall view and the preliminary assembly of the OMV smart hand mechanism and sensors are shown in Figures 2 and 3. The main mechanical design features can be summarized as follows:
o Channel type frame produces a stiff structure.
o One piece aluminum channel frame with integral bevel gear box produces a rigid structure.
o Hardened steel bevel gear drive between motor and ball screws produces a compact and efficient drive train.
o Left hand and right hand ball screw mechanism drives double finger slides in a coordinated fashion.

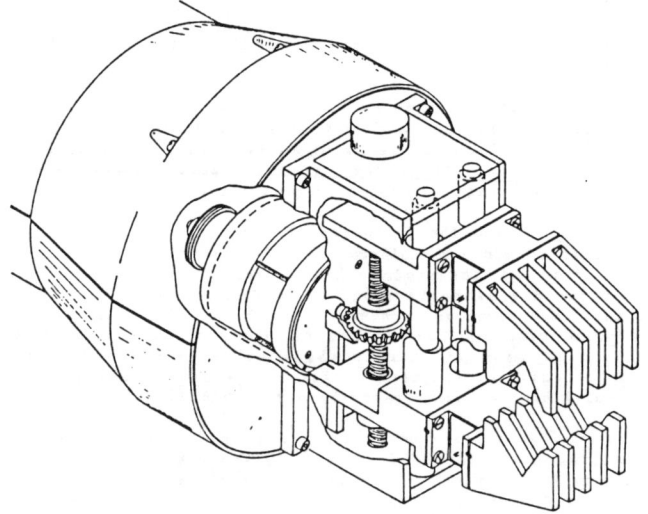

Figure 2. Overall View of OMV Smart Hand Assembly

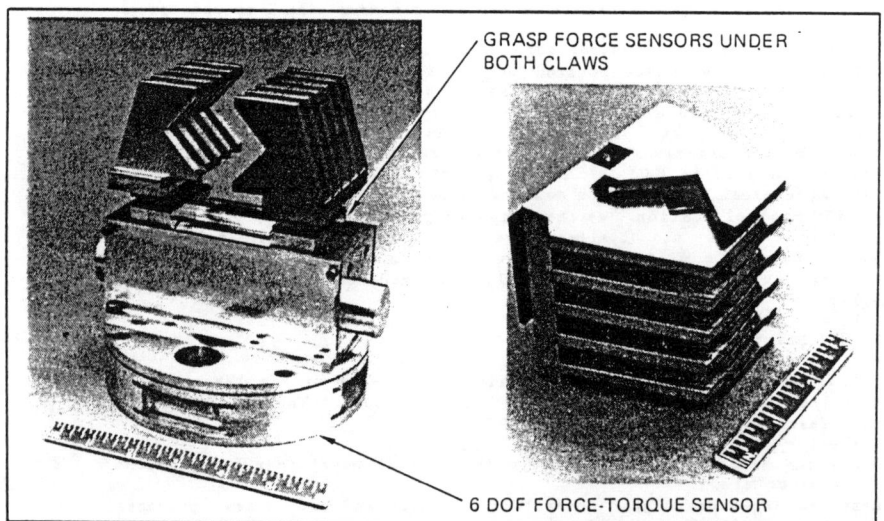

Figure 3. Preliminary Assembly of OMV Smart Hand with Intermeshing Claws and Equipped with Force-Torque and Grasp Force Sensors

o Double slides, each slide on a separate hardened and ground steel rod, are additionally guided by a channel integral with the frame.
o Double slide rods supported at both ends produce a compact design.
o Determinate design and built in adjustment features of slides minimizes tendency to bind and produces a gripper mechanism built of interchangeable parts with few ultraprecision dimensions.
o Partially enclosed drive and slide mechanisms are resistant to damage.
o High efficiency drive will not lock up under any conditions.
o Bevel gear and ball screw drive efficiently matches motor to load.
o Brushless Rare Earth Magnet D.C. motor is used for long life and compact power source.
o Motor can maintain maximum grip force continuously without overheating.
o Fail-safe brake on motor maintains grip in the event of power loss.
o Gripper mechanism attaches with easily accessible screws.
o Claw assembly and grasp force sensor easily changed.

The main mechanical design features of the six dof force-torque sensor mounted to the base of the end effector can be summarized as follows:
o Wrist force sensor resolves all six components of the resultant of forces and moments on the gripper. [Fx, Fy, Fz, Mx, My, Mz]
o Sensor uses a Maltese cross type of design. All sensing is done by strain gage bridges on _bending_ beams. This helps temperature compensation and design has very low tendency to buckle.
o Large bore in center permits gripper motor to extend through sensor to produce a compact sensor/gripper package.
o Flat washer type design keeps wrist/gripper length short for improved work envelope and load capacity.
o Sensing and overload structure is machined from a single piece of high strength aluminum alloy. This produces a low hysteresis, adjustment-free sensing system with good stiffness. There are very few precision machining tolerances.
o Overload protection is built in to withstand unexpected overloads and accidents.
o Six inch diameter of the sensor matches it well to gripper mechanism.
o It is a proven design [2].

The main design features of the grasp force sensors [3] mounted to the base of the intermeshing claws are as follows:
o Sensor senses grip force only on each finger. The sensor is not sensitive to placement of load in claw or other forces or moments applied to or by the claw.
o Parallelogram design produces nearly pure translation type of deflection rather than angular bending typical of many other simple cantilevered beam sensors. Jaws of claws remain parallel or nearly parallel at all times.
o Simple, modular, element "spacer" type design allows that the sensor can be added or removed as needed.
o Strain gage bridge output signals can be read and processed by the same interface and electronics as used by the wrist force sensor.
o Strain gage bridges can be mounted entirely on the inside of the structure and potted in silicone rubber for a robust element.

2.2 OMV Smart Hand Electronics

The electronics contains two major packages: a local electronics package integrated with the end effector mechanism and a central electronics package interfaced to the control station computer and display system. The two major packages are indicated in Figure 4.

The design of the local electronics package incorporates a distributed microcomputer architecture, using advanced integrated circuits, including hybrid and high level multi-functional monolithic packages. This makes it possible to minimize the total chip counts, which are mounted on custom designed printed circuit cards. A single-chip data acquisition system is used as a front end-driver that performs multiple analog signal multiplexing, amplification, sample-and-hold, and finally analog-to-digital conversion. A single-chip microcomputer having its own serial data input/output port is chosen as the heart of the communication process. In fact, two of these are used in the present design. A third microcomputer chip actually performs the function of real-time motor control and sensor turn-on/off control.

As shown in Figure 4, the local electronics is composed of two subsystems, the sensor subsystem, and the end effector subsystem. Interfaced through the slip ring subsystem, this local electronics communicates sensor data, control computer commands, and receives power from the control station, which is remote from the end effector.

The end effector subsystem consists of two CPU's, the Motorola MC68701 and MC68705. The 68701 is basically used as a communication device by virtue of its serial input/output port. It also checks for transmission errors with a 16-bit check sum comparison. It receives the motor drive signals and control modes from the central computer(s) via the slip ring subsystem. This 68701 interfaces with the 68705 which stores and executes the program to control the 3-phase d.c. brushless motor. Pulse width modulation control and winding commutation control is performed in this second CPU. It also receives the clamp force sensor, position and tachometer sensor readings for direct inside-loop control of the motor of the end effector.

The local electronics is installed on three circular and two annular custom designed printed circuit cards behind the force-torque sensor, around and behind the motor. Some of the local electronics is shown in Figure 5 which depicts the hand and sensors on the calibration bench.

Three subsystems and a power supply are the core elements of the central electronics package: the Signal Processing Subsystem, Graphics Subsystem, and the Human Operator Station Subsystem. The signal processing subsystem processes the raw sensor data from the sensor subsystem (of the local electronics), including the conversion of the strain gage readings from the force-torque sensor and the clamp forces sensor into calibrated measurements through scalar and matrix multiplications. In addition, the Signal Processing Subsystem issues primitive graphic commands via an RS232 link to the graphics processor. Vector draw, area fill, and test insertion are typical commands. The graphics processor then generates, at video rate, graphics pictures of the sensor readings and claw configuration on a TV monitor, to be presented to the human operator. The graphics processor is a Parallax 600-M-A unit.

Smart Hand Systems for Robotics and Teleoperation 193

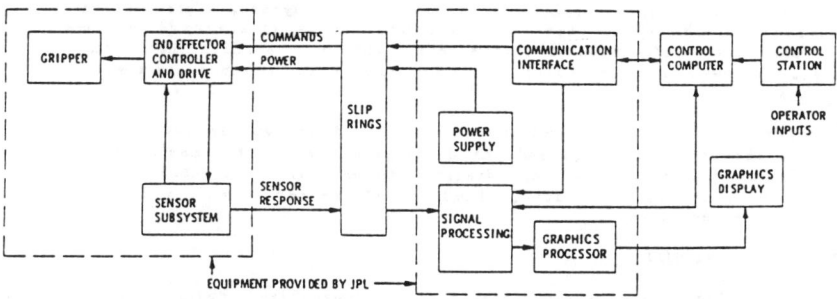

Figure 4. Overall Electronics Organization of OMV Samrt Hand

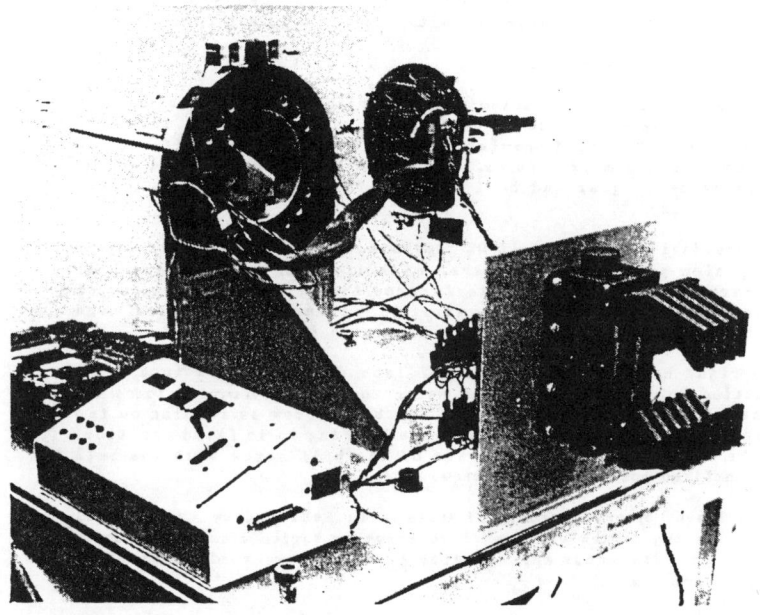

Figure 5. OMV Smart Hand and Sensors on the Calibration Bench

This graphics subsystem is designed to generate signals according to NTSC standards, and will be GENLOCK'ed to the local TV system. This enables the graphics display video signals to be compatible with standard ground based or spacecraft video monitor system nets. This also permits the mixing of one picture over another, which are important design features in a man-machine system.

Integral to this smart hand system are the displays, including the displays of the sensor readings and the displays/menus for the man-machine dialog. Sensor reading displays are designed to provide unambiguous, easy-to-interpret, convention standard, fixed as well as variable formats and presentation perspectives.

3. PUMA 560 SMART HAND

This smart hand system was designed to fit a human-size industrial robot arm, the PUMA 560 arm, for interactive manual and computer control experiments. The main design drivers were two requirements: (i) to provide a versatile object handling capability and (ii) to provide compliance control capability in grasp actions within the limitations of a one dof end effector mechanism. The first objective was accomplished by selecting a contoured inner surface for the claws, and the second objective was accomplished by designing a low friction system together with the use of grasp force sensors.

3.1 PUMA 560 Smart Hand Mechanism

The PUMA 560 smart hand mechanism together with the sensors is shown in Figure 6. As seen, the mechanism is designed with force-torque wrist and grasp force sensors. The hand opening range is 7 cm. The end effector is designed for 140 N maximum grip force. The drive mechanism is based on the non-self-locking bevel gear and ball screw with an anti-friction linear crossed roller bearing.

Power transmission from a small DC torque motor is through a bevel gear system, one pinion gear with two gears, with 1:4 gear ratio to the ball screw. The gears are mounted with interference fit on the shaft for proper torque transfer. Proper gear clearance is achieved by vertical and horizontal adjustment of the motor shaft with pinion gear.

The two right hand ball screws (0.187 inch dia. and 0.062 inch lead) have one function only: to transfer the motor torque for clamping force (no bending force on the ball screw shaft). Each ball screw is running on two "radiax" deep groove ball bearings where one bearing is in fixed position, supporting the axial load. The other end of the ball screw with the ball screw nut is in fixed position in the claw.

Precision linear or one-dimensional motion is achieved by cylindrical crossed roller bearing of the claws with constant or varied clamping force at constant velocity. Stiffness and rigidity to the other two dimensions is attained at the same time.

Claws run on a new type of linear crossed roller bearings which were designed for the special purpose of supporting the claw loads. Bearing have

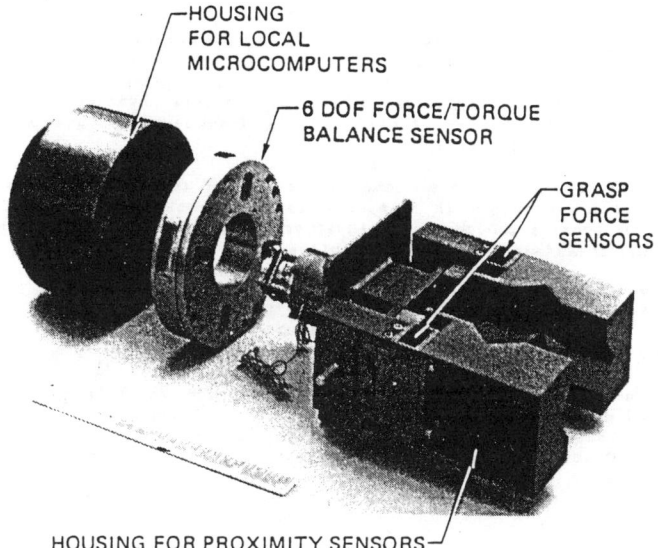

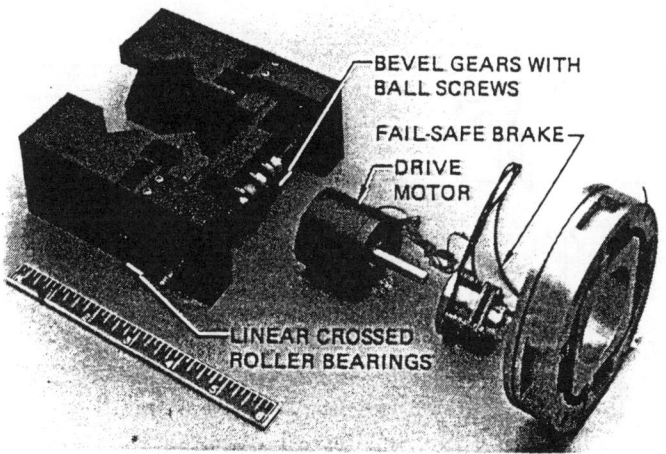

Figure 6. PUMA 560 Smart Hand Mechanism and Sensor System

high load capacity, low friction coefficient ($\mu \sim 0.005$) with long life and no possibility of jamming or caging.

Claw support assembly has threaded holes in the claw mounting surface to allow replacement with different claw sizes and shapes. The grasp force sensor can also be changed easily.

The force-torque sensor is mounted to one side of the gear box frame and the linear roller bearings with claws are mounted to the other side of it. This frame-type housing produces a rigid structure and the gripper mechanism attaches to the frame with easily accessible screws.

The drive motor is equipped with a fail-safe brake. The brake stops the current to the motor or, in the event of power loss, it maintains grip force.

3.2 PUMA 560 Smart Hand Electronics

The organization of the local electronics package which is mounted to the smart hand mechanism is illustrated in Figure 7. The main design driver of this local electronics was to achieve a high density function package with high speed interface to a higher level distributed computer control system. As seen in Figure 7, the local sensor data handling and motor drive utilizes only one microprocessor on a single chip. Local control of grasp force is possible.

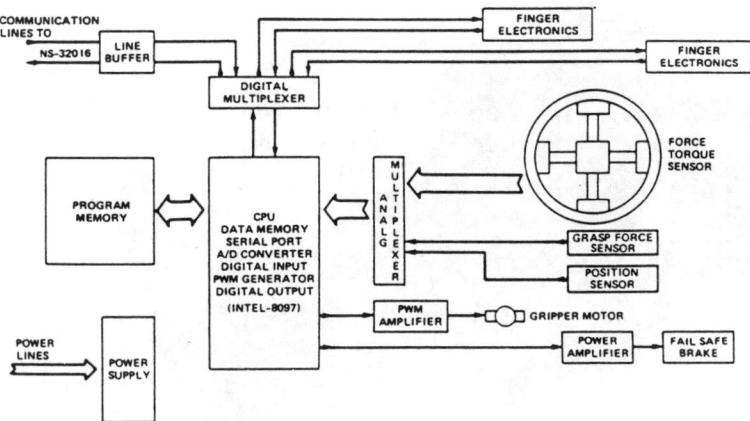

Figure 7. PUMA 560 Smart Hand Local Electronics Organization

4. CONCLUSIONS

The development of smart hand systems is an interdisciplinary task. The development at JPL emphasized an integrated design approach by considering mechanism, sensing, electronics, control and displays within an integrated system design architecture.

The two smart hand systems will be evaluated as soon as they become operationally integrated with the corresponding manipulators. The evaluation will also provide valuable information to be utilized in the design of the next generation hands, the multi-dof smart, dexterous and anthropomorphic hands.

Acknowledgements

This work was carried out at the Jet Propulsion Laboratory, California Institute of Technology, under contract to the National Aeronautics and Space Administration. Several people contributed to this project. The OMV smart hand mechanical design, including wrist force-torque sensor and grasp force sensor is by V. Scheinman of Stanford University. The PUMA 560 smart hand mechanical design is by Z. Vigh. Ideas to electronics and data handling design were contributed by R. Dotson, R. Killion, H. Primus and J. South.

References

[1] Mishkin, A.H., and Jau, B.M., Functional Requirements and Proposed Designs for Space-Based Multifunctional End Effector Systems, JPL Report (in printing), 1986.

[2] Bejczy, A.K. and Dotson, R.D., Force-Torque Sensing and Display System for Large Robot Arms, Proceedings of the IEEE Southeastcon '82, Destin, FL, April 4-7, 1982.

[3] U.S. Patent is pending, CIT File No. 1859.

A Mathematical Model of a Flexible Manipulator of the Elephant's-Trunk-Type

G. Malczyk* and A. Morecki**

*CBKO - Informatics Division, Pruszkow, Poland
**Technical University of Warsaw, Poland

SUMMARY

The paper describes the physical model of an elephant's trunk as well as the mathematical model of its kinematics in static conditions. The model is based upon anatomical and physiological studies of the trunk. Motion capabilities of the trunk and its muscular structure, constituting the power transmission system, have been taken into account.

INTRODUCTION

The technical biomechanics workgroup at ITLiMS PW is occupied with elastic manipulators and robots. This paper describes the kinematics of the elephant's trunk under static conditions. The created model may be used as an aid in the construction of a manipulator arm. A mathematical description is given for the physical model introduced in (6) and (7). The creation of the physical model has been preceded by extensive studies of anatomy and physiology of an elephant's trunk (1,2,6,7,). The aim of the paper is to describe a method of construction of an elephant's-trunk-type manipulator. The literature referring to the subject of snake-like or spine-like constructions (3,4,8,9,10,11) and elastic structures (5) has been investigated.

THE DESCRIPTION OF ANATOMICAL AND PHYSIOLOGICAL PROPERTIES OF THE ELEPHANT'S TRUNK

The description of anatomy will be limited to the motion system (muscular system) and the description of physiology will take into account only the ability to move of the elephant's trunk. The muscular system can be subdivided into two subsystems, namely: the internal subsystem consisting of inner muscles, fat and connective tissue, and the external subsystem comprising an external muscular layer. The trunk does not contain a skeleton, so the structure of movement is continuous. The internal subsystem functions as a "skeleton" and the external subsystem works as a drive mechanism. The internal subsystem functions as a cylindrical core. It is surrounded by a circular muscular layer of variable thickness. The internal subsystem becomes narrower along the trunk and the thickness of the muscular layer increases. The trunk, at its circumference has three callosities, located: one in

front and the other two at either side of the rear part of the trunk. Along the
trunk the callosities become stronger. The structure of the trunk's cross-section
along its length does not change. The fibres in the external muscular layer are
placed along the trunk and the fibres of the muscles in the internal subsystem
are located radially. In the lower end of the trunk the fibres of the rear
callosities, located in the external muscular layer, are placed askew downwards
towards the back of the trunk. The cross-section of the trunk is shown in figure
1, and its model in figure 2.

Certain coordinate systems and nomenclature are introduced to simplify the
physiological studies. The position of the trunk when it is pointing straight
down is considered to be its normal position (figure 3). The figure shows the
axis and planes of reference. The trunk may be subdivided into three parts: the
base, the stem and the tip.

The ability of the trunk to move is immense. Nevertheless, some schematics
of the trunk's motion can be introduced. The base of the trunk can move only in a
sagittal plane. Rotary motion or movement from side to side does not take place or
is negligible. The stem can move in a sagittal plane as well as in a frontal plane.
Here rotary motion is also not possible. The tip of the trunk can move in all
three planes. The motion in the sagittal plane is called the forward and backward
movement. The motion in frontal plane is called left and right bending. The motion
in transverse plane is called turning movement. This nomenclature pertains to
the normal position of the trunk. The possibilities for the trunk's motion are
shown in figure 4. The relation between muscular structure and the nervous system
has been described in (6).

PHYSICAL MODEL OF THE ELEPHANT'S TRUNK CONSIDERED AS AN ELASTIC STRUCTURE

The model is based upon anatomical and physiological studies of the trunk.
Capabilities of motion of the trunk and its muscular structure, comprising the power
transmission system, has been taken into account (6,7). The model (figure 5) consists
of concentrated masses connected with weightless links. Each link is considered to
have a different rigidity in two directions and torsional rigidity about the vertical
axis. The connections between the masses and the links are rigid. The model is
divided into three parts corresponding to the three parts of the elephant's trunk.
The rigidity of each link has been selected so as to represent precisely the
possibilities of the trunk's motion. In the base part of the trunk the rigidity
in the direction of the transverse axis and the torsional rigidity are much larger
than the rigidity in the direction of the saggital axis. Torsional rigidity is
much greater than in the other two rigidities of the stem. All three rigidities
of the tip are of the same order. The rigid connections between masses and links

assure a continuous structure of motion. The model is equipped with a power transmission system of a muscular type (with unilateral action). The power transmission consists of 43 actuators. The two chains located at both sides of the rear contain 21 actuators each. The last actuator is located in front, along the full length of the trunk. These locations correspond to the positions of the three callosities in the external muscular layer. In the tip of the trunk the actuators are positioned askew to the vertical axis. This enables the tip to perform rotary motion. The structure with such power transmission can execute 39 independent movements.

A MATHEMATICAL DESCRIPTION OF THE KINEMATICS OF THE ELEPHANT'S TRUNK

This description pertains to the elastic structure modelling the elephant's trunk, not to the trunk itself. The following assumptions have been made: the segments, corresponding to the elements of the structure, are described by a vector function $r_i = r_i(l_j) = [x_i(l_j), y_i(l_j), z_i(l_j)]$; each of its components being a function of class C^3 for their argument in the range $l_j(0, l_o)$; in transition points between segment i and i+1 the derivatives r_i', r_i'', r_i''' are bilaterally equal; each segment has a constant length. Moreover the range of motion has been restrained, namely: the angle between the normal planes in the starting and ending points of the segment can not be greater than 30° (6).

Several coordinate frames have been introduced to simplify the description (figure 6). The inertial coordinate frame $X_o Y_o Z_o O_o$, having its origin in O_o, is located so that axis Z_o is colinear with the axis of the model in its normal position; axis X_o is defined by a vector with its origin in O and its end at the point of attachment of the front actuator to the base A_o; axis Y_o is constructed in such a way that vectors X_o, Y_o, Z_o will form a dextrarotary orthogonal coordinate frame. Moreover, 21 movable coordinate frames have been fixed to the elements of the model. The origin of each coordinate frame coincides with the origin of each element 1,2,3,...21. The axis of each frame is defined by Frenet's trihedron located in such a way that normal versor is colinear with axis x_i, binormal versor is colinear with axis y_i, and tangential versor is colinear with axis z_i. The axis z_o and Z_o of frames $x_o y_o z_o$ and $X_o Y_o Z_o$ are colinear and the angle of rotation about Z_o between x_o and X_o or y_o and Y_o depends on the operation of the actuators.

Knowing the configurations of all elements in local coordinate frames the configuration of the model can be found. Because all the elements are similar, we will describe only one of them in a local coordinate frame. The 3rd Sarett - Frenet's equation describes the curvature of the element in a local coordinate frame:

$$\frac{d\vec{n}_i(l_j)}{dl_j} = \tau_i(l_j) \cdot \vec{b}_i(l_j) - |\mathcal{H}_i(l_j)| \cdot \vec{t}_i(l_j) \quad , \tag{1}$$

where:

t_i, n_i, b_i — are the vectors: tangential, normal and binormal to the curvature of the element at the point delimited by the length l_j.

τ_i — is the relative torsion at the point delimited by the length l_j.

$\mathcal{K}_i$ — is the relative curvature of the element at the point delimited by the length l_j.

l_j — is the independent variable — the length of the element measured along the curvature, $l_j \in (0, l_{oi})$

l_{oi} — is the total length of the element.

Now, $r_i(l_j)$ can be defined as a vector function, where $r_i(l_j)$ is the vector joining the origin of the local coordinate frame and the consecutive points on the curve (figure 6).

$$\vec{t}_i(l_j) = \frac{d\vec{r}_i(l_j)}{dl_j} \tag{2}$$

$$\vec{n}_i(l_j) = \frac{d^2\vec{r}_i(l_j)}{dl_j^2} \Big/ \frac{dl_j^2}{d^2\vec{r}_i(l_j)} \tag{3}$$

$$\vec{b}_i(l_j) = \vec{t}_i(l_j) \times \vec{n}_i(l_j) \tag{4}$$

where $\times$ denotes the vector cross product.

The curvature and torsion can be derived from (1):

$$\tau_i(l_j) = \frac{(\vec{M}_{si}(l_j))dl_j}{G_i \cdot J_{oi}} \tag{5}$$

$$|\mathcal{K}_i(l_j)| = \frac{|\vec{M}_{qi}(l_j)|}{E_i J_i} \tag{6}$$

where:

J_{oi} — axial moment of inertia of the cross section of the elephant's trunk in segment i.

J_i — moment of inertia of the cross-section in the direction of axis x_i or y_i (axial symmetry of the cross-section is assumed $J_{xi} = J_{yi} = J_i$. Moreover J_i is considered constant along the length of the element).

G_i — modulus of rigidity (coefficient of transverse elasticity).

E_i — longitudinal modulus of elasticity (Young's modulus).

(E_i and G_i are assumed constant along the length of the trunk).

$\vec{M}_{si}(l_j)$ - torque moment acting on an element at the point delimited by the length l_j.

The value of the function $\vec{M}_{si}(l_j)$ describes the distribution of torque moment along the elements of the model. The direction and sense of the vector are determined by the tangential versor at the point l_j.

$\vec{M}_{gi}(l_j)$ - bending moment in two planes acting on an element at the point delimited by the length l_j. The value of the function $\vec{M}_{gi}(l_j)$ determines the distribution of bending moment in space along the model's element. The direction and sense of the vector are determined by the sum of normal and binormal vectors.

Substituting (2,3,4,5,6) into (1) we obtain a vector differential equation describing the vector function $r_i(l_j)$.

$$\frac{d}{dl_j} \frac{\vec{r}_i''(l_j)}{|\vec{r}_i''(l_j)|} = \frac{\vec{M}_{si}(l_j) dl_j}{G_i J_{oi}} \vec{r}'(l_j) \times \frac{\vec{r}''(l_j)}{\vec{r}''(l_j)} - \frac{\vec{M}_{gi}(l_j)}{E_i J_i} \vec{r}'(l_j) \quad (7)$$

Now, the torque and bending moment will be determined from simple relations derived from statics (figure 6). The resultant moment about point j can be denoted as:

$$\vec{M}_{wi} = \vec{S}_{A1}(\vec{a}+\vec{ij}) + \vec{S}_{B1} \times (\vec{b} + \vec{ij}) + \vec{S}_{c1} \times (\vec{c} + \vec{ij}) + \\ +\vec{M}_{Ri} + (m_i\vec{g} + \vec{R}_i) \times \vec{ij} \quad (8)$$

where:

$\vec{S}_A, \vec{S}_B, \vec{S}_C$ - are the muscular forces in the front and both actuators located at the rear.

a,b,c - the distance of the points of application of forces S from the point i+1.

m_i - the mass of element i.

$\vec{R}_i$ - the force of reaction at the point i+1 caused by the elements with a number greater than i.

$\vec{M}_{Ri}$ - the moment of reaction at the point i+1 caused by the elements with a number greater than i+1.

$\vec{ij}$ - vector $\vec{ij} = \vec{r}_i(l_i) - \vec{r}_i(l_j)$.

The resultant moment M_{wi}, expressed in a mobile coordinate frame x_i, y_i, z_i, can be resolved into components expressed in a coordinate frame fixed to the Frenet's trihedron constructed at point "j".

$$M_{wi} = \left[M_{gx_i}, M_{gy_i}, M_{gs_i} \right] \quad (9)$$

where:

M_{gx_i}, M_{gy_i} — are the bending moments in two planes at the point j of element i.

M_{s_i} — torque moment at the point j of element i.

$$M_{g_i} = M_{gx_i} + M_{gy_i} \tag{10}$$

Using equation (7), the following problem can be solved: if the configuration of consecutive elements of the model is given as

$$\vec{r}_i(1_j) = \left[x_i(1_j), y_i(1_j), z_j(1_j) \right] ;$$
$$i = 0, \ldots, 20; \tag{11}$$
$$j = 0, 1, \ldots, n;$$

where:

    i — the number of elements of the model

    j — the number of points at which the element's configuration has been determined in discrete form,

the forces S (which have to be exerted by the actuators, so that this configuration can be obtained) can be determined.

The inverse problem can not be solved using this description, because the boundary conditions are not defined.

CONCLUSIONS

A mathematical description of the elephant's trunk, considered as an elastic structure, has been given. It can be seen, from the analytical description, that the inverse problem can not be solved, if it is stated by the above method, because the boundary conditions are not defined. As a continuation of this work, the verification of the assumptions and the method of description will be performed.

The following conclusions can be derived from the work we have been doing. The continuous mass must be introduced to the mathematical description of the model's kinematics and statics. Moreover global coordinate frames must be employed to determine the boundary conditions. After verification the model will be simulated on a computer.

REFERENCES

1. Boas, J.E.V., S. Pauli: The elephant head, part 2. Published at the cost of the Calsberg-fund., Copenhagen 1925.
2. Чеботарёв И.Т. - Взаимоотношение венозных образований и мышц в хоботе слона. Труды Московской Ветеринарной Академии. Том 27. Москва,1963.

3. Hirose, S., K. Ikuna, Y. Umetani: A new design method of servo-actuators based on the shape memory effect. Tokyo Institute of Technology, Tokyo 1984.
4. Hirose, S., T. Umetani: An active cord mechanism with oblique sivilvel joints and its control. Proc. of the 4th RoManSy-81, Zaborow, Poland 1981, pp. 327-340.
5. Hemani, A.: Studies on a high Weight and Flexible Robot Manipulator. "Robotics" No 1. Novoth-Holland 1985.
6. Malczyk, G.: Model i propozycja rozwiazania manipulatora typu traba slonia. Praca magisterska, Warszawa 1984.
7. Malczuk G., Morecki, A.: Elastyczny manipulator typu traba slonia. Praca Naukowa Instytutu Cybernetyki Technicznej Politechniki Wroclawskiej. I Krajowa Konferencja Robotyki, Tom 2. Wroclaw 1985.
8. Prospekt - Spine-Robotics, Sweden 1983.
9. Prospekt - Spine News, Sweden 1983.
10. Prospekt - Spine News, Sweden 1984.
11. Roth, B., J. Rastegar, V. Scheinman: On design of computer controlled manipulators. Proc. 1st CISM-IFToMM "RoManSy" Udine, Italy, Sept 5-8, 1973, Springer Verlag 1974.
12. Sokolowska-Pietuchowa, J.: Anatomia czlowieka. PZWL, W-Wa, 1983.

A Mathematical Model of a Flexible Manipulator 205

Fig.1 Cross-section of elephant
trunk at a level of the base

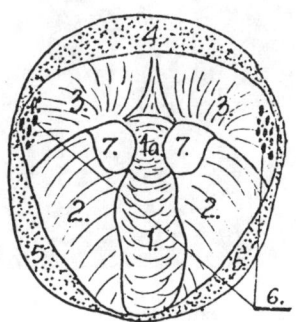

Fig.2 Model of the cross-section
1-5 conventional regions
6-nerves, 7-snout tubes

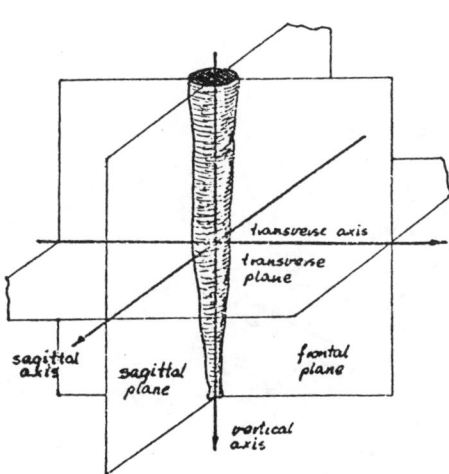

Fig.3 Normal position of elephant
trunk and reference coordinates

Fig.4 Some examples of the elephant trunk movements

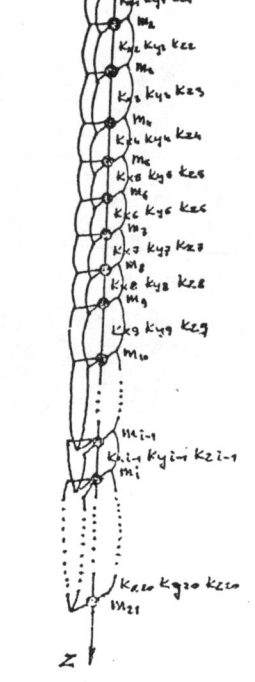

Fig.5 Continous model of elephant trunk movements

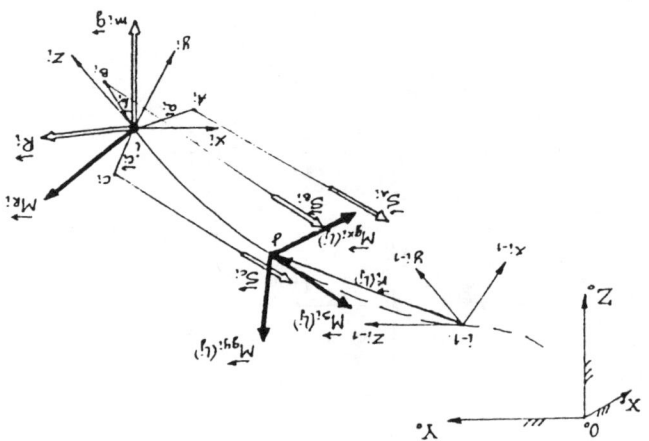

Fig.6 Statical model of the basic element

Analytical Design of Two-Revolute Open Chains

B. Roth

Department of Mechanical Engineering, Stanford University, Stanford, U.S.A.

ABSTRACT This paper deals with determining the dimensions of open chains composed of two moving links connected in series by means of revolute joints. Many manipulators, mechanical hands, walking machine legs and other mechanical devices contain this basic configuration. Equations are presented which can be used to numerically determine linkage dimensions according to specific design requirements. The design specifications treated by these methods can be on position, orientation, error sensitivity, velocity, acceleration, and force transmission characteristics. A numerical example is presented to illustrate the basic theory.

INTRODUCTION The two-revolute open chain shown in figure 1 is a basic component of most manipulators and other robotic type mechanical systems. It is also quite commonly used in closed-loop mechanisms with one or more degrees-of-freedom. This chain can be used as a complete system, as in some pick and place devices and mechanical fingers and legs, or it can be part of a more complex device. In this paper we develop a set of design equations to determine the dimensions of two-revolute open chains in accordance with a predefined set of design constraints on their kinematic and geometric capabilities.

It is well known that as the joint angles θ_1 and θ_2 vary, any point fixed to the outermost moving link, such as point p in figure 1, moves on a cubic surface known as a torus. From the geometry of the figure it follows that the parametric equations of the locus of point p are

$$x = a_2 \cos\theta_1 \cos\theta_2 - a_2 \cos\alpha \sin\theta_1 \sin\theta_2 + s_2 \sin\alpha \sin\theta_1 + a_1 \cos\theta_1$$
$$y = a_2 \sin\theta_1 \cos\theta_2 + a_2 \cos\alpha \cos\theta_1 \sin\theta_2 - s_2 \sin\alpha \cos\theta_1 + a_1 \sin\theta_1 \quad (1)$$
$$z = a_2 \sin\alpha \sin\theta_2 + s_2 \cos\alpha$$

Where, the coordinate system is taken so that z is along axis 1, and the origin, O, is at the point where the normal from axis 2 intersects axis 1. The sense of z can be chosen arbitrary. The x and y axes are any two orthogonal lines in the plane, normal to axis 1, through O. The distance between axis 1 and 2 is the positive quantity a_1, and the angle between them, α, is positive or negative according to the right-hand rule, in going from axis 1 to axis 2, for some arbitrarily assigned positive sense along axis 2. a_2 is the perpendicular distance from point p to axis 2, it is always positive. s_2, the offset, is taken as positive or negative depending upon if it is along the positive or negative sense of axis 2 in measuring from a_1 toward a_2. The angle θ_1 is the rotation of the first moving link about axis 1, it is measured from the x-axis to the link-line a_1 according to the right-hand rule with the thumb along the positive z-axis. The rotation of the second moving link about axis 2 is θ_2, which is measured from the extended common normal (obtained by extending line a_1 beyond axis 2) to the direction of line a_2 according to the right-hand rule with the thumb pointing along the assumed positive sense of axis 2.

Eliminating the angles θ_1 and θ_2 from equation (1) (by solving the last equation for $\sin\theta_2$, and substituting this into the sum of the squares of the first two equations) and rationalizing the resulting equation yields the normal form of the implicit equation in terms of the principal coordinate system:

$$[(x^2 + y^2 + z^2) - (a_2^2 + a_1^2 + s_2^2)]^2 - 4a_1^2[a_2^2 - (z - s_2\cos\alpha)^2/\sin^2\alpha] = 0 \quad (2)$$

This equation has been used by Fichter and Hunt [1] to classify the several different forms of the torus and to derive some of their properties. For our purposes it is important to notice that equation (2) depends on structural parameters: a_1, a_2, s_2 and α.

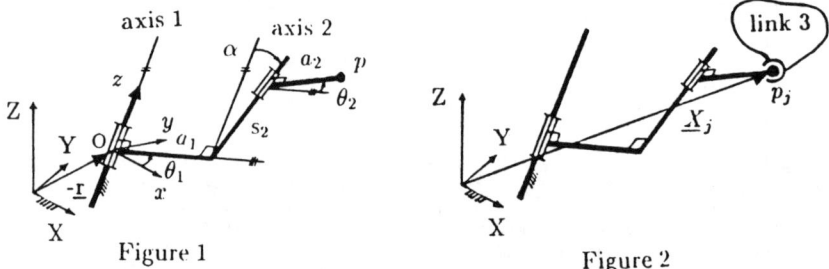

Figure 1 Figure 2

DESIGN EQUATIONS In this paper, we rule out several special cases which simplify matters to such an extent that the design can best be accomplished by much simpler means: (1) We assume $a_2 \neq 0$, i.e., p is not coincident with axis 2, since then its locus would be a circle. (2) We assume $\alpha \neq 0$, i.e., axes 1 and 2 are not parallel, since then p would generate a planar rather than a toroidal surface. (3) We assume $a_1 \neq 0$, i.e., the joint axes do not intersect, since then p lies on the surface of a sphere. If these three restrictions are met equation (2) will not degenerate and can be rewritten in the form

$$[x^2 + y^2 + z^2 - A]^2 + [Cz + D]^2 + B = 0 \quad (3)$$

where:

$$A = a_1^2 + a_2^2 + s_2^2, \quad B = -4a_1^2 a_2^2, \quad C = \frac{2a_1}{\sin\alpha}, \quad D = -2a_1 s_2 \frac{\cos\alpha}{\sin\alpha} \quad (4)$$

Because of our restrictions A, B and C cannot be zero. D=0 is allowed if it occurs due to $s_2=0$ or $\cos\alpha = 0$ (i.e., $\alpha = \pm 90$) or both of these conditions simultaneously.

In a previous paper [2], it was shown that the equation of the torus written in terms of an arbitrary coordinate system can be obtained using

$$(x,y,z)^t = R(X,Y,Z)^t + \underline{d} \quad (5)$$

where, X,Y,Z are the coordinates of p, measured in an arbitrarily chosen coordinate system. R is the 3x3 rotation matrix that takes the directions of the arbitrary system into the principal directions used in deriving (3). $\underline{d}$ is the position vector of the origin of the X,Y,Z system as measured in the principal x,y,z system. We will be interested primarily in the z-component of $\underline{d}$, we label it d_z. Furthermore we will need to know the vector $\underline{d}$ in terms of its measures in the X,Y,Z system; this is given by the vector $\underline{r}$, where $\underline{r}=R^t\underline{d}$. The components of $\underline{r}$ are r_1, r_2, r_3. The superscript t denotes the transpose.

Substituting (5) into (3) yields, after simplification,

$$[X^2 + Y^2 + Z^2 + 2(r_1 X + r_2 Y + r_3 Z) + r^2 - A]^2 + [(n_1 X + n_2 Y + n_3 Z + d_z)C + D]^2 + B = 0 \quad (6)$$

Analytical Design of Two-Revolute Open Chains 209

Here, n_1, n_2, n_3 are the elements of the third row of the matrix R, and $r^2 = \underline{r} \cdot \underline{r}$.
In [2] it was pointed out that (6) depends on twelve independent parameters.
These are $X, Y, Z, N_1, N_2, N_3, r_1, r_2, r_3, A, B$ and D, where $N_1 = Cn_1$, $N_2 = Cn_2$ and $N_3 = Cn_3$.
This conclusion makes use of the following facts: $n_1 r_1 + n_2 r_2 + n_3 r_3 = d_z$,
$r_1^2 + r_2^2 + r_3^2 = r^2$, and $n_1^2 + n_2^2 + n_3^2 = 1$. Our interest here is to use equation (6) to design
linkage chains of the type shown in figure 1. In order to do this we need to be
sure that equation (6) is satisfied at each position of point p that we are
designing for. Hence, we need equation (6) with X, Y, Z replaced by X_j, Y_j, Z_j,
which are the coordinates of point p in design position j. Making this
substitution and writing (6) in vector form yields

$$[\underline{X}_j \cdot \underline{X}_j + 2\underline{r} \cdot \underline{X}_j + A^*]^2 + [\underline{N} \cdot \underline{X}_j + D^*]^2 + B = 0 \quad j = 1, 2, \ldots, m \quad (7)$$

where $A^* = r^2 - A$, $D^* = D + d_z C$, $\underline{N}$ is the vector with components N_1, N_2, N_3, and $\underline{X}_j$ is the
vector with components X_j, Y_j, Z_j in the general coordinate system.
If we subtract the first equation of (7) from all the rest we obtain

$$\begin{aligned}
(\underline{X}_j^2 + \underline{X}_1^2) \cdot (\underline{X}_j^2 - \underline{X}_1^2) &+ 4\underline{r} \cdot (\underline{X}_j - \underline{X}_1) \, \underline{r} \cdot (\underline{X}_j + \underline{X}_1) \\
&+ 4[\underline{X}_j^2 (\underline{r} \cdot \underline{X}_j) - \underline{X}_1^2 (\underline{r} \cdot \underline{X}_1)] + \underline{N} \cdot (\underline{X}_j - \underline{X}_1) \, \underline{N} \cdot (\underline{X}_j + \underline{X}_1) \\
&+ 2A^*[\underline{X}_j^2 - \underline{X}_1^2 + 2\underline{r} \cdot (\underline{X}_j - \underline{X}_1)] + 2D^*\underline{N} \cdot (\underline{X}_j - \underline{X}_1) = 0 \quad j=2,3,\ldots,m
\end{aligned} \quad (8)$$

These are the equations we will use to design the open loop chain to carry some
as yet undetermined point p through m arbitrarily specified positions. If we a
priori specify the point p, then $m \leq 9$ in equation (8).

APPLICATION Figure 2 is helpful in visualizing one important way to use
equation (8). We have repeated the structure shown in figure 1 and added an
additional link, link 3, which is attached to the end of the previous chain by a
ball-and-socket joint centered at point p. This new structure represents many
important practical devices and situations.

The introduction of link 3 allows the designer to specify twelve design
positions of link 3 (which contains point p) while leaving point p unspecified.
In order to specify the twelve positions of link 3, the designer can provide
eleven 3x3 rigid body rotation matrices U_{j1} and eleven displacement vectors $\underline{t}_{j1}$
such that any point, say q, in link 3 will have its position vectors in position
1, say $\underline{Q}_1$, and position j, $\underline{Q}_j$, related by the rigid body transformation
equation: $\underline{Q}_j = U_{j1}\underline{Q}_1 + \underline{t}_{j1}$ where $j = 2, 3, \ldots, 12$. Here, the displacement of link 3 is
measured relative to the arbitrary X, Y, Z system, and therefore $\underline{Q}_j$, $\underline{Q}_1$ and $\underline{t}_{j1}$
are all vectors in the X, Y, Z system. Now instead of using an arbitrary point in
link 3 we use the point p, and we have

$$\underline{X}_j = U_{j1}\underline{X}_1 + \underline{t}_{j1} \quad j = 2, 3, \ldots, 12 \quad (9)$$

The vector $\underline{X}_j$ can thus be expressed as a linear function of $\underline{X}_1$ even though $\underline{X}_j$
and $\underline{X}_1$ are as yet unknown.
If we substitute (9) into (8) we obtain after some algebra

$$\begin{aligned}
&(2\underline{X}_1^2 + 2\underline{t}_{j1}^t U_{j1}\underline{X}_1 + \underline{t}_{j1}^2)(2\underline{t}_{j1}^t U_{j1}\underline{X}_1 + \underline{t}_{j1}^2) + 4\underline{r}^t[(U_{j1}-I)\underline{X}_1 + \underline{t}_{j1}]\underline{r}^t[(U_{j1}+I)\underline{X}_1 + \underline{t}_{j1}] \\
&+4[(\underline{X}_1^2 + 2\underline{t}_{j1}^t U_{j1}\underline{X}_1 + \underline{t}_{j1}^2)\underline{r}^t(U_{j1}\underline{X}_1 + \underline{t}_{j1}) - \underline{X}_1^2 \underline{r}^t \underline{X}_1] + \underline{N}^t[(U_{j1}-I)\underline{X}_1 + \underline{t}_{j1}]\underline{N}^t[(U_{j1}+I)\underline{X}_1 + \underline{t}_{j1}] \\
&+ 2A^*\{2\underline{t}_{j1}^t U_{j1}\underline{X}_1 + \underline{t}_{j1}^2 + 2\underline{r}^t[(U_{j1}-I)\underline{X}_1 + \underline{t}_{j1}]\} + 2D^*\underline{N}^t[(U_{j1}-I)\underline{X}_1 + \underline{t}_{j1}] = 0 \quad j = 2, \ldots, 12
\end{aligned} \quad (10)$$

In this form all products are matrix multiplications, and the superscript t indicates the vector is transposed. In (10), I represents the 3x3 identify matrix, $\underline{X}_1^2 = \underline{X}_1^t \underline{X}_1$ and $\underline{t}_{j1} = \underline{t}_{j1}^t \underline{t}_{j1}$.

Obviously each equation in (10) is a quartic (it is cubic in $\underline{X}_1$, quadratic in $\underline{r}$ and $\underline{N}$, and linear in A* and D*). Because of the nonlinearity of the system we would expect a large number of possible solutions, and this turns out to indeed be the case. We find the solutions of equation (10) in the following manner. For a specified set of U_{j1} and $\underline{t}_{j1}$ (j=2,3,...,12) we choose an initial guess for $\underline{X}_1$, $\underline{r}$, $\underline{N}$, A* and D*. We then apply the Newton-Raphson iteration method to equation (10). When the iteration converges we use a new initial guess and seek to converge to a different set of values for $\underline{X}_1$, $\underline{r}$, $\underline{N}$, A* and D*. This procedure can be repeated until a number of different solutions to (10) are obtained. Each such solution represents either four distinct designs (of which two are closely related) or no real designs. In practice we have found that this iteration converges fairly rapidly and that it is relatively easy to obtain a variety of different solutions for a given set of U_{j1} and $\underline{t}_{j1}$. Once an iteration converges the linkage parameters can be obtained using the results of the following analysis.

We obtain A from $A = r^2 - A*$, which follows from the definition of A*. If we know $\underline{N}$ then C follows from $C = \sqrt{\underline{N} \cdot \underline{N}}$. n_i (i=1,2,3) then follow from $n_i = N_i/C$. Knowing n_i and $\underline{r}$ allow us to calculate d_z from $d_z = n_1 r_1 + n_2 r_2 + n_3 r_3$. We now can calculate D from the definition of D*, viz. $D = D* - d_z C$. Finally, B follows from equation (6), since we can set $\underline{X}_1 = (X,Y,Z)^t$ and we know all the other quantities except B. Thus we have shown that A,B,C,D and n_i (i=1,2,3) follow uniquely from a solution to (10). (It is noted in passing that if we chose instead the negative square root for C, this would simply change the signs of n_i and d_z and the result would be exactly the same physical system as with the positive root). The values of n_i and $\underline{r}$ uniquely determine the location of axis 1 and the origin of the principal coordinate system relative to the general coordinate system. What remains to be determined are the actual linkage dimensions.

Since we now know A,B,C and D we can use these four quantities to determine the linkage parameters a_1, a_2, s_2 and α. This requires that we invert the set of equations (4). If we let $\sin^2 \alpha = u$ it follows that

$$a_1 = C\sqrt{u}/2, \quad a_2 = \sqrt{-B/(uC^2)}, \quad s_2^2 = D^2/[C^2(1-u)] \tag{11}$$

Substituting (11) into $A = a_1^2 + a_2^2 + s_2^2$ yields after some algebra

$$(C^2/4)u^3 - [A+(C^2/4)]u^2 + [A-(B/C^2)-(D^2/C^2)]u + B/C^2 = 0 \tag{12}$$

Solving (12) for u allows us to determine the lengths from (11) and the twist angle from

$$\alpha = \pm \sin^{-1}\sqrt{u} \tag{13}$$

If we use values from (13) such that $-90^0 < \alpha < 90^0$, it follows from (4) that the sign of s_2 must always be opposite the sign of D when $C > 0$.

Since (12) is a cubic it will always have at least one real root. However we require that the roots be not only real but that they be equal to or less than 1, i.e., u<1. It will turn out, as we will see shortly, that (12) will yield either two valid roots or none. For each valid root of (12) we have two structural variants differing only in the sign of the twist angle, α, as given by the plus and minus signs in (13).

If we write (12) in terms of the original structural parameters, the cubic can be factored as follows

$$(u-\sin^2\alpha)[(a_1^2/\sin^2\alpha)u^2 - (a_2^2+ s_2^2+ a_1^2/\sin^2\alpha)u + a_2^2] = 0 \qquad (14)$$

This means that if we have one solution so that $u=\sin^2\alpha$ and the lengths are a_1, a_2 and s_2, the other solutions (if they exist) must be given by the roots of the quadratic factor in square brackets, i.e.,

$$u = [(a_2^2+ s_2^2)\sin^2\alpha + a_1^2]/2a_1^2 \pm \sqrt{[\{(a_2^2+ s_2^2)\sin^2\alpha + a_1^2\}/2a_1^2]^2 - a_2^2\sin^2\alpha/a_1^2} \qquad (15)$$

It can be shown that using the plus value for the radical will yield $u>1$ and using the minus sign will yield $u<1$. In some special cases both roots will coincide at $u=1$. The conclusion then is that whenever our cubic (12) yields one good solution there will always be a second valid solution, and the value of the root for that solution will be exactly determinable in terms of the first solution by using the minus radical of (15). It therefore also follows that the second link lengths, denoted here by a_1', a_2' and s_2', can be given in terms of the first set as

$$a_1'=a_1\sqrt{u}/\sin\alpha, \qquad a_2'=a_2\sin\alpha/\sqrt{u}, \qquad s_2'=s_2\cos\alpha/\sqrt{1-u} \qquad (16)$$

Thus any time a numerical solution gives one valid mechanism it will automatically determine another solution as indicated by equations (15) and (16). In addition each of these variants will work if the negative of their twist angle, α, is used. So we have four linkages in all (perhaps more properly described as two pairs of two solutions) every time the iteration of equation (10) converges to a set of values which leads to a real mechanism. The four-fold generation implied by equations (15) and (16) has previously been discovered by Fichter and Hunt [1] who arrived at these same equations by using a purely geometric argument regarding the bitangent planes of a torus.

NUMERICAL EXAMPLE In order to illustrate the preceding ideas a numerical example is presented: Consider the problem of designing a five-degree-of-freedom manipulator which is required to move an object through the twelve spatial positions shown in figure 3 and tabulated in Table 1. The table gives the vectors $\underline{t}_{j1}$, which are in fact the jth design position coordinates of the point q, in the moving system, which in position 1 coincides with the origin of the (fixed) general coordinate system. The elements for the rigid body rotation matrices U_{j1} are obtained in the standard manner [3] from the specified rotation angles (in degrees) listed in the "rotation" column and the corresponding rotation axes (direction-numbers) listed in the "rotation axis" column. It should be noted that $U_{j1}=I$ for $j=2,3,4,5,6$, since, for this example, the first six design positions are parallel.

TABLE 1. DESIGN POSITIONS

j	t_{j1}	rotation $1 \rightarrow j$	rotation axis $1 \rightarrow j$
2	1.0, 0.0, 0.0	0	—
3	2.0, 0.5, 0.0	0	—
4	3.0, 0.0, 0.0	0	—
5	4.0, 0.5, 0.0	0	—
6	5.0, 0.0, 0.0	0	—
7	4.5, 0.5, 0.0	60	0.0, 0.5, 1
8	4.0, 1.5, 0.5	70	0.0, 1.0, 1
9	3.5, 1.8, 0.0	80	0.5, 1.0, 1
10	2.5, 1.9,-0.5	90	1.0, 1.0, 1
11	2.0, 1.4, 0.0	100	0.0, 1.0, 0
12	1.5, 1.2,-0.5	110	0.0, 1.5, 1

Figure 3

If in figure 2 we replace the ball-in-socket joint and link 3 with a series chain of three links connected by revolute joints having consecutively perpendicular axes which all intersect at point p, we obtain the five-degree-of-freedom manipulator shown in figure 4. If an object is held firmly in the gripper, point q, the reference point in the object, can also be considered a point fixed to link 5. Since q and every other point in link 5 remains at a fixed distance from p, the distances between point p and all the points in link 5 (and the grasped object) are exactly the same as if the points were in link 3 in figure 2. The orientation of the grasped object can be made to coincide with that of link 3 in figure 2, by appropriate rotations in joints 3,4,5 (in figure 4).

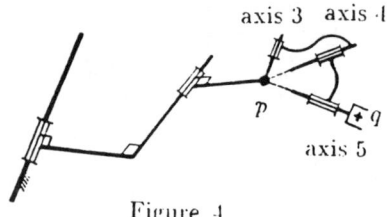

Figure 4

If we choose point q to be along the 5th revolute axis (which often coincides with the center-line of the gripper), once points q and p are in one of the design positions all that remains in order to obtain the specified orientation is to rotate the object about line pq, which is axis 5. So if p is in a design position, rotations about axes 3 and 4 can always bring q to its corresponding position, and then a rotation about axis 5 brings the object into its design position. Hence the design requires that we determine the first two links in accordance with the specifications of Table 1. Once p is known, the length pq and the object size determines the size and location of the gripper. The only restrictions on axes 3,4,5 is that they pass through point p and be sequentially perpendicular.

Using the values from Table 1, 24 different solutions have been obtained for the eleven unknowns in (10). These values are given in Table 2. Finally, in Table 3 we give the linkage parameters corresponding to these 24 solutions. We see that 18 of these solutions yield real linkages. For each of these 18 solutions we list the four different linkages (counting the + and − α variants as separate linkages). All four of these have the same axis 1, which has direction numbers N and passes through the point given by r (these vectors are listed in Table 2.) The designer is free to choose any one of these 72 linkages since they all satisfy the requirements of Table 1. If none of these linkages are suitable (because of design constraints which have not been included in Table 1), the designer must seek additional roots of (10).

Analytical Design of Two-Revolute Open Chains 213

TABLE 2. SOLUTIONS OF EQUATION (10)

No.	X_j			$\underline{t}$			$\underline{N}$			A^*	D^*
1	2.4,	-1.6,	1.4	-4.7,	4.4,	3.8	-0.1,	0.3,	9.4	16.2	10.1
2	12.3,	11.7,	11.1	-14.6,	-12.0,	0.6	17.4,	0.5,	-7.3	63.2	-177.6
3	1.8,	8.2,	9.7	-4.0,	-8.4,	-5.2	-10.4,	-0.9,	-3.6	35.8	83.4
4	-3.2,	3.5,	0.3	1.0,	-0.7,	-2.7	0.1,	-0.5,	6.4	-8.5	-14.9
5	4.4,	5.4,	6.5	-6.7,	-2.7,	0.8	-0.0,	0.3,	1.5	-12.8	15.4
6	5.9,	7.7,	7.6	-8.2,	-7.9,	-1.3	-7.3,	-1.2,	1.8	59.9	55.6
7	-1.9,	-16.1,	1.6	-0.3,	15.9,	11.0	60.3,	0.1,	32.6	-1609.3	-72.2
8	1.1,	1.9,	0.3	-3.3,	-1.7,	-2.4	-2.5,	-3.2,	2.2	8.6	14.0
9	-0.3,	7.3,	-0.6	-2.0,	-4.4,	-2.2	-0.2,	1.2,	-1.3	6.4	-3.2
10	0.4,	1.5,	1.3	-2.7,	-1.6,	0.1	4.1,	2.0,	3.8	-4.5	-19.5
11	-1.4,	1.9,	-0.1	-0.9,	0.9,	-0.7	0.2,	-0.9,	7.3	-10.1	-5.4
12	-1.9,	2.0,	-2.1	-0.4,	-1.9,	0.5	2.5,	3.1,	0.7	-5.5	-6.4
13	-1.7,	13.4,	1.0	-0.5,-10.6,	-7.1	0.1,	-0.3,	2.5	115.0	-17.6
14	-3.3,	-0.3,	0.6	1.1,	0.2,	-0.3	-5.0,	-1.7,	4.5	-14.9	-8.1
15	4.3,	6.0,	7.0	-6.6,	-3.2,	0.4	0.1,	-0.5,	-2.1	-11.5	-3.6
16	0.1,	-0.4,	-1.4	-2.4,	0.6,	0.1	2.4,	3.3,	-3.3	-2.3	-9.5
17	1.9,	-4.3,	-5.2	-4.1,	4.0,	10.6	-10.9,	-0.8,	-13.2	52.3	-26.6
18	-0.0,	1.3,	-1.0	-2.2,	1.5,	-0.8	-0.3,	1.3,	-8.7	-7.2	-4.7
19	-1.2,	1.8,	-1.4	-1.0,	-1.6,	0.2	2.2,	3.4,	0.5	-3.9	-8.3
20	-3.1,	5.8,	0.1	0.9,	-3.1,	-2.3	-0.1,	0.5,	-4.5	-0.8	10.9
21	0.6,	0.8,	-0.6	-2.8,	2.0,	-5.4	0.2,	-0.8,	32.0	-6.3	10.9
22	-2.5,	14.1,	-0.0	0.2,-11.3,	-7.3	0.1,	-0.3,	3.0	116.4	-20.3
23	1.5,	4.1,	-3.5	-3.7,	-1.3,	-3.0	0.1,	-0.6,	1.6	-29.1	-2.9
24	-3.3,	2.4,	-2.6	1.0,	0.4,	-0.3	0.1,	-0.4,	5.3	-18.3	-3.3

TABLE 3. LINKAGE PARAMETERS

a_1	a_2	a_2	α	s_1	s_2	s_2	α
3.2	3.6	4.0	±42.4	2.6	4.6	3.5	±32.6
8.7	8.9	-11.8	±67.1	5.0	15.5	-5.4	±32.0
5.2	5.5	-4.5	±71.4	3.7	7.8	-1.9	±42.4
3.1	2.4	-1.3	±76.0	2.1	3.6	-0.4	±41.5
No linkages							
3.7	4.1	6.4	±76.3	1.9	8.0	1.7	±29.7
30.2	30.2	12.7	±61.6	26.2	34.8	9.4	±49.6
1.9	1.6	-2.2	±56.3	1.1	2.8	-1.4	±28.6
0.8	3.9	2.0	±68.1	0.8	4.1	1.6	±61.8
2.3	2.6	1.4	±51.0	2.1	2.8	1.3	±45.3
3.3	1.2	-0.2	±62.9	1.2	3.3	-0.1	±19.6
2.0	1.6	-1.7	±88.3	1.1	2.9	-0.1	±33.1
No linkages							
3.0	2.6	0.3	±60.6	2.6	3.1	0.2	±46.7
No linkages							
1.9	1.6	1.5	±45.0	1.4	2.2	1.2	±31.1
5.5	5.8	-5.4	±39.7	4.6	6.9	-4.9	±32.6
3.0	0.9	2.3	±43.4	0.8	3.4	1.7	±10.9
2.1	1.4	1.2	±82.9	1.2	2.5	0.2	±34.1
No linkages							
3.2	1.4	-5.9	±11.7	1.3	3.5	-5.8	±4.6
No linkages							
0.8	6.7	-2.9	±71.9	0.8	6.9	-2.2	±65.6
No linkages							

EXTENSIONS Although this paper has considered only specifications on positions and orientations of an object, essentially the same equation can be used in many other ways. Firstly, it is possible to take derivatives of equation (7) and then specify required velocity constraints on positions and orientations. By using both displacement and velocity equations, the designer can specify design positions and the velocities at those same positions. The specification on velocities can also be used to indirectly impose design requirements on force transmission and error sensitivity. This can be done because, respectively, (1) force is inversely proportional to velocity, assuming no frictional losses, (2) the velocity at the free-end (relative to the input angular velocities) is a measure of the error sensitivity in that design position. It is also possible to take higher derivatives of equation (7) at one or more design positions. By specifying both velocity and acceleration the designer obtains a priori control of the inertia loads in the design positions. In general all techniques for mixed position and derivative specifications that have been developed for other types of joints [4] can be applied to (7).

It is also possible to use (10) (or (7)) to design other types of linkages than the one illustrated in the foregoing example. Any linkage which uses the basic chain shown in figure 1 can be treated with the equations developed in this paper. For a 2R manipulator we would generally specify point p. Hence the $\underline{X_j}$ would be known and equation (10) would be used for a nine position design. 6R manipulators can be easily designed by using the Z coordinate axis as a new first axis for the system shown in figure 3, and exploiting the fact that the Z axis then becomes an axis of revolution for the entire 5R chain shown in figure 3. For linkages where $\underline{t}_{j1}$ and U_{j1} (or their equivalents) cannot be completely specified, the methods described in [4] can be applied to deal with the so-called incompletely specified positions.

In all cases the designer is free to specify less than the maximum number of design positions. If in (10) j=2,3,...,m (m<12), the designer is free to arbitrarily fix 12-m of the linkage's location and structural parameters. By reducing the number of design positions it becomes possible to use these design equations with other constraint equations, such as the ones which deal with the maximum reach of a manipulator [5] or its working volume [6]. Alternatively it is possible to specify specific configuratonal constraints. For example if we want $\alpha = 90°$, we take m=11 and use the condition $D^* = N \cdot r$ to eliminate D^* from (10). This would give us the same special configurations studied by Tsai and Soni [6]. In all cases we can have more than 12 design positions by using incompletely specified design positions [4], or by giving up the requirement for exact precision and determining approximate solutions to the overconstrained set of equations one obtains by setting j>12 in (10).

ACKNOWLEDGEMENT The financial support of the Systems Development Foundation and the National Science Foundation is gratefully acknowledged. The computer programming was done by Madhusudan Raghavan who also drew the figures. Donalda Speight also provided some programming assistance.

REFERENCES
[1] Fichter, E.F. and Hunt, K.H., "The Fecund Torus, and its Bitangent-Circles and Derivative Linkages," Mechanism and Machine Theory, Vol. 10, 1975, pp.167-176.
[2] Roth, B., "Analytic Design of Open Chains," Proceedings of the Third International Symposium on Robotics Research, MIT Press, 1986.
[3] Bottema, O. and Roth, B., Theoretical Kinematics, North Holland, Amsterdam, 1979.
[4] Tsai, L-W. and Roth, B., "Incompletely Specified Displacements: Geometry and Spatial Linkage Synthesis," Journal of Engineering for Industry, Trans. ASME, Vol. 95, 1973, pp. 603-611.
[5] Shimano, B. and Roth, B., "Dimensional Synthesis of Manipulators," Proceedings Third CISM-IFToMM Symposium on Theory and Practice of Robots and Manipulators, Elsevier, Holland, 1979.
[6] Tsai, Y.C. and Soni, A.H., "Workspace Synthesis of 3R, 4R, 5R, and 6R Robots," Mechanism and Machine Theory, Vo. 20, 1985, pp. 555-564.

On Fundamental Study of Micro Mechanical Gripper Using Shape Memory Alloy (SMA) Actuator

K. Matsushima* and N. Usui**

*Institute of Engineering Mechanics, University of Tsukuba, Tsukuba, Japan
**Graduate School of Engineering, University of Tsukuba, Tsukuba, Japan

SUMMARY

In this paper we propose a new method for control of the temperature of the SMA actuator using a Peltier effect device. The Peltier effect is a physical phenomenon in which heat generation or absorption at the junction of two different electric conductors or a conductor and a semi-conductor is induced by an electric current flowing into the junction. In this paper, firstly, the temperature control characteristics of the SMA (Cu-Al-Ni single crystal) due to the Peltier effect device are discussed. Secondly, a newly devised micro mechanical gripper using the SMA actuator controlled by the proposed temperature control method is introduced. Finally, the static and dynamic characteristics of the micro mechanical gripper are described.

1. INTRODUCTION

Recently, a new servo actuator using the shape memory alloy, SMA, (i.e. Ti-Ni or Cu-Al-Ni) has been developed. Several kinds of robots actuated by the SMA have been constructed (1), especially in Japan.

The SMA actuator has the following advantages: 1) no mechanical sliding part, 2) the possibility of miniaturization, because the conversion of the thermal energy to mechanical energy is achieved directly, 3) it possesses many degrees of freedom, and 4) the electric resistance of the SMA changes with its deformation, so that SMA has two functions, that of actuator, and sensor. On the other hand, there are some problems which have to be overcome when using the SMA as the servo actuator. One of them is the fact that the SMA has large hysteresis between its deformation and its temperature. The other one is the fact that the time response of the SMA is poor because its control arises out of the thermal process. The latter problem is significant when cooling of the SMA takes place.

S. Hirose and others (2) investigated in detail the behaviour of the SMA from the point of view of the servo actuator design.

In this paper a new temperature control method for the SMA using the Peltier effect device, (the thermo-module), is proposed and a micro mechanical gripper using the SMA actuator controlled by the proposed temperature control method is introduced. Then, the experimental results concerning the gripping force control of a small object using the newly devised micro gripper are presented.

2. TEMPERATURE CONTROL OF SMA USING A PELTIER EFFECT DEVICE

The Peltier effect is a phenomenon in which the heat generation or absorption at the junction of two different electric conductors or a conductor and a semiconductor is induced by the electric current flowing through the junction, and the heat generation or absorption depends on the direction of the current.

The Peltier effect device is called the thermo-module, consists of a junction of p-semiconductor+metal+n-semiconductor, and its size is 4.4 x 4.4 x 2.2 mm^3 as shown in figure 1.

In figure 1, if the temperature of one side of the thermo-module is held constant, the temperature of the other side can be controlled by the magnitude and the direction of the current flowing into the thermo-module.

The static characteristic of temperature versus control current of the thermo-module used here is shown in figure 2. However, the Joule loss caused by the inner resistance of the thermo-module should be released in some way, for instance, by a heat sink.

Figure 3(a) shows the block diagram of the temperature control systems of the SMA actuator. In our case, a single crystal Cu-Al-Ni alloy is used as the SMA, because it has high heat conductivity.

Figure 4 shows the newly devised micro mechanical gripper actuated by the SMA actuator. When the SMA is cooled by the thermo-module, the control current is set as shown in figure 3(b) in order to improve the time response of the SMA.

The waveforms of the control current and the temperature of the SMA are shown in figure 5.

The upper two waveforms are the ones under control and the time responses of the SMA using natural cooling are shown in the lower two (in figure 5) for comparison.

3. GRIPPING FORCE CONTROL OF MICRO MECHANICAL GRIPPER USING SMA ACTUATION

In figure 4 is shown the micro mechanical gripper. This gripper is actuated by the SMA actuator. Its temperature is controlled by the thermo-module. The gripper consists of two plate springs as shown in figure 4. The movement of one of the plate springs is controlled by the SMA actuator.

The gripping force when the finger tips grip a handled object is detected by the strain gauges.

The block diagram of the gripping force control system is shown in figure 6. The system consists of the gripper mechanism, the temperature control loop of the

SMA actuator and the gripping force controller. The controller is designed as a dynamic compensator with which the system has three real poles. Because of that the gripping force should avoid overshooting when the gripper grips an object. However, some nonlinearities, for instance, the current saturation in the temperature control loop and the nonlinearity of the deformation process of the SMA occur in the system.

In our case, in order to improve the input magnitude dependency of the system's response caused by the nonlinearities, the following adaptive control strategy was employed: 1) the feedback gain of the temperature control loop, f, is assumed to be the function $f = f(R)$ of the input command R, and 2) if the function f is determined in some way, then the other parameters K and a can be obtained from equations (1),(2) and (3) under the condition of the dead beat response in the vicinity of the steady state condition:

$$\alpha = \frac{a_1 + a_2 + K_2 K_3 f(R)}{-3} \tag{1}$$

$$K = \frac{3a^2 - a_1(a_2 + K_2 K_3 f)}{K_1 K_2 K_3 K_4} \tag{2}$$

$$a = -\frac{\alpha^3}{K_1 K_2 K_3 K_4 K} \tag{3}$$

where α is the system's pole.

The function f(R) was obtained experimentally as

$$\begin{aligned} f &= 0.14 R - 0.035 \ [V/^\circ C] \quad \text{if } R \geq 0.25 \ [N] \\ &= 0 \quad \text{if } 0 < R < 0.25 \ [N] \end{aligned} \tag{4}$$

In this case, the value of the system's pole varies depending on the magnitude of the input command, but the dead beat response is preserved.

The static characteristics of the gripping force of the system are shown in figure 7, and examples of the dynamic response are shown in figure 8.

The experimental results showed that our system has a resolution of gripping force of less than 0.025 (N) and the settling time of dynamics response of about 3 sec in the range of gripping force between 0 (N) and 0.5 (N).

4. CONCLUSION

In this paper, the new temperature control method of the SMA by means of the Peltier effect device was proposed.

Next, the micro mechanical gripper using the SMA actuator controlled by the proposed method was constructed.

Gripping force control of the gripper was tested and it was shown that fine force control of less than 0.025 (N) using the SMA actuator is possible.

Acknowledgement

We should like to express our thanks to Prof. Dr. K. Ohtsuka (University of Tsukuba) who made valuable suggestions concerning the application of the SMA and made available to us the SMA samples which were made in his laboratory.

REFERENCES

|1| Miwa F.: Shape Memory Alloy Actuator; J. of Robotics Society of Japan, 2.4. 330/337 1984 (in Japanese)

|2| Hirose S. et.al.: A New Design Method of Servo-Actuators based on the Shape Memory Effect; Ro-ManSy '84, Italy, 1984

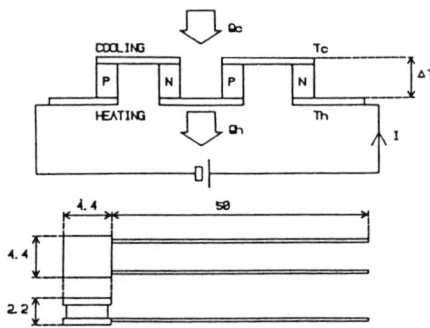

Figure 1 Peltier's effect device (Thermo-module)

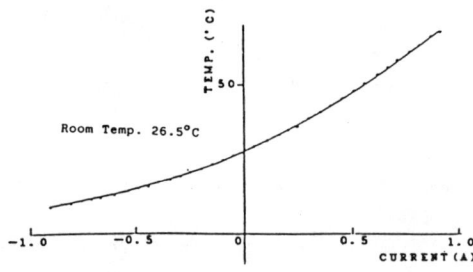

Figure 2   Temperature v.s. Current Characteristic of Thermo-module

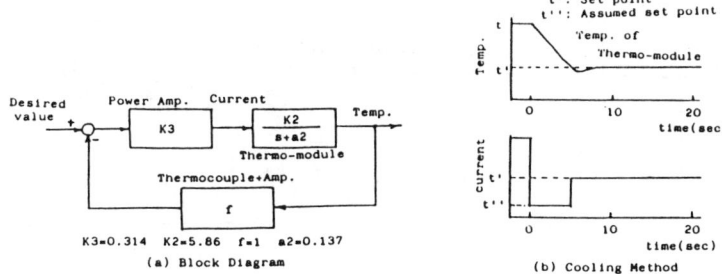

Figure 3 Temperature Control Loop

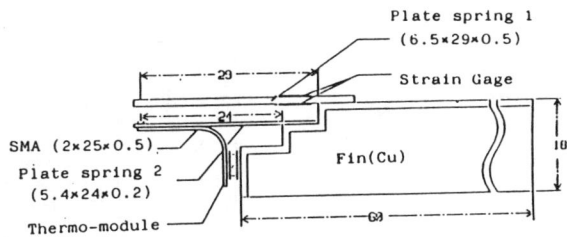

Figure 4 Micro mechanical gripper

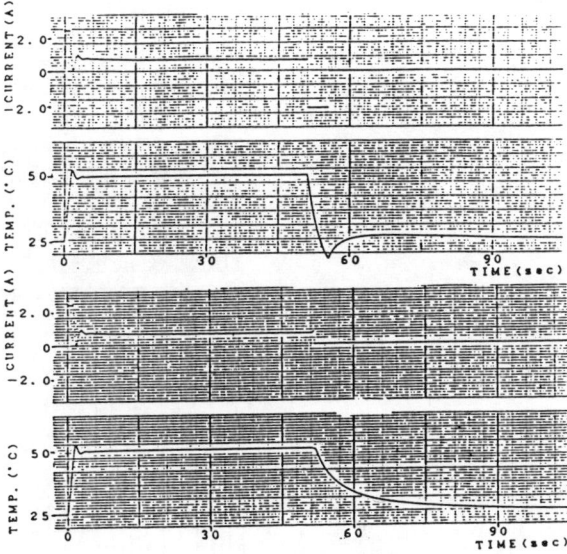

Figure 5 Dynamic response of Temperature control loop

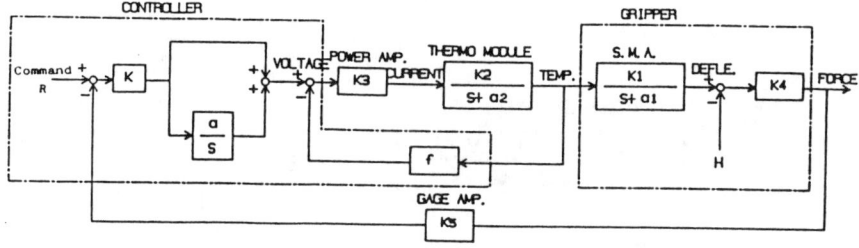

Figure 6 Block diagram of Gripper control system

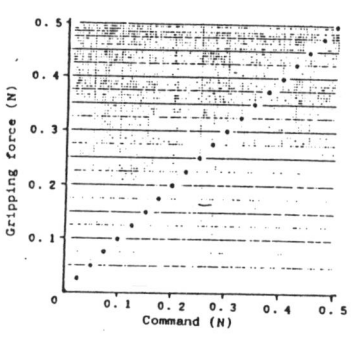

Figure 7 Static characteristic of system

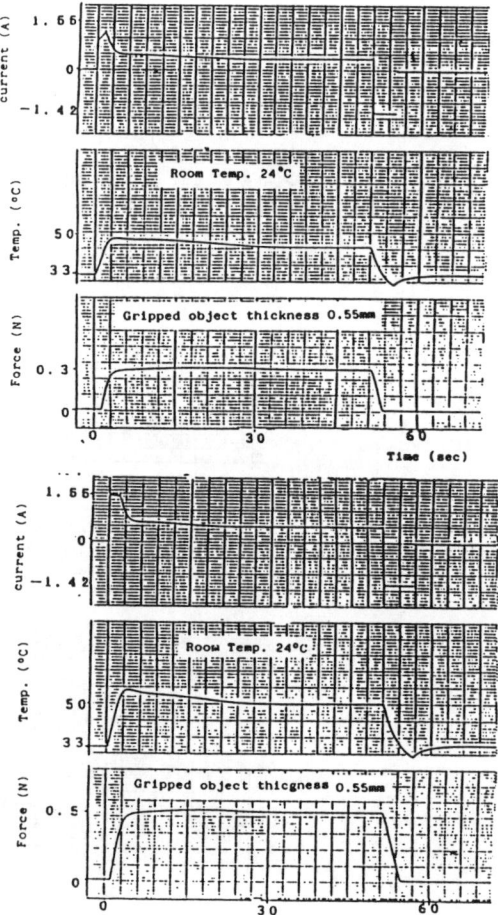

Figure 8 Dynamic response of System

The Kinematic Design and Mass Redistribution of Manipulator Arms for Decoupled and Invariant Inertia

H. Asada

Department of Applied Mathematics & Physics, Kyoto University, Kyoto, Japan

Abstract

A manipulator design theory for reduced dynamic complexity is presented. The kinematic structure and mass distribution of a manipulator arm are designed so that the inertia matrix in the equation of motion becomes diagonal and/or invariant for an arbitrary arm configuration. For the decoupled and invariant inertia matrix, the system can be treated as linear, single-input, single-output system with constant parameters. As a result, the control of the manipulator arm is simplified, and, more importantly, control performance can be improved due to the reduced dynamic complexity.

First, the problem of designing such an arm with the decoupled and/or configuration-invariant inertia matrix is defined. The inertia matrix is then analyzed in relation to the kinematic structure and mass properties of the arm links. Necessary conditions for the manipulator arm to possess a decoupled and/or configuration-invariant inertia matrix are obtained. Using the necessary conditions, we find the kinematic structure and mass properties for which the inertia matrix reduces to a constant, diagonal form. For 2 and 3 degree-of-freedom arms, possible arm designs for decoupled and/or invariant inertia matrices are then determined.

1. Introduction

Dynamic complexity such as coupling and nonlinearities are major concern in the control of manipulator arms. Particularly for high speed manipulators, the coupled and nonlinear dynamics become more prominant, hence the problem to cope with the dynamic complexity becomes more critical. Also, in direct-drive arms and other manipulator arms in which gear ratios of reducers are low, the complicated dynamics have more direct influence upon the drive mechanisms[1]. Again the control problem becomes more complex and difficult because of the prominant coupling and nonlinearities.

A number of control methods have been developed and adopted to the manipulator control. Applications of model-referenced adaptive control[2],[3], nonlinear feedback control[4], sliding mode control[5], for example, have significant contributions to the improvement of control performance. Dynamic computation algorithms[6],[7] and parameter estimation methods[8],[9] allowed us to compensate for the complicated arm dynamics in real time and in an intelligent manner.

Another approach to coping with the dynamic complexity and improving the dynamic performance is to devise the mechanical design along with the controller design. Here, the mechanical construction of a manipulator arm is modified so that the resultant dynamic behavior can be improved or becomes desirable for control[10]. In general, the dynamics of a manipulator arm depends upon mass properties of individual arm links and the kinematic structure of the arm linkage. By the redistribution of mass and the modification of the arm structure, the manipulator dynamics can be changed and improved.

The goal of the paper is to explore the arm linkage design for reducing the dynamic complexity, hence reducing the control difficulties. First, some particular forms of dynamic equations in which the inertia matrix is reduced to a diagonal and/or configuration-invariant form are presented. Then, design theory is developed to accomplish the desired manipulator dynamics. Finally, possible arm structures and mass distributions to reduce the dynamic complexity are determined.

2. Decoupling and Configuration-Invariance of Manipulator Dynamics

A few desirable forms of manipulator dynamics are formulated in this section. The significance of each of the forms is discussed: how the control problem can be simplified, and how much the computational complication is relaxed for the special form of manipulator dynamics.

Let θ_i and τ_i be the joint displacement and torque of the i-th joint, respectively, then the equation of motion of the manipulator arm is given by [11]

$$\tau_i = H_{ii}\ddot{\theta}_i + \sum_{j\neq i} H_{ij}\ddot{\theta}_j$$

$$+ \sum_j \sum_k \left(\frac{\partial H_{ij}}{\partial \theta_k} - \frac{1}{2}\frac{\partial H_{jk}}{\partial \theta_i}\right)\dot{\theta}_j\dot{\theta}_k + \tau_{gi} \quad (1)$$

where H_{ij} is the i-j element of the inertia matrix, and τ_{gi} is the torque due to gravity. The first term on the right hand side represents the inertia torque generated by the acceleration of the i-th joint, while the second term is the interactive inertia torque caused by accelerations of the other joints. The interactive inertia torque is linearly proportional with acceleration. The third term is nonlinear velocity torques due to Coriolis and centrifugal effects. In general, the inertia matrix varies depending upon the arm configuration, hence the nonlinear velocity torque arises.

The inertia matrix is determined by the kinematic structure of the manipulator arm and mass properties of individual links. The design problem we discuss in this paper is to modify both the kinematic structure and mass properties of the arm linkage so that the inertia matrix reduces to particular forms.

Consider the inertia matrix that reduces to a diagonal matrix for an arbitrary arm configuration, then the second term in (1) vanishes and consequently no interactive inertia torques appear. The manipulator inertia matrix in this case is referred to as a decoupled inertia matrix. The significance of the decoupled inertia matrix is that the control system can be treated as a set of single-input, single-out sub-systems associated with individual joint motions. The equation of motion reduces to

$$\tau_i = H_{ii}\ddot{\theta}_i + \sum_k \left(\frac{\partial H_{ii}}{\partial \theta_k}\dot{\theta}_i\dot{\theta}_k - \frac{1}{2}\frac{\partial H_{kk}}{\partial \theta_i}\dot{\theta}_k^2\right) + \tau_{gi}$$
$$(2)$$

where the second term represents the nonlinear velocity torques resulting from the spatial dependency of the diagonal elements of the inertia matrix. Note that the number of terms involved in (2) is much smaller than the original nonlinear velocity torques, because all the off-diagonal elements are zero for all $\theta_1,\ldots,\theta_n$. This reduces the computational complexity of the nonlinear torques.

Another significant form of the inertia matrix that reduces the dynamic complexity is a configuration-invariant form. The inertia matrix in this form does not vary for an arbitrary arm configuration. In other words, the matrix is independent of joint displacements, hence the third term in (1) vanishes and the equation of motion thus reduces to

$$\tau_i = H_{ii}\ddot{\theta}_i + \sum_j H_{ij}\ddot{\theta}_j + \tau_{gi}. \tag{3}$$

Note that the coefficients H_{ii} and H_{ij} are constant for all the arm configurations. Thus the equation is linear except the last term, that is, the gravity torque. The inertia matrix in this form is referred to as an invariant inertia matrix. The significance of this form is that linear control schemes can be adopted, which are much simpler and easier to implement.

When the inertia matrix is both decoupled and configuration invariant, the equation of motion reduces to

$$\tau_i = H_{ii}\ddot{\theta}_i + \tau_{gi}. \tag{4}$$

The system is completely decoupled and linearized except the gravity term. We can treat the system as single-input, single-output systems with constant parameters.

The decoupled and configuration-invariant inertia was first accomplished in reference [1], in which a special five-bar-link parallel drive mechanism was devised for a direct-drive arm. The feasibility and practical usefulness were also demonstrated through the development of the special arm. Yet, it is not made clear how the decoupled and configuration-invariant inertia can be achieved for general manipulator arms. In this paper, we discuss the general design problem to attain the desired dynamics; what are conditions required for the kinematic structure and mass properties of the manipulator arm to possess the decoupled and/or configuration-invariant inertia.

3. Modeling

In this section, we derive inertia tensor matrices of manipulator arms discussed in this paper. As shown in Figure 1, the manipulator arm is assumed to be an open kinematic chain consisting of all revolute joints. The joints are numbered 1 through n from the proximal joint to the distal one. The link between joints i and i+1 is called link i. The direction of the axis of joint i is represented by a unit vector $\underline{b}_i$ and the displacement of link i is denoted by θ_i which is the angle of rotation about the unit vector. The centroid of link k is shown by point c_k in the figure, while the velocity vector of the centroid is denoted by $\underline{v}_k$ and the angular velocity vector by $\underline{\omega}_k$. Let m_k and $\underline{I}_k$ be the mass and the inertia tensor of link k with respect to the O-xyz inertial reference frame, then the total kinetic energy stored in the arm links from 1 to n is given by

$$T = \sum_{k=1}^{n} \frac{1}{2}(m_k \underline{v}_{ck}^T \underline{v}_{ck} + \underline{\omega}_k^T \underline{I}_k \underline{\omega}_k) \qquad (5)$$

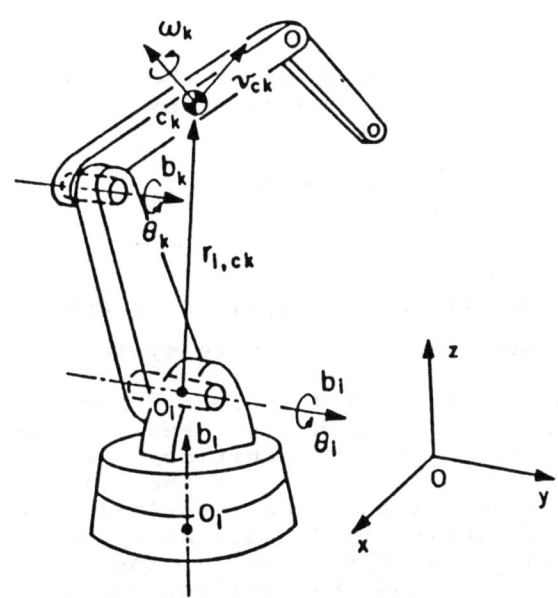

Fig. 1

The motion of link k is generated by the preceding joint motions. The angular velocity $\underline{\omega}_k$, for example, is given by

$$\underline{\omega}_k = \sum_{i=1}^{k} \underline{b}_i \dot{\theta}_i \qquad (6)$$

To represent the linear velocity of the centroid c_k, we

denote the position vector from an arbitrary point on the i-th joint axis to the centroid c_k by vector $\underline{r}_{i,ck}$. Then,

$$\underline{v}_{ck} = \sum_{i=1}^{k} \underline{b}_i \dot{\theta}_i \times \underline{r}_{i,ck} \tag{7}$$

where $\times$ means a vector product. Substituting (6) and (7) into (5) yields

$$T = \frac{1}{2} \sum_{i=1}^{n} \sum_{j=1}^{n} H_{ij} \dot{\theta}_i \dot{\theta}_j \tag{8}$$

where H_{ij} is the i-j element of the $n \times n$ inertia matrix involved in (1), and is given by

$$H_{ij} = \sum_{k=\text{Max}[i,j]}^{n} \{m_k (\underline{b}_i^T \underline{b}_j \cdot \underline{r}_{i,ck}^T \underline{r}_{j,ck} - \underline{b}_j^T \underline{r}_{i,ck} \cdot \underline{b}_i^T \underline{r}_{j,ck}) + \underline{b}_i^T \underline{I}_k \underline{b}_j\} \tag{9}$$

Note that the inertia matrix is symmetry, hence $H_{ij} = H_{ji}$. We also derive a different expression for the inertia matrix, which we need in the following analysis. Let l be an arbitrary joint number. We consider the case in which all the joints between l+1 and n are immobilized. The last (n-l) links are then treated as a single rigid body. As shown in Figure 2, let $M_{l,n}$ and $\underline{p}_l$ be ,respectively, the total mass of the last (n-l) links and the position vector from point O_l to the centroid $C_{l,n}$ that is the centroid of the last (n-l) links. Then,

$$M_{l,n} = \sum_{k=l}^{n} m_k \tag{10}$$

$$\underline{p}_l = \sum_{k=l}^{n} m_k \underline{r}_{l,ck} / M_{l,n} \tag{11}$$

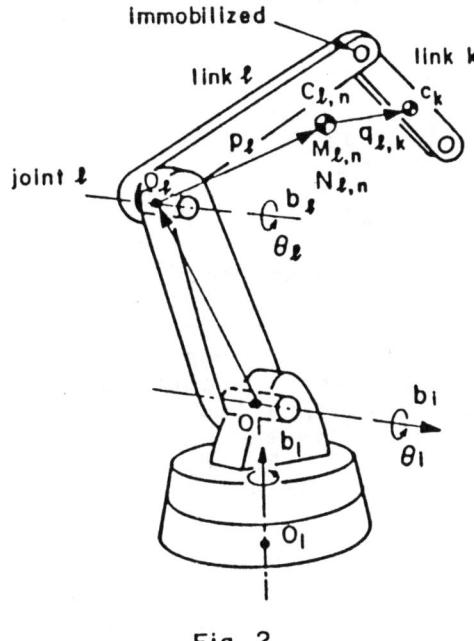

Fig. 2

Let $q_{1,k}$ be the position vector from the centroid $C_{1,n}$ to the centroid of link k, then the following is also derived from the definition of the centroid:

$$\underline{r}_{i,ck} = \underline{r}_{i,1} + \underline{p}_1 + \underline{q}_{1,k} \tag{12}$$

$$\sum_{k=1}^{n} m_k \underline{q}_{1,k} = 0 \tag{13}$$

Substituting (12) and (13) into (9), we can split the i-j element of the inertia matrix into two parts: one is the terms associated with the last (n-1) links and the other is the terms for the other links.

$$H_{ij}^{(1)} = \sum_{k=\text{Max}[i,j]}^{1-1} \{m_k (\underline{b}_i^T \underline{b}_j \cdot \underline{r}_{i,ck}^T \underline{r}_{j,ck}$$
$$- \underline{b}_j^T \underline{r}_{i,ck} \cdot \underline{b}_i^T \underline{r}_{j,ck}) + \underline{b}_i^T \underline{I}_k \underline{b}_j\}$$
$$+ \underline{b}_i^T \underline{N}_{1,n} \underline{b}_j + M_{1,n} (\underline{b}_i^T \underline{b}_j (\underline{r}_{i,1} + \underline{p}_1)^T (\underline{r}_{j,1} + \underline{p}_1)$$
$$- \underline{b}_i^T (\underline{r}_{i,1} + \underline{p}_1) \cdot \underline{b}_i^T (\underline{r}_{j,1} + \underline{p}_1)) \tag{14}$$

where $\underline{N}_{1,n}$ is the composite inertia tensor of the last (n-1) links with respect to the centroid $C_{1,n}$, which is given by

$$\underline{N}_{1,n} = \text{diag.}\{\sum_{k=1}^{n} m_k g_{1,k}{}^T g_{1,k}\} \\ + \sum_{k=1}^{n} \{\underline{I}_k - m_k g_{1,k} \cdot g_{1,k}{}^T\} \quad (15)$$

4. Analysis

In order to eliminate the coupling and nonlinear torques, the inertia matrix must be diagonalized and made invariant for all the arm configurations. In this section, we analyze the inertia matrix and derive the necessary conditions for the kinematic structure and mass properties to allow the inertia matrix to reduce to a diagonal and/or invariant form.

[a]. Off-diagonal elements

We first discuss the necessary conditions for an off-diagonal element, $H_{ij} = H_{ji}$ ($i<j$), to be kept zero or invariant for all the arm configurations. We derive the necessary conditions by considering a special situation where only one joint, say joint 1, rotates 360 degrees, whilst the other joints are immobilized. This simple analysis provides useful conditions that are used to determine the kinematic structure and mass properties for decoupled and/or configuration-invariant inertia.

As shown in Figure 3, we immobilize all the joints beyond joint 1, and regard the last (n-1) links as a single rigid body. Let $OC_{1,n}$ be the perpendicular from the centroid $C_{1,n}$ to the joint axis $\underline{b}_1$. We locate the inertial reference frame O-xyz at the point O, which is coincident with point O_j or O_1 on the joint axis. Without loss of generality, we can direct the z axis in the same direction as the joint axis $\underline{b}_1$. The displacement of joint 1 is then represented by the angle from the x axis to the vector $\underline{p}_1$, which is the position vector of the centroid $C_{1,n}$ relative to the origin O. Denoting the length of the vector $\underline{p}_1$ by $L_{1,n}$, we obtain,

$$\underline{p}_1 = \begin{pmatrix} L_{1,n} \cos\theta_1 \\ L_{1,n} \sin\theta_1 \\ 0 \end{pmatrix} \quad (16)$$

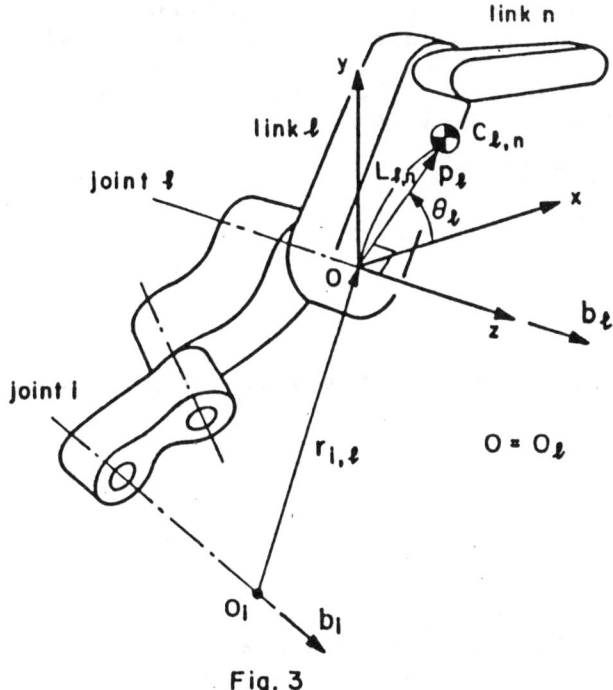

Fig. 3

The inertia tensor $\bar{\underline{N}}_{1,n}$, that is the composite inertia tensor of the last $(n-1)$ links, varies in its matrix expression, as the joint angle θ_1 changes. Let $\underline{N}_{1,n}$ be the composite inertia tensor when $\theta_1 = 0$, the matrix $\underline{N}_{1,n}$ is given by

$$\underline{N}_{1,n} = \underline{A}\bar{\underline{N}}_{1,n}\underline{A}^T . \qquad (17)$$

when $\underline{A}$ is the 3×3 rotation matrix given by

$$\underline{A} = \begin{pmatrix} \cos\theta_1 , & -\sin\theta_1 , & 0 \\ \sin\theta_1 , & \cos\theta_1 , & 0 \\ 0 , & 0 , & 1 \end{pmatrix} \qquad (18)$$

For simplicity, the subscripts 1 and n are neglected unless the notation gets confusing. The elements of the inertia tensor $\underline{N}$ are then written as

$$\underline{N} = \begin{pmatrix} \bar{N}_{xx} & \bar{N}_{xy} & \bar{N}_{xz} \\ \bar{N}_{xy} & \bar{N}_{yy} & \bar{N}_{yz} \\ \bar{N}_{xz} & \bar{N}_{yz} & \bar{N}_{zz} \end{pmatrix} \qquad (19)$$

When the joint axis $\underline{b}_i$ is not orthogonal to the joint axis $\underline{b}_1$, the i-th joint axis intersects with the x-y plane. If we define the intersection as the representative point on the i-th joint axis, the position vector $\underline{r}_{i,1}$, that is the vector from points O_i to O_1, can be written as

$$\underline{r}_{i,1} = \begin{pmatrix} r_x \\ r_y \\ 0 \end{pmatrix} \qquad (20)$$

The direction consines of the i-th joint axis are denoted by

$$\underline{b}_i = \begin{pmatrix} b_x \\ b_y \\ b_z \end{pmatrix} \qquad (21)$$

Substituting (16)-(21) into (14) yields the expression relating the off-diagonal element of the inertia matrix H_{i1} to the joint angle θ_1,

$$H_{i1}^{(1)} = (ML^2 + \bar{N}_{zz})b_z$$
$$+ (b_x \bar{N}_{xz} + b_y \bar{N}_{yz} + MLb_z r_x)\cos\theta$$
$$+ (b_y \bar{N}_{xz} - b_x \bar{N}_{yz} + MLb_z r_y)\sin\theta \qquad (22)$$

Note that the subscripts l and n to be attached to M and L are neglected. The following proposition is derived directly from (22).

[Proposition 1]
The necessary condition for the off-diagonal element of the inertia matrix, H_{i1}, to be invariant for an arbitrary θ_1 are given by

$$b_x \bar{N}_{xz} + b_y \bar{N}_{yz} + MLb_z r_x = 0 \qquad (23)$$

and

$$b_y \bar{N}_{xz} - b_x \bar{N}_{yz} + MLb_z r_y = 0 \qquad (24)$$

In addition to the above conditions, the first term in (22) must be zero when the decoupled inertia matrix is required. However, the off-diagonal element H_{i1} does not vanish unless $b_z = 0$, since the diagonal element of the inertia tensor $\bar{N}_{zz}$ is always positive.

When $b_z = 0$, the two joint axes are perpendicular to each other, then we conclude the following.

[Proposition 2]
For manipulator arms with an open kinematic chain structure, the inertia matrix cannot be decoupled unless the joint axes are orthogonal to each other.
When $\underline{b}_i{}^t\underline{b}_1 = 0$, namely the two joint axes are orthogonal, we can direct the x axis of the inertial reference frame along the joint axis $\underline{b}_i$, without loss of generality. As shown in Figure 4,

$$\underline{b}_i = \begin{pmatrix} 1 \\ 0 \\ 0 \end{pmatrix} \qquad \underline{b}_1 = \begin{pmatrix} 0 \\ 0 \\ 1 \end{pmatrix} \qquad (25)$$

Let O_i be the representative point of joint i that is located at the intersection of the joint axis and the y-z plane as shown in the figure, then the position vector $\underline{r}_{i,1}$ can be expressed as

$$\underline{r}_{i,1} = \begin{pmatrix} 0 \\ r_y \\ r_z \end{pmatrix} \qquad (26)$$

Using the above equations in eg.(14), we obtain

$$H_{i1}^{(1)} = (\bar{N}_{xz} - MLr_z)\cos\theta - \bar{N}_{yz}\sin\theta \qquad (27)$$

The following is derived directly from the above results.

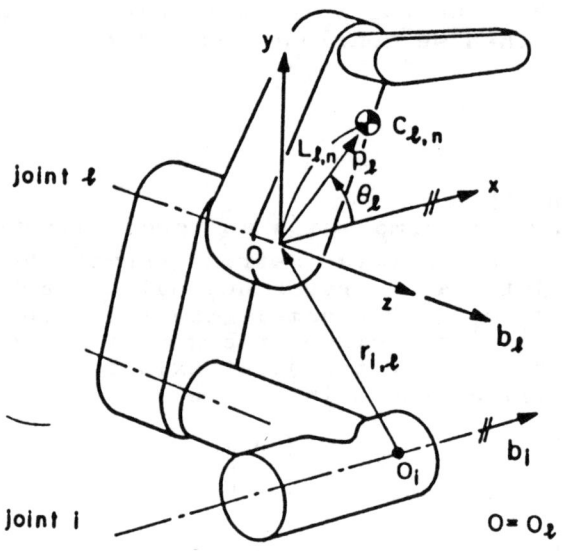

Fig. 4

[Proposition 3]
When joint axes $\underline{b}_i$ and $\underline{b}_l$ are orthogonal, the necessary conditions for the off-diagonal element H_{il} to vanish for an arbitrary joint angle θ_l are given by

$$MLr_z = \bar{N}_{xz} \tag{28}$$

and

$$\bar{N}_{yz} = 0 \tag{29}$$

[b]. Diagonal elements

Next, we consider diagonal elements of the inertia matrix, H_{ii}. Let l be an arbitrary joint number greater than or equal to i. We investigate the change in H_{ii} when the joint angle θ_l varies from 0 to 360 degrees. As in the case of Figure 3, we immobilize all the joints beyond l and treat the last (n-l) links as a single rigid body. Again we use the inertial reference frame O-xyz located at the foot of the perpendicular from the centroid $C_{l,n}$ to the joint axis $\underline{b}_l$. From (14),

$$H_{ii}^{(1)} = \bar{H}_{ii} + \underline{b}_i^T \underline{\underline{A N A}}^T \underline{b}_i$$
$$+ M[\ |\underline{r}_{i,1} + \underline{p}_1|^2 - (\underline{b}_i^T(\underline{r}_{i,1} + \underline{p}_1))^2] \tag{30}$$

where $\bar{H}_{ii}$ is a constant parameter independent of θ_l, and is given by

$$\bar{H}_{ii} = \sum_{k=i}^{l-1} [m_k(|\underline{r}_{i,ck}|^2 - (\underline{b}_i^T \underline{r}_{i,ck})^2] \qquad (31)$$

We first derive the conditions that the diagonal element H_{ii} is invariant for an arbitrary θ_1, when the joint axes, $\underline{b}_1$ and $\underline{b}_i$, are not orthogonal. Then we discuss the case where the axes are orthogonal to each other. Substituting (20) and (21) into (30) yields

$$H_{ii}^{(1)} = a_0 + a_1 \cos^2\theta + a_2 \cos\theta \sin\theta + a_3 \sin^2\theta + a_4 \cos\theta + a_5 \sin\theta \qquad (32)$$

where the coefficients $a_0, \ldots, a_5$ are given by

$$a_0 = \bar{H}_{ii} + M[L^2 + r_x^2 + r_y^2 + (b_x r_x + b_y r_y)^2] + b_z^2 \bar{N}_{zz} \qquad (33)$$

$$a_1 = ML^2 b_x^2 + b_x^2 \bar{N}_{xx} + b_y^2 \bar{N}_{yy} + 2 b_x b_y \bar{N}_{xy} \qquad (34)$$

$$a_2 = 2[ML^2 b_x b_y + b_x b_y (\bar{N}_{xx} - \bar{N}_{yy}) - (b_x^2 - b_y^2) \bar{N}_{xy}] \qquad (35)$$

$$a_3 = ML^2 b_y^2 + b_x^2 \bar{N}_{yy} + b_y^2 \bar{N}_{xx} - 2 b_x b_y \bar{N}_{xy} \qquad (36)$$

$$a_4 = 2[b_z(b_x \bar{N}_{xz} + b_y \bar{N}_{yz}) + M\{b_x L(b_x r_x + b_y r_y) + Lr_x\}] \qquad (37)$$

$$a_5 = 2[b_z(b_y \bar{N}_{xz} - b_x \bar{N}_{yz}) + M\{b_y L(b_x r_x + b_y r_y) + Lr_y\}] \qquad (38)$$

for the $H_{ii}^{(1)}$ to be invariant for an arbitrary θ_1, it is necessary that the coefficients a_i satisfy all of the following conditions:

$$a_1 = a_3 \qquad (39)$$

$$a_2 = 0 \qquad (40)$$

$$a_4 = 0 \qquad (41)$$

$$a_5 = 0 \qquad (42)$$

The above conditions can be further reduced to simpler equations when combined with (23) and (24), which are the necessary conditions for the off-diagonal element H_{i1} to be invariant for an arbitrary θ_1. Substituting (23) into (41),

$$(2b_x^2 + b_y^2)r_x + b_x b_y r_y = 0 \qquad (43)$$

From (24) and (42),

$$(b_x^2 + 2b_y^2)r_y + b_x b_y r_x = 0 \qquad (44)$$

Eliminating r_x in the above two equations, we obtain

$(b_x^2 + b_y^2)r_y = 0$, which is satisfied if, and only if, one of the following two conditions is met;

(i) $b_x = b_y = 0$, namely, joint axes $\underline{b}_i$ and $\underline{b}_1$ are parallel, or

(ii) $r_x = r_y = 0$, namely, the joint axes intersect at the origin O.

In the same way, we can derive a similar expression,

$(b_x^2 + b_y^2)^2 \bar{N}_{xy} = 0$, from (39) and (40). This leads to the following two conditions;

(i) $b_x = b_y = 0$, or

(ii) $\bar{N}_{xy} = 0$ and $\bar{N}_{xx} + ML^2 = \bar{N}_{yy}$

Similarly, necessary conditions can be derived for two joints which are orthogonal to each other. Summarizing the above results, the following proposition is obtained.

[Proposition 4]
The necessary condition for the diagonal element H_{ii} as well as the off-diagonal element H_{i1} to be invariant for an arbitrary θ_1 is that the kinematic structure and mass properties of the manipulator arm satisfy one of the following four conditions;

(i) $\underline{b}_i = \underline{b}_1$; namely the two joint axes are parallel,
$$(45)$$

(ii) $\underline{b}_i^T \underline{b}_1 \neq 0$ and $\qquad (46-1)$

$r_x = r_y = 0$, namely $O_i = 0$, and (46-2)

$\bar{N}_{xy} = 0$, and (46-3)

$\bar{N}_{xx} + ML^2 = \bar{N}_{yy}$. (46-4)

(iii) $\underline{b}_i^T \underline{b}_1 = 0$ and (47-1)

$L = 0$ and (47-2)

$\bar{N}_{xy} = 0$ and (47-3)

$\bar{N}_{xx} = \bar{N}_{yy}$. (47-4)

(iv) $\underline{b}_i^T \underline{b}_1 = 0$ and (48-1)

$r_y = 0$ (48-2)

$\bar{N}_{xy} = 0$ (48-3)

$\bar{N}_{xx} = \bar{N}_{yy} + ML^2$. (48-4)

5. Design

5.1 Arm Design for Decoupled Inertia

Based on the previous analysis on the necessary conditions, we determine the kinematic structure and mass properties of a manipulator arm that possesses decoupled and/or invariant inertia matrices.

First, we consider the decoupled inertia. From Proposition 2 it follows that all the joint axes must be orthogonal for an arbitrary arm configuration in order to accomplish the decoupled inertia. The kinematic structure in which all the joint axes are orthogonal at all time is limited only to two degree-of-freedom arms. For the inertia matrix of the 2 d.o.f. arm, the off-diagonal element to be considered is only H_{12}, and the joint displacement that might cause changes in H_{12} is θ_2. Then we can derive the necessary conditions for invariant inertia by applying

Proposition 3 only to $H_{12}^{(2)}$.

Conversely, if the necessary conditions on $H_{12}(2)$ are substituted into (22) to compute the off-diagonal element H_{12}, one can find that the off-diagonal element vanishes for an arbitrary θ_2. Consequently, the necessary conditions also provide the sufficient conditions for decoupled inertia. Therefore the following theorem is obtained.

[Theorem 1]
An open-kinematic-chain manipulator arm possesses a decoupled inertia matrix for an arbitrary arm configuration if, and only if, its kinematic structure and mass properties satisfy the following conditions:

Structure   :   two degree-of-freedom arm with orthogonal joint axes (Figure 5),

Mass properties: $m_2 L r_z = \bar{N}_{xz}$ and   (49-1)

$\bar{N}_{yz} = 0$ (49-2)

where the notations are defined in Figure 5.

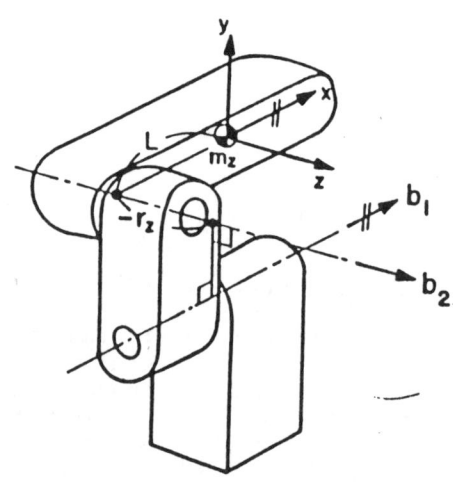

Fig. 5

Special cases, which are practically useful, are shown in Figure 6. In the first arm design, the centroid of link 2 lies on the joint axis b_2, namely, L = 0, and in the second arm design the perpendicular from the centroid of link 2 to the joint axis $\underline{b}_2$ intersects with the common normal of the two joint axes, namely, $r_z = 0$. For each of the two cases, the left hand side of (49-1) is zero, hence $\bar{N}_{xz}$ must be zero. As a result, the principal axis of link 2 must be coincident with the joint axis $\underline{b}_2$, since $\bar{N}_{yz}$ is also zero from condition (49-2).

5.2 Arm Design for Decoupled and Invariant Inertia

Manipulator arms with decoupled and configuration-invariant inertia matrices are discussed in this subsection. To be decoupled, the conditions given in Theorem 1 must be held. At the same time, the diagonal element H_{11} must be constant to meet the requirement for the configuration-invariant inertia. It is therefore necessary that the conditions in Proposition 4 are satisfied for the diagonal element. The first two cases involved in Proposition 4 are conflicting to the decoupling conditions, because the manipulator arm that can possess a decoupled inertia matrix is limited to the ones with orthogonal joint axes. Thus, the set of conditions, either (iii) or (iv), must be met. Conversely, substituting eqs.(47) or (48) as well as the conditions in Theorem 1 into (22) and (30), one can find that the 2 × 2 inertia matrix becomes diagonal and constant, hence that these conditions are sufficient conditions as well.

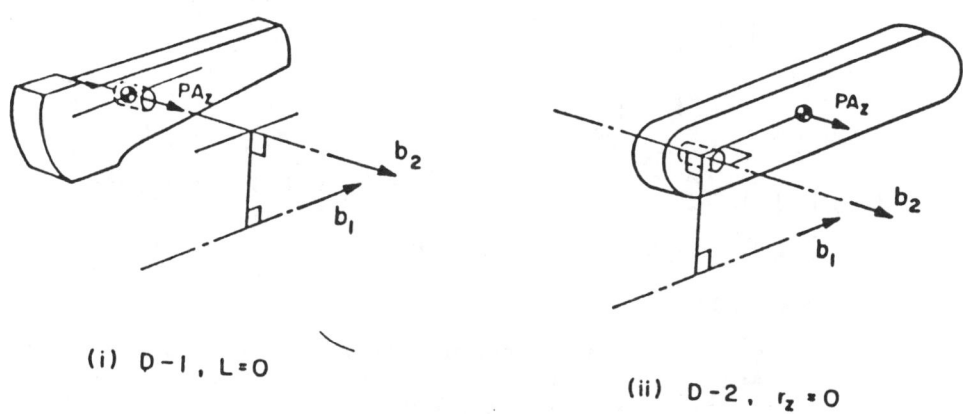

(i) D-1, L=0

(ii) D-2, $r_z = 0$

Fig. 6

[Theorem 2]
The necessary and sufficient conditions for an open-kinematic-chain manipular arm to possess a decoupled and configuration-invariant inertia matrix are given by:

Structure : two degree-of-freedom arm with orthogonal joint axes,

Mass properties: one of the following two sets of conditions must be held;

(i) $L = 0$ and (50-1)

$\bar{N}_{xx} = \bar{N}_{yy}$ and (50-2)

$\bar{N}_{xy} = \bar{N}_{yz} = \bar{N}_{xz} = 0$ (50-3)

(ii) $r_y = 0$ and (51-1)

$\bar{N}_{xx} = \bar{N}_{yy} + ML^2$ and (51-2)

$\bar{N}_{xy} = \bar{N}_{yz} = 0$ and (51-3)

$m_2 L r_z = \bar{N}_{xz}$ (51-4)

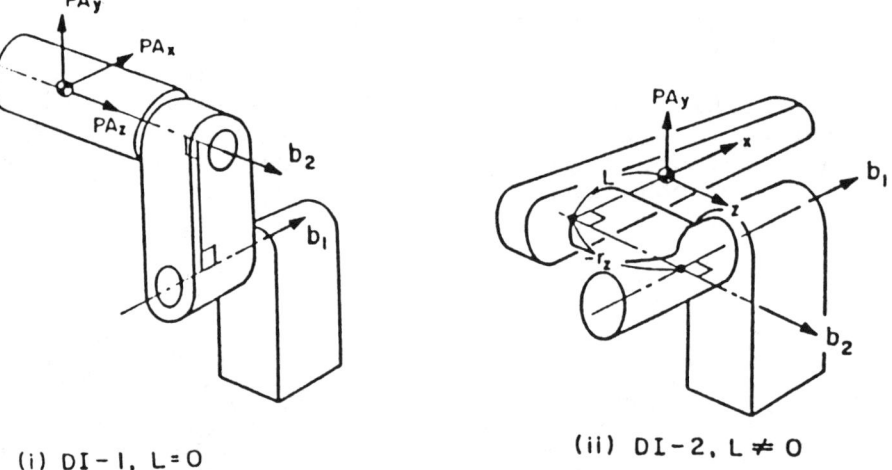

(i) DI-1, L=0   (ii) DI-2, L≠0

Fig. 7

The arm designs corresponding to the above two cases are shown in Figure 7. For convenience, the two arms are called model DI-1 and model DI-2, respectively. From (50-3), the second link of model DI-1 has its principal directions in the same directions as joint axes b_1 and b_2. The direction perpendicular to both b_1 and b_2 is then the principal direction as well. For model DI-2, on the other hand, one of the principal directions of link 2 is along the PA_y axis shown in the figure.

5.3 Arm Design for Invariant Inertia

In this subsection we discuss the arm design when only the configuration invariance of the inertia matrix is required. Since the orthogonality condition is not needed in this case, the arm design can be extended to more than 2 degrees of freedom. However, we first discuss 2 d.o.f. arms.

[a]. 2 d.o.f. arms

We investigate the elements H_{11} and H_{12} in relation to θ_2. Conditions involved in Propositions 1, 3 and 4 are now listed in Table 1, from which conditions for the invariant inertia are to be derived. The upper half of the table is for the conditions on $H_{12}^{(2)}$, while the lower half is for $H_{11}^{(2)}$. Since both elements must be constant for all θ_2, the upper and lower sets of conditions are led to the "AND" gate representing both conditions must be held at the same time. For each of the conditions, either the upper quarter or the lower one must be held, thus the two sets of conditions are led to the "OR" gates representing either one set or the other must be satisfied. These conditions can be reduced to the following.

Table 1 Necessary conditions for 2 d.o.f. arm to possess invariant inertia matrix

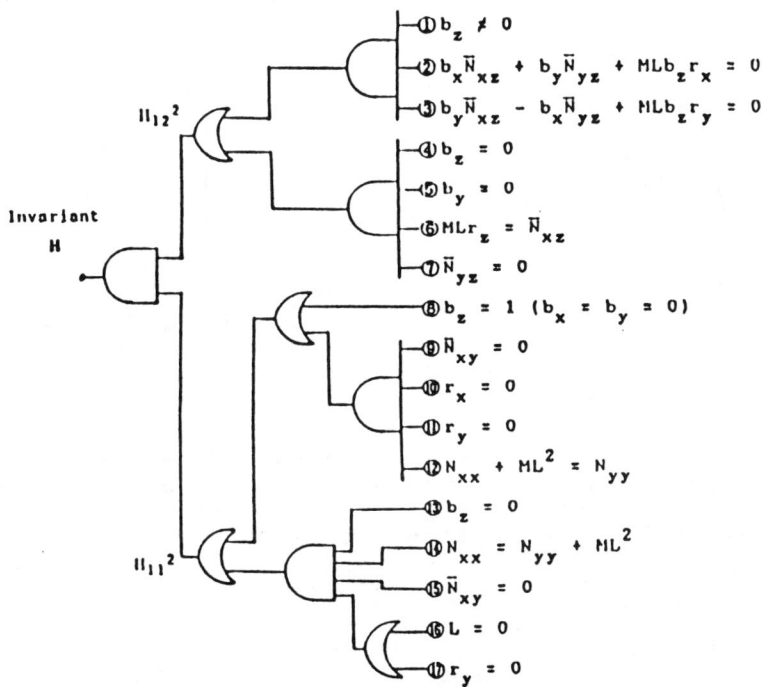

When $b_z = 0$, the two joint axes are orthogonal. Therefore the conditions for the invariant inertia are the same as in Theorem 2. We then focus on the case where $b_z \neq 0$. From the table, two cases arise; (i) one is the set of conditions ①, ②, ③ and ⑧, (ii) the other is conditions ①, ②, ③, ⑨, ⑩, ⑪ and ⑫. In case (i), substituting condition ⑧ as well as its equivalent conditions, $b_x = b_y = 0$, into conditions ② and ③, we find that $Lr_x \stackrel{\triangle}{=} 0$ and $Lr_y = 0$. If we assume that $L \neq 0$, then r_x and r_y must be zero. What this means is that the two joint axes are identical, which is a trivial case, hence neglected. Thus the distance L must be zero for the case (i). In case (ii), substituting conditions ⑩ and ⑪ into ② and ③ yields that $(b_x^2 + b_y^2)\bar{N}_{yz} = 0$. Since b_x and b_y are not zero at the same time, N_{yz} must be zero. In the same way, we can derive that $N_{xz} = 0$. From these results along with condition ⑨, it follows that the three principal directions of link 2 are, respectively, along joint axis b_2, the perpendicular from the centroid C_2 to the joint axis b_2, and the perpendicular to both axes. The above investigations can be summarized in the following theorem.

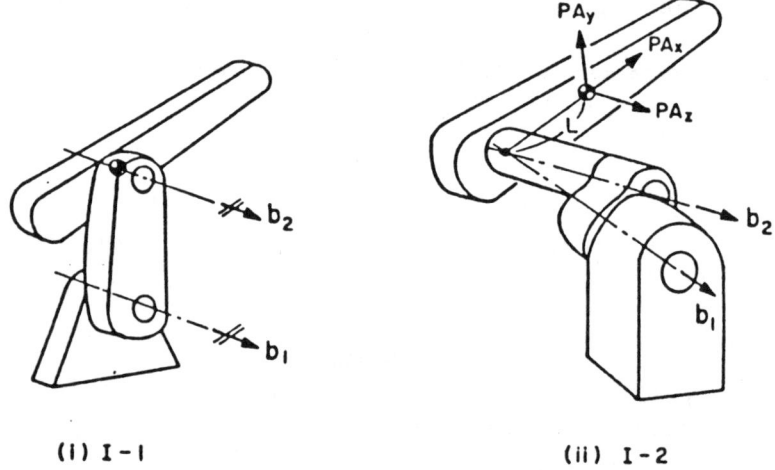

Fig. 8

[Theorem 3]
The two degree-of-freedom manipulator arm consisting of an open kinematic chain possesses a configuration-invariant inertia matrix if, and only if, one of the following four conditions is satisfied.

(i) $b_z = 1$, $(b_x = b_y = 0)$ and   (52-1)

      $L = 0$ (52-2)

(ii) $r_x = r_y = 0$ (53-1)

$\bar{N}_{xy} = \bar{N}_{yz} = \bar{N}_{xz} = 0$ (53-2)

$\bar{N}_{xx} + ML^2 = \bar{N}_{yy}$ (53-3)

(iii) the same as condition (i) in Theorem 2

(iv) the same as condition (ii) in Theorem 2

The manipulator arms corresponding to the first two cases (i) and (ii) are shown in Figure 8. These arm designs are called model I-1 and model I-2, respectively.

[b]. 3 d.o.f. arms

Applying the above theorem, we extend the arm design to three degrees of freedom. A 3 d.o.f. arm reduces to a 2 d.o.f. arm, when one of the three joints is immobilized. According to Theorem 3, the possible designs for 2 d.o.f. arms are limited to the four cases to accomplish the configuration-invariant inertia. Among the four models, model I-2, which includes an oblique joint, is seldom used in practice. We therefore do not adopt this model to our arm design. We build the 3 d.o.f. arm with the combination of the three models only. Table 2 is the list of possible combinations of the three models. For 3 d.o.f. arms, the inertia matrix must be invariant for an arbitrary combination of joint angles, θ_2 and θ_3. We investigate each of the arm models listed in Table 2 whether or not the invariance can be held for all θ_2 and θ_3.

We first consider the case where joints 2 and 3 are the model DI-2 type. As shown in Figure 7-(ii), the centroid of the tip link moves along a circle of radius L as the tip link rotates. As a result, the centroid of the composite rigid body consisting of links 2 and 3 in the 3 d.o.f. arm is not stationary, but moves along a circle when the model DI-2 is adopted to the last two joints. For such a centroid, the mass property conditions required for the first two joints cannot be satisfied no matter which model the first two joints are. Therefore the 3 d.o.f. arms of the invariant inertia cannot be built with the DI-2 type adopted to the last two joints. Thus we eliminate cases 3, 6, and 9 in Table 2.

Table 2 Possible combinations of joint types
for 3 d.o.f. arms

Case	Joints 1 & 2	Joints 2 & 3
1	I-1	I-1
2	I-1	DI-1
3	I-1	DI-2
4	DI-1	I-1
5	DI-1	DI-1
6	DI-1	DI-2
7	DI-2	I-1
8	DI-2	DI-1
9	DI-2	DI-2

Next we consider case 4. The arm construction corresponding to this type is shown in Figure 9. The mass property conditions associated with the model DI-1 used at the first two joints must be satisfied for an arbitrary configuration of the last link. We now investigate this condition. As shown in Figure 9, the inertial reference frame O-xyz is centered at the centroid of the composite rigid body $C_{2,3}$. Let $\underline{I}_3$ be the inertia tensor of link 3 when $\theta_3 = 0$, then the composite inertia tensor of the last two links is given by

$$\underline{N}_{2,3} = \underline{I}_2 + m_2 [\text{diag.}(g_{2,2}^T g_{2,2}) - g_{2,2} g_{2,2}^T]$$
$$+ \underline{\Delta}(\theta_3) \underline{I}_3 \underline{\Delta}(\theta_3)^T + m_3 [\text{diag.}(g_{2,3}^T g_{2,3})$$
$$- g_{2,3} g_{2,3}^T] \tag{54}$$

On the other hand, the centroid of link 3 must lie on the joint axis $\underline{b}_3$, since the last two joints are model I-1. Thus

$$L_{2,3} = 0 \tag{55}$$

Therefore, position vector $g_{2,3}$ is independent of θ_3. Similarly, the centroid of the composite body, $C_{2,3}$, must lie on the joint axis $\underline{b}_2$, because the first two joints are the DI-1 type. Thus,

$$m_2 g_{2,2} + m_3 g_{2,3} = 0 \tag{56}$$

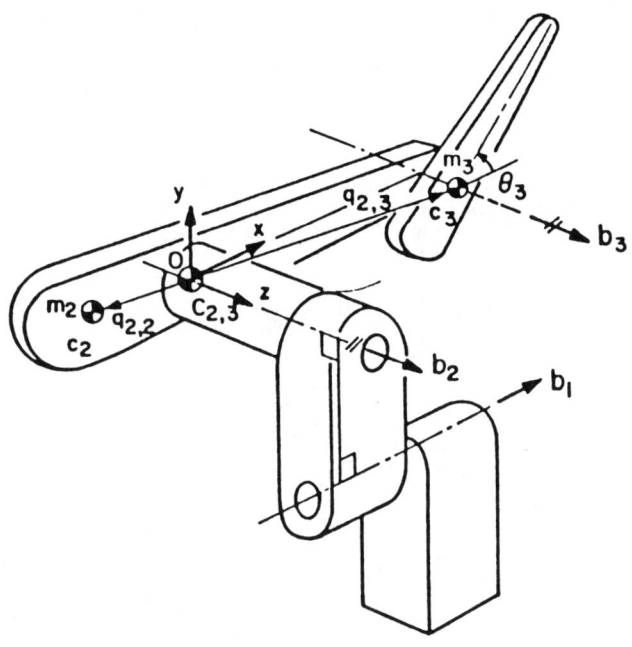

Fig. 9

In addition, according to (50-3), which is a mass property condition for model DI-1, all the off-diagonal elements of the composite inertia tensor $\bar{N}_{2,3}$ must be zero for an arbitrary θ_3. For this to be true, the right hand side of (54) is at least independent of θ_3. Since all the terms except the third term in (54) are independent of θ_3, we need to evaluate only the third term. This yields additional conditions for the inertia tensor of link 3 given by

$$I_{3xy} = I_{3yz} = I_{3xz} = 0 \qquad (57-1)$$

$$I_{3xx} = I_{3yy} \qquad (57-2)$$

Substituting .(55),(56) and (57) into .(54) and evaluating the composite inertia tensor $\bar{N}_{2,3}$, we can derive additional conditions on link 2,3 that are necessary to satisfy (50-2) and (50-3) for an arbitrary θ_3.

$$I_{2xy} = q_x q_y m_2 (1 + \frac{m_2}{m_3}) \qquad (58-1)$$

$$I_{2yz} = q_y q_z m_2 (1 + \frac{m_2}{m_3}) \qquad (58\text{-}2)$$

$$I_{2xz} = q_x q_z m_2 (1 + \frac{m_2}{m_3}) \qquad (58\text{-}3)$$

$$I_{2xx} = I_{2yy} + (q_x^2 - q_y^2) m_2 (1 + \frac{m_2}{m_3}) \qquad (58\text{-}4)$$

As before, the subsitution of the above equations into the general inertia matrix in (9) reveals that the above necessary conditions provide the sufficient conditions as well. Thus the set of conditions we derive are necessary and sufficient for the 3 d.o.f. arm consisting of the DI-1 and I-1 joint types. Similarly the other types of arms are analyzed by using (54), and additional mass property conditions are determined. Thus possible designs for 3 d.o.f. arms with configuration-invariant inertia are all found.

6. Conclusion

This paper first introduced the concepts of decoupled and/or configuration-invariant inertia. The design theory to accomplish these particular forms of inertia were then presented. To this end, first the necessary conditions for the kinematic structure and mass distribution to possess the decoupled and/or configuration-invariant inertia were derived. The necessary conditions provided useful guidelines for the arm linkage design to accomplish the goals.

For decoupling, all the joint axes must be orthogonal. Consequently, manipulator arms with more than two degrees of freedom cannot be dynamically decoupled by design. Based on this design guideline, all of the arm constructions that yield the decoupled inertia matrices were determined.

For the configuration-invariant inertia, there is no limitation regarding the degrees of freedom. In this paper, 2 and 3 degree-of-freedom arms were considered. For more than three degrees of freedom, the same approach can be applied. However, the higher the number of d.o.f. becomes, the more difficult finding arm designs becomes, which are practically useful.

References

[1] H.Asada and K.Yocef-Toumi, "Analysis and Design of a Direct-Drive Arm with a Five-Bar-Link Parallel Drive Mechanism", ASME Journal of Dyn. Sys. Mea. Cont., Vol.106-3, pp.225-230, 1984

[2] S.Dubowsky and D.T.Des Forges, "The Application of model-Referenced Adaptive Control to Robotic Manipulators", ASME Journal of Dyn. Sys. Mea. & Cont., Vol.101-3, pp.193-200, 1979.

[3] M.Takegaki and S.Arimoto, "An Adaptive Trajectory Control of Manipulators", Int. Journal of Control, Vol.34-2, pp.219-230, 1981.

[4] F.Freund, "Fast Nonlinear Control with Arbitrary Pole Placement for Industrial Robots and Manipulators", Int. Journal of Robotics Research, Vol.1-1, pp.65-78, 1982.

[5] J-J.E.Slotine, "Sliding conttroller Design for Nonlinear Systems", Int. Journal of Control, Vol.40-2, 1984.

[6] J.Y.S.Luh, M.W.Walker, and R.P.Paul, "On Line Computational Scheme for Mechanical Manipulators", ASME Journal of Dyn. Sys. Mea. & Cont., Vol.102, 1980.

[7] J.M.Hollerbach, "A Recursive Lagrangian Formulation of Manipulator Dynamics and a Comparative Study of Dynamics Formulation Complexity", IEEE Tran. on Sys.,Man. & Cybernetics, pp.730-736, 1980.

[8] Christopher G. Atkeson, Chae H. An, John M. Hollerbach, "Estimation fof Inertial Parameters of Manipulator Loads and Links", Third Inernational Symposium of Robotics Research, pp.32-39, 1985.

[9] H.Mayeda, K.Osuka and A.Kangawa, "A New Identification Method for Special Manipulator Arms", 9th World Congress of the International Federation of Automatic Control, pp.74-79, 1984.

[10] H.Asada, "A Geometrical Representation of manipulator Dynamics and Its Application to Arm Design", ASME Journal of Dyn. Sys. Mea. & Cont., Vol.104-3, 1983.

[11] H.Asada and J-J.E.Slotine, "Robot Analysis and Control", John Wiley & Sons, 1986.

Graphical-Interactive System for CAD and Simulation of Manipulation Systems

L. Lilov, B. Bekjarov, M. Lorer and V. Atanasov

Institute of Mechanics and Biomechanics
Bulgarian Academy of Sciences, Sofia, Bulgaria

SUMMARY

CAMS is a graphical-interactive software system for CAD and simulation of manipulation systems, based on an originally developed analytical theory. It deals with fundamental problems concerning structure, kinematics, dynamics and accuracy.

CAMS could be applied both in the analysis of already existing manipulation robots for full utilization of their resources in a specific technological process, or in the preliminary design stage for synthesis of systems with optimal characteristics.

CAMS could also be applied at the design stage of robot-automated work places and FMS elements, to select the optimal configuration.

Using CAMS for design and engineering one should model the future system with the aid of the computer and, in this way avoiding the need for experiments, should choose the optimal technological decision.

The CAMS software system is available in several versions for mini-computers PDP 11-34, SM-4 and microcomputers PERQ and IBM - PC/XT.

The proposed software system is intended to bring together the designer's intuition with powerful mathematical methods and algorithms.

BASIC FUNCTIONS

CAMS solves the following fundamental problems, which appear in the process of evaluation and design of manipulation robots.

- direct kinematical problem. This program module provides determination of position, linear and angular velocity and acceleration or an arbitrary link in conformity with given generalized coordinates, velocities and acceleration;
- movement simulation (animation) of the manipulation system (figure 1);
- work space volume. This program module computes the volume of the work space, defined as a set in 3-D space which could be reached by a fixed gripper point;
- approach coefficient computation and visualization (1). This coefficient characterizes the facility of the manipulator to position the gripper point with

different configurations in an arbitrary point of the work space;
- mobility coefficient computation and visualization (1). This coefficient characterizes the possibilities of the manipulator to cause different linear velocities of the gripper point in an arbitrary configuration;
- inverse kinematic problem. This program module determines the generalized coordinates, velocities and accelerations realizing a prescribed motion of the gripper;
- general intertia coefficient computation and visualization (2). This coefficient characterizes the intertial properties of the manipulator in an arbitrary configuration (figure 2);
- accuracy analysis. This program module estimates statistically and determines the extreme values of errors in positioning and orientation of the gripper. All important error sources are taken into account, namely errors in the generalized coordinates execution, clearances in the joints and errors arising from imprecise realization of the links geometry (3) (figure 3);
- direct dynamical problem. This program module computes the generalized coordinates, velocities and accelerations and joint reaction forces and torques in conformity with a given external loading and with servo forces and torques (4, 5);
- inverse dynamical problem. This program module computes the servo forces and torques in conformity with external loading and prescribed motion of the gripper (4, 5);
- actuators test. This program module estimates the capabilities of chosen actuators, to realize the motion of the gripper with prescribed linear velocity.

It promises to include some new modules in CAMS, concerning strength analysis of manipulator links and accuracy tolerance synthesis.

THE CAMS STRUCTURE

CAMS is an interactive system designed on the module principle with a three-level-hierarchical structure.

The first level modules represent the HELP information for the user, allocate the computer memory and organize the dialogue and the work of the second level modules.

The second level modules solve the mechanical problems.

The third level modules illustrate the results of the second level modules' work.

The full HELP information is stored in peripheral devices (8 - inch floppy or hard disk).

INPUT DATA STRUCTURE

CAMS requires the minimum input data, which defines the considered system, such as the kinematic scheme, the link geometry and control parameters. The CAMS input data may be conditionally separated into three groups.

The first group consists of all the input data concerning the mechanics of the evaluated system.

The second group consists of control data for the second level programs.

The third group consists of control data for the third level programs.

CAMS offers powerful techniques for data input and correction. e.g. it is possible to enter analytical functions including standard functions sin, cos, exp, ln, sqrt and the constants π and exp as symbol strings. Also there exists the possibility of describing in an interactive mode only the geometry and material of the links, instead of entering the masses, centers of mass coordinates and the inertia tensors which are defined automatically by CAMS. The user can obtain HELP information about every concrete CAMS module input data. An additional utility is the possibility of using implied data and standard data files for some typical manipulation systems.

INTERACTION WITH CAMS

There are three modes of interaction with CAMS

When the first mode is activated the user may select from several possible variants how to continue his work. This is done with the aid of the CAMS menus.

In the second mode the user must answer "yes" or "no" to questions asked by CAMS.

In the third mode CAMS puts a question to the user about the value of some input parameters or for input/output file-name. In this case the desired value or name should be entered from the key-board.

GRAPHICAL OUTPUT

The results of CAMS' work can be illustrated using the graphical programs VSL, HST, FNC and TDG.

The VSL program presents graphically the values in a rectangle defined by the user (2-D scalar field) of an arbitrary scalar characteristic for functional capabilities of manipulation systems (figures 2, 3).

The FNC program displays on the screen an arbitrary one-argument function graphics.

The HST program displays on the screen the histogram (distribution function) of an arbitrary scalar characteristics for functional capabilities of manipulation systems.

The TDG program gives the user a 3-D axonometric view of a scalar field behaviour.

Varying the input data (axonometric angles) the user obtains only the visible, from different angles parts, of the considered scalar field.

REFERENCES

[1] Lilov L., Bekjarov B., Geometrical and kinematical qualitative characteristics for functional capacities of manipulation systems. V CISM - IFToMM Symposium, Udine, 1984.

[2] Asada H., Geometrical representation of manipulator dynamics and its application to arm design. Journal of dynamic systems, Measurement and control, IX, 1983.

[3] Lilov L., Bekjarov B., Accuracy of systems with tree-like structure and arbitrary joints. Theoretical and applied mechanics. N 1, 1983.

[4] Lilov L., Lorer M., Dynamic analysis of multirigid-body system based on the Gauss Principle. ZAMM 62, 1982.

[5] Lilov L., Structure, kinematics and dynamica of systems of rigid bodies. Advances in mechanics. N 6, 1983.

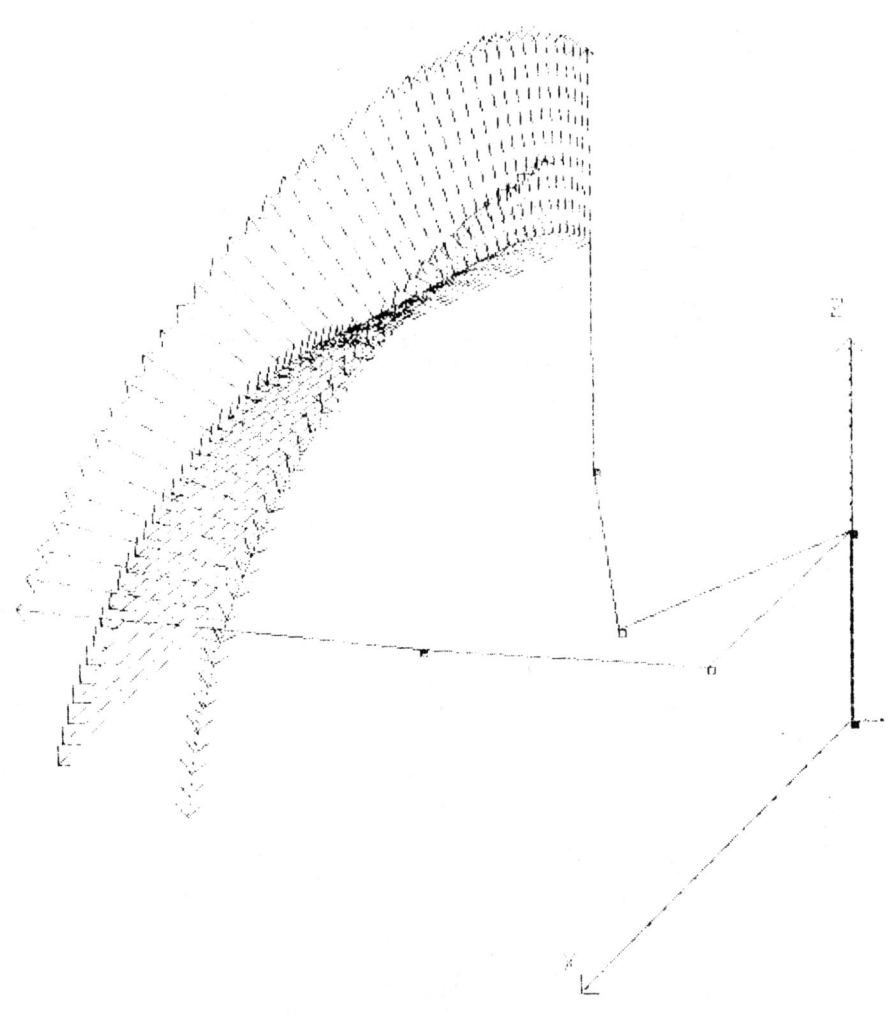

Fig. 1. Movement simulation for UNIMATE 2030 industrial robot.

CAD and Simulation of Manupulation Systems 251

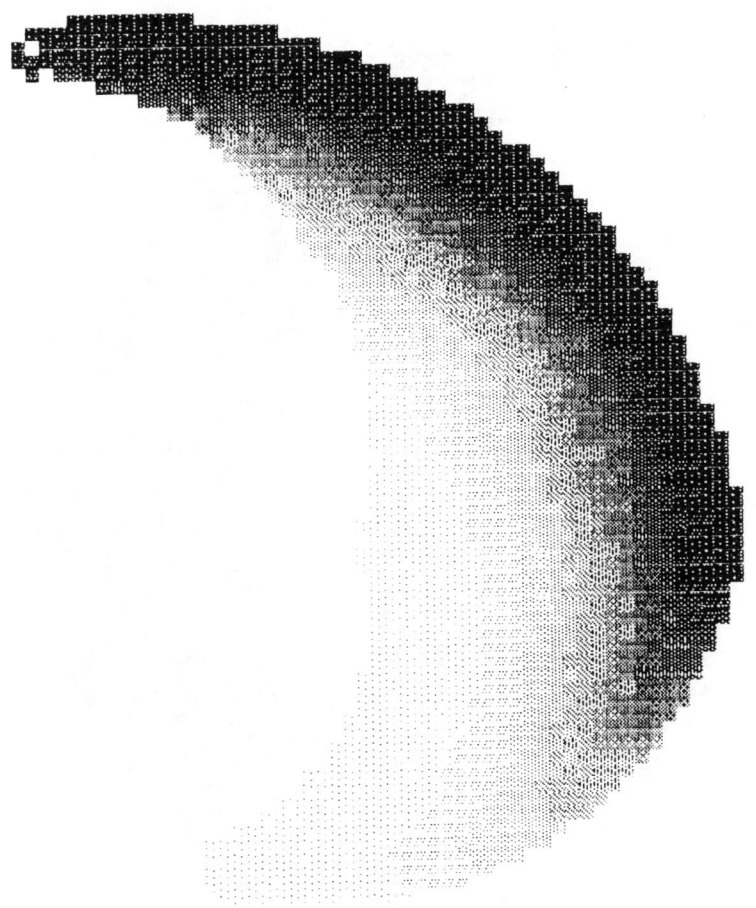

Fig. 2. Visualization of the general inertia coefficient in the working space of COAT-A-MATIC industrial robot.

Fig. 3. Visualization of the maximal positioning errors in the working space of COAT-A-MATIC industrial robot.

Part 5
Sensing and Machine Intelligence 1

Force Feedback in Telemanipulators

R. Bicker and L. Maunder

University of Newcastle-upon-Tyne, England

SUMMARY

The paper gives a brief review of the development of force-feedback in telemanipulators as they have progressed from mechanically linked assemblies to computer augmented generalized control systems.

An experimental facility is described based on three planar three degree of freedom linkages consisting of a master arm articulated at shoulder, elbow and wrist, a kinematically similar slave arm, and a second slave arm incorporating a kinematically dissimilar pantograph mechanism. Force sensors strategically positioned provided feedback in each of the two modes of operation via computer-based transformations between the measured forces and the master arm.

A series of tests based on the peg-in-the-hole task was conducted to investigate the effect of the leading parameters on performance as measured by time to completion. It was found that the force-reflection ratio could be optimized for the task in hand.

REVIEW

Early telemanipulators were based on direct mechanical connections which provided intrinsic force feedback to the operator, and which still account for a high proportion of existing systems in nuclear installations. Their performance, however, is limited by the characteristics of the transmission, and increasingly difficult requirements in space and other hostile environments has led to advances in electrically-linked computer-based methods of force-feedback.

The force-reflecting master-slave manipulators developed by Goertz (1,2) provided a firm foundation for progress in remote system technology. Examples of more recent products include replica manipulators for remote maintenance fitted with improved mechanical components (3,4); the Bilarm

83 for a reflecting system which utilizes direct drive at the joints (5), and a micro-processor controlled bilateral servo-manipulator (6). Where direct kinematic correspondence between master and slave arms is prohibited by restrictions in work-space, the coordination between input and output motions has led to alternative arrangements. Resolved Motion Rate Control (RMRC) described by Whitney (7) used multi-axis joysticks instead of the master-arm: motions of the joysticks are resolved to drive the joints of the slave via transformation from world to tool coordinates and a microcomputer to determine in real time the individual joint rates.

The terminal pointer system described by Saenger and Pegden (8) extended RMRC to a three degree of freedom hand-controller and a terminal mounted camera having rate-control along the camera viewing axis. Booker and Smith (9) recommended a controller for the Shuttle manipulator having force-feedback, position indexing and variable position gain ratios, and Whitney et al (10) reviewed the general relationships between task specification, structural elements and control strategies for industrial and space applications.

Systems developed at the Jet Propulsion Laboratory (11,12,13) include a six degree of freedom controller and kinematically dissimilar slave and master arms, with force-feedback information resolved in real time. Supervisory remote manipulation in which a computer controls sub-tasks on behalf of the operator place emphasis on the ability of the command language to provide smooth interchange between the operator and the computer, as discussed by Ferrell and Sheridan (14,15,16). Distributed microprocessor systems and the advantages of 'smart' displays are being developed at the Jet Propulsion Laboratory (17) and in digital form at the Oak Ridge National Laboratory (18). Areas of promising future development include extensions of supervisory control, applications of teach/playback, automatic camera tracking of end-effectors, and self-adaptive adjustments for compliance and varying loads.

Considerable attention has been given to quantitative measures of performance in attempts to limit the subjective nature of the results, as noted by Sheridan (19). An index of difficulty, I_d, has been proposed by Fitts (20) which relates distance moved by an end-effector to final tolerance in positioning tasks such as fitting a peg into a hole with specified clearance ratios. The index has been applied by McGovern (19,20) to evaluate the performance of two different manipulator systems, where it was found that it provided a valid measure over a small range of operation, and by Ferrell (21), Sheridan (22), McGovern (23) and Hill(24). A dexterity quotient due to Flatau (25) is based upon a well-defined scoring and timing measure which yields a single number: its features include an exact definition of the task in hand and procedures which reduce the influence of acquired skill. Vertut (26) has proposed a dexterity factor which depends on comparing the times to completion of a task carried out manually and with a manipulator. For members of the same family of manipulators, Jelatis (27) refers to the effects of varying mechanical properties including backlash, damping, compliance and mass on responses to standard tasks.

Tests reported by Wilt (28) compared the performance of rate control systems with those employing force-feedback, using the same slave arm but different master arms, and found that for the tasks considered force-feedback offered advantages in terms of task completion time.

EXPERIMENTAL FACILITY

The equipment used in the experimental study consisted of three planar three degree of freedom electrically servoed arms as illustrated in Fig.1. In one combination, the master and slave arms were kinematically identical and in the other generalized control between dissimilar master and slave arms was achieved using a microcomputer to provide resolved position and force signals in real-time. A schematic of the two configurations is shown in Fig.2.

The master arm employed stiff structural elements, and tension cable transmission with high reduction, low backlash, gearboxes mounted so as to reduce inertial mass. Individual joints were servoed using conventional high-speed D.C. servo motors. Strain gauges strategically placed on each link measured joint torques and tight positive torque feedback was incorporated within each servo loop to minimise the influence of friction in the gearboxes. The servo control loop between master and slave arm for a single joint is illustrated in Fig.3.

A dissimilar slave arm, as shown in Fig.4, was based upon a pantograph mechanism developed at AERE, Harwell; it was servoed using brushless D.C. servomotors driving recirculating ball screws for in-plane movements and a conventional D.C. servomotor with a low backlash harmonic gearbox for wrist rotation. Force feedback information from the pantograph slave arm was derived from a two-axis load cell mounted between the X-Y coordinate platform and the input extension arm, together with strain gauges mounted on the terminal joint to measure joint torque. The control scheme was highly dependent upon digital computation to resolve the master arm joint angles into slave arm joint commands, and in the bilateral mode, capable of resolving force/torque feedback information sensed at the slave arm into joint torques at the master arm. Fig.5 illustrates the functional interrelationships.

The implementation requires the master arm joint angles, represented by a vector $\bar{\theta}_m$, to be resolved into the cartesian coordinate frame $\bar{X}_m$, a vector denoting the position and orientation of the terminal joint of the master arm. Where $\bar{X}_m = \bar{L}_m \cdot \bar{\theta}_m$ (see Fig.6)

$$\bar{X}_m = \begin{bmatrix} x & y & \psi \end{bmatrix}^T, \quad \bar{L}_m = \begin{bmatrix} L_1 & L_2 & 1 \end{bmatrix}^T$$

and

$$\bar{\theta}_m = \begin{bmatrix} \cos(\theta_1) & \cos(\theta_1 + \theta_2) & 0 \\ & \sin(\theta_1 + \theta_2) & 0 \\ \sin(\theta_1) & 0 & (\theta_1 + \theta_2 + \theta_3) \\ 0 & & \end{bmatrix}$$

The joint forces/torques measured at the slave arm, denoted by vector $\bar{F}_s$, are resolved into joint reaction torques at the master arm, $\bar{T}_m = \bar{L}_\theta \cdot F_s$, where

$$\bar{T}_m = \begin{bmatrix} T_1 & T_2 & T_3 \end{bmatrix}^T, \quad F_s = \begin{bmatrix} T_\phi & F_x & F_y \end{bmatrix}^T$$

and

$$\bar{L}_\theta = \begin{bmatrix} 1 & 0 & 0 \\ 1 & L_2 \sin(\theta_1 + \theta_2) & -L_2 \cos(\theta_1 + \theta_2) \\ 1 & L_1 \sin(\theta_1) & -L_1 \cos(\theta_1) \end{bmatrix}$$

The control program was written in Fortran and all analogue Input/Output derived using 12 bit resolution. A sine/cosine look-up table was constructed in hardware to eliminate time consuming trigonometric function calls.

EXPERIMENTAL TESTS

The test programme was designed to compare the different control modes employed in the two configurations when applied to the insertion of a peg in a hole. The peg was mounted directly on the end-effector of each slave and two sets of bushes providing a range of tolerances were fitted in the fixed hole. A schematic of the test cycle is shown in Fig.7.

Task completion time was recorded in each test and each test was repeated 25 times in succession to obtain a statistical mean and standard deviation. To limit the effect of operator fatigue, each experimental session lasted for no more than two hours, including rest periods. After each parameter change the operator was allowed a period of time for familiarization with the new conditions, and as far as practical the changes were undertaken randomly to offset operating learning. A full 5 x 5 factorial survey of the 'hole separation distance' and 'peg to hole' clearance was carried out in the bilateral replica master-slave control mode with unity force reflection ratio.

As illustrated in Fig.8, and as predicted by Fitts, an exponential relationship was found to exist between task difficulty (as indicated by the extent of clearance) and task completion time. The standard deviation associated with separation distances of 100 and 200 mm were virtually independent of task difficulty, but because statistical confidence in results obtained at larger distances was much reduced, (probably because of inhibited performance by the operator) all subsequent tests were conducted with a hole separation distance of 100 mm.

Although the importance of force-feedback in remote telemanipulation is well recognised, information on the effect of force-reflection ratios on performance is sparse. From the tests in hand, an interesting relationship, as illustrated in Fig.9, was observed. As the force-reflection ratio was reduced from high values, an improvement in performance as measured by task completion time was observed until an optimum state was reached, after which further reductions in the ratio led to worsening results as unilateral control was approached. Marked improvements compared with unilateral control were found with force-reflection ratios as low as 1:16, and best performance at ratios in the region of 1:8. Although the results clearly depend on the nature of the task, which in the tests so far conducted generate large forces for small misalignments, it would be expected that force-reflection ratios which are well related to the sensitivity of the hand would optimise performance for other tasks as well.

Limited tests using generalized control have been completed, and some useful results have been obtained. As for the replica configuration, the relationship between task difficulty and performance was found to be exponential. The bilateral control mode at a force-reflection ratio of about 1:4 offered significant improvement over unilateral control as shown in Fig.10.

Preliminary tests using the generalized control scheme are not yet completed although some interesting results can be reported. As expected, an exponential relationship existed between task difficulty and time under the different control modes. Bilateral control (having a force/reflection ratio of 1:4), yielded improved performance over unilateral control. However, a passive force compliant device, called the Remote Centre Compliance (RCC)(29), located between the slave arm terminal post and peg produced further improvement when operated in a Unilateral mode (see Fig.10).

Investigation of system parameters, e.g. compliance, inertia and friction are presently being evaluated, based upon a more quantitative performance criteria, which it is anticipated will permit a more critical assessment of those factors influencing performance.

An experimental facility consisting of both similar and dissimilar master-slave arms has shown that the inclusion of force-feedback offers useful advantages in performance, and that the force reflection ratio can be optimized for peg-in-the-hole tasks.

ACKNOWLEDGEMENTS

The authors acknowledge with thanks the support of the Atomic Energy Research Establishment, Harwell.

REFERENCES

1. Goertz, R.C. Fundamentals of general purpose remote manipulators; Nucleonics - Vol.10, No.11, 1952.
2. Goertz, R.C. General Purpose Manipulators; Nucleonics - Vol.12, No.11,1954.
3. Flatau, C.R. et al MA22 - A compact bilateral servo master-slave manipulators; Proc. of 20th Conf. on Remote Systems Technology, 1972.
4. Flatau, C.R. SM229 - A new compact servo master-slave manipulator; Proc. of 25th Conf. on RST, 1977.
5. Yamamoto M. et al Remote maintenance equipment for hot cell facilities; Proc. of 30th Conf. on RST, 1982.
6. Suzuki, M. et al A bilateral - servo manipulator for remote maintenance in Nuclear facilities; Proc. of 30th Conf. on RST, 1982.
7. Whitney, D.E. Resolved Motion Rate Control (RMRC) of manipulators and human prostheses; Trans. of IEEE, Vol. MNS-10, No.2, 1969.
8. Saenger, E.L. and Pegden, C.D. · Terminal pointer hand controller and other recent teleoperator controller concepts; Proc. of 1st Int. Conf. on Remotely Manned Systems, NASA, 1972.
9. Booker, R.A. and Smith, G.W. X - Reference bilateral control for the Shuttle attached manipulator; IEEE Intercon. Int.Conf. & Exposition, Vol.4, 1973.
10. Whitney, D.E. et al Design and control considerations for industrial and space manipulators; Joint Automatic Control Conf., 1974.
11. Bejczy, A.K. and Brooks, T.L. Advanced control techniques for teleoperation in earth orbit; 7th Annual Symp. of Unmanned Vehicle Systems, 1980.
12. Handlykken, M. and Turner, T. Control system analysis and synthesis for a 6 degree of freedom universal force-reflecting hand controller; Proc. of 19th Conf. on Decision and Control, IEEE, 1980.
13. Bejczy, A.K. and Handlykken, M. Generalization of bilateral force-reflecting control of manipulators; Proc. of 4th Romansy CISM IFToMM, 1981.
14. Ferrell, W.R. Command language for supervisory control of remote manipulation; Proc. of 1st Int. Conf. on Remotely Manned Systems, NASA, 1963.
15. Ferrell, W.R. and Sheridan, T.B. Supervisory control of remote manipulation; IEEE Spectrum, 1967.
16. Ferrell, W.R. Adaptive supervisory control of remote manipulation; IEEE Conf. on Decision and Control, 1977.
17. Paine, G. Microprocessors for real time displays and control of space teleoperators; Proc. of 1st Int. Symp. on Mini and Microcomputers in Control, IEEE, 1979.
18. Martin, H.L. et al Distributed digital processing for servo-manipulator control; Proc. of 30th Conf. on RST, Vol.2, American Nuclear Society, 1982.
19. Sheridan, T.B. Evaluation of tools and tasks: reflections on the problem of specifying robot manipulator performance; Performance and Evaluation of Programmable Robots and Manipulators, National Bureau of Standards, Report No. SP459, 1976.
20. Fitts, P.M. The information capacity of the human motor system in controlling the amplitude of movement; J.of Experimental Psychology, Vol.47,No.6, 1954.
21. Ferrell, W.R. Delayed force feedback; Proc. of IRE, Human Factors, Vol.8, Part 5, 1966.
22. Sheridan, T.B. and Ferrell, W.R. Remote manipulative control with transmission delay, IEEE, Vol. HFE4, 1963.
23. McGovern, D.E. Comparison of two manipulators using a standard task of varying difficulty; ASME Paper 74 - WA - B10.4, 1974.
24. Hill, J.W. Study of modelling and evaluation of remote manipulation tasks with force feedback; Jet Propulsion Lab, Report No.95-5178, 1979.
25. Flatau, C.E. et al Some preliminary correlations between control modes of manipulator systems and their performance indices; Proc. of 1st Int. Conf. on Remotely Manned Systems, NASA, 1973.
26. Vertut, J. et al Contribution to define a dexterity factor for manipulators; Proc. of the 21st Conf. on Remote Systems Technology, 1973.
27. Jelatis, D.G. Characteristics and evaluation of master-slave manipulators; NBS Workshop, Report No. SP459, 1976.
28. Wilt, D.R. et al An evaluation of control modes in high gain manipulator systems; IFToMM Mechanism and Machine Theory, Vol.12, 1977
29. Whitney, D.E. and Nevins, D.L. What is the RCC and what can it do? Proc. of 9th ISIR, 1979.

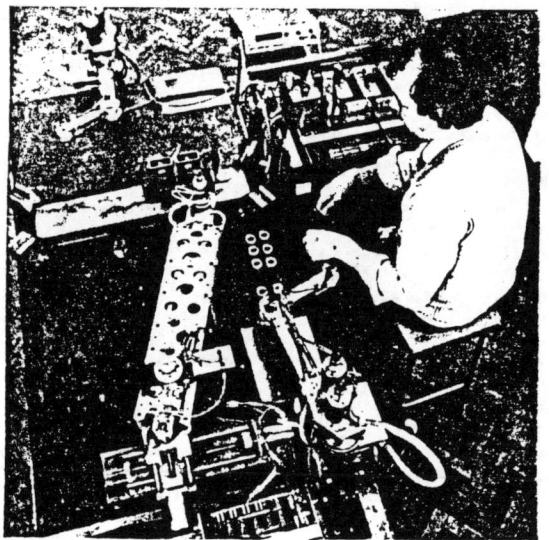

Fig. 1. View of Experimental System

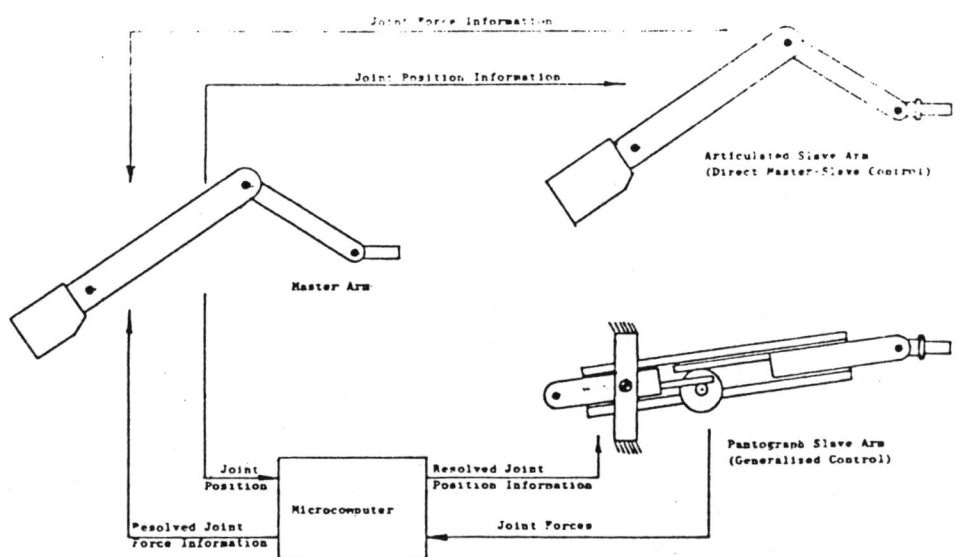

Fig. 2. Schematic of System Configurations

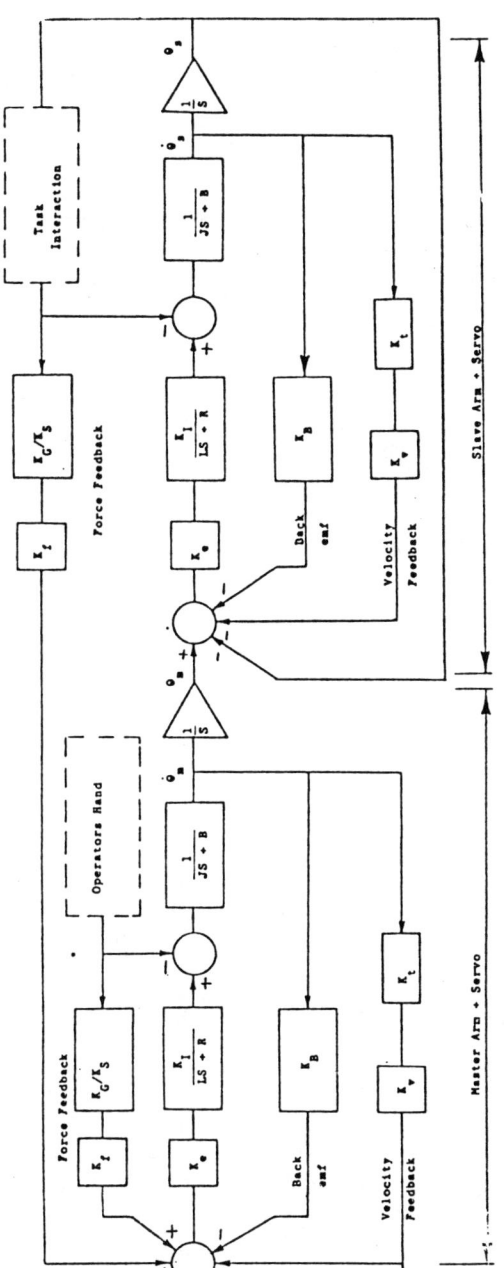

Fig. 3. Servo Control Look Used in Master-Slave Configuration

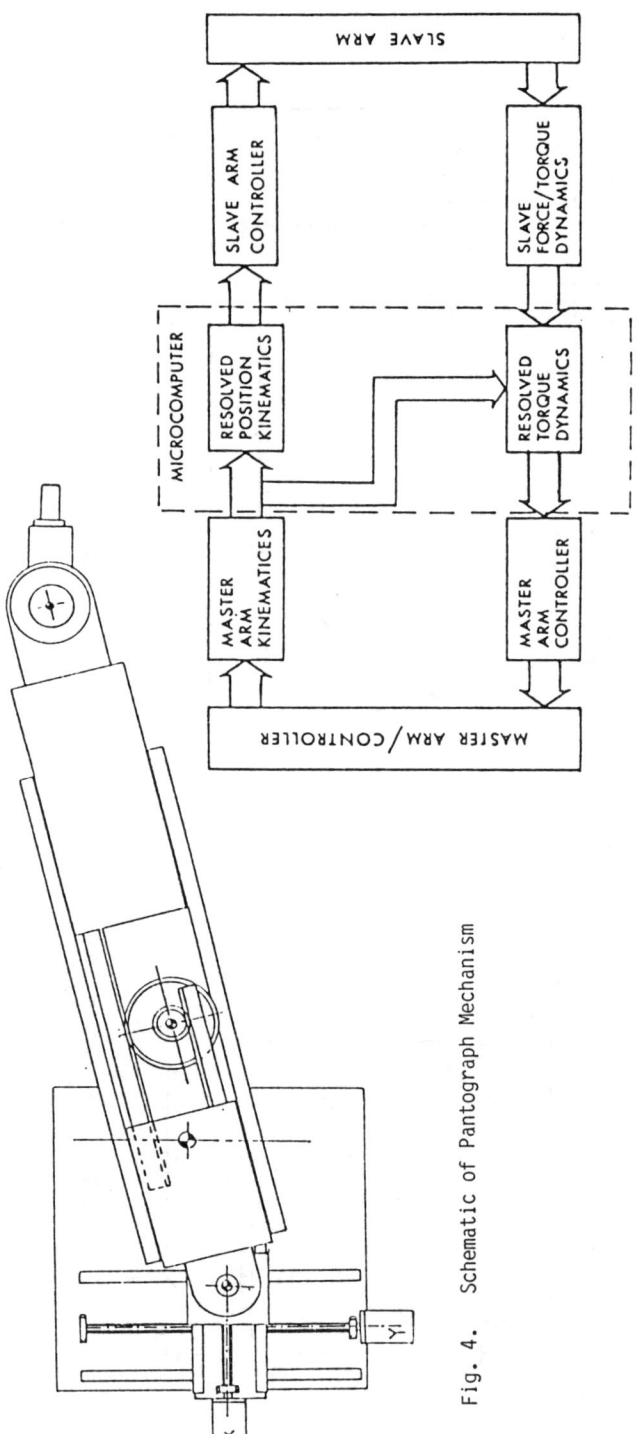

Fig. 4. Schematic of Pantograph Mechanism

Fig. 5. Functional Block Diagram of Generalizated Control

Force Feedback in Telemanipulators 265

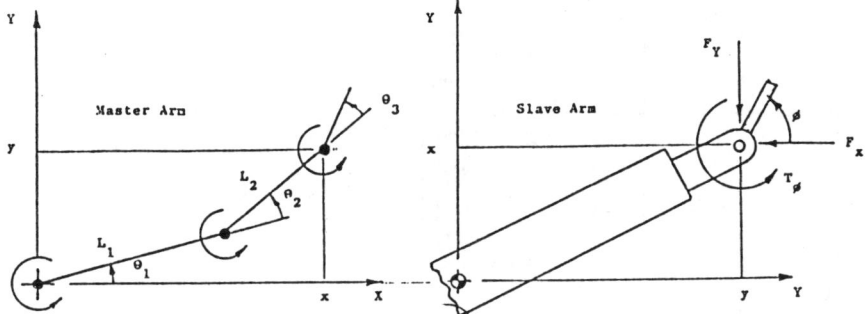

Fig. 6. Kinematic and Dynamic Notation

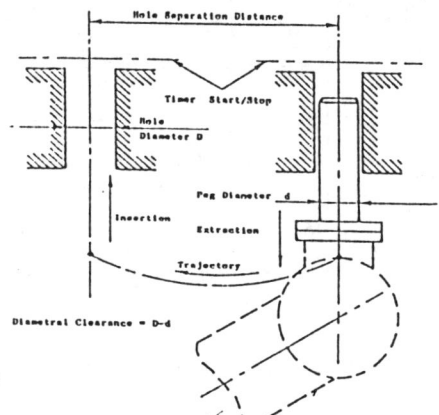

Fig. 7. Schematic of Test Sequence

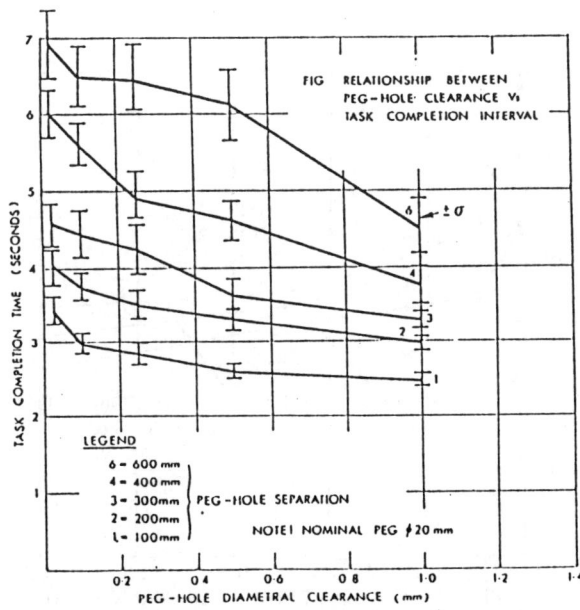

Fig. 8. Relationship between Task Difficulty vs Test Time

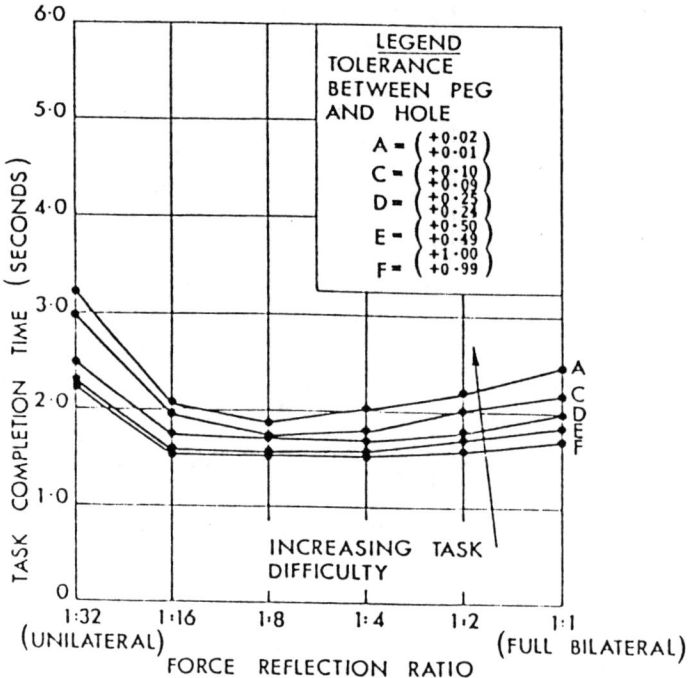

Fig. 9. Relationship between Force Reflection vs Test Time

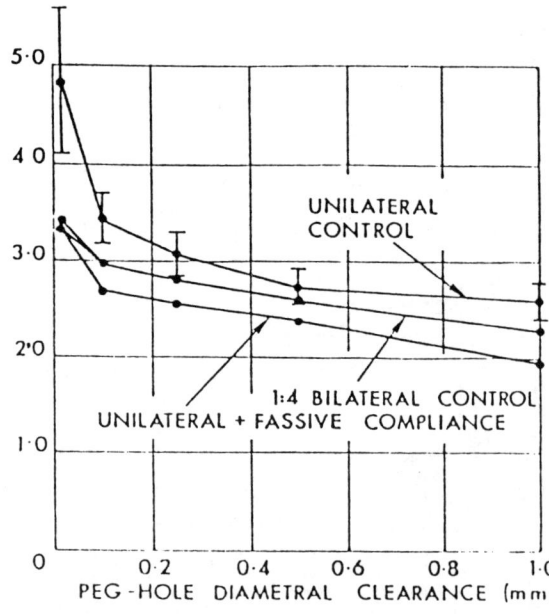

Fig. 10. Generalized Control Modes

Theoretical and Experimental Investigations of Optical Fibre Reflective Sensors for Robotics

E. Marszalec, J. Marszalec* and A. Morecki**

*Polytechnic of Lublin, Poland, and **Technical University of Warsaw, Poland

ABSTRACT

The scale of application of robots in industry or other fields will, to a high degree, depend on the robots' sensing capabilities. Robot sensors are an area of extensive research activity. Over the past few years different types of vision, proximity, touch, slip and force sensors have been developed. Recently there has been rapid development in fibre optics and optical fibre sensors for different industrial applications.

This paper presents two kinds of optical fibre reflective sensors for robotics, namely the optical fibre proximity sensor and the optical fibre colour recognition sensor. Models of both sensors are presented. Investigation of models is described and results obtained are briefly discussed.

Although both sensors were developed for robotics, they can easily be used in other industrial fields.

1. INTRODUCTION

During the last few years we have observed an increasing interest in robots and manipulators. This is due to the needs of automation in different technological processes, development of new technologies as well as labour safety and increased productivity.

In the near future the application of robots in industry will to a high degree depend on the robot's sensing capabilities. Different types of vision, proximity, touch, slip and force sensors have been developed so far. Most of sensors are built with electrical and electronic elements, located in the gripper. However, in many cases building high sensitivity electronic elements into a robot gripper is inadvisable. Electronic elements placed on a working robot gripper may make it impossible for the robot to carry out various operations; moreover they may be damaged during execution of the operation.. Taking this into account it is reasonable to remove electronic elements from the robot gripper to a protected environment. This can be done using optical fibre technology.

Application of optical fibres to robotics has other advantages:
- reducing the dimensions of robot gripper,
- elimination of electrical current from the (sensitive) robot gripper,
- insulated construction, so that the robot gripper with sensors can be used in high voltage, electrically noisy, high temperature, corrosive, or other hazardous environments.

The first type of optical fibre sensors for robotics are reflective sensors. In this type of sensor optical fibres are used to carry the light to and from the transducer. Research on this type of sensors has been described by Bejczy (1,2), Catros and other (3). Optical fibre reflective sensor for robotics, developed at the Technical University of Warsaw, namely optical fibre proximity sensor and optical fibre colour recognition sensor, are described in the next sections of this paper.

2. OPTICAL FIBRE PROXIMITY SENSOR

A proximity sensor is defined as a sensor which measures the distance between the surface of an object and the sensor head, at close range, within 100-200 mm. The general concept of electrooptical proximity sensing is described in (2).

An optical fibre proximity sensor (OFPS) uses two separate clusters of fibres: illuminating fibre, which is connected to a modulated LED source, and receiving fibre, connected to a photodetector (see Fig.1). With the sensor a cone-shaped sensitive volume is associated. It produces an output signal whenever a reflecting surface enters this volume.

Optical power transmitted from the LED Source to the photodetector via a reflecting surface can be described by the formula:

$$P_{FD} = T_1 T_2 T_3 T_4 T_5 P_N \quad , \tag{1}$$

where P_{FD} is the power received at the detector (in mW)
P_N is the power emitted from the LED source (in mW),
T_1 is the transmittance between the source and the illuminating fibre,
T_2 is the transmittance of the illuminating fibre,
T_3 is the transmittance of the sensitive volume with the reflecting surface,
T_4 is the transmittance of the receiving fibre,
T_5 is the transmittance between the receiving fibre and the photodetector.

Transmittances T_1, T_2, T_4 and T_5 are constant for a given construction of the sensor and transmittance T_3 depends on the distance z between the sensor head and the reflecting surface. This transmittance can be written:

$$T_3 = KT(z) \qquad (2)$$

where k is the coefficient determined by geometrical parameters of the sensor head (such as fibre diameter), and its numerical aperture N.A., the angle between the illuminating (receiving fibre and the sensor head's axis) and the reflecting coefficient of the surface,

T(z) is the function which describes variation of the transmittance,

T_3 when the distance z changes.

From formulas (1) and (2) it can be seen that the changes of power P_{FD} with the distance z are just the same as the changes of the function T(z). So the output signal of the sensor U_{FD} will change in the manner described the formula given below:

$$U_{FD} = T_1 T_2 T_3 T_4 T_5 Q_{FD} P_N \qquad (3)$$

where Q_{FD} is the coefficient characterizing the photodetector (in $\frac{V}{W}$). Based on this theoretical approach a computer program for calculating theoretical characteristic curve of the OFPS was written. One of the calculated curves is shown in Fig.2. The distance is a nonlinear function of voltage output.

The OFPS was calibrated in laboratory on an optical bench. The sensor head was fixed in a minipulator on the bench. As a reflective surface a flat sheet of white paper was used, placed in front of the sensor head. The experimental curve of the FOPS was obtained by moving the reflecting surface in relation to the sensor head, e.g. by increasing the distance z from 0 to 300 mm. This curve is shown

in Fig.2. The distance is a double-valued function of the output
signal, but as the maximum output signal is reached at the distance
of about 6 mm we can work on the declining part of the characteristic.
The experimental curve confirms that the distance is a nonlinear
function of the output signal. The sensor has higher sensitivity
at small distances.

The PFPS was mounted on the arm of the robot MA 150 and was
used to control collision avoidance process.

3. OPTICAL FIBRE COLOUR RECOGNITION SENSOR

An optical fibre colour recognition sensor (OFCRS) is a device
which can recognize and or identify objects by their colour. As
a reflective sensor it has an illuminating branch and a receiving
branch. The illuminating branch includes a white light source and
an illuminating optical fibre, while the receiving branch consists
of an optical fibre coupler and a few photodetectors equipped with
narrow-band optical filters. A general diagram of OFCRS is shown
in Figure 3. The flux of light emitted from the light source is
transmitted through the illuminating fibre to a coloured object.
Then the flux selectively reflected from the object is transmitted
through receiving fibres to selective photodetectors. Each fibre
of the coupler can be equipped with optical filter transmitting
light from different parts of the visible spectrum.

Light reflected from objects of different colours, placed in
front of the OFCRS, produces different output signals. The output
signal changes when:
- the distance between the sensor head and object of given colour
changes,
- the distance is constant and the object of a new (different)
colour is placed,

Consequently the output signal is simultaneously a function of the
distance between the sensor head and the reflecting surface and of
the chromaticity of the surface:

$$U_{FC} = RT(z)G(\lambda), \qquad (4)$$

where R is the coefficient which is constant for a given structure
of the sensor

T(z) is the function wich describes variation of the transmittance of the sensitive volume of the FOCRS with the change in distance,

G(λ) is the function which describes the chromaticity of the system's sensor-reflecting surface,

Λ is the wavelength.

For a given construction of the sensor the value of the coefficient R is constant as well as the function T(z) has a constant form. If the coloured objects that have to be recognized are placed at the same distance z the output signal U_{FC} changes like the function G(λ):

$$G(\lambda)\Big|_{\lambda_1}^{\lambda_2} = \int_{\lambda_1}^{\lambda_2} H(\lambda)\tau(\lambda)\rho_K(\lambda)\tau_F(\lambda)\eta(\lambda)d\lambda \tag{5}$$

where H(λ) is the emitted spectrum of the light sources

τ(λ) is the spectral characteristics of transmittance of optical fibre

$\rho_K(\lambda)$ is the spectral characteristics of the reflecting surface

$\tau_F(\lambda)$ is the spectral characteristics of transmittance of the optical filter

η(λ) is the spectral characteristics of photodetector

λ_1, λ_2 are the minimum and maximum wavelengths transmitted through the optical filter.

Narrow band optical filters have a transmittance band of a few nanometers, so the output signals of the sensor can be calculated for the wavelength of the maximum transmittance of the filter λ_{max}. Then the ratio of output signals produced by colour objects K and j using the same filter will have the form:

$$\frac{U_{FCk}}{U_{FCj}} \cong \frac{(\rho_k)_{max}}{(\rho_j)_{max}}, \tag{6}$$

where $(\rho_k)_{max}$ and $(\rho_j)_{max}$ are reflection coefficients respectively of object k and j in the field of maximum transmittance of narrow band optical filter. Two objects k and j can be recognized as different when the ratio, described by equation (6) differs from 1.

The output signal as the functions of the distance between coloured objects and sensor head are shown in Fig.4. Curves were obtained experimentally using one filter and 4 objects of different colour: red, green, blue and yellow.

The OFCRS was mounted in the robot gripper and was used for sorting objects of different colours. Results obtained are quite satisfactory.

4. CONCLUSION

In the paper two types of optical reflective sensors, namely, optical fiber proximity sensor and the optical fibre colour recognition sensor, have been described. Results obtained in the theoretical and experimental investigations of both sensors are quite satisfactory. The development models can be used for real time control of a robot. New models of OFPS and OFCRS are under development. Although both sensors were developed for robotics, they can be also used in the other industrial applications.

ACKNOWLEDGMENT

This work was supported by the Polish Academy of Sciences, key-problem 06.9 and is a part of Ph.D. dissertation prepared both by Elzbieta and Janusz Marszalec in the form of doctoral studies at the Technical University of Warsaw, Poland under the supervision of prof. Adam Morecki.

Authors would like to thank Dr Ryszard Romaniuk, Institute of Electronics Fundamentals, Technical University of Warsaw for many valuable discussions on problems of fibre optics.

References

1. Bejczy A.K. Application of fibre Optics to Robotics, IFDC, vol.1, No 2, Nov. 1980.
2. Bejczy A.K. Smart Sensors for Smart Hands, Progress in Astronautics and Aeronatics, Vol.67, AIAA Publ. New York, 1979.
3. Catros I.Y. at al. Automatic Grasping Using Infrared Sensors, 8 ISIR, Stuttgart, 1978.
4. Marszalec E. at al. Application of Optical Fibre Technology to Robotics, Report of ITLiMS PW, Warsaw, Poland, 1981.
5. Marszalec E. Optical Fibre Colour Recognition Sensor and Its Application for Control of Robot, Its National Conference on Robotics, Wroclaw, 1985, (in Polish).
6. Marszalec E. at al. Fibre Optic Instrumentation of a Robot Gripper III National Symposium on Optical Fibres and Their Application, Jablonna, 1983 (in Polish).
7. Marszalec J., Optical Fibre Proximity Sensor for Control of Robot Gripper, 10th National Conference on TMM, Warsaw, 1984, Poland
8. Yoshimoto K., at al., Development of Colour Information Processing System for Robot Vision, Robot Developments 1983, IFS Publications 1983.

Optical Fibre Reflective Sensors for Robotics 273

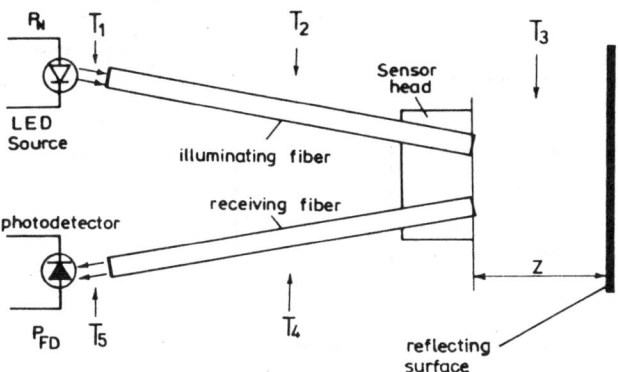

Fig.1. Diagram of optical fibre proximity sensor

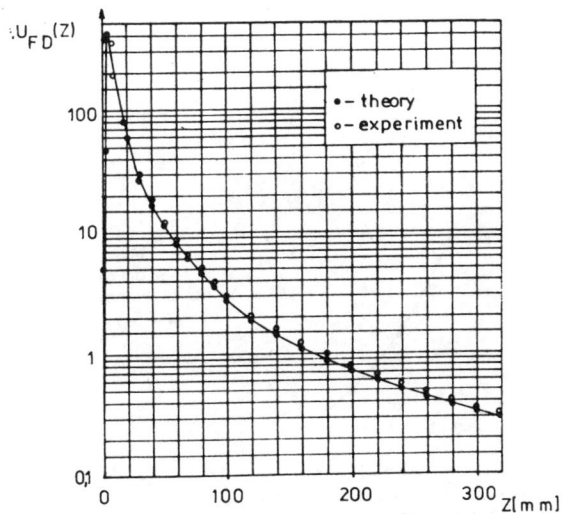

Fig.2. Relative output signal curves of optical fibre proximity sensor

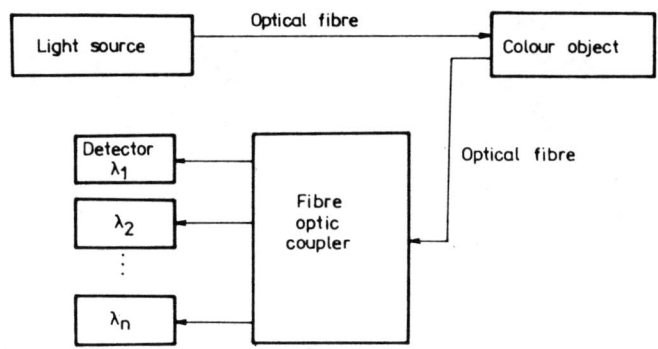

Fig.3. Diagram of optical fibre colour recognition sensor

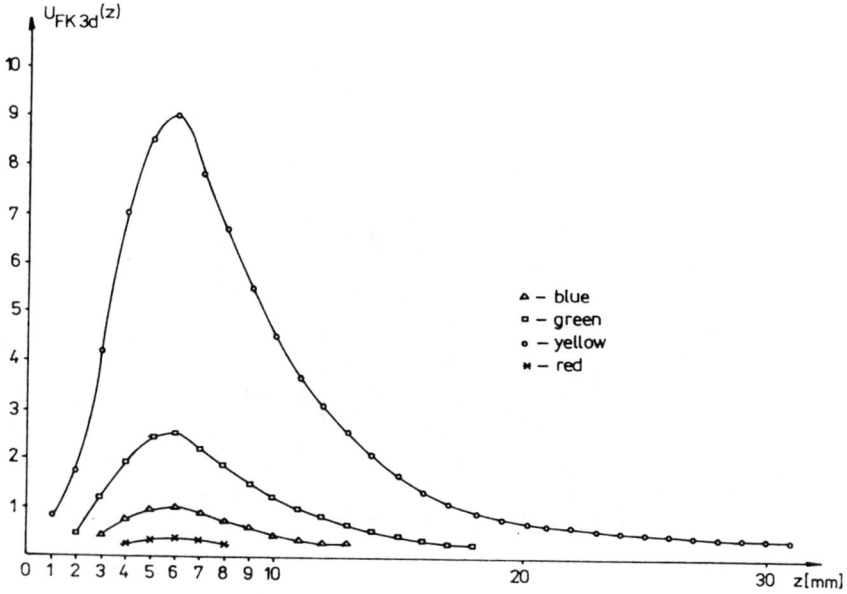

Fig. 4. Relative output signal curver of optical fibre colour recognition sensor.

Task Specification and Closed Loop Control of Manipulators in the Presence of External Sensors

R. Palm, A. Moltmann and H.-J. Horch

Central Institute of Cybernetics and Information Processes
of the Academy of Sciences of the GDR

Introduction

Manipulators in an incompletely described environment require the feedback of the interaction between robot and object world by external sensors.
A formal representation of a sensor guided robot action is characterized by two subtasks. The motion task describes the planned spatial trajectory, and the reaction task represents the interaction between robot effector and object. The control method presented relates to the base equation of manipulation which includes the two subtasks. The closed loop control in spatial coordinates requires an appropriate description of the dynamic behavior of the end effector. Coming from the servocontrolled arm it can be shown that the transient period of each Cartesian coordinate is similar to that one of each manipulator joint. A simple approach leads to an external closed loop control where the robot is modelled as a dead time element in addition to a delay element. To achieve flexible programming we propose a configurator for modules from which a variety of special sensory algorithms (skills) can be constructed. These modules are subdivided into a range of basic and complex functions. An experiment deals with a contouring method by the help of a force sensor. This has already been applied to the process of cleaning of castings.

Task specification

Sensor guided robot actions commonly require the specification both of the planned robot motion and the desired interaction between effector and object /1/. Thus, the robot task can be decomposed into two subtasks:

1. The **motion task** specifies a planned trajectory determining the motion along the contour of an object,
2. The **reaction task** specifies the desired reaction between effector and contour, e.g. reaction forces /2/.

While performing the motion task the reaction task takes care that the planned path in case of inaccuracy of the location and the contour of the object. The formal representation of these subtasks leads to the formulation of an "equation of manipulation" (see fig. 1a)

$$Tx * SEN * TOL = \underbrace{OEN * SJR}_{\text{motion task}} * \underbrace{REC^{-1}}_{\text{reaction task}} \qquad (1)$$

where

Tx	-	Transformation matrix between KH and KX
SEN	-	"	"	"	"	KS	"	KH
TOL	-	"	"	"	"	KT	"	KS
REC	-	"	"	"	"	KF	"	KT
CEN	-	"	"	"	"	KO	"	KF
SUR	-	"	"	"	"	KT	"	KO
KX	-	base frame
KH	-	gripper frame
KS	-	sensor frame
KT	-	tool frame
KF	-	moving object frame
KO	-	fixed frame of the center of the object.

The time discrete representation of (1) yields the motion of the hand

$$Tx,i+1 = (CEN * SUR)d,i+1 * (SEN * TOL * RECd)_{i+1}^{-1} \quad (2)$$

RECd - nominal reaction task.

However, because of the inaccuracies mentioned above and by reason of dynamical effects of the manipulator an actual RECa,i at the time ti is realized. This fact can be considered a difference between the location of the actual and planned object frame KFa,i and KFd,i , respectively. The comparison of the position/orientation vectors $\underline{x}d$ and $\underline{x}a$ of the frame RECd and RECa, respectively, leads, via a control law $\underline{r}(\underline{x}d - \underline{x}a)$ (e.g. PID), to a correction vector

$$\Delta \underline{x}_{c,i+1} = (\Delta \underline{x}, \Delta \underline{\alpha})c, i+1 = \underline{r}(\underline{x}_{d,i} - \underline{x}_{a,i}) .$$

For the next event ti+1 a correction frame RECc,i+1 is computed which results from the actual RECa,i($\underline{x}_{a,i}$) through addition of the correction vector $\Delta \underline{x}_{i+1}$ to the actual vector $\underline{x}_{a,i}$

$$RECc,i+1 (\underline{x}_{i+1}) = REC (\underline{x}_{a,i} + \Delta \underline{x}_{i+1}).$$

The transformation graph in fig. 1b represents the relations among the actual and desired objects frames KFa and KFd, respectively, the tool frame KT, and the base frame KX at ti and ti+1. From this yields the next hand position /orientation matrix

$$Tx,i+1 = (CEN * SUR)d,i+1 * (SEN * TOL * RECc)_{i+1}^{-1} . \quad (3)$$

For (CEN*SUR)d,i = const the algorithm leads to

$$RECa,i \Longrightarrow RECd,i$$

related to the tool frame KT and the actual object frame KFa.

Closed loop control

The servocontrolled robot can be described as a 2nd order system /3/

$$\ddot{\underline{q}} + A * \dot{\underline{q}} + \underline{c} = B * \underline{u}$$
$$\underline{u} = \hat{B}^{-1} * (\hat{A} * \dot{\underline{q}} + \hat{\underline{c}} + \ddot{\underline{q}}_d + P1*(\dot{\underline{q}}_d - \dot{\underline{q}}) + P2*(\underline{q}_d - \underline{q})) \quad (4)$$
$$\text{compensation} \quad \text{feedback-feedforward controller}$$

$\underline{q}$ - vector of joint coordinates
$\underline{q}_d$ - " " " " desired

A, B — matrices of system parameters
$\hat{A}, \hat{B}$ —   "   "   "   "   estimated
$\underline{c}$ — vector of all couplings, and gravitation
$\underline{\hat{c}}$ —   "   "   "   "   ,   "   "   estimated
$P1, P2$ — matrices of control parameters.

All matrices should be diagonal and $\underline{e}_q = \underline{q}_d - \underline{q}$ be sufficiently small. Complete compensation of A, B, $\underline{c}$ yields

$$\underline{\ddot{q}} = \underline{\ddot{q}}_d + P1 * (\underline{\dot{q}}_d - \underline{\dot{q}}) + P2 * (\underline{q}_d - \underline{q}). \tag{5}$$

From the transformations

$$\underline{e}_q = J^{-1} * \underline{e}_x$$

$$\underline{\dot{e}}_q = (J^{-1})^{\cdot} * \underline{e}_x + J^{-1} * \underline{\dot{e}}_x$$

$$\underline{\ddot{e}}_q = (J^{-1})^{\cdot\cdot} * \underline{e}_x + 2*(J^{-1})^{\cdot} * \underline{\dot{e}}_x + J^{-1} * \underline{\ddot{e}}_x$$

J — Jacobian
$\underline{e} = \underline{x}_d - \underline{x}$
$\underline{x}$ — spatial position/orientation vector
$\underline{x}_d$ —   "   "   "   "   desired

and the conditions of identical transfer function poles with

$$P1 = p1 * E$$

$$P2 = p2 * E$$

E — identity matrix

one obtains the system equation of the position/orientation vector in spatial coordinates

$$\underline{\ddot{x}}_x = \underline{\ddot{x}}_{d,x} + p1*(\underline{\dot{x}}_{d,x} - \underline{\dot{x}}_x) + p2*(\underline{x}_{d,x} - \underline{x}_x) + ((J^{-1})^{\cdot\cdot} + p1*(J^{-1})^{\cdot})*\underline{e}_x + 2*(J^{-1})^{\cdot} * \underline{\dot{e}}_x \tag{6}$$

In case of low velocities the nonlinear terms can be neglected so that a linear equation of the error $\underline{e}_x = \underline{x}_{d,x} - \underline{x}_x$ yields

$$\underline{\ddot{e}}_x + p1 * \underline{\dot{e}}_x + p2 * \underline{e}_x = \underline{0} \tag{7}$$

with the same transient function as those of the joint coordinates.
For the design of the controller one has to consider the dead time between sensory input and the output of the computed correction vector to the trajectory generation block. Furthermore, one has to consider the fact that the correction vector is adjusted by linear interpolation on the servo level. On condition of fading transient periods of spatial errors in (7) and introducing the Cartesian transient functions $R(s)$ and $S(s)$ for the robot, and the sensor, respectively, one obtains the control scheme of fig. 2. The control deviation Dx have been computed in the z — domain as follows

$$[Dx]_z = \frac{[\Delta x_d]_z - [x_r \cdot e^{-sT} \cdot R(s) \cdot S(s)]_z + [x_{co} \cdot S(s)]_z}{1 + [e^{-sT} \cdot r(s)]_z \cdot [\frac{1 - e^{-sT}}{s} \cdot \frac{1}{sT} \cdot R(s) \cdot S(s)]_z} \tag{8}$$

where [...] z is the z-transform of the term in parentheses and $r(s)$ is the control law.
Figure 3 shows simulation results regarded to the deviation Dx_i when the effector follows a contour with the ramp function

where

$$x_{co,i} = y1 + A * i * T.$$

$$R(s) = 1$$

$$S(s) = \frac{Cs}{(s-s1)*(s-s2)}$$

$$s1 = AL + BT*j$$
$$s2 = AL - BT*j$$

$$j = \sqrt{-1}$$

$s1, s2$ – complex poles

$Cs = AL^2 + BT^2$ (gain)
T – cycle time
$r(s) = K1 + K2*s$ (control law)

Basic and complex functions

With regard to effective programming of sensor guided control algorithms, basic and complex software modules have been written in PASCAL. These modules can be configurated off line by the user of the robot control system. This set of modules includes simple basic functions for the communication with the sensor interface, the interpreter level and the trajectory generation block. A range of more complex modules contains the control law (e.g PID-controller), geometric operations (e.g. vector and matrix transformations) up to the distance control of the effector perpendicular to a given surface. Table 1 shows a range of functions developed.

Name	Contents
VADD	vector + vector
VSUB	vector - vector
SCAL	vector * vector
CROSS	vector x vector
FACT	vector * scalar
MULT	matrix * matrix
DIAG	diagonal matrix * vector
TRANS	matrix * vector
TRINV	inverse matrix * vector
TRFT	transformation of forces and torques
TRIFT	inverse transf. " " " "
MAV	computation of rotation matrix
ANGL	" " " angles
DELT	transformation of angular velocity
DEIV	inverse transf. " " "
PI	PI - control law
COPF	force control in the object frame

Tab. 1 Basic and complex functions

Experimental results

The following experiment deals with a force adaptive tracing algorithm in the process of cleaning of castings (see fig. 4). Measuring the reaction force by a 3 D - force sensor both the normal force and the orientation of the tool relative to the object surface is controlled.
The correction plane is prescribed by a binormal vector $\underline{o}$ in the coordinate system KX , taught by the operator. At first a velocity vector within the correction plane is also prescribed. With this the 1st motion step can be performed. Each step is divided into n substeps. While performing the 1st n substeps a PI-force control law in the normal direction is realized which leads to the correction of the distance between tool and surface. After finishing the 1st n substeps a regression line is placed through the previous positions which determines the tangent of the contour for the next n substeps. Performing the next n substeps both the force (distance) and the orientation relative to the contour is controlled.

References

/1/ H.Hanafusa, Y.Nakamura: Autonomous Trajectory Control of Robot
Manipulators MIT Press Cambridge, London 1984, p. 863-802

/2/ J.Huebener, R.Palm, A.Holtmann: Programming and Control of 2nd
Generation Robots The Automation of Industrial Processes
Genoa 3-5 dec 1985

/3/ N.Ahlbehrendt, B.Maediger: Algorithms for Point to Point and Continous
Path Control of Industrial Robots
Italian Program on Computer Science Project MODIAC
The Automation of Industrial Processes Torino, Dec 1983

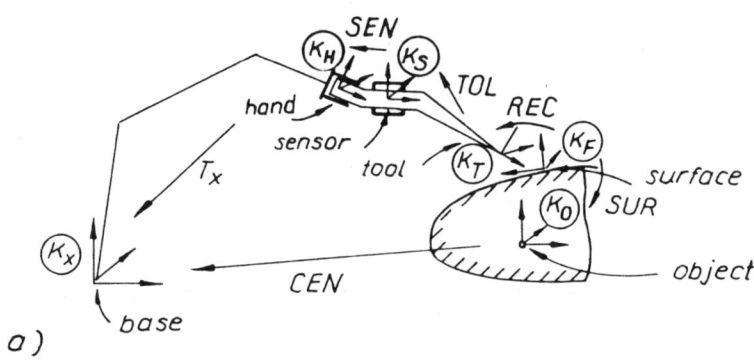
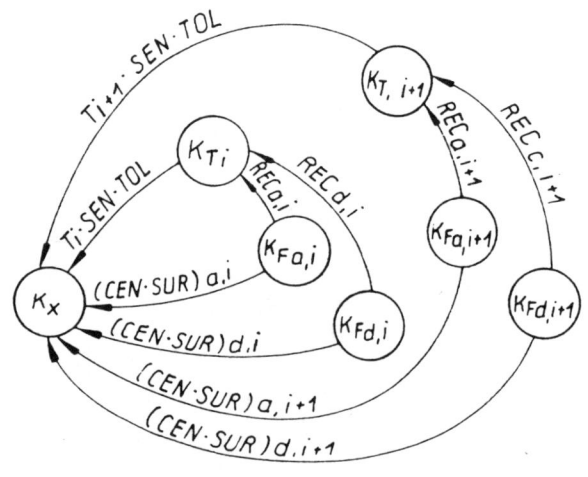

Fig. 1   Frame transformations

Task Specification and Closed Loop Control of Manipulators 281

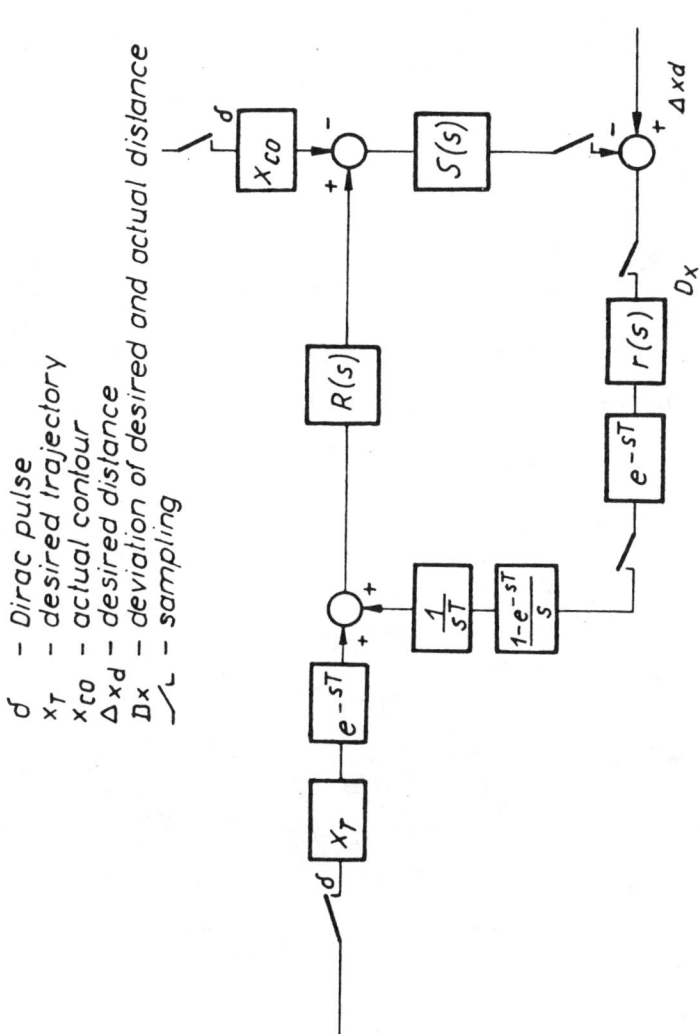

d — Dirac pulse
x_T — desired trajectory
x_{co} — actual contour
Δx_d — desired distance
D_x — deviation of desired and actual distance
$\diagup_L$ — sampling

Fig. 2   External control loop

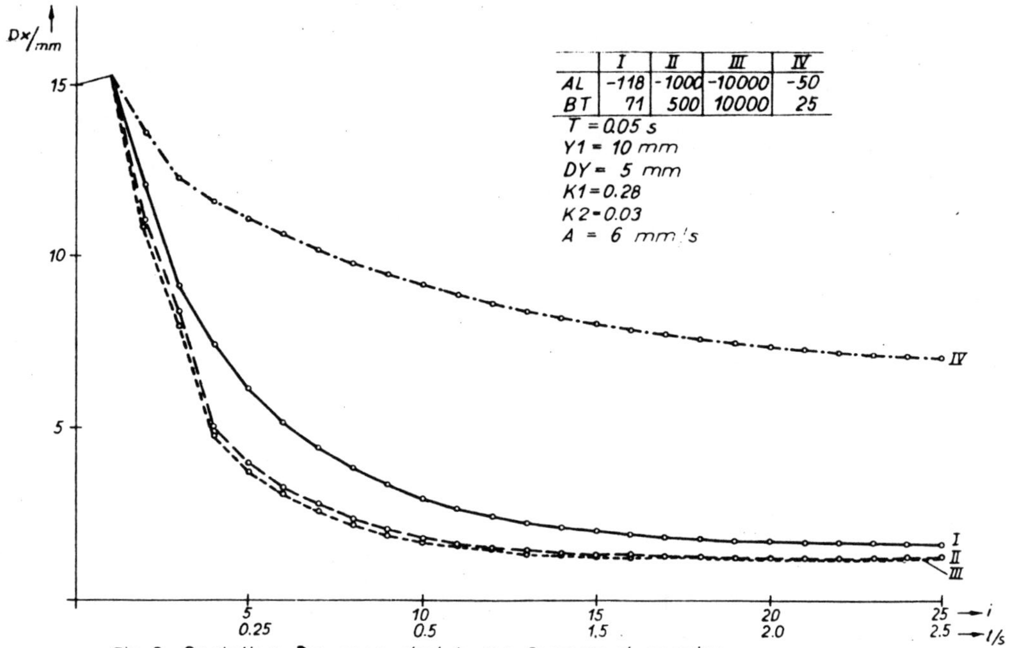

Dx/mm
15
10
5
I
II
III
IV
AL -118 -1000 -10000 -50
BT 71 500 10000 25
T = 0.05 s
Y1 = 10 mm
DY = 5 mm
K1 = 0.28
K2 = 0.03
A = 6 mm/s
5 10 15 20 25 → i
0.25 0.5 1.5 2.0 2.5 → t/s

Fig. 3 Deviation Dx_i regarded to the Sensor dynamics

K1 – Gain
Δf – force error
robot arm
force sensor
milling machine
3rd substep
4th step

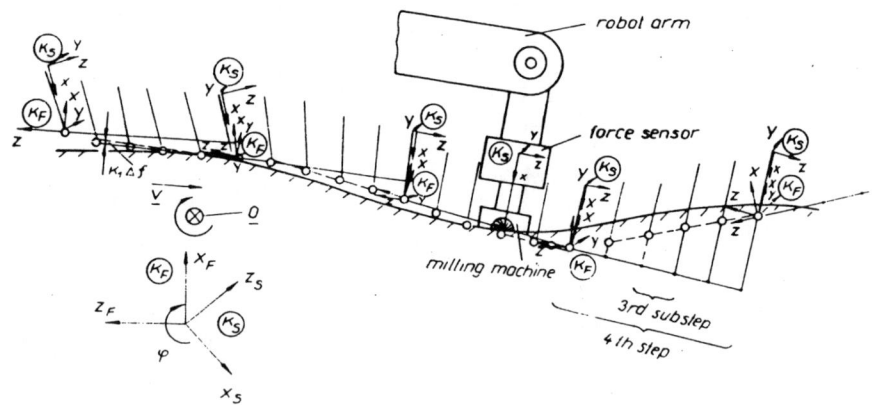

Fig 4 Contouring process controlling force and orientation

Adaptative Force Control of Grippers Taking into Account the Dynamics of Objects

T. Fukuda, N. Kitamura* and K. Tanie**

*The Science University of Tokyo, Japan
**Mechanical Engineering Laboratory, Ministry of International Trade of Industry, Japan

ABSTRACT

In this paper, a force control method for grippers is presented, dynamics of objects based on adaptive control, as demonstrated by experiments being also considered. Since present industrial robotic grippers commonly employ the input/output position control method and sometimes the force control method, it is of great interest to control both the force and position of the gripper against the object simultaneously, However, it is not easy to control the gripper without knowing the nature of the objects, because the dynamics of the object inevitably comes into the overall feedback control system.

Some of the research in this area has been carried out employing steady state position error and various force sensors. Force control is increasingly demanding more sophisticated control for robotic grippers, and this type of gripper requires much more flexible control, such as the handling of various objects adaptively with consideration both of position and force. The adaptive method applicable to say all types of gripper systems, as proposed here, advocates that it is possible to take into account the dynamics of objects by identifying the dynamics via the model reference adaptive system from the viewpoint of force and position control. The proposed method is applied to a gripper handling object. The gripper can hold various objects adaptively with smaller impact forces applied; hence it does not damage objects, unlike conventional grippers. Thus, the range of applicability of force control can be increased with consideration to object dynamics. This can also be shown experimentally in comparison with conventional methods of constant feedback gain.

INTRODUCTION

The Controlled behavior needed for present robots is now becoming more versatile and more complicated. But most of the control systems of present robots, including industrial robots, employ the position input/output control system. However, the control system using position control only cannot meet the requirements for robotic manipulator functions working with force interaction against objects. For example, it is well known that the insertion of work in a parts assembly or crank rotation using robotic arms requires arm controlled motion, adjusting the position and force/torque. Under the present actuator technology, the most often used and practical method to attain the position and force control system is to employ the position servo system for the position control and adjust the position by sensing force/torque employing those sensor outputs or error signals in the servo system (1,2,5). A system for position and force control can be attained using this method, but it is generally difficult to realize the stable force control on all kinds of objects by some particular position control gains, because the desired position control commands corresponding to the force interaction with objects vary according to the dyamics of the manipulated objects in this conventional force/torque control mode. To overcome this problem, the motion is generally stabilized by employing a dead insensitive zone in the force/torque feedback control loop or by decreasing the force control accuracy with some thresholds in force sensing. One stable force control method is the employment of the direct drive(DD) motors as an actuator and the construction of a driving control system for direct commands of force/torque(3,4). Although the DD motor is one of the potential actuators in the next generation, there are cost and performance problems, at present, regarding the motor.

In this paper, the gripper model is employed for control, since the grasping and holding motion of a gripper is one of the typical control systems requiring both position and force control. A new adaptive force control method is proposed here to tackle the problem discussed above, based on adaptive control with regard to the dynamics of objects. We then demonstrate some experimental results. This basic method can be extended to multiple finger systems, which perform holding and grasping manipulations, and also manipulator systems, which perform various tasks such as writing, painting jobs on soft surfaces, contouring jobs, and others.

MODELLING OF THE SYSTEM WITH OBJECT CHARACTERISTICS

Conventional control methods for robotic grippers employ the positional input and output, so that the control system depicted in Fig. 1 takes into account only the dynamics G1 of the manipulator without consideration of object dynamics G2. In this type of control system, force control is necessary to achieve wide flexibility of applicability of handling by manipulators. In this study, the adaptive reference adaptive approach is employed in order to perform hybrid control, including position and force, shown in Figure 2. The output of the proposed control method is combined in the form of hybrid output with position output x(t) and force output F(t), described by

$$y(t) = s1\ x(t) + (1-s1)F(t) \qquad (1)$$

In this proposed method, the adaptive algorithm for the adjustment law of the control gain Kc is derived so that the output error e(t) between the models depends on the characteristics of the objects.

A gripper system holding a wide variety of soft or hard objects, will be considered here. The gripper model employed in this paper is shown in Fig. 3, in which the object G2 is surrounded by broken lines. The assumption is made that both the gripper and the object move symmetrically, so that the center of the gripper is always on the center of the object. This is achieved by controlling the manipulator, at the tip of which the gripper is attached. Thus, the mass effects by objects can be negated. Furthermore, it is assumed that the object can be represented by the linear model with damping C1* and spring constant K1*. This viscoelasticity assumption is generally accepted in many engineering models representing objects. In this figure, kx and cc are the spring constant and the damping coefficient of the gripper respectively, which is known a priori. Therefore, the unknown parameters in this model are K1* and C1*.

Then the state equation and the output equation can be expressed employing the static and dynamic characteristics of the actuator and the load, as follows.

$$\left. \begin{array}{l} \dot{X}(t) = A\ X(t) + B\ u(t) \\ Y(t) = C^T X(t) \end{array} \right\} \qquad (2)$$

where
$$X = (\ x\ \dot{x}\ wx\ )$$
$$y = s1\ x + (1-s1)F$$
$$F = kxK1^*x(t)/(kx+K1^*)+(C1+Cx)\dot{x}(t)$$

$$A = \begin{pmatrix} 0 & 1 & 0 \\ -a2 & -a1 & -a3 \\ -a5 & -a4 & -a6 \end{pmatrix} \quad B = \begin{pmatrix} 0 \\ b1 \\ 0 \end{pmatrix} \quad C = \begin{pmatrix} C1 \\ C2 \\ 0 \end{pmatrix}$$

a0= Ra(Jm n/r + M r /n)/Kt
a1= RarB1/(Ktna0)+nKe/(ra0)-C1*/(C1*-Cx)
a2= RarCx/(Ktna0)-K1*/(C1*-Cx)
a3= Rarkx/(Ktna0)-(K1*-kx)/(C1*-Cx)
a4= C1*/(C1*-Cx),
a5=K1*/(C1*-Cx)
a6= (K1*-kx)/(C1*-Cx),
b1=Ma/a0
C1= S1 + (1-s1)kxK1*/(kx+K1*)
C2= (1-S1)(C1*+Cx)

Ma:	gain of the DC motor,
n:	gear reduction rate,
Jm:	moment of inertia of the motor,
M:	mass of the load,
B1:	frictional resistance of the load,
R:	armature resistance of the motor,
Km:	constant of inverse electric power generation,
Kt:	torque constant of the motor,
K1*:	spring constant of the object,
kx:	spring constant of the gripper,
C1*:	damping coefficient of the object,
Cx:	damping coefficient of the gripper,
r:	radius of the pinion gear in gripper.

It is clear that the conditions for both controlability and observability can be satisfield in this model. It will be clear that this modelling can be extended to the case of robotic manipulators. Hence, the proposed method must be applicable to the manipulator system also.

METHOD OF ADAPTIVE FORCE CONTROL

In this chapter, adaptive force control taking into account the static and dynamic characteristics of objects is derived for the model described above, based on the adaptive control theory (10). Note that this is the position control, if s1 is equal to one, while the other cases become hybrid control with position and force mixed. In particular, if s1 is equal to 0, then it is purely force control, which can be attained by the DD control system (3,4). Since most industrial robots still employ high reduction geared motors, it is assumed in this paper that s1 is not equal to zero.

The transfer function of the model in eq. (2) can expressed by

$$G(s)=(b1C2s^2+(b1C2a4+b1C1)s+b1C1a4)/$$
$$(s^3+(a1+a4)s^2+(a1a4-a3a5+a2)s+a2a4-a3a6), \quad (3)$$

In particular, if s1 is equal to one, then eq. (3) becomes the following transfer function:

$$G(s)=(b1C1(s+a4))/$$
$$(s^3+(a1+a4)s^2+(a1a4-a3a5+a2)s+a2a4-a3a6). \quad (4)$$

Because of the difference of the numerators of the transfer function for both cases, the adaptive algorithm is slightly different.

The dynamical equation for the gripper and the reference model can be expressed by

$$(p^3+a1^*p^2+a2^*p+a3^*)y(t)=b^*(p+a4^*)u(t) \quad (5)$$
$$(p^3+d1\ p^2+d2\ p+d3)ym(t)=(p+\beta1)r(t) \quad (6)$$

where $a1^*=a1+a4$, $a2^*=a1a4-a3a5+a2$,
$a3^*=a2a4-a3a6$, $a4^*=a4$, $b^*=b1C1$, $p=d/dt$.

By introducing the filter $T(p)$.

$$T(p)=p+\lambda1, \quad (7)$$

there exist the following $R(p)$ and $S(p)$, satisfying

$$(p+\lambda1)(p^3+d1p^2+d2p+d3)$$
$$=(p^3+a1p^2+a2p+a3)(p+\lambda2)+p^2+\lambda3p+\lambda4, \quad (8)$$

such that

$$R(p)=p+\lambda2 \quad (9)$$
$$S(p)=p^2+\lambda3p+\lambda4, \quad (10)$$

where $\lambda1,\lambda2,\lambda3,\lambda4$ are all positive constant numbers.
Then the error equation can be derived from equations (5) and (6).

$$e(t)=y(t)-ym(t) \quad (11)$$

$$e(t)=b\ ((p+a4)(p+\lambda2)u(t)+(p^2+\lambda3p+\lambda4)y(t)$$
$$/b-(p+\lambda2)(p+\beta1)r(t)/b)/$$
$$((p+\lambda1)(p^3+d1p^2+d2p+d3)) \quad (12)$$

By introducing the state variable filters $H(p)$, $D(p)$ and $T(p)$,

$$H(p)=p^2+r1p+r2$$
$$D(p)=p+r3$$
$$Q(p)=H(p)D(p)$$
$$=p^3+(r1+r3)p^2+(r2+r1r3)p+r2r3,$$

where $H(p)$ and $D(p)$ are chosen so that the following transfer function is strictly positive real,

$$(p^3+(r1+r2)p^2+(r2+r1r3)p+r2r3)/$$
$$(p+\lambda1)(p^3+d1p^2+d2p+d3) \quad (13)$$

and with some mathematical manipulations, the following equation can be obtained:

$$e(t) = (p^3+(r1+r2)p^2+(r2+r1r3)p+r2r3)$$
$$(u(t)/(p+r3)+ \varphi^T(t) \xi(t))/((p+\lambda 1)(p^3 + \alpha 1 p^2 + \alpha 2 p + \alpha 3)),$$

$$e(t)=(p^3+(r1+r2)p^2+(r2+r1r3)p+r2r3)$$
$$k0(t)(u(t)/(p+r3)-k(t)\xi(t))$$
$$/((p+\lambda 1)(p^3+\alpha 1 p +\alpha 2 p+\alpha 3)), \qquad (14)$$

where $\varphi(t)$, $k(t)$, $\zeta(t)$ and $\xi(t)$ are the vectors of the parameter mismatching, control gains, the outputs of the state variable filter of the reference and the control inputs and the gripper output through $1/H(p)$ and the outputs of (t) of $1/D(p)$ respectively.

The extended error equation can then be obtained as follows:

$$\mathcal{E}(t) = e(t)-\hat{e}(t)$$
$$= ((p^3+(r1+r2)p^2+(r2+r1r3)p+r2r3)$$
$$(\theta 0(t)v0(t)+b\theta^T(t)\xi(t))/ ((p+\lambda 1)(p^3+\alpha 1 p^2+\alpha 2 p +\alpha 3)) \quad (15)$$

where
$$\theta 0(t) = b - k0(t)$$
$$\theta(t) = \varphi + k(t)$$
$$v0(t) = U(t)/D(p) - k^T(t) \xi(t).$$

The equation can be represented by the state variable form:

$$\dot{\mathcal{E}}(t) = F\mathcal{E}(t) + G\, b\theta^*(t)\, \xi^*(t)$$
$$\mathcal{E}(t) = C^T \mathcal{E}(t). \qquad\qquad (16)$$

where $\theta^*(t)$ and $\xi^*(t)$ are the vectors including $\theta_0(t)$ and $v(t)$, respectively.

$$F = \begin{pmatrix} 0 & 1 & 0 & 0 \\ 0 & 0 & 1 & 0 \\ 0 & 0 & 0 & 1 \\ -f4 & -f3 & -f2 & -f1 \end{pmatrix} \quad G = \begin{pmatrix} 0 \\ 0 \\ 0 \\ 1 \end{pmatrix} \quad C = \begin{pmatrix} r2r3 \\ r2+r1r3 \\ r1+r2 \\ 1 \end{pmatrix}$$

Then the constant parameters $\lambda 1$, $r1$, $r2$, and $r3$ can be determined to hold the following Kalman-Yakuvovich lemma,

$$F^T P + P F = -Q$$
$$G^T P = C^T. \qquad\qquad (17)$$

Then the adaptive law can be obtained by

$$\dot{k0}(t) = r0v0(t)\mathcal{E}(t)$$
$$\dot{k}(t) = -\Gamma \xi(t)\, \mathcal{E}(t). \qquad (18)$$

Similarly, the adaptive control law for sl=1 can be derived for these models as follows:

$$(p^3 + a1^*p^2+a2^*p+ a3^*)y(t)$$
$$=b^*(p^2+a4^*p+a5^*)u(t) \qquad (19)$$
$$(p^3+ \alpha 1\, p^2+\alpha 2\, p+\alpha 3)ym(t)$$
$$=(p^2+\beta 1 p+\beta 2)r(t). \qquad (20)$$

For both cases, the adaptive force control input can be given by

$$u(t) = k^T(t)\zeta(t) \ .$$

SIMULATION RESULTS

In order to examine the applicability of the proposed adaptive force control compared with the conventional fixed feed-back gain control method, computer simulations have been extensively carried out. In these works, the stable area by the conventional method is first shown in Figure 4 by continuously changing the spring constant of the object and by obtaining the maximal and minimal feedback gains, between which the gripper can hold the objects. The criterion whether the gripper can hold the object or not is given in this case, so that oscillations in transient holding behaviours can be controlled within the settling time of 2 per cent. The shaded area in this figure gives the stable graspable area performed by the conventional control method. The spring constant and the gear are appropriately normalized by the spring of the reference model divided by that of the object and the control gain divided by the most adaptable gain, respectively. It turns out that the proposed method gives a fourteen times wider stable area than the conventional area and that this adaptive method is very effective taking into account the characteristics of objects.

EXPERIMENTAL METHOD

Photo 1 shows the experimental equipment of the adaptive holding force control. This equipment consists of the DC motor as an actuator and a pair of gripper tips made of aluminum boards of 1 mm thickness, which can be driven by the rack and pinion gear. This gripper can hold the object with a maximal size of 80 mm. The outputs of the position and the holding force are measured by the encoder and the strain gauge attached to one of the gripper tips. The gripper can be controlled by the microcomputer.

A rubber ball is selected as a soft object, such that the ball has a diameter of 64 mm and the stiffness of 160 N/m, while a cylindrical bar made of aluminium represents a hard object, such that the object has the diameter of 60 mm and a stiffness of almost infinity when compared with the former. The adaptive force control method described before is derived under the assumption that the gripper makes contact with the object immediately. However, the size of objects are normally considered to be unknown, so that it is necessary to detect the size of the object first. In this detection process, only the velocity control of gripper is carried out until the force sensor hits the prescribed value of the force. Various experiments of conventional force control by fixed feedback gain and adaptive force control by adjustable gains must be performed from this location of the gripper after the detection of the object.

EXPERIMENTAL RESULTS

Employing the experimental equipment of the gripper shown in photo 1, extensive experiments have been carried out on force control. Throughout these experiments, because of the simplicity of the experiments and the computational time limits for real time control by the microcomputer, the damping effects due to $C1^*$ and Cx have been neglected, so that the dynamical equation can be reduced to two dimensional equations. The simulation studies indicate that there are no large differences under these experimental conditions. Hence, the implementation of the adaptive force control can be much easier.

Some of the experimental results are shown in Figures 5, 6, 7 and 8, in which the solid line shows the output of the gripper, while the broken line shows that of the reference model. In these experiments, the parameters are appropriately chosen, e.g. 1=50, 2=15, 1=0.65. The vertical broken line in these figures shows the location of the object's detection described before.

Figures 5 and 6 show the experimental results using the position input/output only, such that S1 is equal to one. In this case, holding control is carried out employing the constant fixed feedback gains, which are adopted by the presently used industrial robot control. Figure 4 is an example of holding the hard aluminium bar, while Fig. 5 is the one that holds the soft rubber ball with the same feedback gains. In the latter case, it is impossible to hold the object stably, because of the oscillation of the gripper. Thus, it is clear that the applicability of fixed feedback gain is limited to some classes of holding objects, depending upon the characteristics of the objects.

Figures 7 and 8 show the experimental results employing the proposed adaptive force control method, in which the hybrid ratio of the position and the force is 1, so that S1 is equal to half. Figure 7 shows an example of holding the hard aluminium bar, Fig. 8

shows one for holding the soft rubber ball. These figures show that the gripper employing adaptive force control can hold objects stably whether objects have soft or hard characteristics. Therefore, the proposed adaptive force control method is more effective in obtaining stable holding control than the conventional constant fixed feedback gain method and has wider applicability and in obtaining adaptability for force control.

CONCLUDING REMARKS

In this paper, the adaptive force control method is proposed for gripper and manipulator systems with regard to the characteristics of objects and the proposed method was shown to be very effective through the simulation and experimental works.

Furthermore, the proposed method can adapt to the change of object characteristics and can improve the adaptability much more than the conventional control method.

References

1. H. Inoue, "Computer controlled bilateral manipulator," Bulletin of the JSME, Vol.14, no. 69, pp. 199-207, 1971.
2. K. Takeyasu, T. Goto, T. Inoyama,"Precise insertion control robot and its application," Trans. of the ASME, J. of Engineering for Industry, pp.1313-1318, Nov. 1976.
3. H. Asada, H. Yamamoto, "Torque feedback control of MIT direct-drive robot", Proc. 14th ISIR, pp. 663-670, Oct. 1984.
4. T. Suehiro, K. Takase, "Developemnt of a direct drive manipulator: ETA-3 and enchancement of servo stiffness by a second-order digital filter," Proc. 15th ISIR, Vol. 1, pp.479-486, Sept. 1985.
5. M. H. Raibert and J.J. Craig, "Hybrid position/force control of manipulators," Trans. ASME, DSMC, Vol. 103, No. 2, pp. 126-133, 1981.
6. M. K. Brown, "Computer simulation of controlled impedance robot hand," IEEE Int'l conf. on robotics, pp. 442-450, 1984.
7. R.K. Roberts, R.P. Paul, and B. M. Hillberry, "The effect of wrist force sensor stiffness on the control of robot manipulators," IEEE 2nd int'l conf. on robotics and automation, pp. 269-274, 1985.
8. S. Dubowsky, D. T. DesForges, "The application of Model Reference adaptive control to robotic manipulators," Trans. ASME, DSMC, Vol. 101, pp. 193-200, 1979.
9. N. Hogan,"Impedance control: an approach to manipulation, Part I - theory," Trans. ASME, DSMC, Vol. 107, pp. 1-7, Part II-implementation, ibid. pp. 8-16, Part III-applications, ibid, pp. 17-24, 1985.
10. K.S. Narendra, B. B. Peterson, "Recent development in adaptive control," Prepr. of IFAC Int'l symp. on adaptive systems, Ruhr-University Bochum-Germany, Mar. 1980.

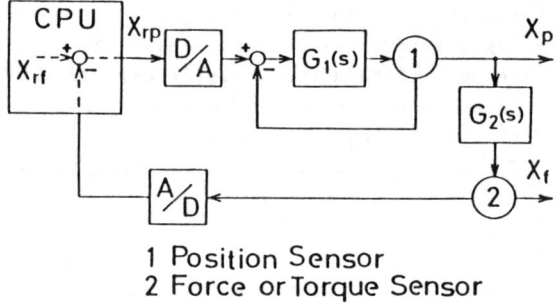

1 Position Sensor
2 Force or Torque Sensor

Fig.1 Conventional position input/output control system

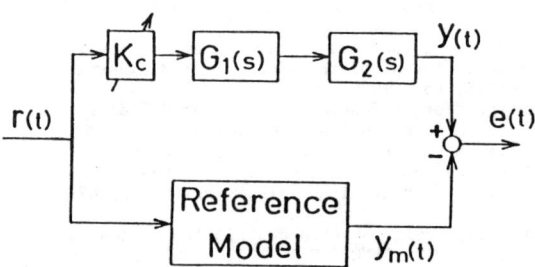

Fig 2 Model reference adaptive hybrid control system

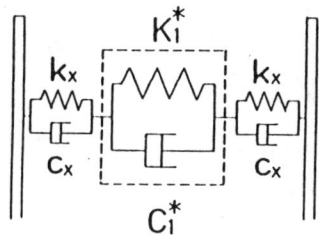

Fig. 3 Gripper model

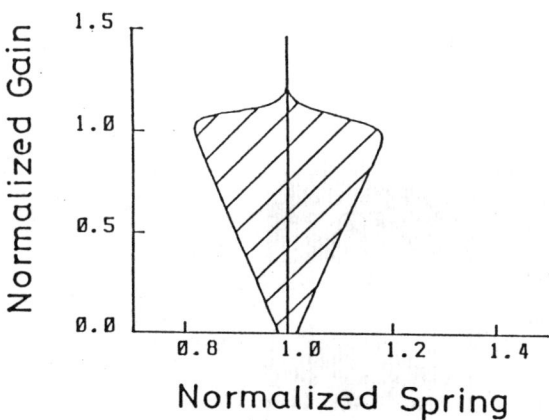

Fig 4 Simulation result for the stable holding limit using fixed feedback gain control

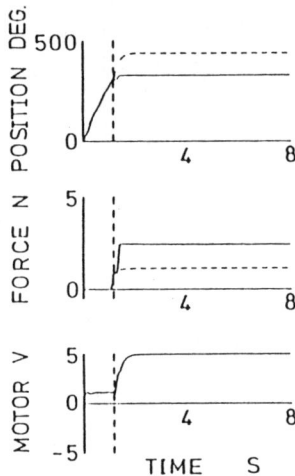

Fig. 5   Experimental result for holding a hard object by the fixed feedback gain control

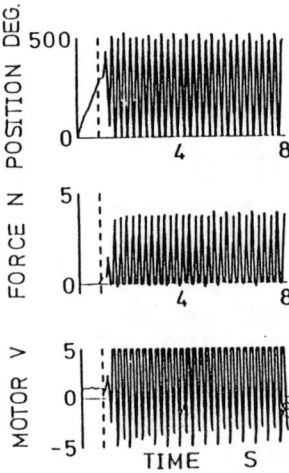

Fig. 6   Experimental result for holding a soft object by the fixed feedback gain control

Adaptive Force Control of Grippers 295

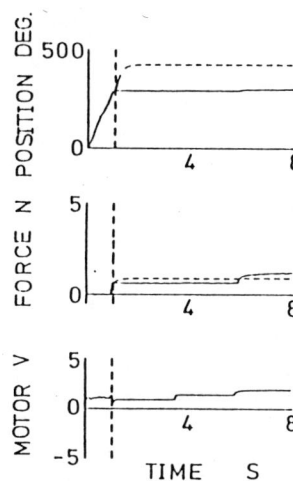

Fig. 7 Experimental result for holding a hard object by the adaptive force control

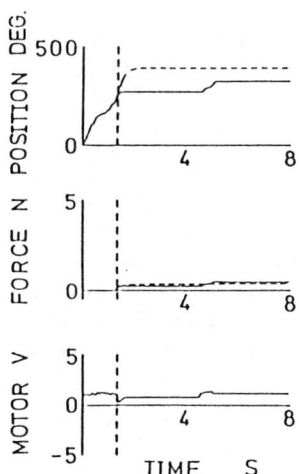

Fig. 8 Experimental result for holding a soft object by the adaptive force control

Bilateral Remote Control with Dynamics Reflection

K. Tanie, K. Komoriya, M. Kaneko, T. Ohno* and T. Fukuda**

*Mechanical Engineering Laboratory, Ministry of International Trade and Industry, Japan
**The Science University of Tokyo, Japan

SUMMARY

This paper describes an effective bilateral master slave remote manipulator system, called object dynamics reflection bilateral control (ODRBC). In the proposed system, the dynamics of the load at the slave side is measured in real-time, and the parameters necessary to describe the dynamics are transmitted to the master side. Using software servo techniques and those parameters, the dynamics of the master mechanism is adjusted. The operator feels the reaction, from the constraints encountered at the slave side, through the change of the dynamics of the master mechanism. The feasibility experiments were carried out using a simulated one axis bilateral system. From the experimental results, it was found that the proposed system worked stably under several load conditions and transmission delays between master and slave sides. The system enabled the operator to feel the compliance of an object.

1. INTRODUCTION

The remote control permits dexterous operation of a human-like manipulator with multi-degrees of freedom (1). There are a lot of applications of remote controlled manipulators in nuclear power plants and in undersea work. Typically, a remote control system can be used in an effective way for work in complex environments.

There are two types of remote control system, unilateral and bilateral (2,3). The unilateral system is one which permits the transmission of position from operator to the load or slave. On the other hand, the bilateral system is one which permits the transmission of force and position from operator to load or slave and simultaneously from load back to the operator or master. The capability of transmission of force and position enables an operator

to feel constraints at the slave side and contributes to reduction
of his/her work load in remote manipulation tasks. So far there
have been developed three kinds of methods to realize efficient
bilateral systems: bilateral position servo type, force reflecting
type and force feedback type. The bilateral character of the system
in bilateral position servo type is achieved by joining in one loop
a pair of position-error servomechanisms and connecting them in
tandem so that their error signals are made common. This common
signal provides a direct means for reflecting the slave forces back
to the operator. Similarly, a slight discrepancy in the spacial
correspondence of master and slave will not only impart a corrective
motion to the slave, but also transmit to the operator a feedback
signal of unbalanced force, proportional to the amount of
desynchronization.

The other two bilateral methods use force or torque sensors
to evaluate constraints at the slave side. In those methods, the
operator at the master side generates position commands to the slave
side to move the slave using position servomechanisms, and feels
reaction force from the slave side through force generating actuators
which are driven by the force or torque sensor signals.

Today almost all practical bilateral master-slave manipulator
systems are of the position servo type. A significant drawback in
those systems is that the force transmitted from the slave side
includes considerable amount of nonlinear friction force of actuator
motion in the slave system (4,5). The operator, therefore, cannot
feel real constraints acting at the slave side. To improve this,
force reflecting and force feedback types have been proposed. However,
the stability problem prevents their practical use. The nature of
the load in a bilateral manipulation system cannot be determined in
advance. It might vary from a near-zero mass, spring and damper
combination to that of an infinite mass, spring and damper combination.
Moreover, the control system has feedforward and feedback lines for
position and force transmission which may have some frequency-
dependent transmission time delay or dead time because of no
dynamic characteristics of slave and master mechanisms or long
distance connection between master and slave locations (6,7,8).

Those factors, the uncertainty of the load characteristics and
the presence of transmission time delay make the system unstable, when
the operator applies appropriate motions through feedforward lines
to the load and feels constraints at the slave side through feedback lines. To solve this problem, the object dynamics reflection
bilateral control method is proposed.

2. METHOD

Fig. 1 shows a block diagram of the proposed object dynamics reflection bilateral control system. The master side has a position sensor and a variable dynamics mechanism, which, for example, can be made, using a direct drive torque motor (DD motor). The slave side includes a position servomechanism with DC motor which has transfer function G(s) and a torque sensor. Assume that an object with transfer function dynamics P(s) is at location Xo. Also, assume that the slave's displacement is Xs, when an operator moves the master by Xm. The reaction force fs generated at the slave side can be written P(s)(Xs-Xo). In the force reflecting system, the force fs is detected by a torque sensor mounted on the slave side and is sent back to the master side. In the proposed system, however, the transmittance P(s) is measured using an appropriate method and is sent back to the master side in lieu of the force fs. At the master side the transfer function P'(s) of the master mechanism is adjusted according to the measured transfer function P(s) so that P'(s) is equal to P(s). In the proposed system, therefore, the operator feels the constraints at the slave side through the force fm= P'(s)(X.-Xo) = P(s)(Xm-Xo). One of the remarkable features in the proposed system is that the force fm does not include dynamic effects, except the load. A block diagram of a force reflecting bilateral system is shown in Fig. 2. In this case the force fm, which an operator feels at the master side can be written as R(s)P(s)(G(s)Xm-Xo), where R(s) is the transfer function included in the force feedback line. The reflecting force is affected by the frequency-dependent characteristics of transmission lines R(s) and G(s) as well as the load P(s). This causes reduction of fidelity of reflection of constraint force at the slave side, and also, causes reduction of stability in force adjustment tasks.

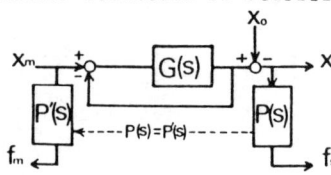

Fig. 1 Object Dynamics Reflection Bilateral Remote Control.

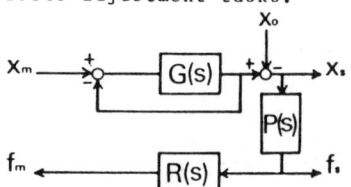

Fig. 2 Force Reflecting Bilateral Remote Control.

3. EXPERIMENTS

3.1. Master and Slave Systems.

In order to confirm the effectiveness of the proposed system, an experimental system was constructed. The system consists of master mechanism and minicomputer system (DEC, LAB11K). The master mechanism includes a joystick connected to a direct drive motor, a potentiometer and a torque sensor (detectable maximum torque 20 kgcm). Outputs of the potentiometer and the torque sensor were transfered to minicomputer via AD converter. The slave manipulator and objects to be handled were simulated on a minicomputer. Also, generation of dynamics of the object and the control of master mechanism were performed in the same computer.

The potentiometer was used to detect the displacement of the joystick handled by an operator. The direct drive motor was a device which generated the force according to the armature current. To achieve the function of varying dynamics of the master mechanism, the armature current was controlled, so that the current kept a specified relation with the angle of joystick measured by the potentiometer. The relation produced a required dynamic condition in the master mechanism. In this experiment, it was assumed that objects to be handled included only spring elements. Therefore, the armature current was controlled according to the following equation: $i(n) = i(n-1) + K(n)$, where $i(n)$ is armature current at time n, $K(n)$ is stiffness of the object at time n, Δ is the difference between angles at time n and n-1. The master system was shown in Fig. 3.

3.2. Experimental Procedures

Two forms of force tracking experiments were performed. The first-form, A, with two variations of the experiment related to the force reflecting bilateral control. Fig. 4 shows the form A experiment block diagram. The motion of the joystick was detected by a potentiometer and was sent to minicomputer via a first order inertial element $1/(1=Ts)$. In the computer, $fs = K(Xs-Xo)$ if $Xs=Xo$, $fs = 0$, where K is a simulated stiffness of the object, and Xs and Xo are the slave displacement and the positon of simulated object, respectively.

Fig. 3 Master System.

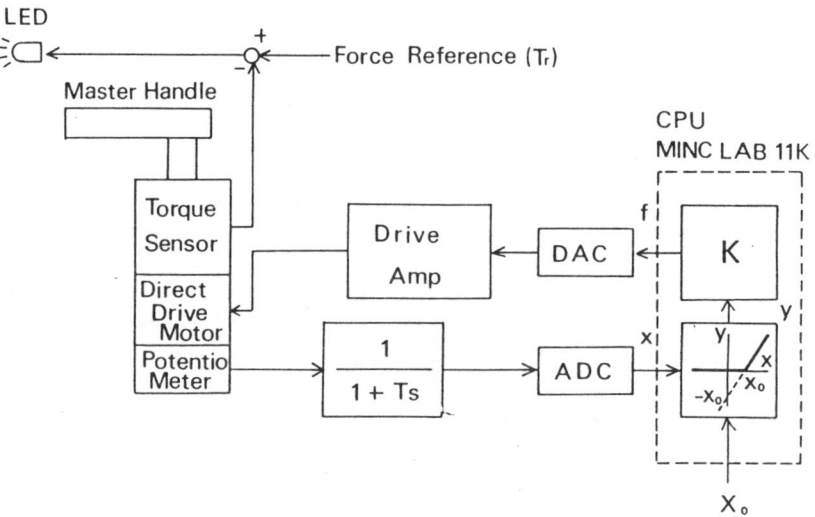

Fig. 4 A Block Diagram of Experiment Form A.

The armature current of direct drive motor at the master side was controlled according to fs and in turn produced the force corresponding to the constraint force at the slave side. An operator was required to manipulate the joystick and to generate a specified constant force during a specified period. In order to make the operator recognize that generating force is at the specified level, a LED was turned on when the difference between the force detected by the torque sensor at the master side and a reference force was less than a specified value. Looking at the LED, each operator performed each requirement.

Form B of the experiment is represented in Fig. 5. It concerns object dynamics reflection bilateral control. In this form, the detected angle of joystick Xm was also transmitted to the microcomputer via a first order inertial $1/(1+Ts)$. In the computer Xs = $1/(1+Ts)$ Xm-Xo and KXs were calculated. KXs corresponds to the torque fs applied to the object with stiffness K at the slave, the simulated slave torque. In order to transmit the information about object's dynamics to the master side, K was estimated using the output of the first order time delay element and the simulated slave torque.

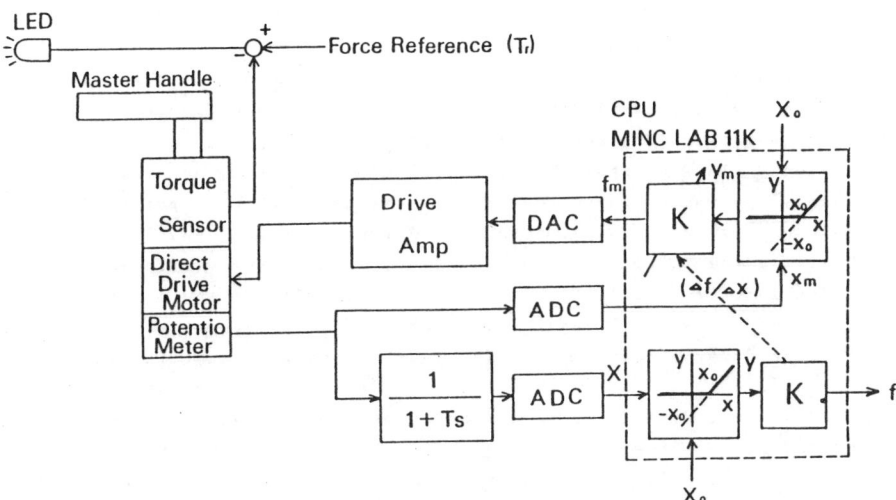

Fig. 5 A Block Diagram of Experiment Form B.

The calculation was carried out as follows: $K(n) = (fs(n) + -fs(n-1)/(Xso(n)-Xso(n-1)))$, if $KSo(n) = Xso(n-1)$; $K(n)=K(n-1)$, if $Xso(n) = Xso(n-1)$, where $K(n)$, $fs(n)$ and $Xso(n)$ are stiffness of the object, simulated slave torque and output of the first order time delay element at time n, respectively. Using $K(n)$ and the above-mentioned equation, the armature current of the direct drive motor at master side was adjusted. Requirements of this experiment were the same as that of the form A. Also a LED was used to make an operator know that generating torque was a specified level.

Experimental parameters were time constant T of the first order time delay element and simulated stiffness K in the two forms of experiments. Two time constants, 0.05s and 1s were used. K was varied from 5.0 to 30.0. The number of subjects employed was one.

3.3. Experimental Results

Figure 6 and Figure 7 show experimental results of form A obtained using time constants T=0.05s and 1s, and simulated stiffness K=5.0 - 30.0. There are three kinds of relation, master torque error-time, simulated slave torque-time and master displacement-time. In those results, Tr is a reference torque described using AD converter unit. The master torque error means the difference between a reference torque and a master torque. The dotted line in a master displacement-time relation indicates the object contact line, where a master displacement arrives at Xo. When a master displacement goes to Xo, the slave torque begins to generate because the simulated slave manipulator hits a simulated object, and this also causes the master torque error to decrease. The configurations of response obtained in Figs. 6 and 7 can explain these behaviours.

Observing the results, it was found that increasing stiffness caused decrease of stability. It is more difficult for the subject to adjust the slave force stably when the longer time constant transmission delay T=1s is used. In order to compare stability in each force adjustment task with each other, a settling time of each simulated slave torque response was calculated. The settling time was defined as the time required for the subject's response to stabilize to a certain level measured from the initiation of the reference torque. When response stayed in the range between 90% and 110% of the reference value during a specified period, the response was judged to become stable. The relations between the settling time and the simulated stiffness was plotted in Figures 8 and 9. The length of the bars indicates that the result of each trail is fluctuating. The area marked by "unstable" in Figure 9 is one where the subject could not stabilize the slave force within a specified range. From the observation of the results, it is evidently found that an increase of simulated stiffness relates to an increase of settling time. Especially, when a longer time constant T=1s and higher stiffness are used, it becomes hard to control slave force stably.

Figure 10 shows experimental results of form A in comparison with the results of form B. Both the results of form A and form B were obtained using time constant T=1s and simulated stiffness K=30.0. From the figure, it is found that in form B the subject can control the slave force stably at T=1s and K=30.0 where the results of form A does not provide a stable response as indicated in Figs. 7 and 9. Fig. 11 shows the relation between the settling time of the simulated slave torque and the simulated stiffness which was obtained from the experiment form B using T=1s and K=5.0 -30.0. Comparing this relation with that of form B, it can be recognized that the settling time of form B at each stiffness condition is remarkably smaller than that of form A. This fact shows the effectiveness of using ODRBC.

4. DISCUSSION

Superiority of ODRBC over FRBC is interpreted as follows. Generally, in force adjusting tasks using a bilateral remote control system, a subject works as a controller in a closed loop system including master mechanism, slave mechanism, transmission lines and object. The characteristics of transfer function of the closed system has an important role to make the subject's performance stable. In FRBC, the force which a subject feels at the master side, will include an object dynamic effect of master mechanisms, slave mechanisms and transmission lines connecting a master side to a slave side as well as object's dynamics, as mentioned in Section 2. This may cause the phase margin of the closed system to decline because of the frequency dependent characteristics of the closed loop transfer function. In ODRBC, on the other hand, the object dynamics is transfered to the master side and is installed in the master mechanism using a software servo technique. From the point of view of dynamics, this enables a subject to manipulate the

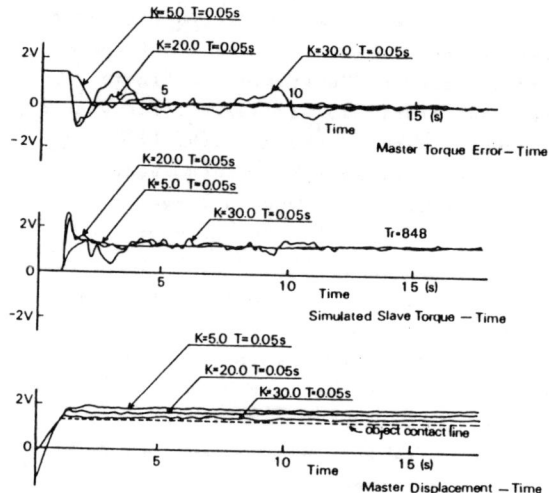

Fig. 6 Experimental Results of Form A
(K=5.0,20.0,30.0:T=0.05s).

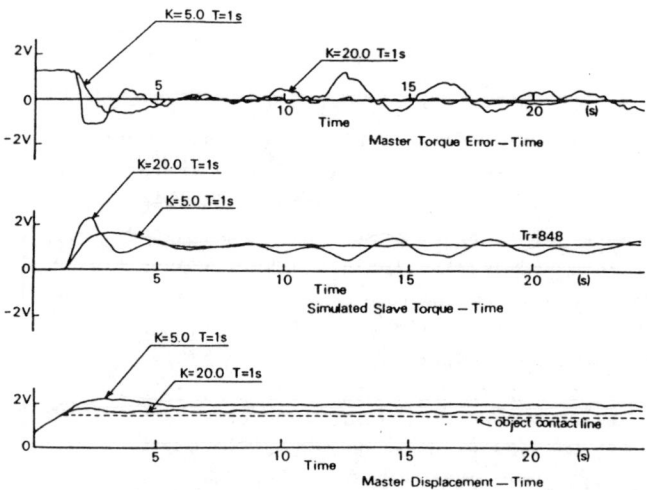

Fig. 7 Experimental Results of Form B
(K=5.0,20.0:T=1s).

Bilateral Remote Control with Dynamics Reflection 305

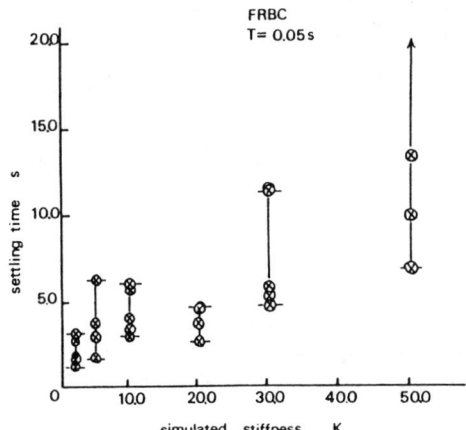

Fig. 8 The Relations Between Settling Time and Simulated Stiffness (FRBC, T=0.05s).

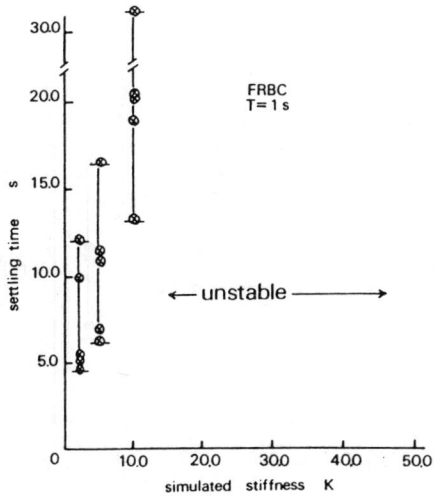

Fig. 9 The Relations Between Settling Time and Simulated Stiffness (FRBC, T=1s).

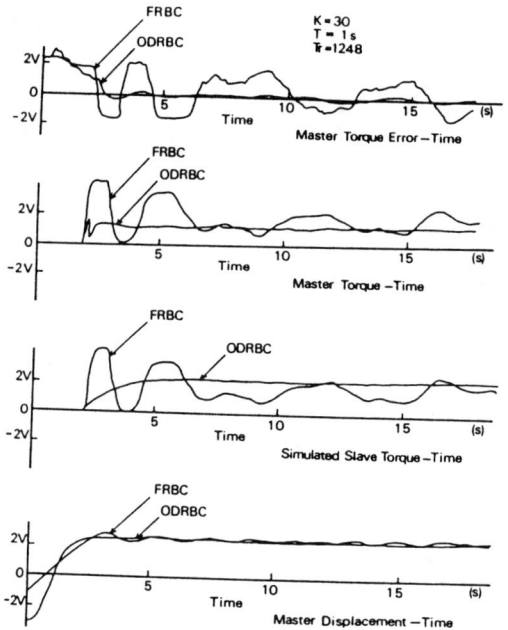

Fig. 10 Comparison of an Experimental Result of Form B with the Result of Form A.

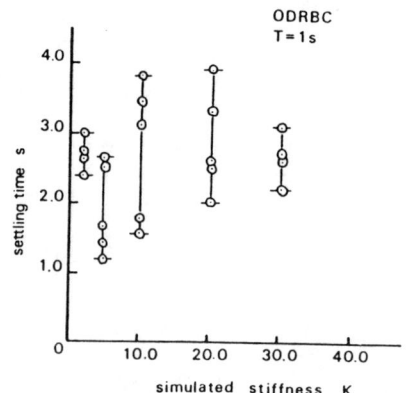

Fig. 11 The Relations Between Settling Time and Simulated Stiffness (ODRBC, T=1s).

object directly. In ODRBC, therefore, the dynamics of slave mechanism and transmission line can be considered to be unity transfer functions, even though a long frequency dependent time delay exists in the slave mechanism and/or the transmission line. This will help to keep the phase margin of the closed loop system higher and in turn to keep the system stable.

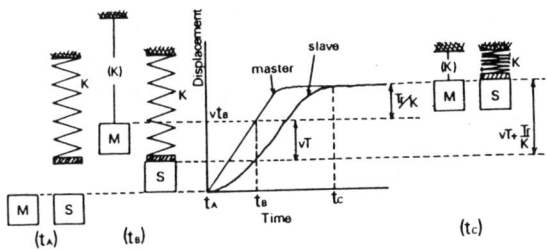

Fig. 12 Force Difference Between Master Side and Slave Side.

There is a drawback in ODRBC. In order to explain this, Figure 12 is used. Suppose that a master arm and a slave arm are located at the initial position at time tA=0. Also, suppose that transmission lines between the master side and the slave side have a inertia described by 1/(1+Ts). There is an object with stiffness K in front of the slave arm. When the master arm moves with a constant velocity v, its displacement will be vtB at time tB. The slave displacement cannot be equal to vtB because of the time delay. The master arm goes vT before the slave arm. Suppose that the slave arm touches an object when the master arm is at vtB. In ODRBC, at time tB the object dynamics will be calculated and the result is transfered to the master side to adjust the master dynamics according to the measured object dynamics. At time tB, also, the subject can begin to feel slave constraints through the master mechanism. If a subject is required to apply a constant force, Tr, to the object, he will move the master arm until its displacement gets Tr/K from the displacement vtB at time tB, where K is the stiffness of the object measured. Because of the delayed motion, the final displacement of the slave arm will be vT+(Tr/K). vT creates a force difference KvT between the master side and the slave side.

Obviously, the force difference depends on the contact velocity of the slave arm with the object. The reduction of the force difference will require that the contact velocity has to be zero. To achieve this purpose, a move and wait strategy may be effectively used (7). Detailed discussion of this is a future topic for this project.

5. CONCLUSIONS

A new type of bilateral control method has been proposed, which is called the object dynamics reflection bilateral control. In order to evaluate the proposed method, the fundamental experiments have been carried out. These experimental results showed that the method enabled an operator to perform a force adjusting tasks stably, even though the system includes a transmission time delay.

References

[1] E.G.Johnsen and W.R.Corliss: Teleoperators and Human Augumentation, NASA, SP-5047, 1967.
[2] J.L.Nevins: Teleoperator Technology - Past, Present and Future, Charles Stark Draper Laboratory, E-2640, February 1972.
[3] R.S. Mosher, B. Wendel: Force-Reflecting Electrohydraulic Servomanipulator, Electro-Technology, pp.138-141, December 1960.
[4] J.R.Burnett: Force-Reflecting Servos Add "Feel" to Remote Controls, Control Engineering, pp.82-87, July 1957.
[5] A.Nagai and K.Matsushima: On the Remote Mini Manipulator-Control of Its Arm and Gripper-, Trans. of the Society of Instrument and Control Engineers, Vol.16, No.1, pp.91-97, February 1980.
[6] W.R.Ferrell and T.B.Sheridan: Supervisory Control of Remote Manipulation, IEEE Spectrum, Vol.4, No.10, pp.81-88, October 1967.
[7] W.R.Ferrell: Delayed Force Feedback, Human Factors, Vol.8, No.5, pp.449-455, October 1966.
[8] G.P.Starr: A Comparison of Control Modes for Time-Delayed Remote Manipulation, IEEE Trans. on System, Man and Cybernetics, Vol.SMC-9, No.4, pp.241-246, April 1979.

Part 6
Control of Motion 1

Finger-Arm Coordination Control Method for Multiple Degrees of Freedom Robot

S. Sugano and I. Kato

Department of Mechanical Engineering, Waseda University, Tokyo, Japan

SUMMARY

This paper proposes a method of controlling the coordination of the fingers and the arm when performing robot work which can be expressed as a sequence of continuous changes in positioning points of robot finger tips.
To establish a plan of coordinated fingers and arm movements, the evaluation function is determined by the change of the angle of each finger and each joint, the change angle of hand orientation and the amount of the wrist movement. In this function, weighting is used so that the change in the finger posture is greater than the change in the arm posture. In the method used for adjusting the finger-arm coordination movements in the robot operation plan, an heuristic function is established for selecting an operation plan where independent quick movements of the robot arm, without coordination with the fingers, is avoided and coordination with the fingers is attained when the arm moves.
The method mentioned above was tested in a case of the keyboard performance work using a multiple degree-of-freedom anthropomorphic robot. As a result, a path including smooth finger-arm coordination movements, similar to a performance conducted by the human, could be created automatically.

1. Introduction

Humans can perform complicated motions by cleverly coordinating the fingers and an arm in certain ways. In the case of the fingers and the arm of a robot, a hand having a multiple degree-of-freedom (DOF) fingers can handle objects of complicated shapes and can efficiently perform delicate operations such as switch operations since the fingers can follow the shapes of objects. But such a hand is not suitable for work extending over a comparatively wide space. However, an industrial robot provided with simple mechanical fingers and able to perform most of the motions necessary for arm joint movement is suitable for work requiring large movements. For accurate work, because the distance between a work point at the arm end and the arm joint actuator is long, strict positional accuracy is required for each joint. Also, strict specifications are required for designing joint drive systems because energy consumption increases due to the drive of large inertia loads.

From the above-mentioned viewpoint, it would be efficient that when a robot performs an operation the fingers and the arm should be used properly utilizing their characteristics in accordance with the operational contents and be coordinated. However, in design of the finger-arm coordination control system, a problem arises since the system has redundant DOF, and it becomes necessary to determine task allocation to fingers and arm as a coordination control standard.

Though many studies on finger and arm control have heretofore been reported, most of the studies on manipulators were on coordination-control among fingers or control of the arm only (1)(2)(3), and very little has been studied on control including finger-arm coordination.

Finger functions are roughly classified into gripping and manipulation. This study takes up continuous finger tip positioning, i.e. manipulation, as the work, and proposes a finger-arm coordination control method for work execution. Also, this study takes keyboard playing as an example of actual work, and reports experimentally verified effects of the algorithm.

2. Finger-Arm Coordination Control Method
2.1 Division of Work Contents

Optimal time-series movements of the fingers and the arm are selected from possible positions and posture thereof so that the value of some index of performance is minimized. However, there are infinite numbers of possible positions and posture of the fingers and the arm, and it is difficult to estimate for all positions and posture at a time. Therefore, a method wherein the process is conducted during the sequence of the work points to be executed is adopted. When positioning is effected sequentially at work points, the point string may be divided into: sections of continuous point string processable only by changing the finger posture; sections positioned between these sections and requiring changes of the arm posture only; and sections requiring simultaneous changes of the finger and the arm posture.

In this study, the string of work points for which the positioning process is possible with changes of finger posture only without changing the arm posture is referred to as the 'block', and the intermediate position between the present work point and the next work point ('block' connection) is referred to as 'CP'.

In this manner, when positioning point strings are given as point strings on a time axis, for example, as shown in Fig.1 (wherein points on the three-dimensional coordinate system are plotted on the time axis), it is possible to express the point strings by a sequence of the blocks defined as described above and the CP representing the boundaries between the blocks. Finger-arm path planning is performed at every block.

2.2 Posture decision evaluating function

Physical amounts concerning motion for evaluating the movements of multi-DOF manipulator include various reference values. However, when the physical amounts including temporal factors are adopted, it becomes necessary to handle constants concerning robot dynamics, the calculation for evaluation becomes complicated, and a long calculation time is required. Therefore, in this study, the process is simplified by adopting the displacement, which does not include the temporal factor, as the evaluation objective.

"Displacement" means the following:
(1) Sum of absolute values of joint angle variation
    amounts of fingers   P1
(2) Absolute value of variation in amount representing
    the hand orientation   P2
(3) Absolute value of estimated variation in amount
    representing the hand orientation   P3
(4) Absolute value of wrist movement distance   P4
(5) Absolute value of estimated wrist movement distance   P5

Item (1) represents the sum of absolute values of joint variation angles during 'block' in a plurality of multi-DOF fingers. The sum increases as the posture of the finger changes. When the function differs among fingers, the sum may be calculated after weighting the values of the fingers.

Item (2) and (3) represent the amount of variation (e.g. sum of absolute values of Euler's angle variations) with respect to reference posture of hand orientation. P2 is at CP before the present block, P3 is at CP after it.

Item (4) and (5) are the length of trajectories along which the wrist moves when the arm posture is changed during work. P4 is at CP before the present block, P5 is at CP after it.

By using P1 to P5, an index of performance for evaluating the finger and arm motion is set as expressed by

$$f = a1*LF*P1 + a2*LW*P2 + a3*LW*P3 + a4*P4 + a5*P5 \qquad (1)$$

Where ai (i = 1-5) denotes the weight coefficient of evaluation parameter Pi. LF denotes the length of each finger part and LW denotes the length from the wrist to the finger tip. The input amounts for joints of the fingers and the arm are decided so that value f is minimized for a given positioning point.

2.3 Task Allocation to Fingers and Arm

Finger and arm movement allocation is fixed by weight coefficient ai set in Equation (1). As mentioned in Section 1, fingers exhibit excellent dynamic characteristics because of a small inertia load and are basically advantageous for delicate movements. However, if the work is burdened only on the fingers during work wherein the work area suddenly extends, it becomes necessary for the arms to change posture quickly, and the arm joint actuators must be controlled quickly. As a result, strict design specifications for the joint driving system are required, and it becomes difficult to execute finger movements during arm movements.

During work execution, the maximum angular velocity and the maximum angular acceleration at the arm joints should be made as small as possible and, when the arm should move, arm movements should be dispersed during the work by considering coordination with the fingers so that the fingers and the arm can move smoothly. Accordingly, joint movement allocation for decreasing the arm joint deflection and comparatively increasing the deflection of the finger part is adopted, and weight coefficients at in Equation (1) are set as a1 < a2,a3 < a4,a5.

2.4 Decision of input amounts to finger and arm joints during work

The joint input amounts are decided by two steps:
(1) The range of work point string processed only with finger movements by fixing the arm posture (wrist position and hand orientation) is

investigated based on the movable range of the fingers. This search is performed in some posture of the arm. The arm posture and finger movement which are able to process the maximum range of work point string is adopted. The maximum range is named a 'temporary block'.
(2) The work point group (temporary block) decided in (1) is re-studied from the viewpoint of introduction of finger-arm coordination, and the blocks and CP are set.

Specifically, step(1) is executed as described below. In Fig.2, work points at which the finger tip should be positioned are indicated by ".", block names are assigned from the start point of the work point string as block(1), block(2), ..., block(n), block(n+1), ..., and the CP between the blocks are designated as CP(1), CP(2), ..., Cp(n), CP(n+1), ... Assuming that block(n-1) and CP(n-1) have already been decided for work contents, the range of positioning work points processable only by finger movements with the arm maintained in possible posture is investigated starting from CP(n-1). That is, an arbitrary finger is sequentially assigned to work points with the arm fixed in a possible posture, and the work points are judged processable only by finger movements when the finger assignment is possible with respect to the movable range of the finger joints. Judgement number Cm is expressed by

$$Cm = \sum_{j=1}^{M} W_j * H_j * F_j \qquad (2)$$

Where M is the number of fingers, Wj, Hj, Fj is the number of types of wrist position, hand orientation and finger posture possible respectively when the tip of j'th (j = 1-M) finger is placed at the work point, assuming that joint angles are converted into discrete values.

Relationship with the work points is investigated for all of Cm. When the maximum number of point strings can be processed only by finger movements with the arm in some position, the set of the point strings is named as block(n)-1, and the intermediate point between the last work point in block(n)-1 and the next work point is named CP(n)-1. Since there are normally a plurality of arm posture and finger movement methods which can process the point strings of block(n)-1, the method minimizing the value of Equation(1) is selected. For the calculation of P3 and P5, the average of coordinates of several work points after CP(n)-1 are used. By determining block(n+1)-1, CP(n+1)-1, ... and so on, it is possible to classify the work point strings into several point string sets, i.e.

temporary blocks, including no change in arm posture. However, only with the decision of finger and arm movements by this method, it is not always possible to expect smooth movements through finger-arm coordination at connection points CP(n)-1, CP(n+1)-1, ... between the divided point strings. This is because nothing is considered for the change in finger and arm posture between one temporary block and the next temporary block. In order to cope with this problem, step (2) is executed to set finally the 'CP' and 'blocks'.

In CP setting, CP(n)-2, CP(n)-3, ... are sequentially set back from CP(n)-1 at intermediate points between work points of block(n)-1 determined in (1). Then, a certain evaluation equation is calculated for CP(n)-i (i=1,2, ...), the CP(n)-i position at which the value of the evaluation equation is the minimum is set as the boundary CP(n) between block(n) and block(n+1), and the set of intermediate work points between CP(n-1) and CP(n) is taken as block(n). The index of performance is determined as described below.

First, when the finger posture and arm posture have been decided up to work point (x-1) and the x'th work point can be processed by using finger-arm coordination movement, parameter fx for identifying the type and posture of the finger possible in finger tip positioning at the point x is expressed as the function of x by

$$fx = coop(x) \tag{3}$$

Parameter fx normally has a plurality of factors. The number of point strings included in the longest temporary block having the start point at the x'th point of the work point string is expressed by

$$Bmax = max\{block(x,fx)\} \tag{4}$$

When parameter fx is not present, Bmax becomes zero.

Then, index of performance hx is expressed by

$$\begin{aligned} hx &= N - \{(x-1) + Bmax\} \\ &= N - [(x-1) + max\{block(x,fx)\}] \\ &= N - [(x-1) + max\{block(x,coop(x))\}] \end{aligned} \tag{5}$$

Where N denotes the total number of work points.

In order to decide block(n) and CP(n), an attempt is made to apply Equation (5) to CP(n)-1. There are many methods (Equation(4)) for moving from the finger and arm posture at work point x-1 prior to CP(n)-1 (left side thereof in Fig.2) to the finger and arm posture for finger positioning at work point x after CP(n)-1 (right side thereof in Fig.2). Such methods are denoted as fx1, ..., fxm, and the number of work points processable only by finger movements from work point x without changing arm posture is investigated for respective fxn (n = 1-m). The maximum among the numbers of work points determined from fxn is taken as Bmax (in the same way as searching for temporary block block(n)-1). Equation (5) is calculated by using Bmax and x to obtain value hx1. In Fig.2, block(n+1)-1 indicates the set of work point strings comprising Bmax, and max-1 indicates the end work point of the set.

This calculation is conducted sequentially for CP(n)-2, CP(n)-3, ..., the point strings corresponding to Bmax are determined as block(n+1)-2, block(n+1)-3, ..., and the end work points are determined as max-2, max-3, ... Also, for the respective cases, hx2, hx3, ..., are calculated. When the minimum value of hx1, hx2, ..., is hxi, CP(n)-i corresponding thereto is selected as CP(n), and the work point set between CP(n+1) and CP(n) is selected as block(n). In Fig.2, CP(n)-3 is selected.

The value hx acts as an index for evaluating the extent of effect obtained on configuration of the next longer block when CP is set between work point strings and the arm posture is changed by introducing finger-arm coordination movements at the CP.

By repeating steps (1) and (2), it becomes possible to automatically make a smooth finger and arm motion trajectory plan minimizing arm movements and utilizing finger-arm coordination movements for given work point strings.

3. Evaluation Experiments

Effects of the algorithm described above were evaluated by applying it to keyboard instrument playing. The keyboard instrument playing may be regarded as the finger tip positioning work, and the work contents are given in the form of a musical score.

3.1 Extraction of control parameter

The anthropomorphic manipulator was used for the evaluation, the arm has 7 DOF, including redundant 1 DOF, and the fingers have 14 DOF (2 DOF for the thumb, 3 DOF for each of the other fingers) (Fig.3)(4)(5)(6).

Since the searching range becomes very large when this algorithm is executed for all DOF of the manipulator, variable control amounts are extracted as described below by considering the characteristics of the keyboard instrument playing work.

First, it was assumed that, when hand coordinate system o-xyz is set for the wrist position and coordinate system O-XYZ for the object (Fig.4), movement and posture change during work are effected only by the wrist movement in the X axis direction, the hand direction (y) is maintained parallel with the Y axis, and rotations around the axis in the hand direction are not used for the work. Also, the X coordinates of the wrist used for the evaluation are taken as discrete values which are set at the centers of white keys and between the white keys. Coordinates of given work point strings (notes) are expressed as values on the X axis. By assuming as described above, it becomes possible to omit P2 and P3 in Equation (1).

For finger posture, the DOF of MP joint flection and extension motion is used for key striking, the DOF of MP joint radial and ulnar flection motion is used for finger tip positioning on the XY plane of the keyboard. In this manner, P1 in Equation (1) can be evaluated only with the angle of MP joint radial and ulnar flection motion.

For estimated wrist position (calculation for P5), the average of coordinates of several work points (five points in this example) is used.

3.2 Finger-arm Coordination

Function coop(x) (Equation(3)) giving fx is calculated with reference to the table (area) which makes it possible to refer to the type of the finger capable of processing the next work point on the basis of the type of the finger currently processing a work point and the distance between the present work point and the next work point. The area indicates the work point processing range possible when the MP joint radial and ulnar flection motion of the finger and the x axis movement of the wrist are coordinated with each other. This area can be calculated easily from the movable range of the MP joint radial and ulnar flection motion DOF of each finger.

3.3 Evaluation result

The results of this algorithm are shown in Fig.5. In this example, though the work point are discrete, finger and arm motion allocation free from wrist jumping was obtained by finger-arm coordination as much as possible.

Further, the process was conducted with the same parameters for approx. 20 scores of medium difficulty electronic organ music. As a result, trajectories approximately coinciding with those of human movements were obtained, and smooth movement was achieved.

Obtaining finger and arm movements equivalent to those of humans using the algorithm obtained by this study verified the efficacy of this algorithm and Equation (1) and (5) used for evaluation.

4. Conclusion

The finger-arm coordination control method for automatically calculating the motion plan in the case where continuous finger tip positioning work is conducted by the multiple DOF robot is proposed. Experimental application of this method to anthropomorphic manipulators verified that it can make the robot motion in the same way as natural human movements and is efficient.

This method can also be used for general tasks by expressing the work contents as changes of continuous finger tip positioning points.

REFERENCE

(1) S.C.Jacobsen et al. "The UTAH/M.I.T. Dextrous Hand: Work in Progress", The International Journal of Robotics Research, Vol.3, No.4, 1984
(2) T.Okada "Computer control of multi-jointed finger system", Sixth Int. Joint Conf. on Artificial Intelligence, 1979
(3) J.K.Salisbury et al. "Articulated Hands: Force control and Kinematic Issues", The International Journal of Robotics Research, Vol.1, No.1, 1982
(4) S.Sugano et al. "Keyboard Playing by an Anthropomorphic Robot" Theory and Practice of Robots and Manipulators (The fifth CISM-IFToMM Symposium), Kogan Page, 1985
(5) S.Sugano et al. "Limb Control of the Robot Musician 'WABOT-2'", Int. Conf. on Advanced Robotics, 1985
(6) S.Sugano et al. "Autonomous Limb Control for Information Processing Robots", Bulletin of Science and Engineering Research Lab. No.112, WASEDA University, 1985

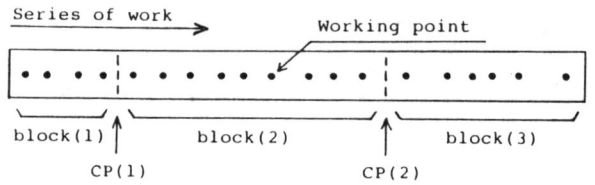

Fig.1 Block and CP

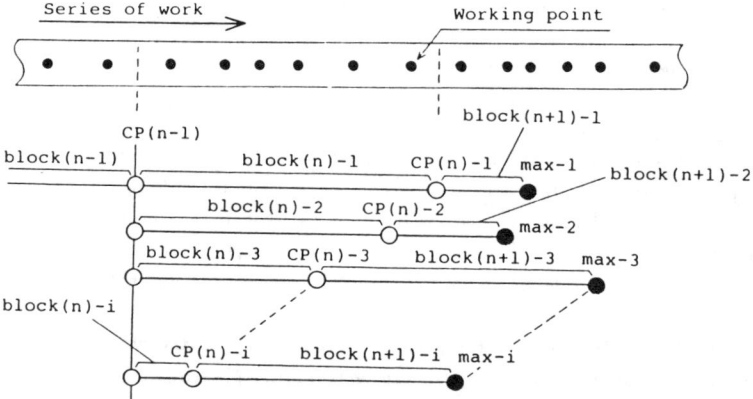

Fig.2 Fingers and an arm coordination method

Fig.3 WAM-8

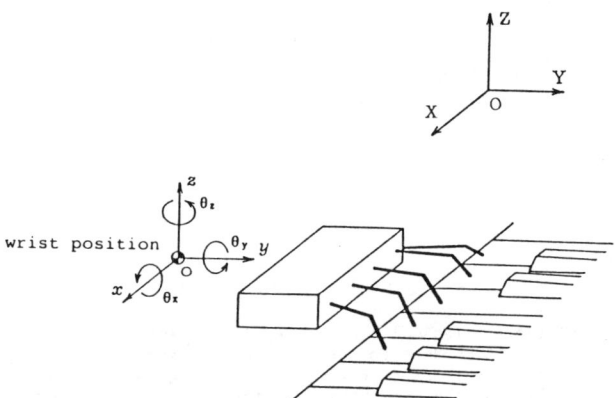

Fig.4 The coordinates of the hand

Fig.5 Example of the results

A Model-Based Expert System for Strategical Control Level of Manipulation Robots

D. Stokić and M. Vukobratović

Mihailo Pupin Institute, Beograd, Yugoslavia

SUMMARY

A new solution of the strategical control level for manipulation robots is presented in the paper. According to our approach the strategical control level includes data base of various approximate models (geometric, kinematic and dynamic) of manipulation robot and various methods which are used for solving various control tasks. The software system for strategical control level is described in global and its operation is presented in detail for one particular control task.

INTRODUCTION

The basic approach in solving complex control problems in robotics, as well in other large-scale systems, is by hierarchical control structure. The usual control levels in robot controllers are [1]: strategical control level which has to solve trajectory planning problem in complex environment; tactical control level has to map external (hand) coordinates of the robot into internal (joint) coordinates and to compute joint trajectories of the robot; executive control level represents the servosystem level which has to ensure tracking of the joint trajectories specified by tactical level.

In our previous papers we have dealt mostly with the two lower control levels - executive [2, 3] and tactical [4]. In this paper we shall consider strategical control level. There are many tasks that have to be solved at strategical control level of robotic systems. Some of them are: minimization of travel time between the given points in space or along a prescribed trajectory in space (or through several points in space), path planning in the presence of obstacles, parts assembly by robots, planning of forces realized by the robot, coordination and sinchronization of two or more robots, etc. All these problems have been focused by many researchers. The problems have been attacked by various approaches involving various models and algorithms. Here, we shall not survey these various approaches which might be found in literature [5-8].

It is obvious that to solve the above mentioned tasks statical control level has to use models of robotic systems and environment. The more precise and exact models we use, the better results might be expected from the point of energy and time required for accomplishing the tasks by the robot. However, the more complex models of robots and environment usually require more computational efforts and/or more powerfull (and more expensive) control computer in order solve the problem in acceptable time period (or, even more, to solve the problems in real time, if necessary). In order to overcome these problems and make a trade-off between the optimality of the solution at the strategical control level and requirements concerning computation time and computer equipment, we suggest a new solution of the strategical control level for manipulation robots.

NEW SOLUTION OF STRATEGICAL CONTROL LEVEL

Our approach to strategical control level [9] attempts to enable solving of trajectory planning problems in real time with the available computer system, and to improve the performance if there is more available time. Actually, our attempt is to find the best solution in the available computation time (and with available computer system). Thus, our strategical control level includes various models and algorithms for solving trajectory planning problems. Actually, it includes data base of various approximative models-geometric, kinematic and dynamic - of manipulation robots and various algorithms which can be used for solving all above mentioned control tasks.

The new solution of strategical control level consists of three sublevels (Fig. 1): (a) The first sublevel represents an expert system which has to identify the robotic task imposed by higher control level or by operator. Next, this expert system has to select the models of robot and environment and the algorithm from above mentioned data base which will be used for solfing the given robotic task. This selection of the model and the method is made in accordance with available computer equipment and computation time which is available for solving the stated control problem; namely, the expert system has to determine the time which is available for computation and solving of the trajectory problem. For each model and algorithm in data base computation time required for their execution is estimated (according to available computer system). The expert system selects the most precise model of robot and the best algorithm which can be executed in available time. (b) The second sublevel has to solve the trajectory planning problems using the models and algorithm selected by the sublevel (a). Actually, the second sublevel represents the process of solving the given control task using the selected algorithm and the models. The solution found on this sublevel is sent to lower control levels of the robot controller (i.e. to tactical control level). (c) The third sublevel has to compare the solution determined by the sublevel (b) and the real behaviour of the robot. On the basis of this comparison the sublevel (c) identifies the parameters of the approximate models of the robot and environment and changes their values in data base. The sublevel (c) gathers the information from sensors and from tactical and executive control levels. By identification of parameters of robot's and environment's models the third sublevel improves these models and algorithms for solving the trajectory planning problems.

The expert system at the first sublevel plays the central role in this solution of strategical control level. This expert system requires two data bases. The first is the knowledge base necessary for identification of robotic task and selection of the appropriate models and algorithm. The second data base is already mentioned base of various models of the robotic systems and environment and of various algorithms for solving problems at strategical control level.

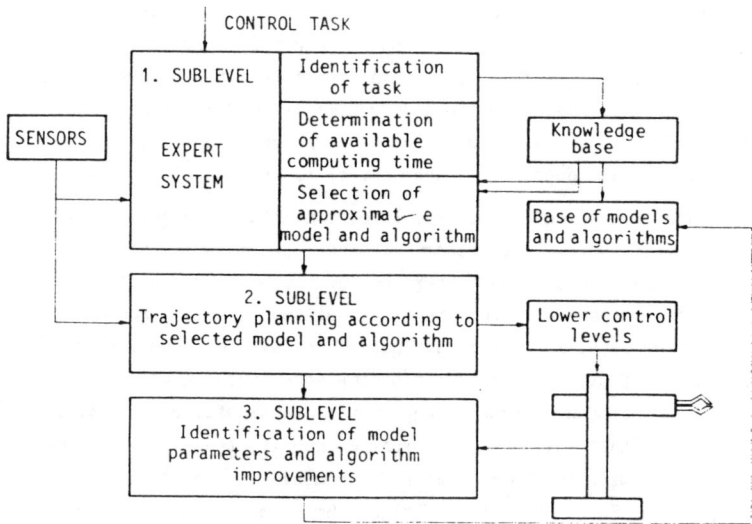

Fig. 1. Strategical control level

The strategical control level operates in the following way. When the robotic task is given by the upper hierarchical control level or by the operator, the expert system identifies the control task, determines the available time for solving the problem and selects (from data base) the best model and algorithm which can be executed in the available time period. If the available time is short and the available computer system is not powerfull, it must select very rough (approximate) model of the system and very suboptimal algorithm for solving of the given task. The control system memorizes the solution obtained by the sublevel (b). If the process repeats (which is very often in industrial practice), the expert system might re-select the model and the method and improve the solution of the same task in the repeated cycles. This means that the expert system has to check the capabilities of the computer system and available computation time. If there is sufficient time to choose better model (more complex and more adequate) and better algorithm (which requires longer computation time but offers less suboptimal solution), the expert system commands the sublevel (b) to re-compute the solution of the same control task with better model. By this we achieve improvement of the path planning solution at the strategical control level. On the other hand this improvement does not require more complex computer equipment, but the expert system always has to determine the least suboptimal solution for the given computer system and for the time period available for computation of solution. Thus, we achieve maximal efficiency of the control computer.

The software system for the strategical control level is so designed that it can be easely adapted various types of manipulation structures and with various parameters values. In the phase of initialization of the software system, the structure, geometric, dynamic and other parameters of the given robot are set, and the system automatically generates various approximate models of the robot: geometric, kinematic and/or dynamic. Also the data base of various algorithms is generated for a specific manipulation system and desired set of control tasks.

From the above description it can be concluded that our strategical control level represents combination of the expert system and model-based algorithms. Actually, the expert system uses the various models and algorithms for solving trajectory planning problems at the strategical control level.

COORDINATION OF THE MULTI-ROBOTS SYSTEM

As we have already explained the strategical control level has to solve various robotic tasks. We shall briefly explain the performance of our solution of strategical control system for one particular robotic task. We shall consider the task of coordination of several robots cooperating in one technological cell (or line). In this, we can distinguish three cases: (a) if robots cooperate in the same cell or line, but they do not share common working space; thus, the robots cannot come into contact; (b) if robots share common working space their paths must be planned to avoid collisions; (c) the robots might cooperate in such a way that they must be in contact with each other (for example if two or more robots have to lift an object together). In the case c) becides the problem of synchronization of the robots in time and in space, the problem of the control system robustness to structural perturbations has to be considered, too. We have considered the case c) in [10].

Let us consider the case b) when two robots share common working space. In this case coordination of robots has to be done in time and in space. This problem might be solved in various ways [11, 12]: (A.1) The simplest but the most suboptimal solution is to forbid the both robots to enter simultaniously into common working space [11], i.e. if one robot enter into this space, this it forbidden for the other. This simple solution, usually implemented in practice, is not acceptable for all applications and it is very time consuming from the standpoint of task execution. (A.2) The better solution of space coordination of two (or more) robots might be obtained by "master-slave" approach [12]: the "master" robot tracks its nominal (prescribed) trajectory as if it were alone, while the "slave" robot has to avoid collision; this means that the "master" robot represents moving obstacle for the "slave" robot; this approach is much less suboptimal than approach 1. but it imposes several problems to be solved. (A.3) The most complete and the least suboptimal solution of collision avoidance problem is by allowing the both robots to enter common working space and to change nominal trajectories of both robots so as to avoid collision between them. Actually, this approach allows to determine the "optimal" solution from the standpoint of time and/or energy required for task execution. However, this approach is the most complex from the standpoint of necessary computation.

The approach 1. does not require additional computation. However approaches 1., 2. require application of at least geometrical models of robots in order to determine collision free paths. In this, we might use various geometric models of robots:

(Ga) Since the robotic tasks are oftenly set in cylindrical coordinates system and since many robotic structures can be approximated by cylindrical structure, we may adopt cylindrical structure as the geometric model of robot: we approximate the robot by one bar which can rotate with angle ψ and which can change its length ρ (Fig. 2). The thikness of the bar is determined by the requirement that the real robot must be "inscribed" in the bar for all allowed joint angles. The lifting of the bar (z-coordinate) is not taken into account but the problem is solved in the plane [12]. The reason for this is that in many practical applications the movement in z-direction is restricted and cannot be used for collision avoidance. By this simple geometrical model the coordination of robots in common working space can be realized in real-time by relatively cheap microprocessors.

(Gb) In order to achieve less conservative solution of collision free trajectories we have to use more complex model in which all three cylindrical coordinates (ψ, ρ, z) are included (Fig. 3). In spite of obstacles in work space in many cases the collision avoidance can be obtained by vertical movement. Thus, we must explore this opportunity, too. However, this considerably complicates the coordination problem since it is necessary to check simultaneously collision avoidance by all three coordinates. The solution oftenly is not unique, so we have to introduce some criterion how to choose the best solution, but this approach is very time consuming. For each robotic structure we might select the best approximative geometric model in the form of some regular geometric body (cylindar, prism, sphere etc.), in which the given robot might be "inscribed" for all allowable joint angles.

(Gc) All above mentioned geometric models of robots offer rough representation of real robot's geometrical structure and parameters, but they allow simple investigations of collision free conditions. Obviously, these models occupy much larger volume than real robots, and thus, require much stronger conditions for collision avoidance than the real robots do. As the consequence much larger deviations from the nominal trajectories are required. For precise investigations of the collision avoidance and computation of "state" trajectories we should consider exact geometric model in which all robot's links are precisely described. However, investigation of such models would be extremely complex. Thus, instead precise description of each robot's link, we may adopt representation of each link by regular geometrical bodies; i.e. for each link we should determine regular body (prism, cylinder, sphere) with minimal volume in which we can "inscribe" the link. In this way we get geometrical model consisting of a serial of regular bodies representing the links of the robot. This model is more complex than (Ga) and (Gb), but it is simpler than precise (and complete), model of robot.

However, in solving coordination problem of multi-robots system at strategical control level it is not sufficient to use only above listed geometrical models of the robots, but it is necessary to take into account kinematic and dynamic features of the robots. This is required in order to predict the movement of the robot i.e. to determine the collision free trajectories in accordance with the robot's kinematic and dynamic capabilities. In other words, the strategical level must synthesize the collision free trajectories which the robots can realize. On the other hand when the strategical control level plans the collision free path it must take some safety factor ε, i.e. ε is the least distance between two robots which might be allowed. This safety factor ε depends on precision of tracking of imposed paths by the lower control levels of the robot controllers. In order to define this safety factor strategical control level must be supplied with models estimating the performance of the robots. These models might be of various complexity and the estimation of robots' precisions ε depends on the selected dynamic models. Let us briefly list some of these dynamic models which might be used at strategical control level:

(Da) The complete dynamic model of the robot is given by the model of mechanical part of the system S^M and the models of actuators S^i which drive the joints of the robot. The model of mechanical part of system is given by set of complex, nonlinear differential equation of the second order [2]:

$$P_i = H_i(q)\ddot{q} + h_i(q, \dot{q}), \qquad \forall i \in I \qquad (1)$$

where q is the n×1 vector of joint angles, P_i is the driving torque in the i-th joint, $H_i(q)$ is n×1 vector of inertia factors, $h_i(q, \dot{q})$ represents centrifugal, Coriolis and gravity forces, n is the number of joints of the robot, $I = \{i: i=1,2,\ldots,n\}$. The models of actuators are given by:

$$S^i: \quad \dot{x}^i = A^i x^i + b^i N(u^i) + f^i P_i, \qquad \forall i \in I \qquad (2)$$

where x^i is $n_i \times 1$ vector of actuator state, n_i is the order of the i-th actuator model, u^i is the scalar input, A_i is $n_i \times n_i$ matrix, b^i, f^i are $n_i \times 1$ vectors, $N(u^i)$ is the nonlinearity of the amplitude saturation type. This complete model

(1), (2) of the robot might be very complex. Thus, its application for prediction of "safe" trajectories and determination of the safety factor ε is very difficult although by this model we way obtain the best estimation of system performance.

(Db) Instead complete dynamic model (Da), we may use approximate decoupled model of the robot. Instead the complex interconnected model of mechanical system (1) we may adopt the model in form:

$$P_i = \bar{H}_{ii}\ddot{q}_i + \bar{h}_i q_i \qquad (3)$$

where $\bar{H}_{ii}$ and $\bar{h}_i$ are constant estimations of $H_{ii}(q)$ and $h_i(q, \dot{q})$. Combining (3) with (2) we get decoupled model of the robotic system:

$$\bar{s}^i: \quad \dot{x}^i = \bar{A}^i x^i + \bar{b}^i N(u^i) \qquad (4)$$

where $\bar{A}^i = (I_{n_i} - f^i T_i \bar{H}_{ii})^{-1} A^i$, $\bar{b}^i = (I_{ni} - f^i T_i \bar{H}_{ii})^{-1} b^i$, T_i is $1 \times n_i$ transformation matrix ($\dot{q}_i = T_i x^i$). In model (4) complex, nonlinear coupling is neglected. We may use model (4) to predict the motion of each joint separately and to estimate the tracking precision for each joint separately. This dynamic model is very convinient is we use geometrical model (Gc): for each link we must predict the collision free trajectory and safety factor. However, this model requires the specification of trajectories in joint coordinates q. Since the robotic tasks are usually set in external (hand) coordinates s, we must map the desired path from external into joint coordinates at the strategical control level; the mapping might be time consuming.

(Dc) In order to avoid the transformation from external into joint coordinates instead dynamic models (1), (2) or (4) in joint coordinates we may consider model in hand coordinates. We adopt m×1 vector of external coordinates s (which might include Descartes coordinates of the gripper tip (or center of mass) and Euler angles of the gripper with respect to absolute coordinates system). The complete dynamic model of the robot in external coordinates in centralized form might be written as:

$$\ddot{s} = JT[diag(\bar{A}^i)\bar{T}^{-1}(g^{-1}(s), J^{-1}\dot{s}) + diag(b^i)N(u) +$$

$$+ diag(f^i)P(s)] - \frac{\partial J}{\partial q}(J^{-1}\dot{s})^2 \qquad (5)$$

where J is m×n Jacobian matrix, $T = \text{diag}(T_i)$, $\tilde{T} = \text{diag}(\tilde{T}_i)$, $\tilde{T}_i$ is 2×n_i transformation matrix $(q_j, \dot{q}_j)^T = \tilde{T}_i x^i$, and g is function transformation from q into s (s = g(q)). J^{-1} represents inverse of J (if m=n), or generalized inverse of J (if m<n).

The model (5) is at least complex as model (1), (2). Thus, instead centralized model (5) we might adopt approximate decoupled model in external coordinates.

$$S_s^i: \quad \dot{x}_s^i = A_s^i x_s^i + b_s^i N(v^i), \qquad i=1,2,\ldots,m \tag{6}$$

where $x_s^i = (s^i, \dot{s}^i)^T$, $v = (v^1,\ldots,v^n)^T$, $v = Ju$, v^i is so-called equivalent input for subsystem in external coordinates S_s^i. The amplitude constraint upon the equivalent input $N(v^i)$ might be defined as

$$v_m^{iD}(s) < v^i < v_m^{iG}(s) \tag{7}$$

where v_m^{iD} and v_m^{iG} depend on the robot's position s. This means that for each point in work space amplitude constraints vary. The decoupled model (6) is very approximate since dynamic coupling between external coordinates is completely neglected. However, this model is very simple for estimation of the robot performance directly in external coordinates in which the problem of space coordination might be directly solved. This model is very convenient if models (Ga) and (Gb) are used. Using model (6) we may directly determine maximal allowable velocities $\dot{s}_m$ and accelerations $\ddot{s}_m$ of the robot external coordinates. However, since amplitude constraints (7) depend on the robot's position, these maximal velocities $\dot{s}_m$ and accelerations $\ddot{s}_m$ also vary with the robot's position s. This highly complicates the estimation of the robot's trajectory and safety factor. Obviously, these maximum values of velocities and accelerations might be memorized in advance for various regions of work space.

(Dd) In order to avoid the problems with variable constraints upon velocities and accelerations of external coordinates, we may adopt constant limits upon the equivalent inputs of subsystem S_s^i (6):

$$\bar{v}_m^{iD} < v^i < \bar{v}_m^{iG} \tag{8}$$

where $\bar{v}_m^{iD} = \max_{q \in Q} v_m^{iD}(s)$ and $\bar{v}_m^{iG} = \min_{q \in Q} v_m^{iG}(s)$, Q is the region in the work space in which the robot can move. Using model (6), (8) we get constant values for maximal velocities and accelerations of external coordinates. We can also estimate constant safety factors ε for all external coordinates. Thus, we get very simple kinematic model of the robot which is usually used on the strategical control level for various tasks [13].

(De) At last, instead kinematic or dynamic models of the robot, we may adopt finite automata approach. This approach, however, is suitable for the considered problem, only if we adopt solution A.1. For solutions A.2. and A.3. we must use geometrical and kinematic and/or dynamic model of the robot.

According to suggested solution of strategical control level for the given task of coordination of two robots the expert system has to select the approach and model for solving of the problem. Let us briefly consider the example of two robots which have common working space (Fig. 4); one robot is of cylindrical while the second is of spherical structure. Let us assume that robot (a) has to move from point A ($-60°$, 0.2m, 0.3m) to point B ($60°$, 1m, 0.5m) (coordinates with respect cylindrical coordinate system in the robot's base), while the robot (b) has to move from point C ($-30°$, $30°$, 0.4m) to point D ($+30°$, $60°$, 0.6m) (with respect to spherical coordinate system connected to the base of the robot (b)). The both robots have to keep their grippers in vertical positions during the motions. The maximal accelerations of the robots are ($480°/s^2$, 3.2 m/s^2, 1. m/s^2) for the robot (a), and ($360°/s^2$, $120°/s^2$, 0.8 m/s^2) for robot (b) (the accelerations correspond to external coordinates as defined above). Obviously, if we do not coordinate the motions, the collision between the robots must appear during these movements. The problem of space coordination might be solved with various approaches leading to various solutions from the standpoint of the time required for execution of the movement. In Table 1. we list several above mentioned approaches and geometrical and dynamical models which might be used for solution of the problem. The rough estimations of the time required at the strategical control level to compute the collision free trajectory at each sampling period by microprocessor INTEL-80-86 is given in the table. The estimations of the times required to complete movements of both robots are also given. It is evident, that the more complicated model we use the longer computation time is required, but the time necessary for execution of the resulting motion is shorter. Depending on the time available on the microprocessor for strategical control level, the expert system selects the approach. For example, if there is no time for off-line (pre calculation) of trajectories, the expert system must adopt approach A.1 or A.2 with geometrical model (Ga) and dynamic model (Dd). If some period of time is available for off-line computations, or the process repeats, the expert system might select better but more complicated models. It should be noted that estimations of computation time are obtained under

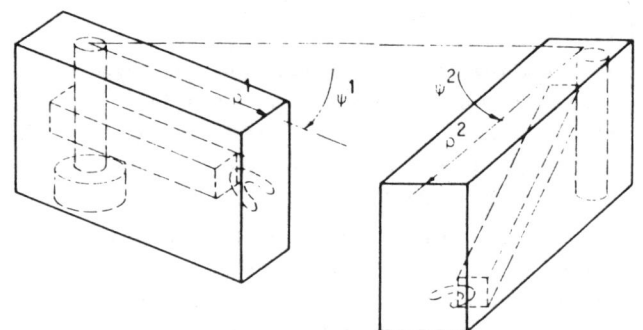

Fig. 2. Geometric models of robots (Ga)

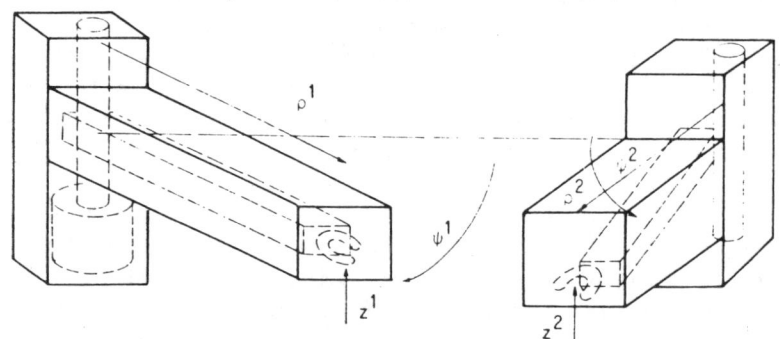

Fig. 3. Geometric models of robots (Gb)

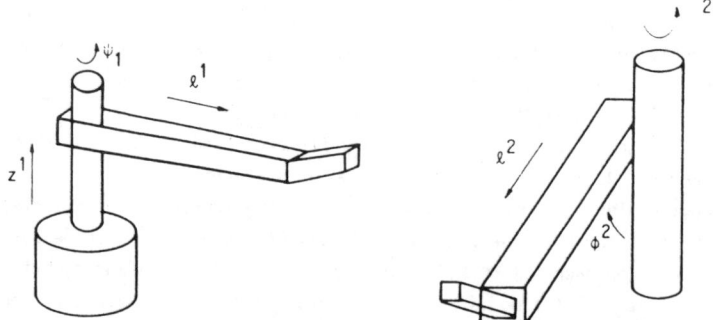

Fig. 4. Robots with common working space-geometric models (Gc)

assumption that the collision free trajectories are achieved in one shoot, which is not always true.

Approach	Geometric model	Dynamic model	Computation time [ms]	Task execution duration [s]
A.1.	-	(De)	-	2.
A.2.	(Ga)	(Dd)	2.	1.3
A.2.	(Ga)	(Dc)	2.3	1.25
A.2.	(Gb)	(Dd)	2.5	2.3
A.2.	(Gc)	(Db)	20*)	1.1

T.1. Comparison of various approaches and models for solving coordination problem

*) Time for computation of inverse kinematic problem has to be added

REFERENCES

[1] Popov E,P., Vereschagin F.A. and Zenkevich S.L., Manipulation Robots: Dynamics and Algorithms, (in Russian), Series "Scientific Fundamentals of Robotics, "Nauka", Moscow, 1978.
[2] Vukobratović K.M. and D.M. Stokić, Control of Manipulation Robots: Theory and Application, Monograph, Springer-Verlag, Berlin, 1982.
[3] Vukobratović K.M., D.M. Stokić and N. Kirćanski, Non-Adaptive and Adaptive Control of Manipulation Robots, Monograph, Series "Scientific Fundamentals of Robotics", Vol. 5, Springer-Verlag, 1985.
[4] Vukobratović K.M., and M.V.Kirćanski, Kinematics and Trajectory Synthesis of Manipulation Robots, Monograph, Series "Scientific Fundamentals of Robotics", Vol. 3, Springer-Verlag, 1985.
[5] Lozano-Perez T., M. Wesley, "An Algorithm for Planning Collision-free Paths among Polyhedral Obstacles", Communications ACM 22(10), pp. 560-570, 1982.
[6] Zapata R., P.Coiffet, A.Fournier, "Trajectory Planning for a Multi-arm Robot in an Assembly Task", Digital Systems for Ind. Automation, Vol. 2, No 2, 1984.
[7] Albus S.J., A.J. Barbera, M.L. Fitzgerald, "Programming a Hierarchical Robot Control System", Proc. 12th ISIR/Robots pp. 505-517, 1982.
[8] Khatib O., "Dynamic Control of Manipulators in Operational Space", 6th IFToMM Congress on Theory of Machines and Mechanisms, New Delhi, 1983.
[9] Vukobratović K.M., and D.M.Stokić, "Application of Robots in Assembly Automation", II Intern. Conference on Manufacturing Science, Technology and Systems of the Future, Ljubljana, 1985.
[10] Vukobratović K.M., and D.M.Stokić, "Control and Stability of the Multi-Robots System", 4th IFAC/IFORS Symp. on Large Scale Systems, Zurich, 1986.
[11] Dunne M., "An Advanced Assembly Robot". Industrial Robots, Vol. 2, Societa of Manufacturing Engineers, Michigan 1979.
[12] Hoyer H., "On-Line Collision Avoidance for Industrial Robots", Preprints Symposium on Robot Control, Barcelona, pp. 477-486, 1985.
[13] Luh J.Y.S., C.S. Lin, "Optimum Path Planning for Mechanical Manipulators", Trans. of the ASME, Journal of Dynamic Systems, Measurement, and Control, Vol. 102, pp. 142-151, June, 1981.

Robot-Task Adaptability by Semi-Local Correction without Contact

A. Jutard, T. Redarce, J.F. Chabrier and G. Liégeois

INSA-LAI, Lyon, France

ABSTRACT

The robotization of tasks which are actually manual requires a permanent adaptation between robot and task.

The problem of adaptability may be expressed as :
- taking into account of environment
- processing of information which is collected at the task level.

Proceeding without contact, we equip the terminal device of robot with proximity sensors, and owing to a judicious choice of their number and their locations, we are able to define quickly and without ambiguity the position and the orientation of terminal device with regard to the task referential.

I- ADAPTABILITY OF THE ROBOT TO THE TASK

Industrial robots now available on the market are capable of performing a great variety of movements with precision which is sufficient for the task they must carry out.

However the analysis of robotisable tasks often shows an obvious inadequacy between Robotic products and certain processes of operation [1] designated as soft tasks [2], which require that dexterity, skill and know-how to be conferred on the robot like to any operator.

These various tasks which for the most part are carried out manually could very well be automatized if the robots, capable of carrying out this type of operation at a high degree of variability, were available at present ; this however is not the case... Making robots adaptable to the task and flexible has become an economic necessity in the past few years. In fact, it is a well-known fact that the time spent on tasks such as : finishing, checking, polishing, mounting, assembling.... can take as much as 30% of production time.

Because of this, the research work undertaken in this direction is plentiful [3] and explores a great number of directions and techniques : artificial vision [4], checking of efforts [5] proximetry [6], active and passive compliance [7] [8],...

With this view in mind our laboratory has studied certain strategies of performance of this robot-task adaptability and has developed some original adaptative devices for passive complianc [9], active compliance [10] and for dimensional control [11].

This is a continuation of our research concerning this last approach which we present here, on the planeoof strategy of control as well as on the plan of the working of the adaptative device.

II- ADAPTABILITY BY DIMENSIONAL CONTROL

We would like to proceed by dimensional control without contact. Therefore in order to carry out the adaptation of the robot's referential of the effector at any moment to the hazardous evolution of the referential of the task (Fig. 1), we put proximity sensors on the terminal organ of the robot which allows us to measure the distances which separates the effector (thus drawn up from the task's object). The locations of the sensors are such that the measures of distance obtained define the position and the orientation of the referential of the task without ambiguity in relation to that of the sensor ; the number of sensors depends on the task which is to be performed.

The coincidence set-up of the two referentials can be developed from the measures of distance made between the sensor and the task object according to three force-ideas which bring about the conception of more or less complex adaptative devices and different control strategies :

- The first consists in creating an active autonomous organ with certain specific mobilities whose features and the number are adapted to the problem presented. At this point the robot is simply a carrier and it is the extra device which develops at any moment locally the required adaptation.

We have studied an active device of the [11] type to develop surface tracking using a six axes robot.This system, which is valid from the point of view of its dynamic properties, has 3 degrees of freedom (DDL) and it was equipped with 3 proximity sensors of the inductive sort. Nevertheless it presented some drawbacks : first it was heavy and bulky, secondly, by adding three more DDL to a robot which already had 6, the whole thing became superabundant.

Unlike the former, the second solution uses all the mobilities of the robot to reposition its terminal organ in accordance with the task which it must accomplish. It is sufficient to add a sensor-carrier on the robot effector which communicates the measures made in the task environment to the control cabin. This method of procedure seems to us to be contrary to the effects required ; in fact, it demands a complete command procedure(the opposite of the robot model, amongst other things) thus it is penalising from the time of response point of view. On the other hand, it leads to more or less bigger movements of the whole robot structure to correct positioning errors of a few millimetres at least, and it does this with the accuracy one associates with robots [12].

- A more detailed analysis of the problem presented leads us to propose a hybrid solution which consists in developing the required adaptation by using both its own specific mobilities at the wrist of the robot and extra mobilities, brought about by yet another device added to the extremity of the terminal organ of the robot, the carrier of the sensors required for the adaptation. The control strategy, which allows the development at any time of the referentials coincidence of the effector and the task is shown in figure 2.

This last solution, which we recommend and which we have developed presents certain advantages :

- The local adaptive device only has a limited number of mobilities. Thus it is light weight and small in volume. Together with the carrier-sensor part, it carries the tool or the effector which is necessary for the robot to accomplish the task for which it has been programmed.

- The adaptation control is just a simple as the one mentioned above, that is the autonomous local device. In fact, the whole set-up formed by the wrist of the robot and the supplementary local organ makes up a local architecture of correction which is compact and the mathmatical model of which is extremely simple. Its inversion, which is essential to reach the articular variables, is unalterable due to the invariability of the mechanical parameters of the structure. The reverse model of the device programmed once and for all, then allows direct control in real time of the correction to be done.

III- SMALL DISPLACEMENTS METHOD

In order to resolve the complex problem of the adaptability of the robot to the object of the task, we have adopted a method of dimensional control using the theory of small wrench displacements. This method, which has been developed in [11], allows the connection of elementary movements which must be done to assure the coincidence of the two referentials R_E and R_T (figure 1) for the measurements of distances carried out by the sensors.

The local adaptive device has as many sensors as it has degrees of freedom. Let C_i one of the sensors, be fixed on the effector at point M_i and the measure of direction $\vec{n}_i$ (figure 3) and let λ_i be the measure made in relation to the object of the task. Since whatever point A of the effector we show [11] that, during the correction stage, the measure λ_i is equal to the component of the two wrenches :

$$\mathcal{P}(\vec{n}_i) \cdot \mathcal{T}(A) = \lambda_i \qquad (1)$$

with $\mathcal{P}(\vec{n}_i)$ the wrench in A of Plucker's coordinates of the vector $\vec{n}_i$:

$$\mathcal{P}_{(\vec{n}_i)} : \left\{ \begin{array}{c} \vec{n}_i \\ \vec{AM}_i \wedge \vec{n}_i \end{array} \right\}$$

and $\mathcal{T}_{(A)}$ the small displacement wrench at point A of the effector.

The wrenches of Plucker's coordinates of each direction of measure n_i only depend on the position and the orientation of associated sensors C_i with regard to the referential of the effector. The processing of composant values of the small displacements wrench, expressed in the R_e reference, from known measures λ_i, leads to a linear equations system with constant coefficients. The isostatism condition of the referential R_E relative to the referential R_T is such that this system always finds a solution. Therefore it is possible, using a local, correction device, to impose the processed displacements from relation (1) at point A of the effector. The wise choice of this point allows us to simplify this relationship. We fix it arbitrarily at the origin of the effector referential.

Thus, we developed a local adaptative device whose variable articular values are directly proportional to wrench composant values $\mathcal{T}_{(A)}$.

If, from a practical point of view this method leads to results which are the same as those obtained by the Jacobian inversion method, it has certain advantages :

- it presents a synthesis of information collected on environment evolutions,
- it uses wrench algebra, and thus simplifies the writing of different relations using the specific properties of wrenches,
- it follows the evolution system step by step, and thus leads to simple processing which is consequently rapid.

IV- CORRECTION ARCHITECTURE ASSOCIATING THE WRIST OF THE AKR ROBOT TO A TRANSLATION MODEL

In order to find out if our approach was valid, we developed an adaptation to surface tracking which uses wrist mobilities of the AKR robot in the laboratory and a translation module, situated at the end of the wrist with 3 proximetry sensors and on which a tool can be adapted (for example a pneumatic machining tool).

Figure 4 shows the diagram of the correction architecture.

By writing all the small displacement wrenches at the same point and using Chasles' relationship to obtain the variations to be applied to the articular variables, we finally obtain the small displacements wrench of the effector in relation to the fixed referential of the robot :

$$\mathcal{T}_{4/0} = \left\{ \begin{array}{l} -\sin\alpha_3 \, \Delta\alpha_2 + \cos\alpha_3 \, \cos\alpha_2 \, \Delta\alpha_1 \\ \Delta\alpha_3 - \sin\alpha_2 \, \Delta\alpha_1 \\ \cos\alpha_3 \, \Delta\alpha_2 + \sin\alpha_3 \, \cos\alpha_2 \, \Delta\alpha_1 \end{array} \right. \left| \begin{array}{l} -(L_2+L_4)\cos\alpha_3 \, \Delta\alpha_2 - \sin\alpha_3(L_1+(L_2+L_4)\cos\alpha_2)\Delta\alpha_1 \\ \Delta L_4 \\ -(L_2+L_4)\sin\alpha_3 \, \Delta\alpha_2 + \cos\alpha_3(L_1+(L_2+L_4)\cos\alpha_2)\Delta\alpha_1 \end{array} \right\}_{R_3} \quad (2)$$

Its composants can be processed for a type of problem owing to information collected on the evolution of the environment.

By expressing it with the following composants :

$$\mathcal{T}_{04} \, 4/0 : \left\{ \begin{array}{l} \theta_1 \\ \theta_2 \\ \theta_3 \end{array} \right. \left| \begin{array}{l} u \\ v \\ w \end{array} \right\}_{R_3} \quad (3)$$

We obtain the system of equations below :

$$\theta_1 = -\sin\alpha_3 \, \Delta\alpha_2 + \cos\alpha_3 \cos\alpha_2 \, \Delta\alpha_1$$

$$\theta_2 = \Delta\alpha_3 - \sin\alpha_2 \, \Delta\alpha_1$$

$$\theta_3 = \cos\alpha_3 \, \Delta\alpha_2 + \sin\alpha_3 \cos\alpha_2 \, \Delta\alpha_1$$

$$u = -(L_2+L_4)\cos\alpha_3 \, \Delta\alpha_2 - \sin\alpha_3 \, (L_1+(L_2+L_4)\cos\alpha_2)\Delta\alpha_1 \quad (4)$$

$$v = \Delta L_4$$

$$w = -(L_2+L_4)\sin\alpha_3 \, \Delta\alpha_2 + \cos\alpha_3 \, (L_1+(L_2+L_4)\cos\alpha_2)\Delta\alpha_1$$

The unknown $\Delta\alpha_1$, $\Delta\alpha_2$, $\Delta\alpha_3$ and $\Delta\alpha_4$ are the four variables of the correction structure. The adaptation is only possible if $\mathcal{T}_{04\ 4/0}$ only imposes four values.

In the case of surface tracking where only three axes are used to develop adaptability, we can express the wrench more freely by :

$$\mathcal{T}_{04\ 4/0} = \left\{ \begin{array}{c|c} \theta_1 & x \\ x & v \\ \theta_3 & x \end{array} \right\} \quad (5)$$

The system then becomes :

$$\begin{cases} \theta_1 = -\sin\alpha_3 \, \Delta\alpha_2 + \cos\alpha_3 \cos\alpha_2 \, \Delta\alpha_1 \\ \theta_3 = \cos\alpha_3 \, \Delta\alpha_2 + \sin\alpha_3 \cos\alpha_2 \, \Delta\alpha_1 \\ v = \Delta L_4 \end{cases} \quad (6)$$

and for $\cos\alpha_2 \neq 0$, we obtain the reverse-model of the wrist :

$$\begin{cases} \Delta\alpha_1 = (\cos\alpha_3 \, \theta_1 + \sin\alpha_3 \, \theta_3) \, \dfrac{1}{\cos\alpha_2} \\ \Delta\alpha_2 = (-\sin\alpha_3 \, \theta_1 + \cos\alpha_3 \, \theta_3) \\ \Delta L_4 = v \end{cases} \quad (7)$$

Through processing θ_1, θ_3 et v, it is easy to obtain $\Delta\alpha_1, \Delta\alpha_2, \Delta\alpha_4$; the prohibited case, where $\alpha_2 = \frac{\pi}{2}$, corresponds to a physical impossibility of the robot's wrist to fit the task.

The composants θ_1, θ_3 et v depend firstly on the position of the sensors, and secondly on the measures made by these.

If the sensors are positionned in a plane at distance a from point θ_4, at 120° one from the other we arrive at :

$$\begin{cases} \theta_1 = \dfrac{1}{3a} (\lambda_1 + \lambda_2 - 2\lambda_3) \\ \theta_3 = \dfrac{1}{a\sqrt{3}} (\lambda_1 - \lambda_2) \\ v = \dfrac{1}{3} (\lambda_1 + \lambda_2 + \lambda_3) \end{cases} \quad (8)$$

V- EXPERIMENTAL DEVICE

In order to increase the possibilities of the AKR robot and to make it apt to execute surface tracking we have been led to design and develop a translation module, tool carrier and proximetry sensors carrier. This supplementary axis is operated by a step by step motor. Figure 5 shows the device set up in our laboratory, fixed on the wrist of the robot.

We process by the relationships 8 from information given by the sensors, the movements to be done and thereby the signals to be given to each of the actuators which carry out the three mobilities required.

The control of the direction is assured by a Mostek Microprocessor Z80. The calculations were given to an arithmetical processor and the data acquisition was made by using A/N converters. The orders of correction were transmitted by N/A converters for the rotation axes of the wrist and by P.I.O. for the translation axis.

Note that the orders corresponding to the control of specific axes of rotation of the robot superimpose on those owing to the control of the trajectory imposed on the robot through experience.

So as to reduce the correction time and to make it compatible with industrial use, we have chosen the translation axis to be first in control.

The tests taking place at the moment have confirmed the proof of our approach ; the simplicity of the control strategy adopted makes the adaptation rapid enough so that its use is absolutely compatible with a robotized work of quality.

VI- ACKNOWLEDGEMENTS

We would like to thank the Centre National de la Recherche Scientifique (C.N.R.S.) who backed this study through our participation in the Automatic and Advanced Robotics program (ARA). We also thank the Agence Nationale pour la Valorisation de la Recherche (ANVAR) who financed a part of our work on the subject.

REFERENCES

[1] G. LIEGEOIS, M. BETEMPS, A. JUTARD, P. ROMAND
Task analysis for industrial robots and design
11th International Symposium on Industrial Robots
TOKYO 7-9 octobre 1981 pp 43/50

[2] A. JUTARD, G. LIEGEOIS
Mise en oeuvre d'un robot en vue de l'exécution des tâches molles nécessitant un haut niveau de variabilité.
Compte-rendu final à l'Agence de l'Information (convention 82/141) juillet 1984

[3] B. ESPIAU
Prise en compte de l'environnement local dans la commande des robots manipulateurs.
Thèse de doctorat d'état, juin 1982 - RENNES

[4] A.G. MAKHLIN
Stability and sensibility of servo-vision systems
ROVISEC 5 - AMSTERDAM 29-31 octobre 1985 pp 79/89

[5] J.P. MERLET
Commande par retour d'efforts en Robotique
INRIA - rapport de recherche n° 351, décembre 1984 - PARIS

[6] P. MARCHAL, J. CORNO, J.M. DETRICHE
Soudage auto adaptatif au moyen d'un système de suivi de joint automatique.
RomanSy 81 - Warsaw, septembre 1981

[7] D.E. WHITNEY
Quasi-static assembly of compliantly supported rigid parts
Journal of Dynamic Systems measurement and control.
mars 1982 - Vol 104   pp 65/77

[8] J.P. ROUGET
Contribution à l'étude structurelle et fonctionnelle d'organes compliants passifs pour robot d'assemblage.
Thèse de 3ème cycle, juillet 1983 - BESANCON

[9] A. FAKRI
Conception, modélisation et simulation de la structure compliante passive DCR pour robot d'assemblage.
Thèse de doctorat (nouveau régime), septembre 1985 - LYON

[10] J.F. CHABRIER, A. JUTARD, G. LIEGEOIS, T.REDARCE
Robot/task adaptability : two examples with and without contact
ROVISEC 5 - AMSTERDAM, 29-31 octobre 1985 PP 119/128

[11] M. PILLET, J.F. CHABRIER, A. JUTARD, G. LIEGEOIS
Robotique et théorie des petits déplacements - Application au suivi de surface.
APII - AFCET Vol 19 n°2 , juin 1985 pp 99/116

[12] B. CARON
Etude et réalisation d'un logiciel d'aide à la modélisation de systèmes mécaniques articulés.
Thèse de 3ème cycle, janvier 1985 - LYON

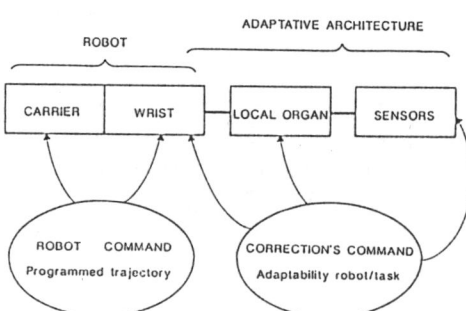

Fig 2 Correction strategy

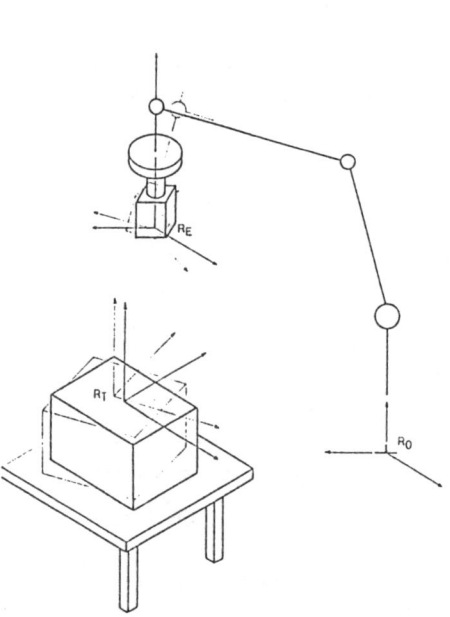

Fig 1 Relative position of robot effector and task referentials

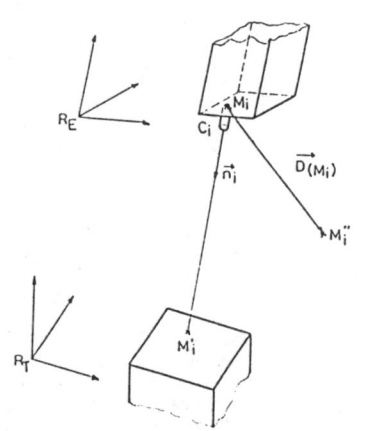

Fig 3 Elementary displacement

Robot-Task Adaptability 341

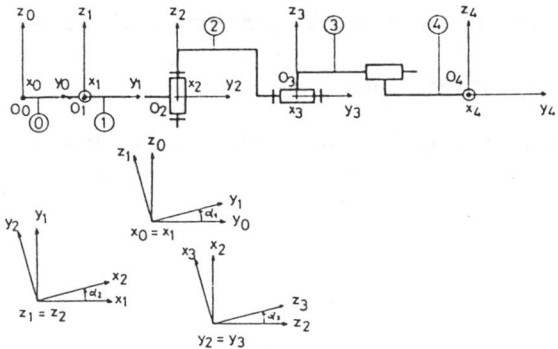

Fig 4 Modelling of an adaptative system

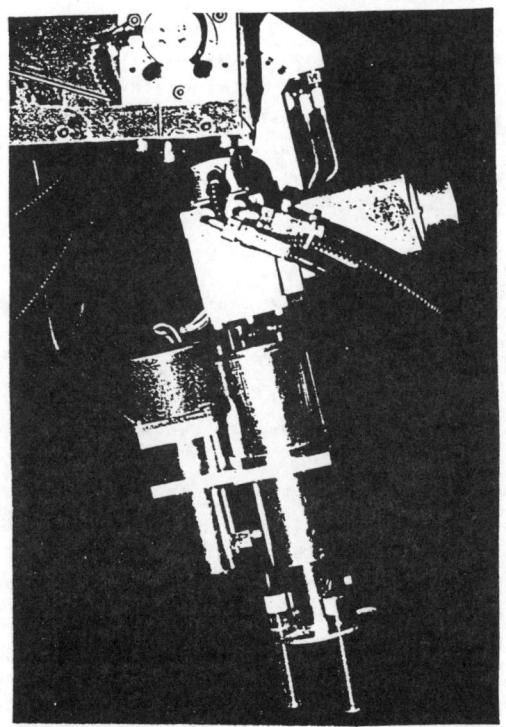

Fig 5 A.K.R. robot's wrist with the translation
system and the proximity sensors

Robot Control Synthesis in Conjunction with Moving Workpieces

P. Kiriazov and P. Marinov

Bulgarian Academy of Sciences, Institute of Mechanics and Biomechanics, Sofia, Bulgaria

SUMMARY

    Many applications of manipulators are connected with work on conveyors and it is necessary to make control synthesis of point-to-point motion in conjunction with moving workpieces. The velocity of the manipulator end-effector should be equal to the conveyor's velocity at the operation point and this fact involves a new shooting technique for exact solution of the corresponding two-point boundary-value problems. Besides the proposed off-line control synthesis, a final adjustment of control parameters can be performed on the manipulator itself as in a self-learning adaptive procedure. The method is verified on a dynamic model of a two-degrees of freedom manipulator using D.C. motors.

1. INTRODUCTION

    In most industrial applications manipulators do not have to follow a prespecified path -- they are given freedom to move along any path between two given intermediate or end points. But now continuously increasing demands for enhanced productivity and improved precision have imposed special requirements on the design of the control system.

    Industrial manipulators are usually controlled by conventional (linear) control techniques, which are based upon simplified models of the mechanisms. The simplified nature of these models, in characterizing the highly coupled and nonlinear manipulator dynamics, results in low speeds or overshoot and unnecessary oscillations of the gripper (1,2).

    Control synthesis of manipulator movement from one point to another, which operation must be performed as fast as possible, leads to complex nonlinear two-point boundary-value problems (TPBVP). Using a shooting technique, several algorithms for the exact solution

of these TPBVPs have been proposed (3,4). In these works, the
TPBVPs are characterized by zero boundary conditions for the initial
and final values of the generalized velocities.

Many applications of manipulators are connected with work on
the conveyors and it is necessary to make control synthesis of
point-to-point motions in conjunction with moving workpieces. The
velocity of the manipulator gripper should be equal to the conveyor's velocity at the operating point which means that at this
point the boundary conditions for the generalized velocities have
to be non-zero.

In view of the preceding fact, a new shooting technique for
the exact solution of the corresponding TPBVPs is proposed. The
control synthesis, which is based on the real manipulator dynamics
and the actuators' maximum capabilities, can be performed on the
manipulator itself for final adjustment of the control parameters.

The method was verified on a dynamic model of a two-degrees
of freedom manipulator using D.C. motors.

2. STATEMENT OF THE PROBLEM
2.1. Real space formulation.

For the sake of a simpler statement of the problem, a manipulator with polar coordinates is considered (Figure 1). The manipulator picks up a workpiece on the table, places it on the conveyor with a given velocity v^c and then goes back. This operation consists in two point-to-point motions: (I) motion from point P.1 ($\dot{x}=0$) to point P.2 ($\dot{x}=v^c$) and (II) motion from P.2 ($\dot{x}=v^c$) to P.1 ($\dot{x}=0$). It is accepted in this case that there is no time-delay at the operation point P.2. Both the operation points P.1 and P.2 have some specified coordinates in the real space.

The problem is to make control synthesis of the motions (I) and (II) based on real manipulator dynamics and the actuators' maximum capabilities.

2.2 Joint space formulation.

Lagrangian formulation of the dynamic behaviour of a manipulator with an acceptable approximation presents the following model:

$$J(q)\ddot{q} + N(q,\dot{q}) = R \qquad (1)$$

where: q is the nx1 generalized coordinates vector, J(q) is the matrix of inertia, $N(q,\dot{q})$ represents non-linear interaction forces, gravity, and the dry and viscous frictions, and R is nx1 vector of the force/torque which the actuators apply at the joints. The magnitudes of the driving forces or torques depend on the control inputs (voltages):

$$R_i = R_i(q_i, \dot{q}_i, u_i), \qquad i=1,\ldots,n \tag{2}$$

We assume that the constraints on the control inputs of the actuators have the form:

$$|u_i| \leq U_i \tag{3}$$

The conversion from the real space coordinates to the joint coordinates yields the following boundary conditions:

(I) motion

$$q(t^o) = q^o, \quad \dot{q}(t^o) = 0 \text{ -- initial state P.1} \tag{4}$$

$$q(t^c) = q^c, \quad \dot{q}(t^c) = r^c \text{ -- final state P.2} \tag{5}$$

(II) motion

$$q(t^c) = q^c, \quad \dot{q}(t^c) = r^c \text{ -- initial state P.2} \tag{6}$$

$$q(t^f) = q^o, \quad \dot{q}(t^f) = 0 \text{ -- final state P.1} \tag{7}$$

In view of the time-optimal concept, a bang-bang control strategy is accepted. Each joint motion has one switching point from acceleration to deceleration when the manipulator moves from P.1 to P.2 and vice versa.

So, we come to the solution of two TPBVPs: (1-2-4-5) and (1-2-6-7), under the above mentioned control laws.

3. SOLUTION APPROACH

3.1. Off-line control synthesis.

Figure 2 shows a joint motion in the phase plane. The values $q_i^{s.I}$ and $q_i^{s.II}$ are the switching values of the joint coordinates corresponding to the (I) and (II) manipulator motions.

We are more familiar (3) with the control synthesis of the second motion because its final values of the joint velocities are to be zero. The control synthesis consists in performing several simulation movements from the point P.2 to some terminal positions converging to the required P.1. Each such movement means that the system (1÷2) with some approximate switching values $q_i^{s.II}$ is integrated from the initial state (6) until satisfying the final conditions (7)$_2$: $\dot{q}_i(t_i^f)=0$, i=1, ..., n. Denoting $q_i(t_i^f)=F_i$, we come to the following vector shooting equation:

$$F(q^{s.II}) = q^o \tag{8}$$

In this way the other final condition (7)$_1$ will be satisfied and thus the second TPBVP (1-2-6-7) will be solved.

If we perform backward integration in time of the system (1÷2) with the initial conditions (5), then, using the above proposed shooting technique, we are able to solve, also, the first TPBVP, finding $q_i^{s.I}$, i=1, ..., n. The obtained final times will be the starting times t_i^o and $t^o = \min t_i^o$, i=1, ..., n. Note that, in general, t_i^o are different as well as t_i^f, where $t^f = \max t_i^f$, i=1, ..., n.

Thus the obtained values of the control parameters t_i^o, $q_i^{s.I}$ and $q_i^{s.II}$ cannot be used directly for the control system of the manipulator itself, because the system (1÷2) cannot present, in practice, exactly the real manipulator dynamics. It is difficult or impossible to take into account the influence of backlash, friction, deflections, etc. That is why we shall use these values of the control parameters like some approximations to the real ones in the next proposed method for on-line control synthesis.

3.2. On-line control synthesis.

This control synthesis procedure consists in performing several cyclic movements of the manipulator: table - conveyor - table. As an initial guess for the control parameters we take the previously obtained ones using the off-line method. It was stated that the corresponding trajectory of the manipulator end-effector will not, in general, pass through the given point P.2 and the terminal position will not coincide with the point P.1 on the table. Therefore, we have not to adjust the control parameters for the exact solution of the first and second TPBVPs on the base of the real manipulator dynamics.

The first TPBVP can be solved using consecutive solutions of two shooting equations. The first one is for the switching values $q_i^{s.I}$ which are varied in order to satisfy the boundary conditions for the point P.2: $q_i(t_i^c)=q_i^c$, where t_i^c are defined from the conditions: $\dot{q}_i(t_i^c) = r_i^c$, $i=1, \ldots, n$. Denoting $q_i(t_i^c) = F_i^c$ and taking into account that with some given initial times t_i^o the vector F^c depends only on the vector $q_i^{s \cdot I}$, this shooting equation may be written as:

$$F^c(q^{s \cdot I}) = q^c \qquad (9)$$

The other shooting equation for the first TPBVP has the initial times t_i^o as variables and the aim is to satisfy the condition of simultaneous approach to the values q_i^c of the joint coordinates, i.e.:

$$t_1^c(t_1^o, \ldots, t_n^o) = \ldots = t_n^c(t_1^o, \ldots, t_n^o) \qquad (10)$$

Once the solutions of the eqs (9) and (10) are obtained, the control synthesis of the first TPBVP thus will be effected, which means that the trajectory of the manipulator end-effector will pass through the point P.2.

Further, the control synthesis of the motion (II) is preformed in the same way as in the off-line method, i.e. we have then to adjust the switching values $q_i^{s \cdot II}$ according to the shooting eq. (8).

Now, one can see that the proposed method, with a modification, will solve the problem when the manipulator end-effector has to move synchronously with the workpiece on the conveyor during some time interval.

4. NUMERICAL EXAMPLE

By way of illustration, a dynamic model of a manipulator with polar coordinates is taken into consideration. Including D.C. actuator systems and taking some specified values of the model parameters, the equations of the coupled joint motion are written as:

$$\ddot{q}_1 = (u_1 - 6.8\dot{q}_1 - 0.34(10.0q_2 + m(q_2+1.0))\dot{q}_1\dot{q}_2)/$$
$$(0.4 + 0.17(20.0 + 10.0q_2^2 + m(q_2+1.0)^2)),$$
$$\ddot{q}_2 = (u_2 - 6.8\dot{q}_2 + 0.17(10.0q_2 + m(q_2+1.0))\dot{q}_1^2)/(0.4 + 0.17(10.0+m))$$

where: $U_1 = 15V$, $U_2 = 10V$, $m=5$ kg (loading), $m=2$ kg (unloading) P.1(0;2), P.2(1;0), $v^c = 0.15$ m/s, $q_1^o = \pi/2$ rad, $q_2^o = 1.0$ m, $q_1^c = 0$ rad, $q_2^c = 0$ m, $r_1^c = 0.15$ rad/s, $r_2^c = 0$ m/s.

The trajectory obtained of the manipulator end-effector and the joint motions are depicted in figures 3, 4 and 5, respectively.

5. CONCLUSIONS

A direct method for off-line and on-line control synthesis of the manipulators' point-to-point motions in conjunction with moving workpieces has been proposed. The method is based on the real manipulators' dynamics and the actuators' maximum capabilities, and can be easily realised by the control system in a self-learning adaptive procedure.

The concept for the singular (bang-bang) control laws has been taken for a simpler presentation of the method but similar control synthesis algorithms have been performed in the case of more general control laws improving the manipulator behaviour in the switching and positioning times (4,5).

References

1   Kahn, M.E., and Roth, B., The Near Minimum-Time Control of Open-Loop Articulated Kinematic Chains, Journal of Dynamic Systems, Measurement and Control, ASME Trans.,Vol. 93, n° 3, Sept.1971, pp.164-171
2   Kim, Byung Kook and Kang G. Shin, Suboptimal Control of Industrial Manipulators with a Weighed Minimum Criterion, Proc. 22nd IEEE CDC, San Antonio, TX., Dec. 1983, pp.1199-1204
3   Marinov, P., and Kiriazov, P., Synthesis of Time-Optimal Control for Manipulator Dynamics, Teor. Appl. Mech., Publ. House Bulg. Acad. Sci., Year 15, $\underline{1}$, 1984, pp.13-19
4   Kiriazov, P., and Marinov, P., A Method for Time-Optimal Control of Dynamically Constrained Manipulators, Prepr. of 5th CISM-IFToMM Symposium on Theory and Practice of Robots and Manipulators, June 26-29, 1984, Udine, Italy, pp.131-138
5   Marinov, P., and Kiriazov, P., A Direct Method for Optimal Control Synthesis of Manipulator Point-to-Point Motion. Prepr. of 9th World Congress IFAC. Vol. IX, Colloquia 14.2, 09.2. Budapest, Hungary, 2-6 July, 1984, pp.219-222.

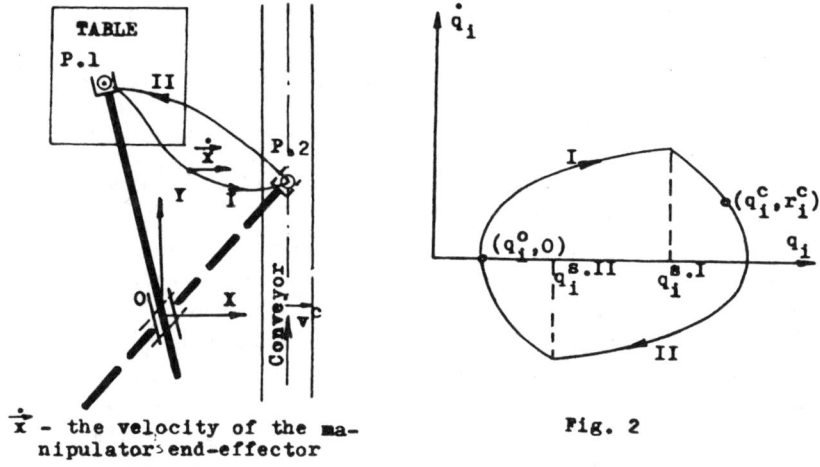

$\dot{x}$ - the velocity of the manipulator's end-effector

Fig. 1

Fig. 2

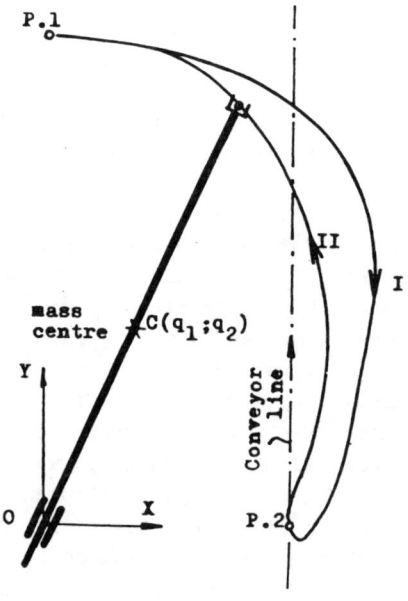

Fig. 3

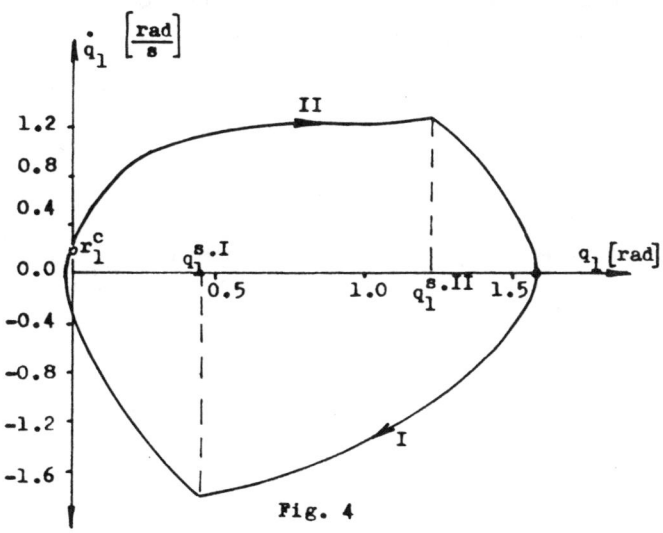

Fig. 4

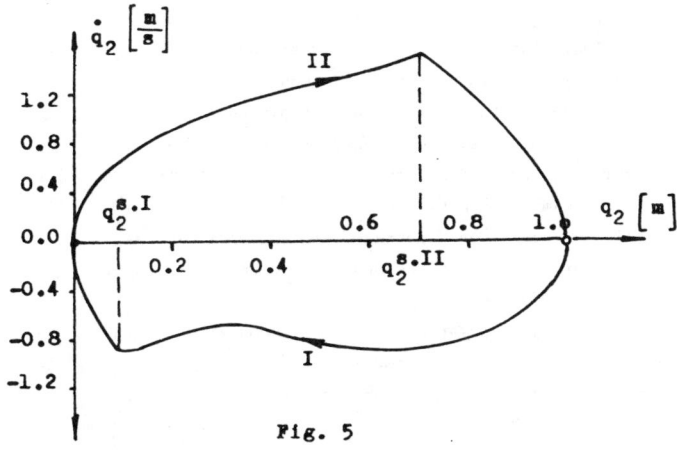

Fig. 5

Dynamic Command Motion Tuning for Robots
A Self Learning Algorithm

G.W. Vernon, J. Rees Jones and G.T. Rooney

Mechanisms and Machines Group Liverpool Polytechnic, Liverpool, England

Summary - Computer control offers a significant advantage in that the motion command can be readily up-dated. The robot's trajectory can be stored as a set of discrete data rather than coefficients of an explicit function. A simple algorithm is derived for tuning this data subsequent to each run. Its use requires minimal knowledge of the dynamics and no additional transducers. Once the required tracking accuracy has been achieved, the up-dating process can be curtailed. Should the system be subject to parameter variations, retention of the process should cope with effects due to these variations. Experimental results are included, the peak dynamic tracking errors being reduced from 10% of the motion range to 0.5% in the best case and 4% to 0.6% in the worst.

Notation

$t, \Delta t$	- time, sample interval	A, B, C	- system, input & output matrices
i	- sample number	P, Q	- difference eqn. soln. matrices
j	- coordinate or axis number	L	- learning gain matrix
k	- repeat cycle or run number	I	- identity matrix
exp	- exponentiation	x, y, r	- state, response & command vectors
λ	- dummy time variable	$\bar{r}$	- reference command vector
		e	- displacement error vector

Introduction

Static accuracies of industrial robots can be very high, typically 0.05% but dynamically they may be over 10%. The tracking accuracy of the robot can be improved by tuning the shape of the command profile. Conventional dynamic tuning requires a mathematical model of the system and is typified by the work of Stoddart 1953 (1) for polynomial motions, Rees Jones 1977 (2) for harmonic motions. The form and complexity of the dynamic model in the manipulator case makes it impractical to use the same approach.

In the robot and controller we have a self contained experimental and data logging system. It is capable of estimating its own dynamic parameters but substantial effort is required to use these in tuning the command, for an arbitrary trajectory. The robot dynamics are inherently contained in the response, so it is appropriate to use this information in a self learning strategy. The objective of the strategy is to progressively increase the trajectory tracking accuracy, until some machine limitations are reached.

Relation to Previous Work
Monitoring response information from a trajectory run has been widely studied, in the parameter estimation of robots inertial components. Most of this work has been simulation, one exception being that of Atkeson 1985 (3). Highlighted are the problems of using actual torque measurements as opposed to simulated ones. Craig 1984 (4) produces a learning scheme which does utilise the robot repeatability, but in constructing a torque function, in a simulation only. It relies on substantial dynamic modelling and the use of a complex control law. Arimoto 1984 (5) proposed a "Betterment Process" in which arbitrarily weighted error derivatives are added into the trajectory. The "simplest structure" proposed here is overlooked, because apparently convergence is not always assured. In his implementation, Arimoto 1985 (6) uses an up-dating algorithm :

$$r_{k+1} = r_k + L \frac{d}{dt}(\bar{r}_k - \dot{y}_k)$$

After 28 runs, the displacement error is reduced to around 10%. The slow convergence is attributed to the low magnitudes of the L matrix elements, which is also chosen to be diagonal.

The dynamic tuning by self learning system presented here requires only a simple feedback control system. The extra computation of the learning algorithm does not need to be in real time; if it is, it only comprises a few multiplications and additions. No dynamic model is used in its application.

Derivation of the Learning Algorithm
If the robot is repeatable, the errors ahead can be accurately estimated, and used in a learning system to negate errors before they arise on the next run. By restricting attention to the corresponding individual samples i and i+1 from each run, an analysis can be based on a linearised form for the manipulator. Thus at any time t; constant matrices A,B and C can be found such that :

$$\dot{x}(t) = A\ x(t) + B\ r(t) \quad \text{and} \quad y(t) = C\ x(t) \qquad (1,2)$$

The A and B matrices include components due to both the feedback controller and the robot's open loop dynamics. In the discrete time case, solutions may be obtained as difference equations :

$$x((i+1)\Delta t) = P\ x(i\Delta t) + Q\ r(i\Delta t) \qquad (3)$$

or in a more convenient form, adding the run number, k :

$$x_{i+1,k} = P x_{i,k} + Q r_{i,k} \quad \text{and} \quad y_{i,k} = C x_{i,k} \quad (4,5)$$

where $P = \exp(A \Delta t)$ and $Q = \int_0^{\Delta t} \exp(A \lambda) B \, d\lambda$ (6,7)

The scheme is shown in figure 1. The change in response produced by the change in command between runs k and k+1; assuming modification to $r_{i,k}$ only takes place at or after the ith sample and is given by :

$$y_{i+1,k+1} - y_{i+1,k} = CP(x_{i,k+1} - x_{i,k}) + CQ(r_{i,k+1} - r_{i,k}) \quad (8)$$

or $\quad \Delta y_{i+1,k} = CP \Delta x_{i,k} + CQ \Delta r_{i,k}$ (9)

Now because the change in command does not take place prior to the ith sample, the response x_i is due to the command r_{i-1} which is unchanged. If the system is repeatable and has been tuned up to this point, then :

$$x_{i,k+1} = x_{i,k} \quad \text{hence (9) becomes} \quad \Delta y_{i+1,k} = CQ \Delta r_{i,k} \quad (10,11)$$

The error on the previous run, at sample i+1 is given by :

$$e_k = \bar{r}_{i+1} - y_{i+1,k} \quad (12)$$

Note $\bar{r}$ is the reference trajectory and is not a function of k. Clearly for there to be zero error in the response at sample i+1 on the next run, k+1 :

$$\Delta y_{i+1,k} = e_k \quad (13)$$

This condition will be satisfied by the learning algorithm :

$$\Delta r_{i,k} = (CQ)^{-1} e_k \quad (14)$$

So for the learning system to work in a single run, the $(CQ)^{-1}$ matrix would require computation at a suitable set of intervals of i. To produce these accurately would defeat the object of the scheme. Referring back to equation 11, note that the relative slopes of response & command are defined by the (CQ) matrix. If the feedback control system is working reasonably well, the change in response can be approximated to the change in command.

hence $(C\ Q)^{-1} \simeq I$; the identity matrix   (15)

so rewriting the learning algorithm as :

$$\Delta r_{i,k} = L\ e_k \tag{16}$$

or $r_{i,k+1} = r_{i,k} + L\ (\bar{r}_{i+1} + y_{i+1,k})$ (17)

It is therefore suggested that L should be diagonal or at least dominantly so and that the diagonal elements should be close to unity for fast convergence.

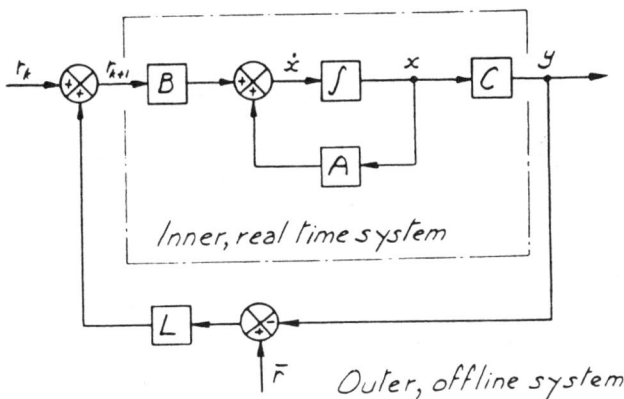

Fig.1 The self learning scheme

System Stability

The learning algorithm does not affect the dynamic response in real time. If the feedback controller is stable the system should remain stable. A condition for stability is that :

$$0 < \frac{\Delta y_{i+1,j,k}}{\Delta r_{i,j,k}} < 1 \tag{18}$$

these being the jth axis elements of the associated vectors. From equation 11 it is clear this is not a function of the learning algorithm. The learning algorithm itself is capable of generating a diverging, or increasingly oscillatory command shape if any of the diagonal elements of (L C Q) are greater than unity. This assumes (L C Q) to be substantially diagonal.

Implementation

The three major axes were used in the tests, being hydraulically driven, under analog position control, comprising proportional gain and a stabilising pressure feedback on the swing and vertical axes only. The experimental scheme was controlled by an i8086 based commercial microcomputer, in Pascal. It is outlined in figure 2. The analog interface unit contained 12 bit DAC's and ADC's for the three axes and a programmable real time clock to control synchronisation. The sample interval was set at 10mS; the highest fundamental natural frequency of the robot being around 16Hz. Control gains were empirically maximised to yield high static and tracking accuracy consistent with stability requirements. Position feedback was provided by 0.1% linear potentiometers. The robot configuration is outlined in figure 3.

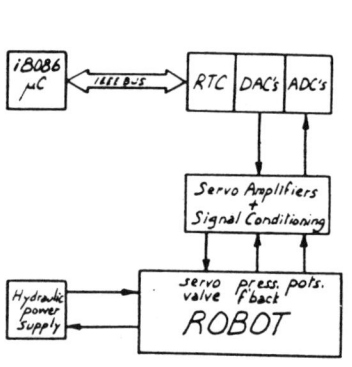

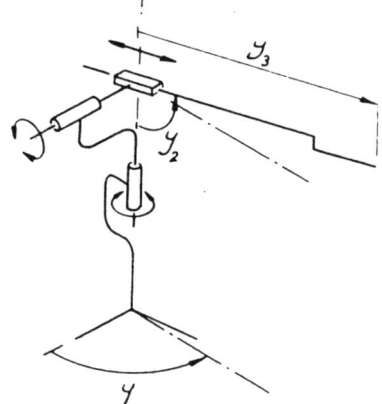

Fig.2 The System Hardware   **Fig.3 The robot configuration**

The nominal or reference motion used comprised 5 segments of durations 1,1.5,1,1.5 and 1 second respectively. With zero velocity and acceleration boundary conditions polynomials were defined with the following displacement end conditions:

AXIS	y1,swing (rad)	(units)	y2,vertical (rad)	(units)	y3,horizontal (m)	(units)
low condition	−0.3927	1663	1.4399	1535	0.8	1671
high condition	0.3927	2431	1.7017	2559	1.0	2751

These values give around 25% of the motion range for each axis. The 'units' of measurement are 12 bit numbers, giving a system resolution of 0.024%. Only the second and fourth segments involve these motions, the remainder are static. The same motion, run at one second duration saturates in velocity on two axes. The un-tuned results of this motion are presented in figures 4,5 & 6, the errors in 7,8 and 9. The horizontal (time) axis 0 to 550 corresponding to the sample number. The vertical axis is in 12 bit integer form (0 to 4095).

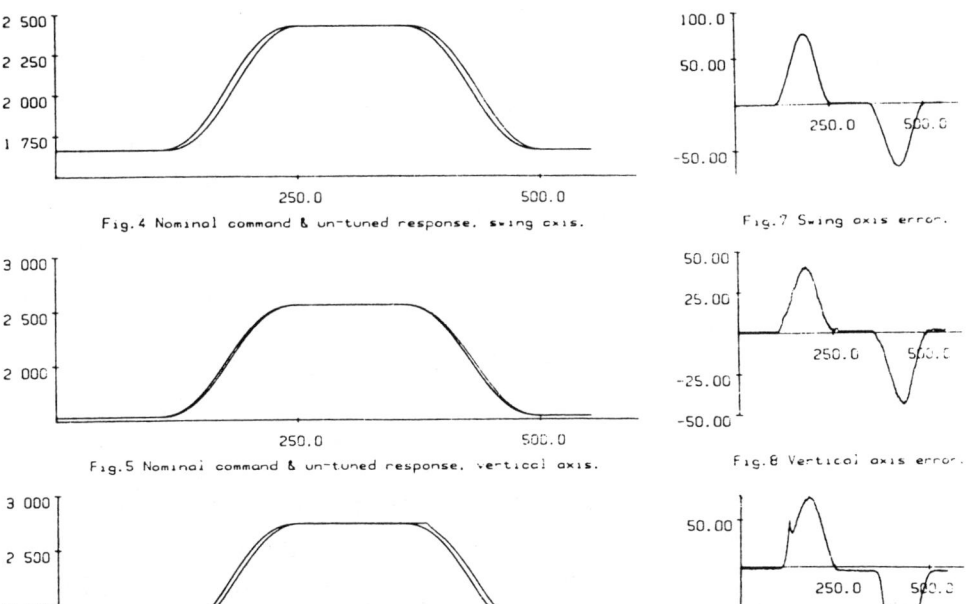

Fig.4 Nominal command & un-tuned response, swing axis.

Fig.7 Swing axis error.

Fig.5 Nominal command & un-tuned response, vertical axis.

Fig.8 Vertical axis error.

Fig.6 Nominal command & un-tuned response, horizontal axis.

Fig.9 Horizontal axis error.

A conservative L matrix was initially chosen with diagonal elements of 0.8. A subsequent change to 0.9 increased the rate of convergence without detriment. The tuned results presented (figures 10 to 15) are based on a run which took 15 cycles to reach the specified accuracy. Repeating the same test took varying numbers of cycles, from 13 to 19. The specified error bounds on the axes were 2,3 and 4; in system 'least significant bit' units. These values were found to be the lowest feasible which allowed tuning to continue without interference by noise. The maximum number of repeat cycles allowed for any one datum sample was 3, which led to one or two response samples being outside the specified error limits. Note that for comparison purposes, tuning was only applied to the second of the five segments, where the lift occurs.

Fig.10 Nominal command & tuned response, swing axis.

Fig.13 Swing axis error.

Fig.11 Nominal command & tuned response, vertical axis.

Fig.14 Vertical axis error.

Fig.12 Nominal command & tuned response, horizontal axis.

Fig.15 Horizontal axis error.

Fig.16 Swing.

Fig.17 Vertical.

Fig.18 Horizontal.

Fig.19 The three tuned commands.

To confirm that the approximation used in equations 11 and 15 is justified, the response of each robot axis is plotted as a function of the corresponding tuned command, in figures 16,17 and 18; the tuned commands in figure 19. The large spike in axis 3 command arises automatically to combat the static lag.

Concluding Remarks

A simple algorithm for improving the trajectory tracking accuracy of a robot, has been presented. It utilises the fact that if a robot motion is dynamically repeatable, previous response data can be used to null out errors before they arise. The algorithm is relatively independent of the manipulator system, its design requires the estimation of few parameters. Extra commands in a robot programming system could be used to enable operator selection of tuned motion. Additional costs amount to approximately 35kbytes/minute of tuneable motion.

To move on, more analysis is needed to aid the setting of error bounds and learning gains. A similar technique could be used to progressively reduce the motion duration until saturation or other limits are reached.

Acknowledgements - This work was supported by research grant GR/D/24081 under the ACME directorate of the Science and Engineering Research Council.

References

(1) Stoddart,D.A.,"Polydyne Cam Design",Mach. Design,Nos.1-3,121,pp.146,1953.
(2) Rees Jones,J., "A Comparison of the Dynamic Performance of Tuned and Non-tuned multiharmonic cams", Second IFToMM Int. Symp. on Linkages and Computer Aided Design Methods, Bucharest,Romania,Vol III-2,paper 47,pp.553-564, 1977.
(3) Atkeson,C.G., An,C.H. & Hollerbach,J.M., "Estimation of Inertial Parameters of Manipulator Loads and Links", 3rd International Symposium on Robotics Research, Gouvieux (Chantilly), France, October 1985.
(4) Craig,J.J., "Adaptive Control of Manipulators Through Repeated Trials", paper FP2, pp.1566-1573, JACC 1984.
(5) Arimoto, S.,Kawamura,S. & Miyazaki,F., "Bettering Operation of Dynamic Systems by Learning : A New Control Theory for Servomechanism or Mechatronics Systems",TP4,pp.1064-1069,Proc. of the 23rd Conf. on Dec. and Control,IEEE,Las Vegas, NV, December 1984.
(6) Arimoto,S., Kawamura,S. & Miyazaki,F., "Can Mechanical Robots Learn by Themselves ?", Proceedings of 2nd International Symposium of Robotics Research, Kyoto, Japan, August 1985

Part 7
Sensing and Machine Intelligence 2

C-Surface Theory Applied to Force-Feedback Control of Robots

J.-P. Merlet

INRIA, Le Chesnay, France

ABSTRACT:

For a robot in contact with its environment the constraints between the positional variables may be described in the parameters space in term of a "C-surface". The robot performs its task by sliding on this surface. We proves a theorem which relates the generalized forces acting on the manipulator to the C-surface normal. A general scheme for force feedback control is then proposed illustrated by simulation results of the contour following problem.

A parallel manipulator is then presented and some kinematic features of this architecture are demonstrated. We show that this architecture is very convenient for force-feedback control.

I - INTRODUCTION

C-surface was first introduced by MASON (1) as a task configuration which allows only partial positional freedom. In this case neither pure position or force control is convenient but rather a hybrid mode where some degrees of freedom are pure position controlled and the others pure force controlled. Freedom of motion occurs along C-surface tangents and freedom of force occurs along the C-surface normal. From this point we may introduce an "hybrid" controller to be one where we may define degrees of freedom to be force or position commanded. Such controlers may be find in (2) (3). The main problem for their use is that the choice of the degrees of freedom must be determinated by the operators and are fixed. But in some case the degrees of freedom may vary (e.g. the surface following problem) or there may be some uncertainty on their real positions (e.g. assembly tasks). We will show first that the C-surface normal may be related directly to the force measurements.

II - C-SURFACE NORMAL THEOREM

First we generalize the notion of C-surface, as the set of points belonging to the robot's configuration space where contact with its surrounding occurs. Figure 1 shows the C-surface for the peg-in-hole problem. According to this definition we may demonstrate that *in the static case the generalized forces measurements and two orientation parameters enable to determinate the C-surface normal whatever is the task to be performed*.

The demonstration of this theorem may be found in the Annex. For example if the parameters are the cartesian coordinates and the angle $(\alpha, \beta, \partial)$ of the rotation around each axis of the absolute frame the components of the C-surface normal N are

$$N \mid F_x, F_y, F_z, -\frac{M_x}{F}, \sin\alpha \frac{M_z}{F} - \cos\alpha \frac{M_y}{F}, \cos(\alpha+\beta)\frac{M_z}{F} - \sin(\alpha+\beta)\frac{M_x}{F}$$

where F is the value of the contact force, F_x, F_y, F_z its components and M_x, M_y, M_z the torques around the axis.

III - FORCE FEEDBACK CONTROL

From this theorem we may deduce a force-feedback controller by defining two "experts", a normal expert and a sliding one.

1) *the normal expert* : according to the force and position measurements it calculates the C-surface normal and then a basis of the tangent hyperplane.

2) *the sliding expert* : this expert generates the displacement of the manipulators : we may control the force magnitude by a displacement along the C-surface normal. Then the others constraints of the task enable to calculate the motion in the tangent hyperplane (e.g. the motion which minimize the distance between the actual position and the goal).

C-surface method is a hybrid control method where the force and position controlled degrees of freedom are automaticaly determinated..

IV - EXAMPLE : SURFACE FOLLOWING PROBLEM

We have here a 2-D problem where a robot move in a plane with a fixed orientation and must follow a contour with a constant applied force. In this case the dimension of the configuration space is two. The forces measurements give the normal to the contour (normal expert). The sliding expert generates a motion along the normal to control the force and along the tangent according to the desired velocity. A PID controller is used to control the force by a displacement along the normal. A simulator for this task has been implemented (4) and figure 2 shows the graphic display. Figure 3 shows the force

C-Surface Theory Applied to Force-Feedback Control

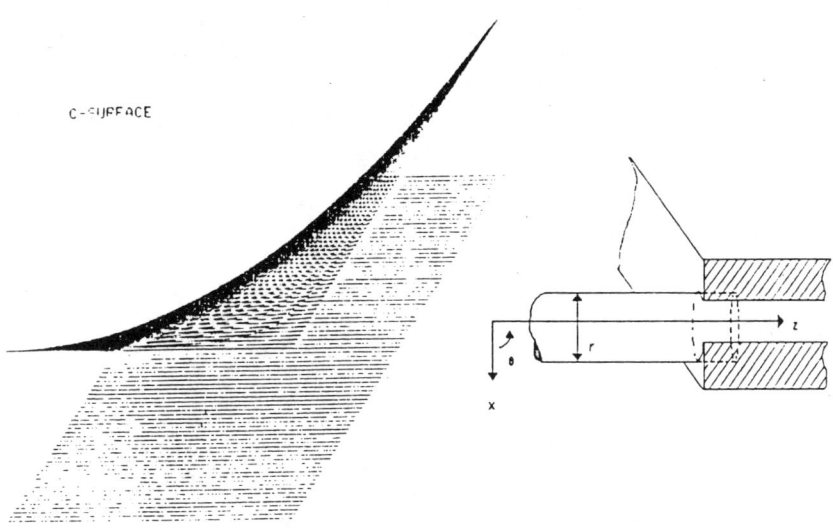

Figure 1 : C-surface for the peg-in-hole $(x \geq 0, \theta \geq 0)$

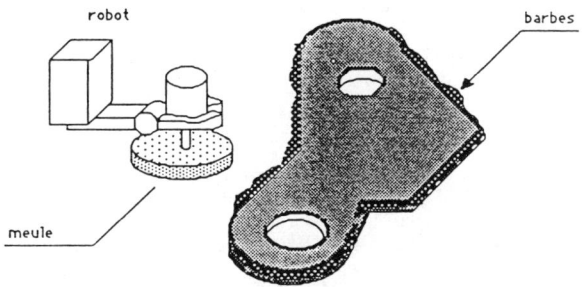

Figure 2 : surface following simulation

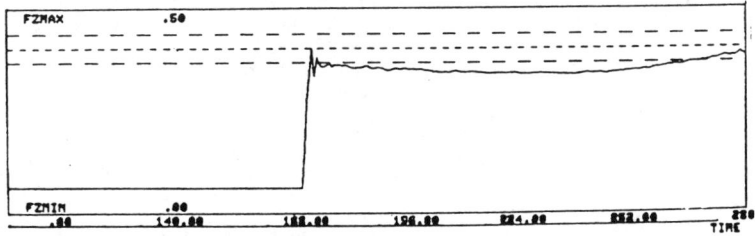

Figure 3 : force record for the surface following problem

records (the dashed lines are the desired value ± 10 %). The stiffness matrix and generalized damper methods (5) have been also tested. The results may be found in Table 1.

method	C-surface	stiffness matrix	generalized damper
force average (desired value : 0.5)	0.4315	0.651	0.5815
force variance	0.0009093	0.005038	0.002093

Table 1: surface following problem, force results (with a ± 0.1 mm accuracy robot)

Table 2 illustrates the computation time for each of these methods.

method	C-surface	stiffness matrix	generalized damper
computation time (μs)	460	350	685

We have also experimented in simulation the turning of a crank. Figure 4 shows the followed trajectory for a manipulator submitted to a random positional noise of ± 1mm. In this case we have only a rough approximmation of the position of the axis so that the manipulator is only force controled. It may be seen that the followed trajectory is very close to the ideal circle.

The C-surface method has been also tested for the peg-in-hole problem in simulation. We have shown that this command enable to insert a 20 mm diameter peg with a clearance of 5 μm using a robot which repeatability is ± 0.1 mm (4). According to the simulation results these method seems to be very insensitive to the variation of the features of the process, and the stability is quite good.

According to the abovementioned good simulation results we are planning to perform experiments, in particular complex assembly using a "left hand".

V - PARALLEL ROBOT

At the moment few complex assembly task has been performed (6). For such assembly we are currently developing a "left hand" i.e. a very accurate manipulator with small mobility range. We use a parallel architecture, this is each actuator is directly connected to the base and to the hand of the robot. This manipulator (figure 5) has been developed with the cooperation of M. REBOULET from CERT DERA, Toulouse, (2), which design the prototype. Basically we have 2 plates linked together with 6 segments. Each segment consists of an electrical jack, a damper, a force sensor and a linear potentiometer to control the length of the jack. With convenient articulation we have a 6 d.o.f. device with a positionning accuracy of about 5/100 mm. The position and orientation of the upper plate is uniquely determinated if we know the length of each segment. Such architecture has interesting features (4) :

1) The inverse kinematic problem is solved analytically in a parallel way i.e. if we know the position and orientation of the plates we can calculate the length of each segment, which does not depend of the length of the 5 others segments.

2) The inverse of the Jacobian matrix can be determined analytically and each line of this matrix is independant from the other one.

3) The force exerted on each segment is axial (this is very convenient for the strain gages sensors) and we can calculate the forces and torques acting on the upper plate on the basis of the force measurements.

C-Surface Theory Applied to Force-Feedback Control 365

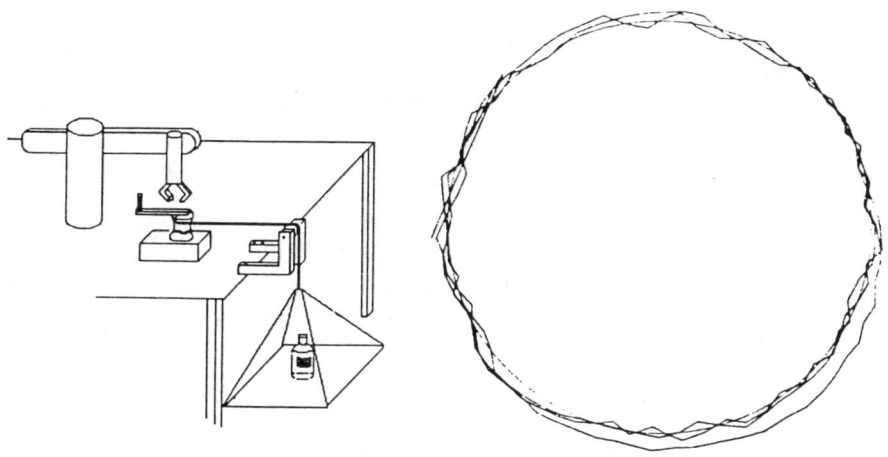

Figure 4 : turning of a crank and the trajectory of the hand

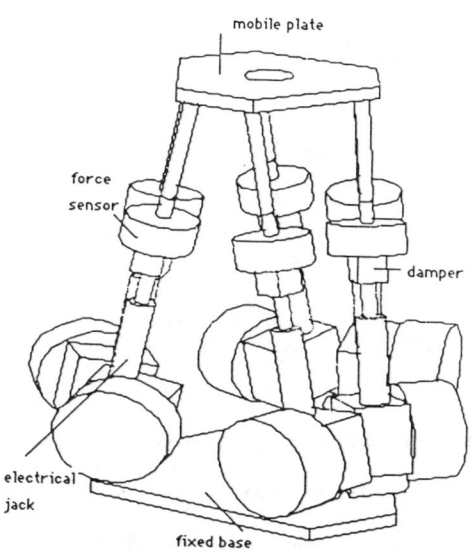

Figure 5 : parallel manipulator

4) The damper give some passive compliance to this manipulator but this compliance is controlled through the linear potentiometers (the other components of the robot are quite infinitly rigid). This compliance may be ajusted very easily. A theoretical study of the mechanics of assembly has shown that in most case it exists some insertion depth limit beyond which no jamming may occur if there is some passive compliance. Thus active C-surface compliance may be used only at the beginning of the assembly so that the cycle time is minimum.

The range of the manipulator is 0 to 5 cm in the vertical direction, ± 10 cm in the lateral one, ±20° for the Euler's angle θ and ± 70° for the Euler's angle ψ, ϕ. The velocity will be about 2 cm/s and the maximum measured force 88 daN with an accuracy of about 0.1 N. The actuators enable to have a maximum load of about 900 kg.

VI - CONCLUSION

The C-surface normal theorem presented here enables to find automaticaly the position and force commanded degrees of freedom for an hybrid force/position controller.

The deduced force-feedback command seems to be very efficient. Simulation for assembly, turning of a crank and surface following has shown interesting results. A parallel manipulator has been presented: this kinematic is very convenient for force-feedback command. Experiments are currently under development.

REFERENCES

(1) MASON M.T.
"Compliance and force control for computer controlled manipulator", IEEE Trans. on Syst., Man and Cybern, Vol. SMC 11, N° 6, June 1981, pp. 418-432.

(2) CRAIG J.J., RAIBERT M.H.
"A systematic method of hybrid position/force control of a manipulator", J. of Dyn. Syst., Meas. and Control, 1981, 103, N° 2, pp. 126-133.

(3) REBOULET C., ROBERT A.
"Hybrid Control of a manipulator with an active compliance wrist", 3th Int. Symp. on Robotics Res., GOUVIEUX, FRANCE, 7-11, Oct. 1985, pp. 76-80.

(4) MERLET J.P.
"Contribution à la formalisation de la commande par retour d'efforts en robotique", Thèse de doctorat de l'Université Paris VI, 1986, to appear.

(5) WHITNEY D.E.
"Force-Feedback control of manipulator fine motions", J. of Syst., Meas. and Control, June 1977, pp. 91-97.

(6) GIRAUD A.
"Generalised active compliance for part-mating with assembly robots", 1st Int. Symp. on Robotics Res., BRETTON WOODS, USA, 25 August-2 September 1983, pp. 949-960.

ANNEX

Let 2 frames R (O,x,y,z), the absolute frame, and $R_2(E,x_2,y_2,z_2)$ (figure 6).

Surface of object 1 can be described by : $U(X) = 0$ where X is a point in frame R.

For object 2 we have : $U_2(X_2) = 0$ where X_2 is a point in frame R_2.
Let $X_e = [x_e, y_e, z_e]$ the coordinates of E in frame R.
We can write : $X_2 = R^T(X-X_e)$ (1) where R is a rotation matrix.

The C-surface may be defined by : $U_2(R^T(X-X_e)) = 0$.

We set $U_2(R^T(X-X_e)) = V_x(X_e,\alpha,\beta,\partial)$ where α,β,∂ denote the orientation parameters.

Let S be a contact point. The normal of object 1 at point S is in the R frame :

$$\frac{\partial U}{\partial X}\bigg|_S \quad (2)$$

and for object 2

$$\frac{\partial U_2}{\partial X}\bigg|_S \quad (3) \text{ and we have } \frac{\partial U}{\partial X}\bigg|_S = - \frac{\partial U_2}{\partial X}\bigg|_S \quad (4)$$

Consider now the normal N of the C-surface which is a 6-dimensional vector

Let $\partial V_x = [\frac{\partial V_x}{\partial x_e}, \frac{\partial V_x}{\partial y_e}, \frac{\partial V_x}{\partial z_e}]$ be the three first components of N

We have $\frac{\partial V_x}{\partial X_e} = \frac{\partial U_2}{\partial X_e} = \frac{\partial X_2}{\partial X_e} \cdot \frac{\partial U_2}{\partial X_2}\bigg|_S \quad (5) \quad \frac{\partial U_2}{\partial X_2} = \frac{\partial X}{\partial X_2} \cdot \frac{\partial U_2}{\partial X}\bigg|_S \quad (6)$

From (1) we get $\frac{\partial X_2}{\partial X_e} = - R^T I \quad (7)$ and $\frac{\partial X}{\partial X_2} = RI \quad (8)$

Equations (7), (8), (5) yields :

$$\frac{\partial V_x}{\partial X_e} = - R^T R \frac{\partial U_2}{\partial X}\bigg|_S \quad (9)$$

R is a rotation matrix, thus $R^T R = I_3$ where I_3 is the identity matrix. Then using equation (4) yields

$$\frac{\partial V_x}{\partial X_e} = \frac{\partial U}{\partial X}\bigg|_S \quad (10)$$

Thus *the three first components of the C-surface normal are the components of the normal to the surface. In the static case the contact force F is also directed by this normal, thus with the force measurements we may determine the 3 first components of the C-surface normal.*

Consider now the torques M. In the frame R we have

$$M = F \wedge (X_S - X_e) = F \frac{\partial V_x}{\partial X_e} \wedge (X_S - X_e) \quad (11)$$

Let $\frac{\partial V_x}{\partial \Omega}$ be the 3 last components of N and M_1, M_2, M_3 the three components of M.

We have $\frac{\partial V_x}{\partial \Omega} = \frac{\partial X_e}{\partial \Omega} \cdot \frac{\partial V_x}{\partial X_e} \quad (12)$

and $\frac{\partial X_e}{\partial \Omega} = [(\frac{\partial R}{\partial \alpha} X_2)^T \; (\frac{\partial R}{\partial \beta} X_2)^T \; (\frac{\partial R}{\partial \partial} X_2)^T]^T \quad (13)$

Equations (1) yields

$$\frac{\partial X_e}{\partial \Omega} = [\frac{\partial R}{\partial \alpha} R^T (X-X_e)^T, \frac{\partial R}{\partial \beta} R^T (X-X_e)^T, \frac{\partial R}{\partial \partial} R^T (X-X_e)^T]^T \quad (14)$$

We have : $R = R_\alpha R_\beta R_\partial$ (15)

with

$$R_\alpha = \begin{bmatrix} 1 & 0 & 0 \\ 0 & \cos\alpha & -\sin\alpha \\ 0 & \sin\alpha & \cos\alpha \end{bmatrix} \quad R_\beta = \begin{bmatrix} \cos\beta & 0 & -\sin\beta \\ 0 & 1 & 0 \\ \sin\beta & 0 & \cos\beta \end{bmatrix} \quad R_\partial = \begin{bmatrix} \cos\partial & \sin\partial & 0 \\ -\sin\partial & \cos\partial & 0 \\ 0 & 0 & 1 \end{bmatrix}$$

We have to estimate $\frac{\partial R}{\partial \alpha} R^T, \frac{\partial R}{\partial \beta} R^T, \frac{\partial R}{\partial \partial} R^T$ from (15) we get :

$$\frac{\partial R}{\partial \partial} R^T = (R_\alpha R_\beta) \frac{\partial R_\partial}{\partial \partial} R_\partial (R_\alpha R_\beta)^T \quad (16) \qquad \frac{\partial R}{\partial \beta} R^T = R_\alpha \frac{\partial R_\beta}{\partial \beta} R_\beta^T R_\alpha^T \quad (17)$$

$$\frac{\partial R}{\partial \alpha} R^T = \frac{\partial R_\alpha}{\partial \alpha} R_\alpha^T \quad (18)$$

Thus

$$\frac{\partial X_e}{\partial \Omega} = \begin{bmatrix} 0 & -(z_s - z_e) & y_s - y_e \\ (y_s-y_e)\sin\alpha - (z_s-z_e)\cos\alpha & -(x_s-x_e)\sin\alpha & (x_s-x_e)\cos\alpha \\ (y_s-y_e)\cos(\alpha+\beta) & -(x_s-x_e)\cos(\alpha+\beta) - (z_s-z_e)\sin(\alpha+\beta) & (y_s-y_e)\sin(\alpha+\beta) \end{bmatrix} \quad (19)$$

which yields :

$$\frac{\partial V_x}{\partial \Omega} = [-\frac{M_1}{F}, \sin\alpha \frac{M_3}{F} - \cos\alpha \frac{M_2}{F}, \cos(\alpha+\beta)\frac{M_3}{F} - \sin(\alpha+\beta)\frac{M_1}{F}]^T \quad (20)$$

Consequently :

$$N = [\frac{F_x}{F}, \frac{F_y}{F}, \frac{F_z}{F}, -\frac{M_1}{F}, \sin\alpha \frac{M_3}{F} - \cos\alpha \frac{M_2}{F}, \cos(\alpha+\beta)\frac{M_3}{F} - \sin(\alpha+\beta)\frac{M_1}{F}]^T \quad (21)$$

Thus we may say that *in the static case the C-surface normal can be expressed in term of the generalised forces and only two orientation's parameters.*

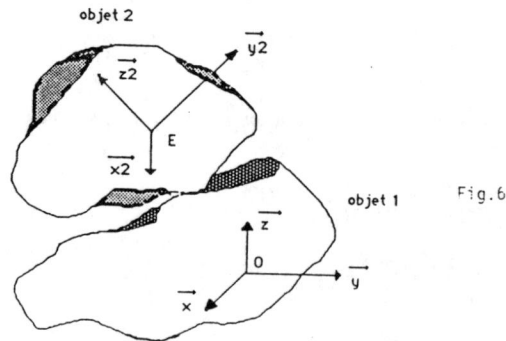

Fig.6

Experimental Investigation of Active Force Control of Robot and Manipulator Arms

G. Galatis, A. Hadzistylis and J.R. Hewit

University of Newcastle-upon-Tyne, England

Recently a new method for effecting robust coordinative control of robot and arm movement has been proposed. The method, Active Force Control, operates by substituting measurement for computation and thus avoids many of the difficulties in obtaining fast robotic movements imposed by the highly non-linear coupled dynamic equations and the need to perform calculations in real time.

To date, however, only theoretical results have been offered backed up by a certain amount of digital simulation.

Limited stability criteria have been obtained which show certain limits to system parameters in order that the method be effective.

Simulation results have indicated that with proper design the method can offer improvements in high speed performance over a wide range of operating conditions.

Here we describe the results of an experimental investigation into this method of robot movement control.

An experimental planar arm with three degrees of freedom is described. This arm has been designed specifically for the validation of Active Force Control and to this end it possesses a number of desirable features including gearless transmissions, high performance actuators, special purpose instrumentation (including transduction of accelerations and motor torques) and a fast control computer with floating point hardware.

A series of experiments is described consisting of specially selected tasks aimed at monitoring the performance of Active Force Control and comparing it with more conventional control techniques.

These tasks include attempting to hold the arm still while applying disturbances of both internally generated and external forces and torques, and attempting to move the arm in a coordinated way in the face of random disturbances.

The results of these tests are given. They show clearly that Active Force Control can provide up to 100% immunity to external forcing although the improvement obtained over all the tests ranges from 15% to 100% depending upon the parameters (such as frequency) of the disturbances.

Reasons for non-ideal control are discussed and methods proposed for extending Active Force Control to applications to robotic arms with flexible transmissions and other indesirable but almost inevitable deficiencies.

1. Introduction

The problem of obtaining high speed coordinated movement of robotic arms and manipulators has proved to be unusually intractable.

Many different schemes have been proposed to deal with this problem, ranging from the deliberate design of direct-drive arms with resulting simplified dynamics,[1] to the implementation of complex adaptive algorithms to force arms of complex geometry to conform to simple models.[2]

Most of these schemes have been supported by simulation studies only and serious questions must remain about the feasibility of their application in the demanding environment of the real-time real-world.

A few years ago, Active Force Control was proposed by one of the authors as an alternative approach to the problem.[3]

In what follows, the results of applying Active Force Control to an experimental arm are described, and it is shown that the initial optimism in the method has been justified although much work remains to be done in implementing it in an actual industrial situation.

2. Active Force Control

The dynamical behaviour of an n-degree of freedom robotic arm can usually be modelled by an equation of the form.

$$\underline{A}(\theta)\underline{\ddot{\theta}} = \underline{B}(\theta,\dot{\theta}) + \underline{T} \qquad 2.1$$

where $\underline{\theta}$ is the arm coordination n-vector (i.e. the vector whose elements are the displacements of the individual links comprising the arm and which are individually activated by the arm motors), $\underline{T}$ is the generalised force/torque vector, $\underline{A}(\theta)$ is the n x n mass/inertia matrix and $\underline{B}(\theta,\dot{\theta})$ is an n-vector of the various internally generated forces and torques such as these due to friction, gravity, centrifugal effects and Coriolis effects (we neglect here external random forces).

In all but the most simple arm geometries it is the presence of the vector $\underline{B}(\theta,\dot{\theta})$ and the complication and non-linear nature of its elements which poses the predominant problem for control.

(A further complication arises from the need, in many cases, to transform between the arm coordinates vector $\underline{\theta}$ and some other n-vector which describes the arm configuration in a coordinate frame suited to the task. This will not be considered here).

Fig.1 shows the form which equation 2.1 takes when displayed as a conventional block-diagram (solid lines).

All methods of coordinated control must generate the torque vector $\underline{T}$ so as to produce the desired coordinate vector $\underline{\theta}$ in the face of the intruding $\underline{F}$ vector. Thus the $\underline{T}$ vector may be considered to consist of two parts so that

$$\underline{T} = \underline{T}_C + \underline{T}_B \qquad 2.2$$

The purpose of $\underline{T}_B$ is to cancel $\underline{B}(\underline{\theta},\underline{\dot{\theta}})$ so that ideally we would aim for

$$\underline{T}_B \equiv -\underline{B}(\underline{\theta},\underline{\dot{\theta}}) \qquad 2.3$$

The purpose of $\underline{T}_C$ is then to control $\underline{\theta}$ in the simpler conditions where $\underline{B}(\underline{\theta},\underline{\dot{\theta}})$ has been nullified by 2.3.

There are two main approaches to the nullification of $\underline{B}(\underline{\theta},\underline{\dot{\theta}})$ via 2.3 and these may be described as feedforward and feedback. Each demands an accurate model $\underline{B}'(\underline{\theta},\underline{\dot{\theta}})$.

In the feedforward implementation, the demanded time functions $\hat{\underline{\theta}}$ and $\hat{\underline{\dot{\theta}}}$ are available (from a so-called 'trajectory planner') and these are used to compute $\underline{B}'(\hat{\underline{\theta}},\hat{\underline{\dot{\theta}}})$ from which $\underline{T}_B$ is calculated as

$$\underline{T}_B = -\underline{B}'(\hat{\underline{\theta}},\hat{\underline{\dot{\theta}}}) \qquad 2.4$$

In the feedback implementation, the actual time functions $\underline{\theta}$ and $\underline{\dot{\theta}}$ are transduced as $\underline{\theta}_M$ and $\underline{\dot{\theta}}_M$ and the values fed to the control compute for the computation of

$$\underline{T}_B = -\underline{B}'(\underline{\theta}_M,\underline{\dot{\theta}}_M) \qquad 2.5$$

Fig. 2 (dotted lines) shows the implementation of the schemes.

Each of these control algorithms suffers from the disadvantage of requiring the computation of $\underline{T}_B$ in real time. Since 'real-time' means, typically a sample period of only 5 ms it is clear that this imposes a severe burden on the control computer.

The method of Active Force Control obviates the necessity for fast computation (or extensive memory look up) by replacing the calculations inherent in equations 2.4 or 2.5 with additional instrumentation.

Effectively $\underline{T}_B$ is obtained from measurement rather than from calculation.

Returning to equation 2.1, we may write

$$\underline{B}(\underline{\theta},\underline{\dot{\theta}}) = \underline{A}(\underline{\theta})\underline{\ddot{\theta}} - \underline{T} \qquad 2.6$$

If $\ddot{\theta}$ is regarded not as "the second derivative of displacement" but as "acceleration", equation 2.6 is purely algebraic.

Thus if measurements $\underline{T}_m$ and $\ddot{\underline{\theta}}_m$ are available and if a model $\underline{A}'$ of the mass matrix $\underline{A}$ is known, the torque vector $\underline{T}_b$ can be computed as

$$\underline{T}_b = \underline{A}'(\underline{\theta}_m)\ddot{\underline{\theta}}_m - \underline{T}_m \qquad 2.7$$

Fig.2 shows the Active Force Control implementation. It should be noted that there is no computation of $\underline{B}$ and, while there is computation of $\underline{A}$ this is a minor problem.

If now $\underline{T}_c$ is generated as

$$\underline{T}_c = \underline{A}(\underline{\theta}) \cdot \underline{C} \qquad 2.8$$

where $\underline{C}$ is a control signal vector then results, in the ideal case, the relationship

$$\ddot{\underline{\theta}} = \underline{C} \qquad 2.9$$

so that the original non-linear coupled set of equations 2.1 has been reduced to a set of 6 uncoupled linear second order equations. These may then be subject to conventional controls.

3. Experimental Investigation

An experimental robot arm with 3 degrees of freedom and of a planar design has been built. (Fig 3) The planar design is typical of industrial SCARA-type arms used extensively for fast insertion operations in automated assembly.

Activation is via torque motors whose torque is transduced by measurement of armature current.

Torque transmission is direct via steel belts to eliminate gear boxes and their associated problems of friction and backlash.

Angle transduction is via precision potentiometers.

Transduction of angular acceleration is via specially designed angular accelerometers. These consist of inertial bars mounted on 'frictionless' flexible pivots. Damping is provided by miniature rotating viscous dampers.

The relationship between angular acceleration $\ddot{\theta}$ of the arm link and the angular displacement d of the inertial bar relative to the arm link is given by

$$\frac{d}{\ddot{\theta}} = \frac{K}{s^2 + 2\zeta\omega_n s + \omega_n^2} \qquad 3.1$$

where γ and ω_n are the damping factor and natural frequency of the inertial bar system and may be chosen by appropriate choice of parameters. In the design described here we chose $\gamma = 0.707$ and $\omega_n = 2\pi.30$ rad/s. This particular value of natural frequency was considered to be well above that of any naturally occurring frequency within the arm structure.

The constant K is a scaling factor dependent upon the method of transducing the angular deflection d.

Here we used an L.V.D.T. to measure the linear deflection of one end of the inertial bar. The resulting sensitivity after filtering and amplification was 40 mV/rad/s. (Fig. 4).

The whole rig is controlled by an LSI-11/23 computer using 12 bit A/D, D/A interface. The control algorithm is written in floating point FORTRAN. The processor is augmented by a floating point microcode option KEF11-A capable of performing a single precision multiple in 80 μsec.

4. Results

Three sets of experimental tests were run to investigate the ability of Active Force Control, as described above, to provide decoupled movement without computation.

The first set of tests consisted of applying a sinusoidal command to the second link and examining the induced vibration in link 1 which was commanded to remain stationary.

Figure 5 shows the effect of Active Force Control and it can be seen that with AFC absent, the oscillation of link 1 was around $6mV$ $(=.21°)$ while with full AFC this oscillation was reduced to zero; (when the amount of AFC was increased beyond unity instability began to manifest itself).

The second set of tests consisted of applying the same test as in the first set but in addition, attaching the tip of link 2 to ground via a stretched spring. The induced offsets of links 1 and 2 were measured.

In the third set of tests the effect of external forcing was investigated by attaching the tip of link 2 to an oscillating table via a stretched spring. The whole arm was commanded to be stationary and the induced offsets and oscillation amplitudes were measured.

A full description of the tests and the results is available elsewhere (4,5). (See Table 1.)

5. Discussion

According to theory it should be possible with perfect transduction and perfect actuation to achieve decoupling without the need for accurate modelling.

Discrepancies in the results from 100 can be attributed to non-ideal implementation of the above functions.

Test No.	Test	Parameters	Parameter Plotted	Results - % improvement	
1	Sine input to link 2	Freq. = 1 Hz 1.5 2.0 2.5	Amp. of Osc. of Link 1 Amp. of Osc. of Link 1 Amp. of Osc. of Link 1 Amp. of Osc. of Link 1	100 92 100 100	100 92 100 99
2	Sine input to link 2	No Spring Freq. = 3 Hz No Spring Freq. = 3 Hz No Spring Freq. = 3 Hz	Amp. of Osc. of Link 1 Offset of Link 1 Offset of Link 2	100 55 44	100 25 33
		With Spring Freq. = 3 Hz With Spring Freq. = 3 Hz With Spring Freq. = 3 Hz	Amp. of Osc. of Link 1 Offset of Link 1 Offset of Link 2	100 40 50	100 25 37
3	External forcing of link 2	Freq. = 3 Hz Freq. = 3 Hz Freq. = 3 Hz Freq. = 3 Hz	Amp. of Osc. of Link 1 Amp. of Osc. of Link 2 Offset of Link 1 Offset of Link 2	100 64 22 50	97 64 15 45
		Freq. = 2 Hz Freq. = 2 Hz Freq. = 2 Hz Freq. = 2 Hz	Amp. of Osc. of Link 1 Amp. of Osc. at Link 2 Offset of Link 1 Offset of Link 2	100 48 30 33	95 40 27 30

Table 1

The accelerometers have a natural frequency of 30 Hz. This may be too low and we are currently designing new instruments with higher bandwidth.

Faster computation would reduce the effects of sample interval. In the above we did not take account of aliasing or other quantisation effects. These should be able to be substantially reduced with a more modern, and hence faster, computer for control.

The method of transducing applied torque by measurement of motor current is less than ideal and it is believed that this could be improved by monitoring motor torque directly via the use of specially placed strain guages.

Finally, and probably most serious of all, it has been assumed that the transmission elements from the motors to the joint are rigid. These elements, being steel cables are of course subject to strain and cannot be considered to be rigid. The resulting anomaly is likely to be the cause of control deficiency.

Work is continuing to add a model of transmission flexibility into the Active Force Control scheme so as to provide coordination in the face of non-rigid transmission. Simulation studies are promising.

References.

1. Asada, H., Kanade, T., Takeyama, I. "Control of a Direct Drive Arm" in Robotics Research and Advanced Applications, Phoenix, Arizona, November 14-19, 1982 (A.S.M.E.).

2. Dubowsky, S. "On the Dynamics of Computer Controlled Manipulators" 4th RO-MAN-SY Symposium, Zaborow, Poland, September 8-12, 1981.

3. Hewit, J. R., Tan, N. "Dynamic Coordination of Robot Movement" ibid.

4. Galatis, G. MSc Thesis, University of Newcastle upon Tyne, 1985

5. Hadzistylis, A. MSc Thesis, University of Newcastle upon Tyne, 1985.

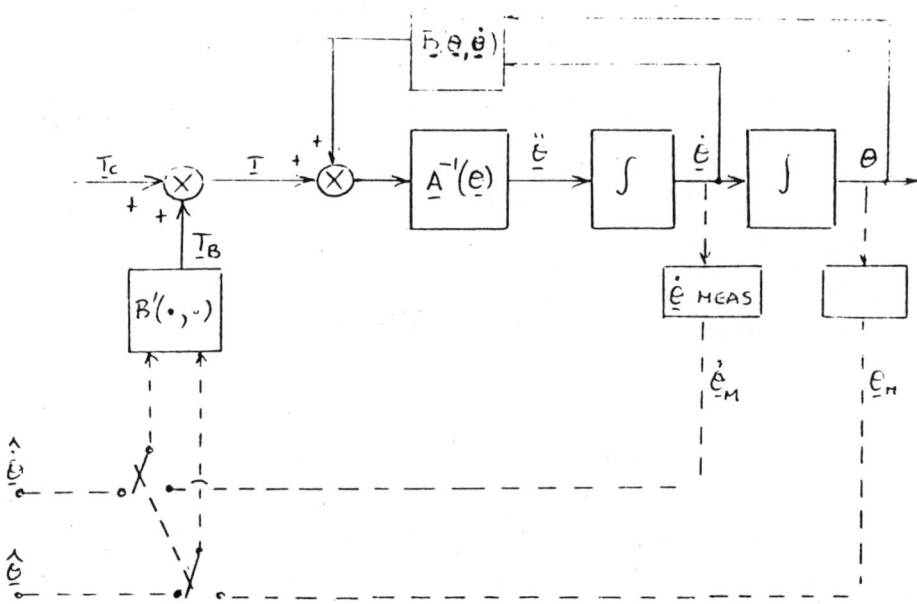

Fig. 1

Active Force Control of Robot and Manipulator Arms

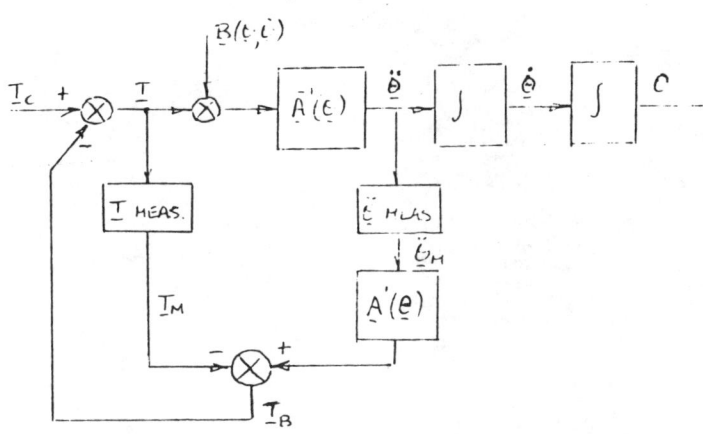

Fig. 2

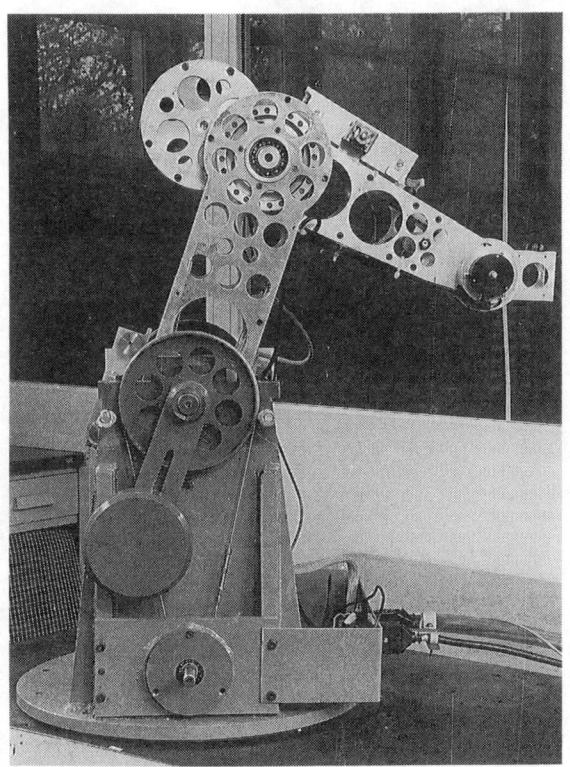

Fig. 3

Fig. 4

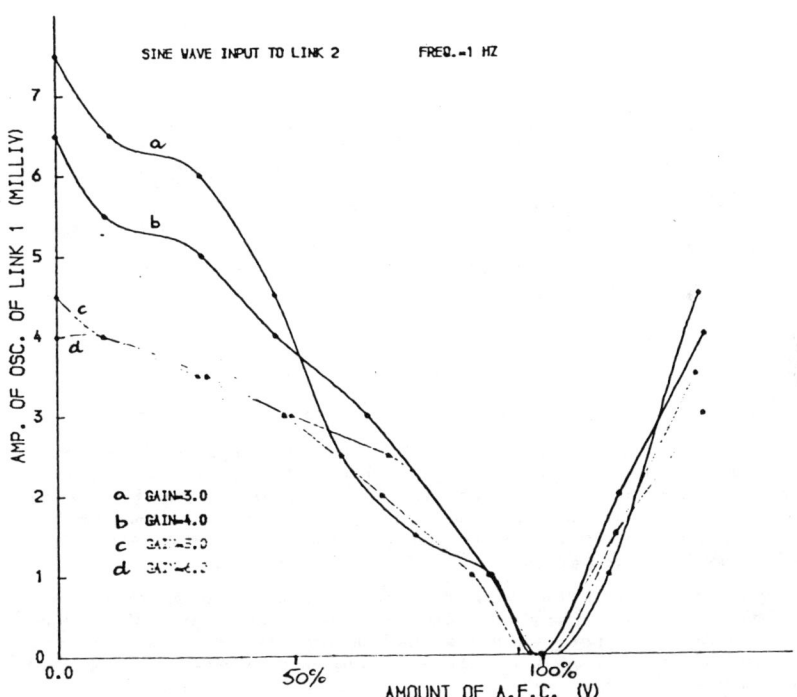

Fig. 5

Automatic Grasp Planning. An Operation Space Approach

M. T. Mason and R. C. Brost

Computer Science Department, Carnegie-Mellon University, Pittsburgh, U.S.A.

Summary

During grasping motions, frictional forces and geometric constraint can combine to eliminate uncertainty in the location of the object. This paper incorporates the mechanics of friction and constraint in automatic planning of grasping motions. Our approach to automatic planning is centered on the use of *operation space*, which can be constructed for any parameterized class of operations. This approach was applied by Brost [1986] to plan planar parallel-jaw grasping motions. We extend Brost's work to include the finite dimensions of the fingers and the finite lengths of the motions. We explore the conditions for a successful grasp, and derive constraints in the operation space. An operation satisfying all of the constraints should produce the required result.

I. Introduction.

An important aspect of a grasping operation is how it deals with uncertainty in the initial location of the object. In many applications, the orientation and position of an object will vary slightly from predicted values. At the least, a grasping operation should succeed despite such uncertainty. At best, we may be able to construct a grasping operation that eliminates the uncertainty. This paper focuses on grasping operations that eliminate the uncertainty mechanically, by the aligning and centering motions that occur when an object is pushed and squeezed between two fingers.

To plan grasping motions, we adopt the operation-space approach of Brost [1986], and extend it to address a broader set of issues. Brost explored the conditions for stability and convergence of object orientation, assuming that the fingers are wide enough, and approach from a great enough distance. To address the finite width and stroke of the fingers, we must construct bounds on the translations of the object, and incorporate these bounds in the planning process.

The central planning mechanism is the *operation space*. For any parameterized class of operations, we can define the operation space to be the space whose dimensions correspond to the parameters of the class. Each point in the space corresponds to a particular operation. A typical operation space constructed by Brost is shown in figure 1. For this diagram, the fingers are modeled as infinite half-planes, and the duration and speed of motions are not addressed. A two-parameter operation results, which can be viewed as a mapping from parameter values to final object configurations. This mapping is represented explicitly in the operation space, simplifying the choice of an appropriate pair of parameter values to obtain a desired outcome.

The operation-space approach involves three steps:

Automatic Grasp Planning 381

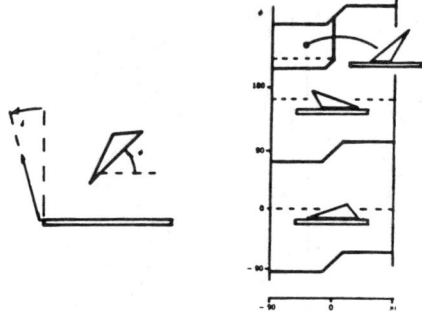

Figure 1. A two-parameter operation space for pushing, from Brost [1986]. The diagram shows the mapping from combinations of finger angle (ϕ) and pushing direction (δ) to resultant configurations of a triangle.

1) Define a parameterized operation. Each parameter defines a dimension of the operation space.

2) Develop the mechanics of the operation, combining models of the task and the operation.

3) Express the results as a mapping from the operation space to task outcomes. For any desired outcome, select a parameter value-set from the region that maps to that outcome.

Some forms of uncertainty are easily incorporated. For example, if uncertainty in one of the parameters is present, we can simply select a point that is not too close to the limits of the feasible region. This is equivalent to shrinking the feasible region before selecting a point. Other forms of uncertainty must be included in the mechanics of the operation.

Brost modeled the fingers as infinitely wide half-planes, approaching from infinity—planning the initial position was not an issue. In this paper, we assume that the two fingers have a finite width, and an initial position must be planned. Figure 2 shows part of the resulting four-dimensional operation space. As we shall see, the constraining relations among the parameters are not as simple as Brost's, but they are still fairly simple.

I.A. Previous work.

This paper draws primarily on research in the mechanics of pushing, and planning of pushing and squeezing motions. Mason [1982, 1986] explored the mechanics of pushing, and demonstrated automatic planning of push–grasp operations. Fearing [1984] incorporates the mechanics of pushing, and explores acquisition and stability of grasping with point fingers. Mani and Wilson [1985] address planning of a sequence of pushes to orient objects, and construct an operation space for pushing. Brost [1985, 1986] constructed operation spaces for three different grasping operations. Peshkin and Sanderson [1986] improved on Mason's analysis of pushing.

A number of other issues arise in the context of grasping, including control of a grasped object, mechanical properties of grasp configurations, and finding feasible grasp configurations. A short survey of this area should include Hanafusa and Asada [1977], Lozano-Pérez [1976, 1981], Salisbury [1982], Cutkosky [1985],

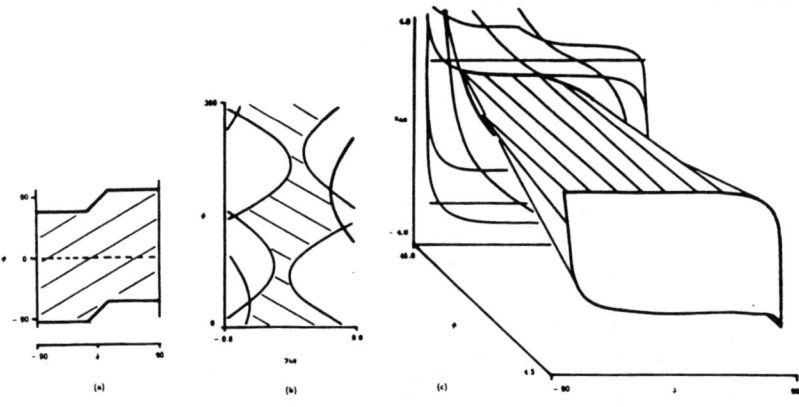

Figure 2. A four-parameter operation space for a push-grasp. Besides the finger angle ϕ and pushing direction δ used by Brost, we include the initial position of the reference finger (x_{h0}, y_{h0}). Constraints arise (a) in ϕ–δ space, (b) in y_{h0}–ϕ space, and (c) in x_{h0}–ϕ–δ space. The region representing successful operations is the intersection of regions shown in (a), (b), and (c).

Asada and By [1985], Kerr and Roth [1986], Abel, Holzmann, and McCarthy [1985].

There are a number of precedents in the use of operation space. Our operation space is a direct extension of the spaces constructed by Brost [1986]. Mani and Wilson [1985] independently constructed a pushing space. Kerr and Roth [1986] defines a space whose dimensions are the magnitudes of internal forces in a grasp; and the jamming and wedging diagrams of Simunovic [1975] and Whitney [1982] have the same character.

The operation space approach is similar to the use of configuration-space for robot path-planning [Lozano-Pérez 1981]. If we view "being there" as an operation, then configuration-space is the corresponding operation space. A more useful operation is a straight-line motion, which is characterized by the initial and final configurations. The operation space would be $C \times C$, where C is the configuration space.

Similarities also hold with recent work in fine-motion planning [Lozano-Pérez, Mason, and Taylor 1984, Mason 1984, Erdmann 1984]. Assuming a simple, fixed, model of compliance, and a known task geometry, the robot path for a single motion command is completely characterized by an initial configuration and a nominal velocity.

II. Analysis of the operation.

Brost [1986] addresses three operations: push-grasp, squeeze-grasp, and offset-grasp. We will extend his results for the push-grasp, then briefly consider variations required for the other two operations. The push-grasp (see figure 3) pushes the object until it is stably aligned with the finger, then squeezes the object with the second finger. The general form is:

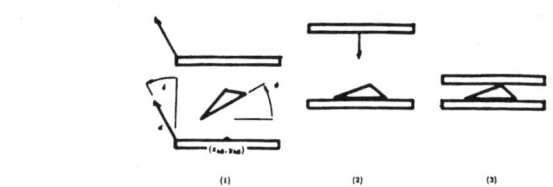

Figure 3. The push-grasp. The first finger pushes a distance d in direction δ, until the object is aligned with the finger. Then the second finger squeezes the object.

1) Move the hand to an initial position (x_{h0}, y_{h0}) and orientation ϕ, above the object;

2) Move down close to the table;

3) Translate the hand in direction δ for a distance d with velocity v;

4) Close the fingers, and translate the hand to hold the reference finger still.

We will treat the push-grasp as a four-parameter operation $(x_{h0}, y_{h0}, \phi, \delta)$. A lower bound for the pushing distance d arises as a by-product of our analysis. See Mason [1985] for an approach to producing bounds on the pushing velocity v. The initial position and orientation of the hand are expressed in a coordinate system fixed in the object, i.e. they represent *relative* position and orientation. Uncertainty in the object's initial location are equivalent to uncertainty in the parameters of the operation.

We will require the planner to verify that the following necessary conditions are satisfied:

- *geometrical conditions*: fingers clear object at initial position and orientation;
- *convergence and stability*: object converges to desired orientation relative to the fingers, and is stable in the final orientation; and
- *finite motion limits*: object does not roll or slide past the face of a finger.

II.A. Geometrical conditions.

First we address the problem of preventing undesired collisions, and producing the desired collision, between the fingers and the object. For this problem, the operation-space reduces to configuration-space. Using the notation of figure 4, the constraints are:

$$r_i \sin(\theta_i + \phi) - s < y_{h0} < r_i \sin(\theta_i + \phi)$$

for all i. This means that y_{h0} must fall between two sets of sine waves, as shown in figure 2(b).

II.B. Convergence and stability of orientation.

This is precisely the problem addressed by Brost [1986], based on the earlier work of Mason [1982, 1986]. Assuming negligible inertial forces and Coulomb friction, the direction of rotation is determined by the coefficient of friction, the center of gravity, the direction of the finger motion, and the contact geometry. Stability and convergence are obtained for any operation inside the simple constraints shown in figure 2(a).

II.C. Finite fingers and finite pushing distance.

The finite translations of the object are important in two respects. First, if our previous analysis is to be

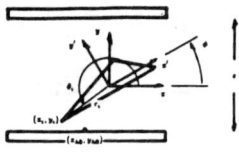

Figure 4. Notation for the push-grasp. Two coordinate frames are located at the object center of gravity. The $x'-y'$ frame is fixed with respect to the object. The $x-y$ frame is aligned with the fingers.

valid, we must ensure that each finite-width finger has the effect of an infinite half-plane. This means that the object must never make contact with an edge of the finger-face. Second, we must have an upper bound on the distance traveled during the rotation, so we know how far to push.

The primary complication is indeterminacy. We do not expect to know the distribution of support forces between the object and the table. Although the rotation direction is determined, the rotation rate, and the translational components of the motion, are not determined. Our first requirement, then, is to obtain at least some bounds on the possible motions. Figure 5 shows the bounds that we will use.

Given some bound on the possible rotation centers, we can proceed as follows. Circumscribe a rectangle, aligned with the finger face, about the set of possible rotation centers. Let the coordinates of the rectangle edges be $x_r^+, x_r^-, y_r^+, y_r^-$. It is easily shown that

$$y_r^- \le \frac{dx}{d\phi} \le y_r^+,$$
$$-x_r^+ \le \frac{dy}{d\phi} \le -x_r^-.$$

where (x,y) describes the trajectory of the object vertex. If we compute these bounds for some fixed δ, letting ϕ vary over some interval, we can bound the variations in x and y:

$$y_r^+ \Delta\phi \le \Delta x \le y_r^- \Delta\phi,$$
$$-x_r^- \Delta\phi \le \Delta y \le -x_r^+ \Delta\phi.$$

for $\Delta\phi < 0$, as in the example. Let the origin be placed at the initial location of the contact vertex, then the final location is in the bounding rectangle $[x^-, x^+] \times [y^-, y^+]$, where

$$x^- = y_r^+ \Delta\phi,$$
$$x^+ = y_r^- \Delta\phi,$$
$$y^- = -x_r^- \Delta\phi,$$
$$y^+ = -x_r^+ \Delta\phi.$$

for $\Delta\phi < 0$. Although the details are not presented here, the bounds on the rotation center, y_r^- etc., are simple trigonometric functions of δ. Hence the bounds on the contact vertex, x^- etc., are trigonometric functions of δ, and linear in ϕ.

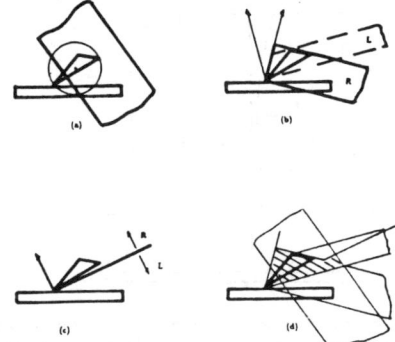

Figure 5. In order to bound the translational motion of the object, we require bounds on the instantaneous rotation center. We will use bounds arising from a variety of considerations. (a) shows the rotation center cannot fall inside the perpendicular bisector of the line connecting the contact point with the center of gravity. Nor can it fall outside a parallel line of distance $\frac{(p+r)^2}{p}$ from the contact, where r is the radius of a circle circumscribed about the support, and p is the distance from the contact to the center of gravity. (Peshkin and Sanderson [1986] have improved this bound close to $p + \frac{r^2}{p}$.) For a fixed ϕ, this defines a band—for ϕ varying over some interval, this defines a "bow-tie". Constraint (b) arises from considering the angle of force at the contact. Assume that the force is in the direction γ_R, a rotation center outside band R would give rise to net forces in a direction orthogonal to the hypothesized direction of force. If the contact force can actually lie anywhere inside the friction cone, we sweep the band to obtain another bow-tie. (b) also includes the result from [Mason 1982] that the rotation center cannot fall inside the friction cone. For constraint (c), we construct a line perpendicular to the finger velocity. Rotation centers above that line must be in region R of (b), and rotation centers below the line must be in region L of (b). (d) shows an example combining the constraints for a fixed δ.

We can be confident of $\Delta\phi$ rotation by pushing at least as far as y^+ in the y-direction. However, it is almost certain that the rotation will be completed before the motion is complete, so we must also consider the pure translation of the object after rotation ceases. It is easily shown that if $|\delta| \le \alpha$, where $\alpha = \tan^{-1}\mu$, the object will translate in direction δ, otherwise it will translate in direction α.

These observations are summarized in figure 6. The path of the contact vertex, and the contact edge after rotation is complete, is bounded by a polygonal region of the plane. Letting (x_i, y_i) be a vertex of the bounding polygon, we can write

$$x_i - w < x_{h0} + y_i \tan\delta < x_i + w.$$

The x_i are 0, x^-, or $x^+ + l$, where l is the length of the edge, and the y_i are 0, y^-, or y^+. These are all simple trigonometric functions of δ, and linear in ϕ. In general, we obtain constraints of the form,

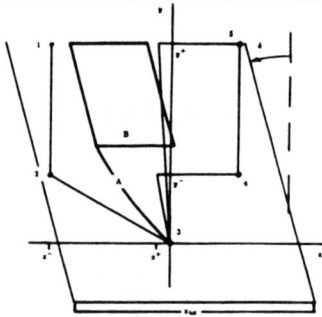

Figure 6. The region 12345 bounds the possible locations of contact of the object with the finger. A sufficient condition is that the area swept by the finger include this region. Also shown is the contact region for a typical motion. A shows the locus of the contact vertex as the object rotates, and B shows the region of contact after the rotation is complete.

$$f_i(\delta) \Delta\phi + c_i - w < x_{h0} < f_i(\delta) \Delta\phi + c_i + w$$

where the $f_i(\delta)$ are trigonometric functions, and the c_i are constants. Figure 2(c) shows an example of these constraints plotted in the operation-space.

This analysis suffices for simple push grasps involving rotation about a single contact vertex. A more general analysis is required for complex grasps where the object may roll from one vertex to the next. The basic idea would be to repeat the above analysis for each vertex, and splice the envelopes together.

II.D. Related operations: squeeze–grasp and offset–grasp.

The analysis of the push–grasp almost encompasses two similar operations, the squeeze–grasp and the offset–grasp. Both motions dispense with the preliminary pushing motion, and rely on pure squeezing for alignment and centering of the object. The offset–grasp requires that the reference finger makes first contact with the object, while the squeeze–grasp is indifferent to which finger hits first. The main difference in constructing the operation spaces is that a different set of rotation center bounds are required. Rotation center bounds for an example problem proved to be somewhat simpler than the bounds described in figure 5.

References

Abel, J. M., Holzmann, W., and McCarthy, J. M. 1985. "On Grasping Planar Objects with Two Articulated Fingers," *Proceedings of the 1985 IEEE International Conference on Robotics and Automation*, St Louis.

Asada, H., and By, A. B. 1985. "Kinematics of Workpart Fixturing," *Proceedings of the 1985 IEEE International Conference on Robotics and Automation*, St Louis.

Brost, R. C. 1985. "Planning Robot Grasping Motions in the Presence of Uncertainty," CMU-RI-TR-85-12, Robotics Institute, Carnegie-Mellon University.

Brost, R. C. 1986. "Automatic Grasp Planning in the Presence of Uncertainty," *Proceedings of the 1986 IEEE International Conference on Robotics and Automation*, San Francisco.

Cutkosky, M. R. 1985. "Grasping and Fine Manipulation for Automated Manufacturing," PhD thesis, Carnegie-Mellon University.

Erdmann, M. A. 1984. "On Motion Planning with Uncertainty," TR-810, Artificial Intelligence Laboratory, Massachusetts Institute of Technology.

Fearing, R. S. 1984. "Simplified Grasping and Manipulation with Dextrous Robot Hands," American Control Conference, San Diego.

Hanafusa, H., and Asada, H. 1977. "A Robotic Hand with Elastic Fingers and its Application to Assembly Process," *IFAC Symposium on Information and Control Problems in Manufacturing Technology*.

Kerr, J., and Roth, B. 1986. "Analysis of Multifingered Hands," *International Journal of Robotics Research*, 4(4).

Lozano-Pérez, T. 1976. "The Design of a Mechanical Assembly System," TR-397, Artificial Intelligence Laboratory, Massachusetts Institute of Technology.

Lozano-Pérez, T. 1981. "Automatic Planning of Manipulator Transfer Movements," *IEEE Transactions on Systems, Man, and Cybernetics*, SMC-11(10).

Lozano-Pérez, T., Mason, M. T., and Taylor, R. H. 1984. "Automatic Synthesis of Fine-Motion Strategies for Robots," *International Journal of Robotics Research* 3(1).

Mani, M., and Wilson, W. R. D. 1985. "A Programmable Orienting System for Flat Parts," *Proceedings, NAMRI XIII*, Berkeley.

Mason, M. T. 1982. "Manipulator Grasping and Pushing Operations," AI-TR-690, Artificial Intelligence Laboratory, Massachusetts Institute of Technology.

Mason, M. T. 1984. "Automatic Planning of Fine Motions: Correctness and Completeness," *Proceedings, IEEE International Conference Robotics*, Atlanta.

Mason, M. T. 1985. "On the Scope of Quasi-Static Pushing," *3rd International Symposium on Robotics Research*, Gouvieux.

Mason, M. T. 1986. "Mechanics and Planning of Manipulator Pushing Operations," *International Journal of Robotics Research*, 5(1).

Peshkin, M., and Sanderson, A. 1986. "Manipulation of a Sliding Object," *proceedings, 1986 IEEE International Conference on Robotics and Automation*, San Francisco.

Salisbury, J. K. 1982. "Kinematic and Force Analysis of Articulated Hands," PhD thesis, Department of Mechanical Engineering, Stanford University.

Simunovic, S. N. 1975. "Force Information in Assembly Processes," *Proceedings, 5th International Symposium on Industrial Robots*.

Whitney, D. E. 1982. "Quasi-Static Assembly of Compliantly Supported Rigid Parts," *Journal of Dynamic Systems, Measurement, and Control* 104:65.

A Method of Optical Processing in the Robot Vision

B. Macukow

Institute of Mathematics, Technical University of Warsaw, Poland

SUMMARY

Real-time object recognition is, of course, one of the most important and attractive requirements in robot vision. Many methods have been developed for pattern recognition. Among these, correlation methods are of great importance. Shift and rotation invariances are the basic problem in recognition. A conventional matched spatial filter system is shift invariant but its shortcoming is that only objects in known orientations can be recognized.

The method presented in this paper preserves shift invariance while adding rotation invariance. Digital as well as optical implementations are presented.

INTRODUCTION

Pattern recognition is important in almost any field of human or intelligent machine activity - in photo interpretation, in automatic devices for character recognition in civil and military applications. However, in robotics, in classical application of industrial robots for mechanical assembly and specially in the higher generation of robots, for adaptional robots with machine intelligence this problem is exceptionally important (5).

Rotation and shift invariances are the basic problem in pattern recognition. The ideal recognition system ought to be shift, scale and rotation invariant. Many efforts have been made to solve these problems. A conventional matched spatial filter method (MSF), well known in optics, was partially described in previous papers (2,4). This method can search for objects in a certain area, but it can not find targets of unknown orientations. Differently oriented targets do not trouble man but they do trouble a machine - this is known as the rotation problem. The system is said to be shift-and-rotation invariant if, whenever the input image (object) is shifted and rotated, then all the recognized target locations simply shift and rotate by the same amounts as those in the object plane.

SHIFT AND ROTATION PROBLEM

In this part we use circular harmonic expansion to analyze the shift and rotation problem for the conventional matched spatial filter system.

Let us assume a system of a coherent optical correlator (1) with the impulse response matched to the object described by $f_1(x,y)$. When the object is rotated through an angle α, the function $f_\alpha(x,y)$ appears at the input plane of the system and the system output is determined by the correlation function

$$R_\alpha(x,y) = \int\!\!\!\int_{-\infty}^{+\infty} f_\alpha(\xi,\eta) f_1^*(\xi-x,\eta-y)\, d\xi\, d\eta \tag{1}$$

This output varies with the angle α. For an object with the same orientation as the filter reference (i.e. $\alpha = 0$), the autocorrelation function

$$R_0(x,y) = \int\!\!\!\int_{-\infty}^{\infty} f_1(\xi,\eta) f_1^*(\xi-x,\eta-y)\, d\xi\, d\eta \tag{2}$$

has the peak value

$$R_0(0,0) = \int\!\!\!\int_{-\infty}^{\infty} |f_1(\xi,\eta)|^2\, d\xi\, d\eta \tag{3}$$

The value of $R_\alpha(0,0)$ is called the center correlation value (3). Only for a small α is the modulus value high enough to be the peak of the correlation function. In general $R_\alpha(0,0)$ will decrease too much to be a peak due to the large orientation error α.

In our further investigation we will use two coordinate systems because the convolution achievable by the optical system is convenient to be described in Cartesian coordinates, whereas the rotation phenomenon is more easily studied in polar ones. The origin of those two systems coincide (that is $r = 0$ corresponds to $x = 0$ and $y = 0$).

The short-hand notation in polar coordinates of (1) and (2) are respectively

$$R(r,\theta) = f(r,\theta) * f(r,\theta) \tag{4}$$

$$R'(r,\theta) = f(r,\theta) \star f(r,\theta) \tag{5}$$

where $f(r,\theta)$ denotes the function $f_1(x,y)$ and $f(r,\theta+\alpha)$ denotes the function $f_\alpha(x,y)$.

An object function $f(r,\theta)$ can be expanded into a Fourier series called the circular harmonic expansion (2,3).

$$f(r,\theta) = \sum_{M=-\infty}^{\infty} f_M(r) \exp(jM\theta) \tag{6}$$

where

$$f_M(r) = \frac{1}{2\pi} \int_0^{2\pi} f(r,\theta) \exp(-jM\theta)\, d\theta \tag{7}$$

Therefore, an object rotated by an angle α can be expressed as

$$f(r,\theta+\alpha) = \sum_{M=-\infty}^{\infty} f_M(r) \exp[jM(\theta+\alpha)] \tag{8}$$

The center correlation values (see eq. (3)) in polar coordinates has the form

$$C(\alpha) = \int_0^\infty r\,dr \int_0^{2\pi} f(r,\theta+\alpha)f^*(r,\theta)\,d\theta \qquad (9)$$

where $C(\alpha)$ denotes the value $R_\alpha(0,0)$

After some transformations one obtains

$$C(\alpha) = 2\pi \sum_{M=-\infty}^{\infty} \exp(jM\theta) \int_0^\infty r\,|f_M(r)|^2\,dr \qquad (10)$$

Equation (10) expresses the center correlation value in the form of a sum of all the circular harmonic components. This value does not remain constant when α varies.

Hsu and Arsenault (3) proved that to realize the system output modulus (peak value) which shifts and rotates along with the input (without changing), the input response must be one single circular harmonic component. If more than one component is present the system becomes rotation-variant.

The above discussion suggests using the reference

$$f_r(r,\theta) = f_M(r)\exp(jM\theta) \qquad (11)$$

and then the center correlation value between this reference and object becomes

$$C_M(\alpha) = A(M)\exp(jM\alpha) \qquad (12)$$

where

$$A(M) = 2\pi \int_0^\infty r\,|f_M(r)|^2\,dr \qquad (13)$$

From (11) A(M) is the autocorrelation peak value of the reference and is a constant $\sqrt{s}$, the target orientation α, so it can determine the presence of the object.

EXPERIMENT AND RESULTS

Most experiments were performed for both the digital and the optical systems. Each time the results were compared and agreed very well. The digital system has its advantage in adjusting to the multiple changes of tasks and very precise results. On the other hand, optical systems provide real-time recognition. The optical system processes the image in a parallel way that makes recognition much faster than the digital processing, specially for large-size inputs.

The digital system is divided into two parts as shown in figure 1. The first one refers to the preliminary operation on the standard object and is carried out in advance. The standard object described by $f(r,\theta)$ is analyzed to find the strongest circular harmonic component $f_r(r,\theta)$. This component is used as the reference. The Fourier transform of $f_r(r,\theta)$ is calculated (by the subroutine FFT), then conjugated and stored in the system.

The second part is the recognition system. The input image is read into the system and is Fourier-transformed. This Fourier spectrum of the input is multiplied by the stored in-memory conjugated spectrum of the reference (see first part of the system). The inverse Fourier transform of this product gives the spatial function of the correlation R(0,0) between the input and the reference. The largest square modulus value of R(0,0) indicates the target location. From the detected location the system draws a value of $C_M(\alpha)$ and next calculates the orientation angle α from $C_M(\alpha)$. It took about 30 secs on the IMB-370 to complete the recognition tasks for the input image of 192 x 192 pixels. This is much slower than the action expected from a robot. For larger objects, for example of 512 x 512 pixels, it will need about 4.5 minutes of CPU time.

The optical recognition system is the standard Fourier transform system with the computer generated hologram filter matched to one of the circular harmonic components of the object.

The filter was prepared in advance. Data was calculated using the IBM-370 computer and then converted by the photomation into the filter pattern into the film. The filter was put at the Fourier plane of the optical system shown in figure 2.

Lenses L1 and L2 form a telescope to match the scale of the input image with the scale of the filter. L3 is the Fourier transform lens and L4 performs the inverse Fourier transform. L5 and L6 lenses form the second telescope to match the output scale to the size of detector. After optical thresholding every target has been recognized.

CONCLUSIONS

Since recognition by conventional MSF is rotation variant (but of course shift invariant), we seek to solve recognition using partially matched filters. One of the circular harmonic components of the target is chosen to be the reference of the filter. The choice of order of circular harmonic component strongly depends on the object. Both the optical and digital implementations of the method coincide very well. The experiments involving searching for certain objects in the specified area (objects like characters, airplanes at airfield etc.) show very high efficiency of the system. However, the optical systems provide real-time recognition, practically delay-free. Although the optical system works only with the coherent light, transducers such as the liquid crystal light valve which converts the image of incoherent light into that of coherent light can be used.

This work has partially been made during the author's stay at Laval University, Quebec, Canada. The author would like to thank H. H. Arsenault and Y. N. Hsu for their cooperation.

REFERENCES

1. Goodman J.W., Operations Achievable with Coherent Optical Information Processing Systems. Proc. IEEE 65/1,29-38,1977
2. Hansen E.W., Goodman J.W., Optical Reconstruction from Projections via Circular Harmonic Expansion, Opt.Comm. 24/1, 268-272,(1978)
3. Hsu Y.N., H.H.Arsenault, Optical Pattern Recognition Using Circular Harmonic Expansion, Appl.Opt. 21/22,4016-4019,(1982)
4. Macukow B., New Approach to Robotic Visual Processing, in: Theory and Practice of Robots and Manipulators, Proc. RO.Man.Sy, Kogan Page, London eds. 7,A.Morecki, G.Bianchi, K.Kędzior.1985, 255-259
5. Macukow B., Some Problem of Control of a Two-Link Anthoropomorphic Manipulators. Proc. Symp.SyROCO'85, IFAC, Barcelona, 467-469

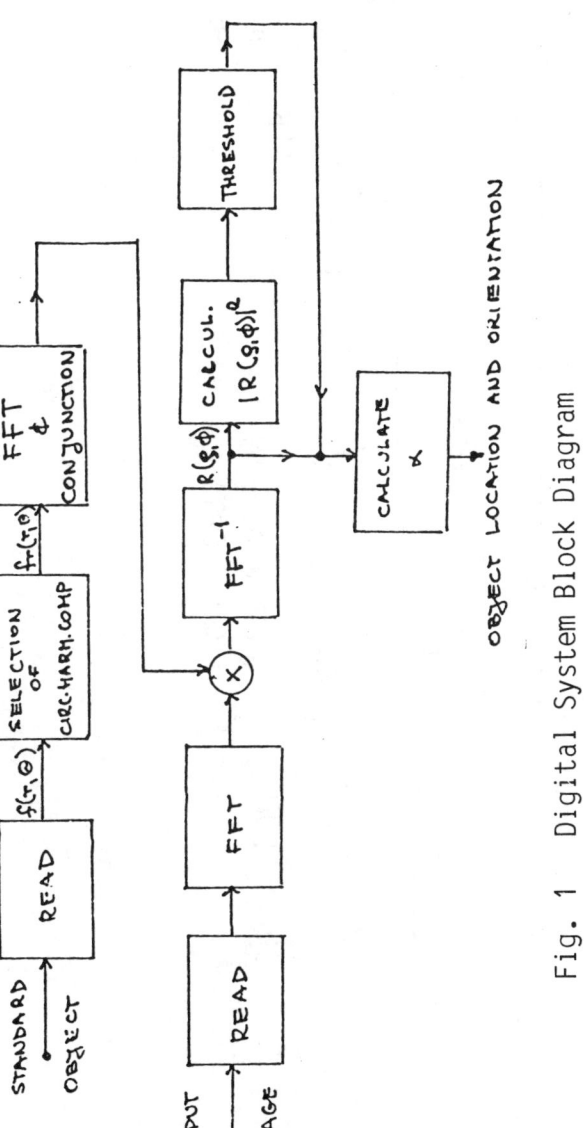

Fig. 1 Digital System Block Diagram

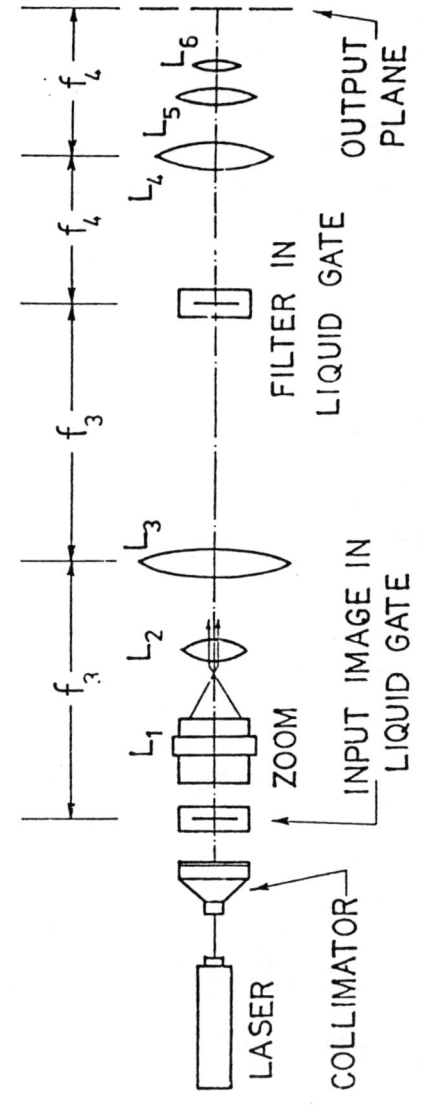

Fig. 2 Experimental Optical recognition System

Tridimensional Optical Syntaxer

A. Oustaloup and P. Melchior

E.N.S.E.R.B., Université de Bordeaux I, Talence, France

SUMMARY

In tele-control, the control unit or syntaxer produces, in real time, trajectories to be followed by the slave arm. "Dynamic" syntaxers (master arms or mini-master arms) and "static" syntaxers (isometric control handles with strain gauges) produce at each moment a target position. "Push button" syntaxers produce a target speed, which goes to zero when the pressure on the button is relaxed.

The "optical, three dimensional" syntaxer described here produces a target position. This dynamic syntaxer differs completely in nature and design from existing models, since the moving part comprises neither mechanical articulation, nor any kind of mechanical or electrical junction with a fixed frame. It has none of the disadvantages of current syntaxers. The most obvious advantages occrue from simple design, easy construction of the moving part and the absence of mechanical play, whip, inertia and wear.

The dynamic design allows the operator to follow the movement, so that the tele-commanded manipulation is an exact replica of the operator's movement.

The principle of this tridimensional optical syntaxer, is the movement of a hand-held stylus in the field of view of two cameras with perpendicular optical axes.

It gives control over seven degrees of freedom.

The sets of points on the stylus are distinguished by the cameras. A calculator commands the different degrees of freedom by reference to the coordinates of these points. Several stylus configurations are possible of course.

I - INTRODUCTION : THE STATE OF THE ART OF SYNTAXERS

Syntaxers currently used to drive robots are of two principal kinds :

I.1 - "Dynamic" syntaxers

These, the more common kind, are articulated mechanical systems. They may be similar to the manipulator or slave arm being driven (case of command arms and master arms), or not (handle-like). They carry sensors of position, potentiometers in most cases, of which there are as many as there are degrees of freedom, both position and speed, to be controled. Their disadvantages are :
- Constructional difficulties related to complicated design.
- Purely mechanical difficulties, such as poor intrinsic accuracy, due in part to mechanical play and in part to overall slackness which appears when the motion, or that of the operator, is too fast. A further cause of inaccuracy, particularly at high speeds, is an inertial effect which perturbs the natural reactions of the operator.
- Electromechanical problems : wear of the contacts of the potentiometers at different articulations. This leads to crackle.

Portable syntaxers (command handles or mini-master arms) minimise the strictly mechanical problems. The electromechanical problems generally are solved by use of other means, such as numerical sensors.

I.2 - "Static" syntaxers

These are in general isometric command handles based on strain gauges. They are less common than the first kind of syntaxer. Their disadvantage is that the operator's wrist does not move. This makes it difficult to translate the desired movement of the robot into a sequence of pressures to be applied to the handle. Operators must gain practice in order to synchronise and coordinate their fingering in order to execute rapidly a given trajectory.

II - SUGGESTED SYNTAXER

II.1 - Introduction

Although the syntaxer described here is of the dynamical kind, it is of completely different design, since the mobile part comprises neither any articulations nor any mechanical or electrical connexion, to the fixed part. It has none of the short comings of present day syntaxers. Its most obvious advantages are due to its simplicity - no mechanical play, no slackness, no inertial effects, no wear. Being a dynamical syntaxer, it allows the operator to accompany the movement of the manipulator, so there is no problem of transcribing the required movement of the manipulator into a set of movements by the operator. The two movements are identically similar. These advantages are acquired at the cost of more sophisticated and complex electronics.

II.2 - Principles of operation

In the tridimensional optical syntaxer, an operator moves a "stylus" in the common field of vision of two cameras set at a right angle one to the other.

The syntaxer commands the position of seven degrees of freedom and the speed of rotation about one of them.

The stylus carries a set of reference points which are discriminated by the cameras (figure 2). The positions and the components of their speeds are used by a calculator to command the degrees of freedom of the manipulator.

Several shapes of stylus are possible, for which we make some suggestions below.

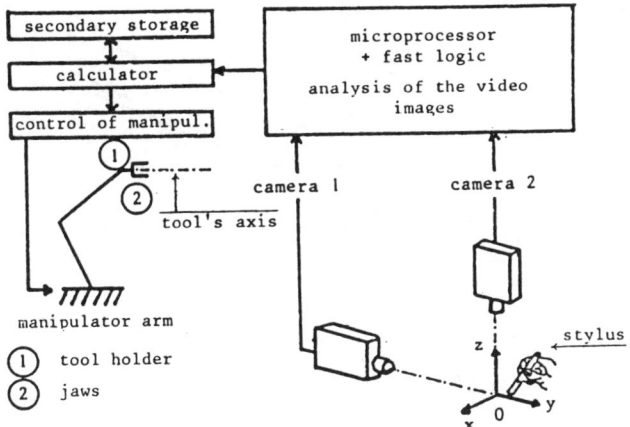

Figure 1 - Block diagram of the remote control system.

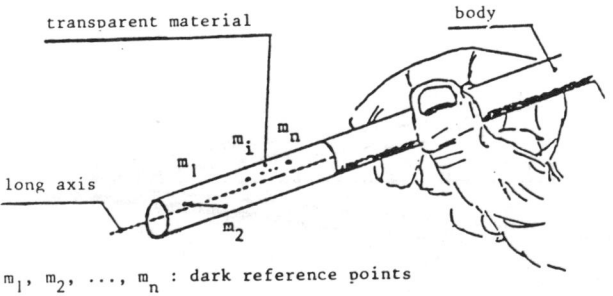

$m_1, m_2, \ldots, m_n$: dark reference points

Figure 2 - Design of the stylus transparent material.

1 - **First example**

The reference points are distributed as in figure 3 on a pencil shaped stylus.

m_1, m_2 and m_3 are fixed points ; m_4 is movable by hand on a cursor on the body of the stylus.

- The coordinates of m_1 command the position of the tool-holder on the manipulator (three degrees of freedom).
- The components of vector $\overrightarrow{m_1m_2}$ define the longitudinal orientation of the tool (two degrees of freedom).
- The vector $\overrightarrow{m_1m_3}$ defines the rotation of the stylus about its long axis and commands the rotation of the tool (one degree of freedom).
- Finally, vector $\overrightarrow{m_1m_4}$ commands the opening of the jaws (one degree of freedom).

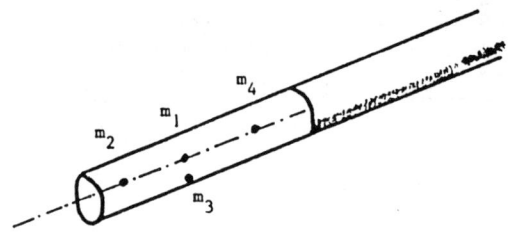

Figure 3 - A pencil shaped "stylus".

2 - Second example

This stylus is designed like a pair of tweezers, particularly useful for delicate work. It avoids use of a cursor to command the opening of the jaws. Figure 4 shows a suitable arrangement of the reference points.

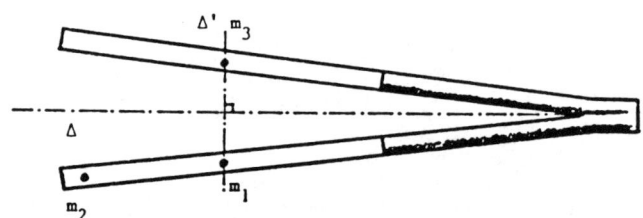

Figure 4 - A "stylus" like a pair of tweezers.

- The coordinates of the intersection of axes Δ and Δ' (algebraic means of the coordinates of points m_1 and m_3), command the position of the tool-holder (three degrees of freedom).
- The components of vector $\overrightarrow{m_1 m_2}$ command the orientation of the longitudinal axis of the tool (two degrees of freedom).
- The vector $\overrightarrow{m_1 m_3}$, which defines the rotation about axis Δ, commands the rotation of the tool (one degree of freedom).
- The length of the vector $\overrightarrow{m_1 m_3}$ commands the opening of the jaws (one degree of freedom).

3 - Notes

a) - Automatic rotation of the tool could be controlled, for example, by manual movement of m_2 (figures 3 and 4) on a cursor. The length of vector $\overrightarrow{m_1 m_2}$ can be used to control rotational speed.

b) - Problems related to trembling by the operator when picking up objects or screwing them, can be overcome by decreasing the sensitivity of the command of the position of the tool-holder with respect to fluctuations of position of point m_1.

II.3 - Feedback

1 - Feedback of grasping force

Feedback of grasping force may be obtained from a linear motor in the stylus whose moving part exerts a force on the cursor of m_4 in example one, or on the spread of the jaws in example two.

2 - Feedback of screwing torque

* Hand screwing

This feedback is difficult to obtain since the stylus is completely free (figures 3 and 4).

One solution would be to control screwing not by rotation of vector $\overrightarrow{m_1 m_3}$, but by a knob and a radio control. A rotating motor could be applied to the knob to provide feedback.

* Automatic screwing

When controled by a cursor m_2 (cf. note 3-a), automatic screwing can have the same kind of feedback as grasping.

3 - Note

Feedback could be delivered to some other device accessible to the free hand of the operator. This would also provide feedback of the pressure applied by the tool as well as feedback of the grasping force.

III - EXTENSION

Various options could be added to the stylus which we have presented here in its simplest form as a holder for the reference points discriminated by the cameras.

The object of these options, specific to different applications, is to avoid the operator having to learn by practice or to translate the desired robot movement into his own different movements. Two levels of options may be distinguished :

* The first would be to try to reproduce as closely as possible in the syntaxer' space the movements to be accomplished in the work space, particularly a hostile environment. This implies two conditions :
- The first one follows from the fact that a screw driver is not used in the same way as an adjustable spanner, which means the operator must transcribe his movements in order to drive a screw with a tool more resembling an adjustable spanner than a screw driver, and vice-versa. The answer to this is to let the operator use real tools, such as screwdrivers, adjustable spanners, drills, hammers and so on in the syntaxer space between the cameras. The tools obviously must carry reference points identifiable by the cameras. This solution is to us a debatable one, useful only if the operator cannot go without familiar tools.
- The second point is that the operator should use these real tools on a workpiece placed in the field of the cameras and as far as possible identical to that worked on by the manipulator. It would be advisable to add a third or even a fourth camera, covering specific areas, since the workpiece or the operator could at times hide some reference points from just two cameras. When competition occurs between cameras, information is taken from the one which can see the most reference points.

* The second level exploits the remarkable flexibility of the wrist. One solution would be to carry out a partial recognition of the hand, assimilated to a pair of pliers with the thumb as one jaw. This kind of application could be specially useful for handicapped people by doing away with master harnesses. One can also envisage a set of light cameras attatched to the waist of an able person. They would record the movement of his legs when walking, to guide a handicapped person.

IV - GENERAL CONFIGURATION OF THE SYNTAXER

The camera holder of our syntaxer was built from tubular framing as in figure 5. The bridge carries the cameras while luminous screens are fixed to the chassis opposite them. We are thus looking here at dark reference points on a light background.

For cheapness, we used tube cameras, but later use of CCD cameras is envisaged.

V - CONCLUSION

In tele-control, the control unit or syntaxer produces, in real time, trajectories to be followed by the slave arm. "Dynamic" syntaxers (master arms or mini-master arms) and "static" syntaxers (isometric control handles with strain gauges) produce at each moment a target position. "Push button" syntaxers produce a target speed, which goes to zero when the pressure on the button is relaxed.

The "optical, three dimensional" syntaxer described here produces a target position. This dynamic syntaxer differs completely in nature and design from existing models, since the moving part comprises neither mechanical articulation, nor any kind of mechanical or electrical junction with a fixed frame. It has none of the disadvantages of current syntaxers. The most obvious advantages occrue from simple design, easy construction of the moving part and the absence of mechanical play, whip, inertia and wear.

The dynamic design allows the operator to follow the movement, so that the tele-commanded manipulation is an exact replica of the operator's movement.

The principle of this tridimensional optical syntaxer, is the movement of a hand-held stylus in the field of view of two cameras with perpendicular optical axes.

It gives control over seven degrees of freedom.

The sets of points on the stylus are distinguished by the cameras. A calculator commands the different degrees of freedom by reference to the coordinates of these points. Several stylus configurations are possible of course.

The resolution reached with such a system is better than one millimeter if we use 625 lines of 1024 points to describe a cubic stylus workspace of side 50 cm.

The technic used here, optical acquisition of the coordinates of carefully distributed reference points and careful use of them may be put to other uses than that of a syntaxer. One example is direct measurement of the coordinates of the terminal organ of a robot arm, which need only carry a few reference points discriminated by sight. Indeed, present measurement of the position is indirect, relying on a kinetic calculation based on data from the sensors at each degree of freedom. The inaccuracy of this indirect measurement easily may be imagined when one thinks of the mechanical play and elasticity of the manipulator.

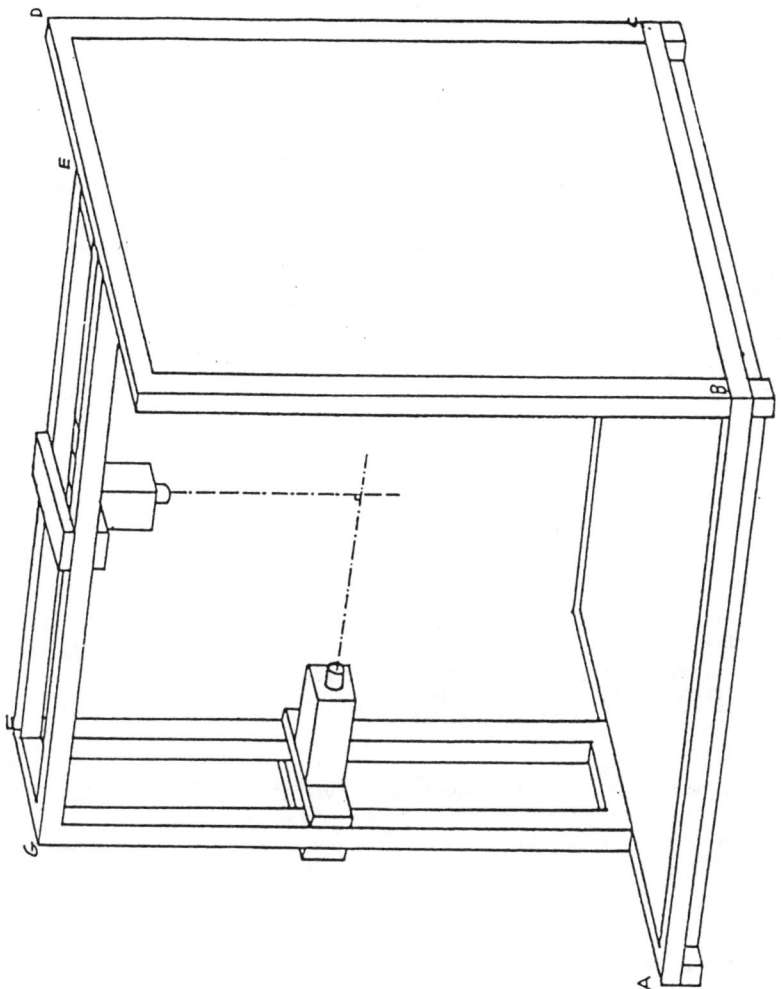

Figure 5 - Experimental arrangement of the cameras of the optical, three dimensional syntaxer.

The above syntaxer could also be useful in off-line programmed manipulators, where a first level of programming is accomplished without the manipulator followed by a second, finer, level with it : the first level may then be done very easily with a considerable gain of time. A particular case which comes to mind is programming techniques for welding.

BIBLIOGRAPHY

(1) T. JAWAD

"Etude et réalisation d'un banc de mesure de coordonnées spatiales tridimensionnelles par caméra opto-électronique".

Thèse de Docteur-Ingénieur, Université de Bordeaux I, 16-11-1984

(2) B. EL HADJ AMOR

"Etude et réalisation d'un système de détection et de localisation tridimensionnelle d'objets. Application à un robot agricole ramasseur d'asperges".

Thèse de Docteur-Ingénieur, Université de Bordeaux I, 09-12-1982

(3) P. JEUNET

"Commande en boucle fermée avec retour visuel d'un robot manipulateur destiné à tailler automatiquement la vigne".

Thèse de 3ème cycle, Université de Bordeaux I, 15-11-1984

(4) J. VERTUT, P. COIFFET

"Les robots".

Tome 3A : Téléopération, évolution des technologies, Hermès Publishing, 1984

(5) Rapport d'activités A.R.A. : "Téléopération avancée"

"Automatisation et robotique avancées", Journées de Poitiers, 28-30 Septembre 1982

    a - J. VERTUT, R. HUGON, R. FOURNIER, A. MICAELLI
    b - A. LIEGEOIS, H. VILAMOSA, R. ARDEBILI, M. CHIROUZE
    c - J. GUITTET, J.P. GAILLARD, N. QUETIN

(6) S. LELANDAIS

"Réalisation et première exploitation d'un système de numérisation de formes tridimensionnelles".

Thèse de 3ème cycle, E.N.S.T. - Université de Technologie de Compiègne, 17-04-1984

(7) S. LELANDAIS et H. MAITRE

"Analyse et prétraitement de données tridimensionnelles".

C.E.S.T.A., Premier Colloque Image, Biarritz, France, 21-25 Mai 1984

(8) B. GORLA, M. RENAUD

"Modèles des robots manipulateurs. Application à leur commande".

C.E.P.A.D.U.E.S.

Part 8
Locomotion and Walking Machines

Towards Generalized Concepts and Tools for Unconventional Mobile Robots General Languages, Mobility Modes

J.-J. Kessis, J. Penne, J.-P. Rambaut and N. Mattar

Laboratoire de Robotique et Intelligence Artificielle
Université de Paris VII

SUMMARY

Unconventional, i.e. non-wheeled, mobile robots (UMR), e.g. walkers, reptile, submarine, were seen until recently as pure research subjects. Now in several fields, such as maintenance, exploration, intervention, in hard or hostile environments, they, including advanced robots, i.e. fully autonomous ones, are in demand.

In previous RoManSys, we described general architecture (R'81) and intelligent systems (R'84) as developed in our laboratory for walking robots. This paper describes computer science applications for system and languages for mobile robots.

After our experiments on two walking prototypes (H1, H2) we found it useful to develop two levels of dedicated languages: first an assembly-level one (LRA: LP 4.5), producing an object code interpreted by the robot as its own machine language; second, a high-level language (LRE: MULTICIBLE) which compiles into LRA or other microprocessor code. The pair LRE/LRA allows easy programming of adaptive autonomous behaviour. On the other hand, intelligent planning generates LRA directly.

The generalization of various mobile robots, including UMRs, is currently in development. The main concept is mobility mode, an extension of the gait concept. A declarative part of the future mobile robot language (MRL) should be the description of the particular MR in use, with its physical properties and performance.

INTRODUCTION

Mobile robots will be used in otherwise unaccessible places or uncongenial environments, places without roads or smooth sites. Man-built unhealthy places, such as nuclear plants or other industrial severe environments, are full of stairs, pipes and other pitfalls, quite unfriendly for wheeled devices.

Consequently, it is of interest to define, as opposed to mechanically simple and overstudied rolling devices, the category of non-wheeled Unconventional Mobile Robots (UMRs) and focus our research on the specific general concepts involved, since a lot of robots in practical use will be UMRs. These are distinguished by the complexity and rich diversity of the possible means of mobility, including those of animals.

Complexity - the number of moving parts and configurational degrees of freedom make it necessary to design and implement reducing concepts, such as gaits for Walking Robots, UMRs. Our incremental approach to gaits (1, 12) makes possible generalization to other UMRs, the mobility modes concept, as described below.

The diversity of the means used by living creatures to move, and of those possible is really impressive. To our knowledge very few were implemented or at least studied in robotic research: besides walking, snake-like creeping (9, 10) which is itself a world, hopping (11)... submarine true robots should be of great interest (7, 8). There is again a need to order and manage such a diversity. We propound hereafter means to satisfy this need, using above-mentioned general concepts, along with architectural and software "robot-independent" design.

Table 1 shows (left) the state of implementation and experiments on our hexapods H1 and H2; and (right) the generalization of similar architecture, concepts and tools to general mobile robots. Compared with our initial design as shown at RoManSy'81 (1) one may remark the subdivision of logical levels 1 and 3, leading to a 1.5 "real-time" level and a 3.5 "advanced language" level. The software levels - 1.5 and above - may be materially implemented on separate processors. In addition to that hierarchical organization, "satellite" software+hardware nodes cope with sensing and/or non-locomotional actuators. Hierarchical organization is encountered now in several mobile devices, e.g. A.S.V. (5).

The underlying idea is to avoid "ad hoc" programming of robots. As a matter of fact, the short history of computer science shows that the need for general software concepts and works is always largely underestimated. Moreover, computer systems ideas (e.g. machine and high-level languages, compilation, real-time, concurrent programming, device-independence...) may be applied to mobile robots whose complexity is somewhat similar. Hereafter we follow the ordering of levels as shown in Table 1. We first overview the state of the experiments on walking robots at the laboratory. Then we emphasize the new language and compiler developments, both at gait level and in the new 3.5 structured language level; these developments make the subject of N. MATTAR's thesis (3). Finally, we examine the new category of UMRs, and how to generalize the above works.

CURRENT STATE OF HEXAPODS H1 AND H2

Legs of both robots are planar cartesian (i.e., with separately controlled x and z degrees of freedom) devices, of an original pantograph-type design on H1 (1), direct cartesian for the 40-kg-payload H2. Due to proper compliance (1) the robots can turn on the spot with only 12 controlled configurational degrees of freedom, instead of the usual 18.

Level 1.5 implements continuous generation of command ("Tonus"), reflex contact-avoidance and safety moves, and more generally, all "real-time" needs of a robot, as opposed to "background" software tasks.

Our conception of gaits (12) is wider than usual, first by the usage of amplitude, sign and cycle-shape parameters that lead to a combinatorial multiplicity of gaits, including backward, turning-on-the-spot, etc. Second, by an incremental (differential) implementation of the gait generator allowing slicing of moves in elementary-time gait parts. Third, due to the complexity of specification of such "generalized gaits," we find it useful to gave a gait compiler (ALLCMP) as shown below.

The assembly-type "robot's machine language" LP4.5 (level 3), with robotic primitive functions such as (sliced) gaits, was from the beginning an originality of our approach; its current state is shown in the book "Recent Advances in Robotics" (12).

A higher level language and its compiler (MULTICIBLE), filling the new 3.5 level, opens the way to general UMR software; these are detailed in the following.

The intelligent level 4, which combines algorithmic parts (path-finding A*-type system) and expert-system parts (universe-modelling and decision-making systems), was described at RoManSy'84 (2).

LANGUAGE AND COMPILERS DEVELOPMENTS

The need for high-level language tools for easy specification, modification, portability, etc., is now well known among computer scientists, e.g. L.M. (4) For mobile robots we may cite OWL, a language for walkers with concurrent abilities (6).

1. - ALLCMP

This is, to our knowledge, the first gait compiler ever made. It allows easy specification of fully parametrised, generalized gaits, with a variable amount of detail. First, a number of usual gaits are predefined, thus declared by the programmer by simply using their reserved name. Second, one may specify individually every parameter, including leg cycle shape and/or crawling style. A previous declaration of variables and arrays, with possible initial values gives a style facility by avoiding repetitions. The whole generates a table for every gait, for use by the level-2 gait operator.

2. MULTICIBLE

This structured language and compiler (the latter using the newest and most efficient computer-science compiling techniques) (3), allows easy programming of adaptive robot behaviours. The advanced structure statements may embrace machine-code parts in various processors' own code. This ability is new in the field of computer languages. Possible machine-codes are those of microprocessors in the hardware organization of the robot, and the LP4.5 robot's level-3 language, seen as a machine code among others.

After these recent improvements the levels 3 and above offer a choice for the decision-making part of a robot, depending on situations and tasks: either an artificial intelligence (L4) plan generator produces new LP4.5 code (L3) as a robot's action progresses; or a general MULTICIBLE program (L3.5) is compiled once for all into a LP4.5 robot monitor program. The first solution is mandatory for an unknown-universe outside intelligent tasks such as exploration. The second may be good for inside routine tasks, e.g. periodic maintenance or cleaning. A combination is possible, i.e. a monitor for routine jobs, or plan generator otherwise.

PROSPECT: TOWARDS A GENERAL MOBILE ROBOTS LANGUAGE (MRL)

To become more than a word, the category of unconventional mobile robots (UMRs) needs common methods and tools, and first a common programming system: MRL. Walking robots are complex enough to require reducing concepts such as gaits. We think it possible to generalize the gait concept, in the improved form we designed, into a differential mobility modes concept. Our first test on the subject will be the mobility of the Amiens modular axial-symmetric reptile robot (10). Using such concepts it is possible to design in the spirit of MULTICIBLE, a MRL compiler and language exhibiting a "device-independence" such as usual input/output devices as seen by a modern computer operating system. That is to say, a MRL computer program may run on various UMRs (and on - easier - wheeled robots) with no changes or by changing only a declarative part describing the idiosyncrasies of the particular robot in use. Nowadays, MULTICIBLE runs on both H1 and H2 although these are different in size and organization. The adaptation to wheels, which should be easy, is in development. The conceptual analysis needed by complex UMRs such as reptile robots is harder; is has been currently started by the Amiens team.

REFERENCES

(1) KESSIS J.J., RAMBAUT J.P., PENNE J.
"Walking Robot multi-level Architecture & Implementation"
4th Symposium on Theory and Practice of Robots and Manipulators
(RoManSy'81), Warsaw, Poland, Sept. 1981
PWN Polish Scientific Publishers, Warszawa, 1983, pp. 297-304

(2) KESSIS J.J., RAMBAUT J.P., PENNE J., WOOD R., MATTAR N.
"Hexapod walking Robots with Artificial Intelligence capabilities"
5th Symposium on Theory and Practice of Robots and Manipulators
(RoManSy'84), Udine, Italy, Sept. 1984
Hermes Publishing, Paris, 1985, pp. 395-401

(3) MATTAR N.
"Contribution à la conception et à la mise en oeuvre de Robots Mobiles : implantation d'un langage évolué spécialisé; systèmes sensoriels et de controle"
Thesis, Paris 7 University, 1985

(4) MAZER E., MIRIBEL M.
"Le Langage LM"
Cepadues, Toulouse, 1984

(5) MCGHEE R., ORIN D.E., PUGH D.R., PATTERSON M.R.
"A Hierarchically Structured System for Computer Control of a Hexapod Walking Machine"
RoManSy'84, Udine, Italy, Sept. 1984, pp.375-381

(6) DONNER M.
"The design of OWL, a language for walking"
Proc. SIGPLAN'83 Symp. on prog. languages
SIGPLAN Notices, 18(6),pp. 158-165

(7) GUERRIER G.
"Projet RAM Océanique"
Journées ORIA'85 (Offshore, Robotics & A.I.), Marseille, June 1985
Actes, IIRIAM Institute, Marseille, 1985

(8) KESSIS J.J., MATTAR N.
"Langages et Architectures logicielles pour Robots Mobiles"
Journées ORIA'85 (Offshore, Robotics & A.I.), Marseille, June 1985
Actes, IIRIAM Institute, Marseille, 1985

(9) HIROSE S., UMETANI Y.
"An active cord mechanism with oblique swivel joints & its control"
RoManSy'81, Warsaw, Poland, Sept.1981, pp. 327-340

(10) DEMARCQ Y., LAMBERT M., MALLEJAC P.
" R.A.M.S.O.M. : Un projet de Robot Rampant"
Séminaire "Les Robots Automoteurs", Nov 1983, Paris
Actes, Agence de l'Informatique, Paris, 1984

(11) RAIBERT M.H.
"Dynamic Stability & Resonance in a one-legged hopping machine"
RoManSy'81, Warsaw, Poland, Sept.1981, pp.352-367

(12) KESSIS J.J., RAMBAUT J.P., PENNE J., WOOD R
"Six-Legged Walking Robots"
in "Recent Advances in Robotics" , G.BENI & S.HACKWOOD Edtrs
John Wiley & Sons, New York, 1985, pp. 243-260

TABLE I : A GENERAL ARCHITECTURE FOR MOBILE ROBOTS

LEVEL	WALKING ROBOTS (LARIA's H1 , H2)	GENERAL MOBILE ROBOTS including UMRs
1	Cartesian Legs linear servomotors	**Mechanics & Servocontrol level** Mobility organs w. motors & local control
1.5	"Tonus" Contact & collision avoiding	**"Real-Time" level** Continuous Control Local "reflex" moves Safety
2	Gait concept implementation: incremental gait generator Gait compilator:ALLCMP	**Co-ordination level** Mobility Modes concept Differential M.M. generation
3	Assembly-level language LP 4.5 gaits as primitive fns	**First level of language** **"LRA"** "Robot's machine language" mobility modes as primitive fns
3.5	Structured advanced Language MULTICIBLE compiles any processor machine language,including Hexapodes' LP 4.5	**Second level of language** **"LRE"** General Mobile Robot Language MRL "robot-independent": declarative
4	Path finding system "P2A*" Universe modelling & Decision Expert System with variables "W3V"	**Intelligent Systems level** Perception: Universe models... Decision: Plan Generation, Path finding...

Concepts and Tools for Unconventional Mobile Robots 411

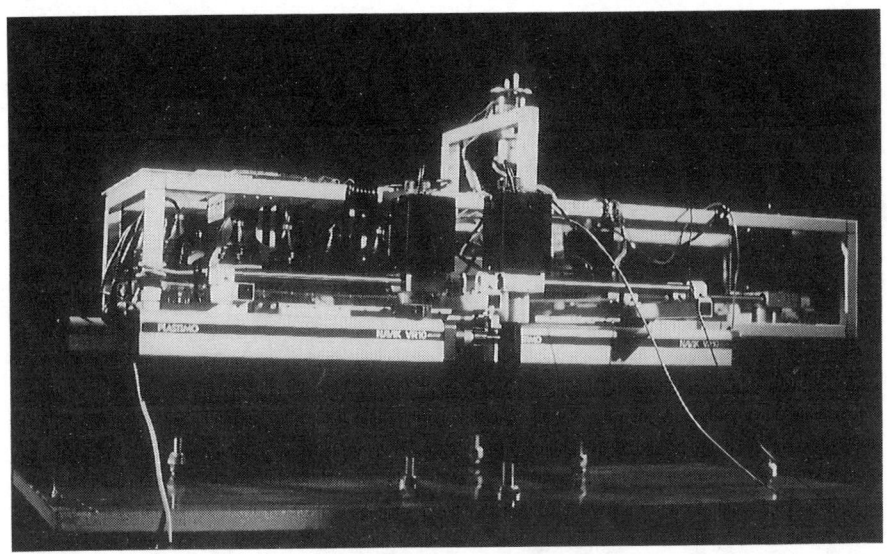

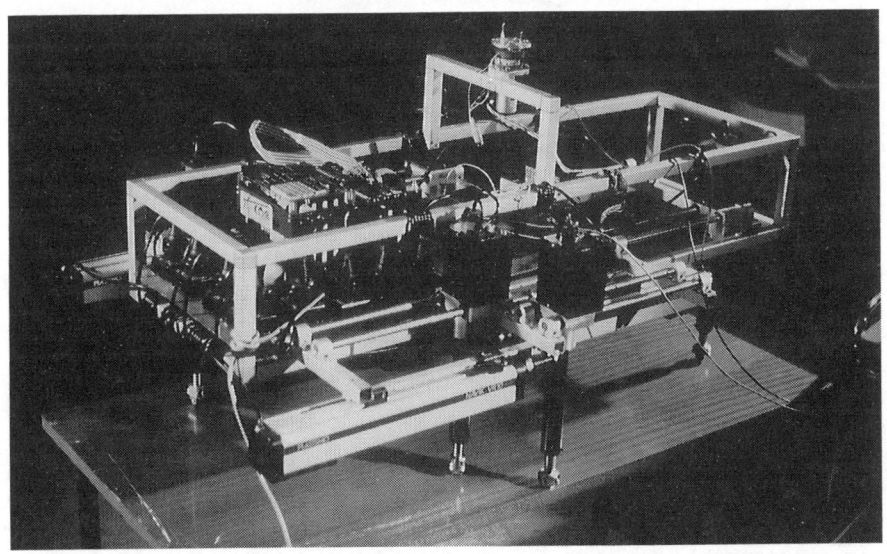

THE H2 HEXAPOD ROBOT

Mobile Robotic Systems for Use in Unstructured Terrain

K.J. Waldron and S. Agrawal

Department of Mechanical Engineering, The Ohio State University, Colombus, U.S.A.

Abstract

   Autonomous and teleoperated mobile platforms which have, so far, been constructed for use in unstructured terrain are of two types. One of these is a platform with wheels or tracks having a passive suspension. The other is a fully terrain adaptive legged locomotion system. However, the use of active coordination of multiple joints makes other configurations feasible, and attractive. A simple example is a wheeled system with an active suspension.

   In this paper, unconventional locomotion element configurations will be explored, and a theory for their design to optimize their performance over both small and large amplitude terrain variations will be presented. An important point is that the characteristics desirable in a mobile robotic platform are somewhat different from those desirable in conventional vehicles. Another is that the performance characteristics required of the locomotion system are determined by sensing and computation capabilities of the system, and vice-versa.

1. Introduction

   The suggestion is frequently made that wheels should be added at the ends of the legs of the Adaptive Suspension Vehicle [1] (Figure 1). The reasoning is that the machine could then act as a conventional wheeled vehicle with an actively controlled suspension on paved or relatively smooth terrain, and could walk in rough terrain. In fact systems have been designed before with wheeled legs [2,3], and the proposition of replacing the suspension of a conventional vehicle with an active system is also raised from time to time. Nevertheless, little progress with these ideas has been reported in the literature. One reason is that, in the pure feedback mode of operation, which is usually envisaged for the active suspension, the idea is basically unsound. Regardless of stiffness and weight distribution, the suspension and wheel assembly acts as a spring-mass oscillator. The power required to drive a spring-mass oscillator at frequencies above its natural frequency increases with the cube of frequency for constant amplitude. That is, for a given locomotion speed, the power increases inversely as the cube of wavelength of terrain variations. Of course, in real terrain, amplitude is not independent of wavelength. However, even if amplitude is proportional to wavelength, which is a reasonable approximation, the power required increases as the inverse of the square of the wavelength. Further, at short terrain wavelengths, or high frequencies, computational bandwidth becomes a problem. Thus, attempting to follow short wavelength terrain irregularities with an active system is not very productive.

Also, as far as the use of a walking action in rough terrain is concerned, a wheel is a rather poor foot, particularly in soft soil. The action by means of which a foot generates traction is quite different from that of a wheel. Broadly speaking, soil failure decreases wheel traction, but increases foot traction [4]. In experiments in our laboratory, this has been demonstrated to be due to a rather complex soil failure sequence which results in a tendency for the foot to dig in deeper as a consequence of soil failure under combined normal and shear loading. Wheels and tracks are most effective at low ground pressure. Feet, at least on the basis of biological system evidence, are most efficient at relatively high ground pressures.

However, in order to operate in unstructured terrain, mobile robotic systems, such as the Adaptive Suspension Vehicle, must have sophisticated means of sensing and internally modelling the terrain, including terrain preview capability. This capability is essential, regardless of the locomotion principle used. In conventional vehicles, it is provided by the human operator. By taking advantage of the preview capability, and by making proper use of the respective capabilities of active and passive suspension systems, it is possible to devise a configuration which promises superior mobility.

2. Small Amplitude Performance

As was pointed out in reference [5] it is possible to adequately characterize locomotion system performance for design purposes by considering performance at two scales. These are small terrain variation amplitudes; defined as those for which a passive suspension is not subject to bottoming, or other saturation effects, and large amplitudes; defined as those for which the suspension saturates, or for which there is interference between the vehicle body and the terrain. For small amplitudes, the performance of either an active or passive suspension system can be quantitatively studied by means of its attenuation of the power spectral density of the terrain elevation which is viewed as a spatial random variable. Figure 2 shows the attenuation factors of a three dimensional model of a wheeled system, and of an idealized legged system of similar scale. Further details can be found in reference [6].

Although there are substantial differences in the mechanics of the wheeled and legged systems, it can be seen that, at this scale, their performances are not substantially different. The legged system behaves as a sampled data system sampling the terrain at a spatial frequency which is the inverse of the leg stroke. In principle, a fully terrain adaptive legged system can completely isolate its body from terrain variations [7]. In practice, the body does respond to terrain variations. Its response is determined by the algorithm used to control body attitude. The response plotted is for an algorithm which maintains the body parallel to an average terrain gradient determined by the last three foot-falls, and maintains constant average walking height. The body is maintained level with the horizon about the roll axis.

The most significant difference is the complete lack of response of the legged system to very short wavelength variations. However, in a passive suspension system, the presence of a compliant tire (not included in the model of Figure 2) tends also to produce enhanced attenuation at very short wavelengths.

In summary, whatever mobility advantages may be attainable by using an actively controlled walking, rather than a passively suspended rolling, locomotion action, are not very evident at small terrain variation amplitudes.

3. Large Amplitude Performance

In reference [8], a characterization of large amplitude performance was presented which offers a meaningful compromise between the over-simplification of characterization using simple obstacle geometries, such as steps and ditches [9], and the complexities of terrain interference studies [4]. This characterization makes use of an obstacle which has both height and width. It is a very natural mode of characterization for a walking system since all that is important for such a system is the locations of the take-off and landing areas. The geometry of the terrain in-between is immaterial. It is less suited to characterization of wheeled or tracked systems since, in those cases, the geometry of the intervening terrain is important. Nevertheless, an adequate characterization can be constructed using just two cases: the "convex case in which the take-off and landing areas are joined by a linear ramp; and the "concave" case in which there is a void between them. Figures 3 and 4 show the feasible values of obstacle width A, versus height B plotted for these two cases for a four-wheel drive wheeled system and the Adaptive Suspension Vehicle [1]. A variety of other cases may be found in reference [6]. In these diagrams, any combination of obstacle width, A, with height, B, is negotiable provided it falls within the envelope for that vehicle. As may be seen, the legged system exhibits substantially superior mobility on this basis. This is true for both the convex and concave cases. The mobility of the wheeled system in the concave case is very poor.

For the legged system, the critical position is that in which the center of mass is just over the landing point. It is necessary that this position be achievable while a foot is still placed on the take-off point. This leads to the geometry shown on Figure 5. Here H is the distance from the bottom of the leg working volume to the center of mass of the vehicle, R_x is the maximum leg stroke and R_z is the depth of the working volume, P is the pitch, or distance between legs. Note that the working volume has a rectangular cross-section, at least for the ASV, by virtue of the pantograph leg mechanism [9]. Then:

$$A = (P + R_x/2) \; C\alpha - H \; S\alpha$$
$$B = (P + R_x/2) \; S\alpha + H \; C\alpha - (H-R_z)/C\alpha \tag{1}$$

Elimination of α from these equations results in a quartic equation relating A and B which gives the bounding curve of the legged vehicle area shown in Figures 3 and 4. It is readily plotted directly from the equations (1) by substituting varying values of α. For the convex case, there is a limiting gradient for the ramp. Either the legged or wheeled system can locomote indefinitely on gradients lower than this limit. This is the linear boundary on the top right side of Figure 3.

The wheeled case turns out to be remarkably complex, with several regimes of operation to be considered. Full details of the models used may be found in reference [6].

4. Design of an Active Suspension Configuration

From the discussion above, it may be concluded that an active suspension offers little advantage in mobility over small amplitude, short wavelength obstacles and, in fact, it is energetically quite undesirable to use an active system in this manner. However, it also suggests that there may be mobility advantages to using an active system for handling large amplitude terrain variations. Further, if wheels are used on an actively operated legged system, the walking action will be impaired by the wheel's deficiencies as compared to a properly designed foot. However, the mobility advantage demonstrated over terrain variations with large amplitudes is only partially due to the walking action. A large part of it is simply due to the long vertical travel of the locomotion elements.

This suggests the use of a system with actively controlled, long travel, suspension elements in series with a passive compliance. The passive compliance can then be used to take care of small amplitude terrain variations. The active system can be used to adapt to large amplitude variations under the control of the terrain preview system, at quite low bandwidths. At least 8 wheels are needed to maintain stability while lifting a front wheel in anticipation of an obstacle. The suspension travel is in the vertical direction only. All wheels are driven, preferably under individual active control. Either articulated or skid steering may be used. In the latter case the end wheel pairs would be raised when changing direction. The arrangement of this system is shown in Figure 6.

As might be expected, the mobility of this system is intermediate between that of a conventional wheeled system and that of a walking system. Mobility envelopes are plotted on Figures 3 and 4. Figure 7 shows the model used for the concave obstacle case. Here the second last wheel pair, fully retracted, must be at the landing point and produce a resultant force which passes through the center of mass at the instant the last wheel pair leaves the take-off point. Then:

$$A = R\, S\phi + P\, C\alpha - E\, S\alpha$$
$$B = E\, C\alpha + P\, S\alpha + R(1 - C\phi) \tag{2}$$

Here R is wheel radius, P is wheel pitch, E is suspension travel and ϕ is friction angle. The bounding contour can be computed in the same way as for equations (1) by substituting varying values of α.

Summary

We have presented a configuration for a robotic vehicle with actively controlled suspension which is consistent with the mechanics of the suspension. As compared to conventional wheeled systems it offers significant mobility advantages. As compared to walking systems, it offers relative simplicity and superior speed in easy terrain.

Acknowledgement

The authors wish to acknowledge the support of DARPA, contract No. DAAE 07-84-K-R001 during the course of this work.

References

1. Waldron, K.J. and McGhee, R.B., "The Adaptive Suspension Vehicle," to appear in IEEE Control Systems Magazine.

2. Anon., Kaiser Spyder Model X5M, Industrial and Municipal Engineering Corp, Galva, Illinois, 1982.

3. Anon., Tomorrow's Mobile Man, Film, Defense Advanced Projects Agency, circa 1970.

4. Bekker, M.G., Introduction to Terrain-Vehicle Systems, University of Michigan Press, Ann Arbor, Michigan, 1969.

5. Waldron, K.J., "Mobility and Controllability Characteristics of Mobile Robotic Platforms," Proceedings of 1985 IEEE International Conference on Robotics and Automation, March 25-28, 1985, St. Louis, pp. 228-236.

6. Agrawal, S.K., Mobility Characterization of Robotic Platforms, M.S. Thesis, The Ohio State University, 1985.

7. McGhee, R.B., "Vehicular Legged Locomotion," Advances in Automation and Robotics, Vol. 1, pp. 259-284, JAI Press, 1985.

8. Waldron, K.J., "The Mechanics of Mobile Robots," *Proceedings of 1985 International Conference on Advanced Robotics*, September 9-10, 1985, Tokyo, pp. 533-544.

9. Waldron, K.J., Song, S.M., Wang, S.L. and Vohnout, V.J., "Mechanical and Geometric Design of the Adaptive Suspension Vehicle," *Theory and Practice of Robots and Manipulators*, Ed. Morecki, A., Bianchi, G., and Kedzior, K., Hermes, London, 1985, pp. 396-306.

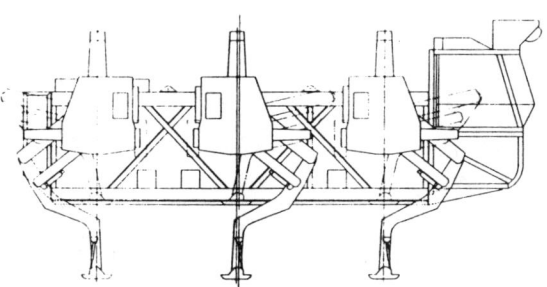

Figure 1: The Adaptive Suspension Vehicle (ASV). This machine is presently under test by The Ohio State University and Adaptive Machine Technologies Inc.

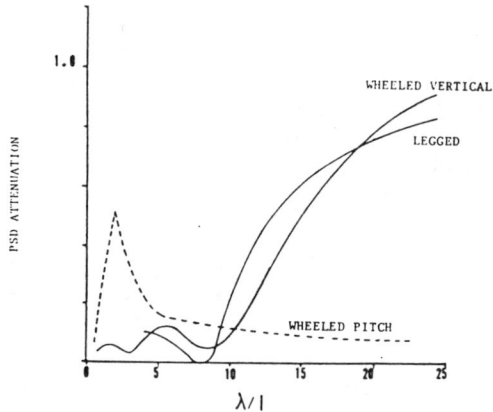

Figure 2: Power spectral density attenuation characteristics for eight-wheeled passive suspension system (vertical and pitch motions), and the ASV (vertical motion).

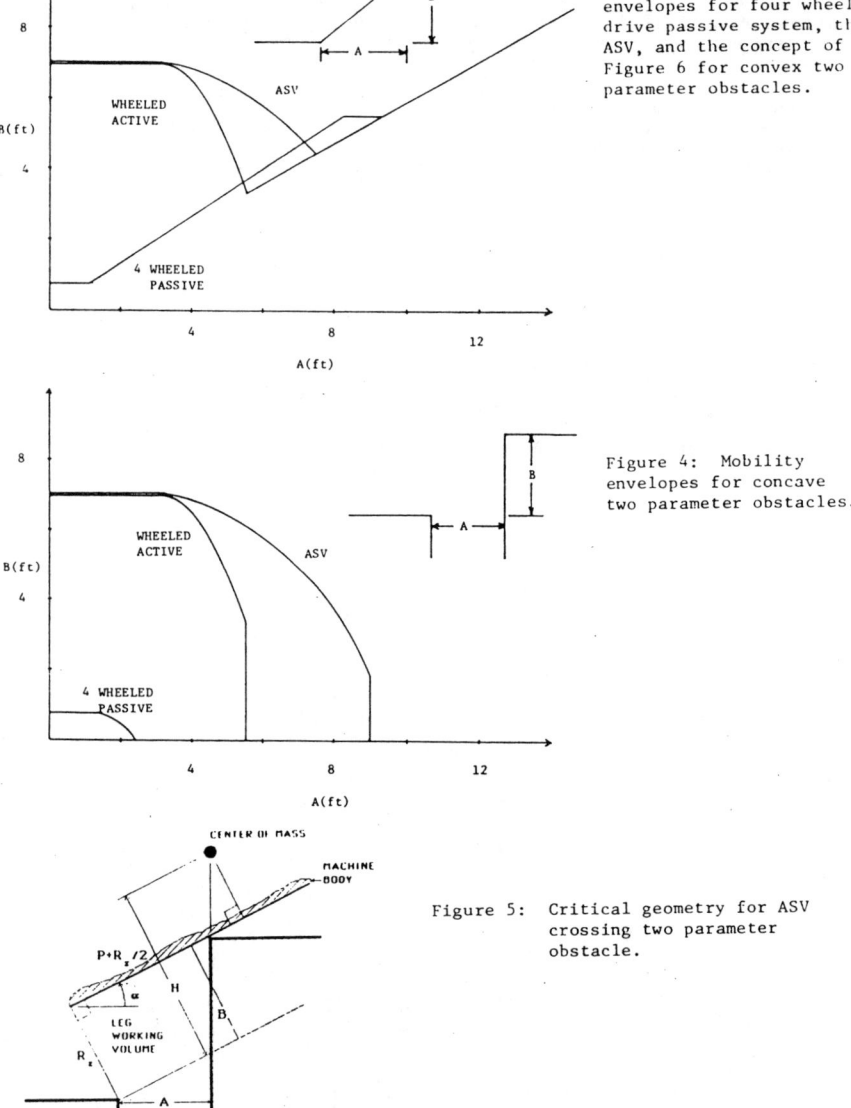

Figure 3: Mobility envelopes for four wheel drive passive system, the ASV, and the concept of Figure 6 for convex two parameter obstacles.

Figure 4: Mobility envelopes for concave two parameter obstacles.

Figure 5: Critical geometry for ASV crossing two parameter obstacle.

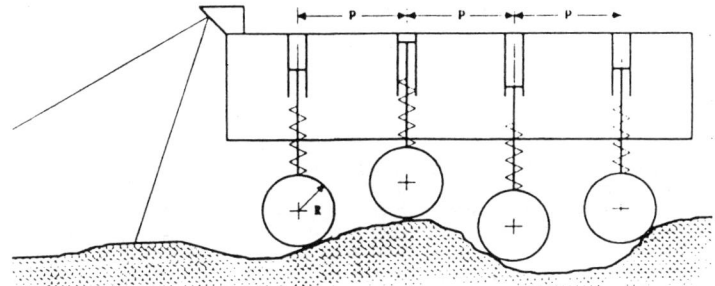

Figure 6: Vehicle with a combination of active and passive suspension elements. A means of pre-viewing the environment is essential to the control of the active elements.

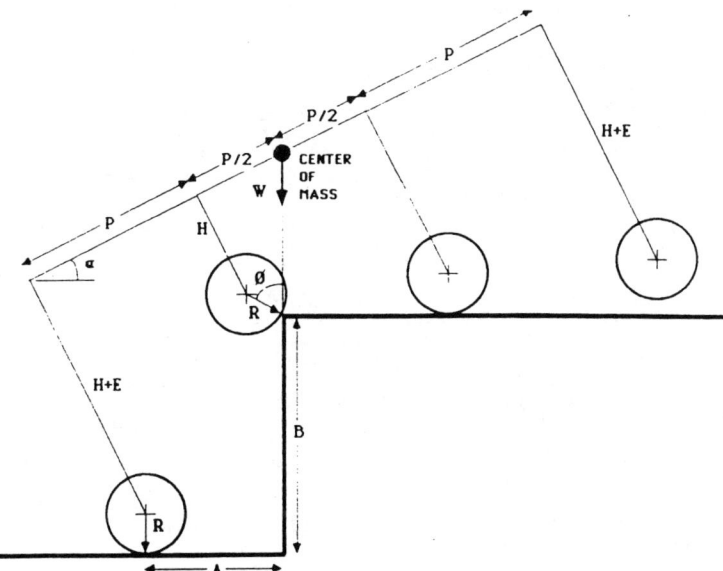

Figure 7: Critical position for two parameter obstacle crossing by the vehicle shown in Figure 6.

Wall Climbing Vehicle Using Internally Balanced Magnetic Unit

S. Hirose

Tokyo Institute of Technology, Tokyo, Japan

Summary

The fields on which a robot has to work are now expanding, from inside of a factory to outer environments. Robots are even expected to walk on vertical surfaces of buildings and structures to perform various types of works. This paper discusses a new mechanism of such robot which moves around the iron walls of the steam boiler of a nuclear power plant or sides of ships sustaining itself with magnetic sticking devices.

At first, a new magnetic sticking device is introduced. The device, named "Internally-Balanced Magnet" (abbreviated "IB Magnet") consisting of permanent magnet, spring, control rod and frame, enables to produce strong sticking force of the permanent magnet itself and at the same time can easily release it from the fixing surface. The validity of this mechanism is successfully demonstrated by human experiments where a man wearing four IB Magnets with manual operating levers on hands and feet walks around iron wall and ceiling.

Subsequently, a magnetic sticking vehicle using six IB Magnets is constructed. It is about 60 kg in weight, 700 x 700 x 300 mm in scale and 100 mm/sec in maximum locomotion speed. Stable and relatively fast linear motion as well as turning motion on walls is experimentally demonstrated.

1. Introduction

The current study intends to construct a robot which can magnetically stick to and move on vertical surfaces of steel sheets often used for bulky structures such as atomic reactors and ships, etc. and will make the following observations:

(1) This paper, at first, proposes a new internally-balanced magnet (abbreviated IB magnet) which can produce strong sticking force by itself and can easily stick to and release itself from iron walls and subsequently, clarifies the characteristics thereof and discusses the design method.

(2) A device which enables a man not only to continuously stick on vertical walls but also to move on such surfaces is experimentally produced, making most of the IB magnets. The experimental locomotion on steel walls demonstrates that a robot

of wall-surface-mobile type using the IB magnets will be able to perform highly adaptable locomotion on a vertical wall of iron/steel sheets in the future.

(3) A robot which can magnetically stick to and move on a flat surface is specifically materialized. At first, the mechanism is discussed and subsequently, two robot, experimentally created are explained. The experiment on controlled sticking walks proves the high probability of practical use of a sticking mobile machine based on the IB magnet.

2. Proposal of Internally-Balanced Magnet

At first, the author would like to propose a new magnet unit named IB (internally-balanced) magnet which can produce strong magnetic sticking force and simultaneously, can release itself from the fixing surface with a zero force in principle.

The IB magnet, as shown in Fig. 1, consists of such components as permanent magnet, spring of non-linear characteristic, a control stick for operation and a non-magnetic frame housing them. In principle, the releasing action of IB magnet is exerted in the following steps: at first, an internal mechanism executes a stroke motion to release the permanent magnet from the surface on which it is completely fixed ($x=\alpha$), in such distance as to nearly nullify the sticking force ($x=\beta$), and subsequently, the whole IB magnet unit lifts itself from the surface. Initiating this stroke motion directly by an actuator, however, requires a large force to release the permanent magnet from the surface, to make the concept unrealistic in many cases. Therefore, the idea of IB magnet is intended to nullify the force by internally balancing it with a spring. Specifically, a non-linear spring designed to have characteristics in direct opposition to the characteristics of the sticking force F_m and the displacement x of the permanent magnet, as shown in Fig. 2, is placed between the frame and the permanent magnet, as illustrated in Fig. 1. As the result of this structure, whatever position x ($\alpha \leq x \leq \beta$) the permanent magnet might take up, the magnetic sticking force F_m exerted on the permanent magnet always balances with the recovering force F_s created by the spring, and consequently, the driving force F_c of the joystick becomes nil, that is,

$$F_c = F_m + F_s = 0 \qquad (1).$$

This structure allows to lift up the permanent magnet toward the frame with zero force because the internal force always remains in balance with the force working on the permanent magnet. When $x = \alpha$, as the magnet unit is sticking on a steel wall, the permanent magnet is producing strong magnetic force in a way to press the frame to the steel wall by a spring power. This force, as it is, acts as the sticking force on outer material. When $x = \beta$, the force of the permanent magnet exerted on the steel plate is negligibly small so that the IB magnet can

be easily lifted from the steel wall. In this manner, the IB
magnet is a device enabling to control large sticking force F_m
acting on external materials with a very small operating force F_c
of the joystick.

The action principle of the IB magnet unit is explained in
terms of energy as follows: When the permanent magnet is sticking, the released amount of the magnetic potential energy is
accumulated in the spring as the strain energy, which is used
afterwards as the pulling-off energy of the permanent magnet.

At this stage of the discussion, the characteristics of the IB
magnet unit are summarized as follows:

(1) The characteristic enables a strong permanent magnet to
stick/release by very weak force (Zero force, if perfectly
adjusted). A permanent magnet usually has 0.5 to 1.0MPa of
sticking force per unit area, as will be concretely presented in
later passage. For comparative purpose, it should be noted that a
sucking disk which makes use of negative pressure for fixation,
cannot produce any more than 0.1MPa of sticking force at best.
In this respect, the current system depending on a permanent
magnet may be said to have outstanding advantage of strong
sticking force although its movable environment is limited to
iron/steel sheets and members. Besides, the switching system
between sticking and releasing actions may adopt either electrics
or simple pneumatic cylinder or levers. Such flexibility is
another merit of this IB magnet unit.

(2) The characteristic allows simplifying the configulation of
the magnet unit so that the production is easy. In addition,
leaking magnetic flux within the magnetic circuit can be
minimized to enable efficient use of the magnetic power.

(3) If layered leaf-spring is used for the non-linear spring, the
whole unit can be designed compactly.

(4) The mechanism is simple and highly reliable.

The key point of designing such IB magnet unit is the precise
analysis of the relations between the magnet sticking force and
the distance with the sticking surface, and designing the nonlinear spring having opposite characteristic. While the gist of
such designing technique will be presented here, the details will
be explained in another paper[1] It is recommendable to design the
magnet based on a modified permeance method, which is known to
facilitate considerably precise work.

For constructing the non-linear spring, the author has newly
introduced a concept of stepped-multiple-spring system. The
system consising of linear leaf-springs each having different
motion-initiating point, can cancel magnetic attraction force of
non-linear characteristic, as shown in Fig. 3. The compensating
characteristic is presented in Fig. 4.

3. Man-Walking Experiment Using IB Magnet Unit

For the purpose of testing the effectiveness of the IB magnet
as a robot sticking system, the devices were experimentally made

(Fig. 5) to be worn by a man on his hands and feet, because the main robot was not actually assembled by the time of this experiment. Frame dimensions were 74 x 90 x 62.6 mm. Each unit weighed 1 kg, and was able to generate approx. 1400N of sticking force. The lever can be controlled very lightly, and can be mounted/dismounted easily. Although the control force can be adjusted to near zero (less than 10N), the experiment set the force at approx. 30 to 50N by means of spring washers for sure control operation.

Fig 6 shows a man wearing the devices was actually walking on vertical walls and ceiling made of iron sheets (about 10 mm thick). The experiment proved that adaptive walking on such surfaces could be easily executed by repeating the sticking/releasing actions with hands and foot-tips.

4. Experimental Assembly of Wall-Climbing Robot (MAVERICK)

Figs. 7, 8 & 9 show the wall-climbing robot MAVERICK (Magnet Vehicle) experimentally assembled using the newly developed IB magnet. The robot demonstrated sure performance of straight-forward and turning locomotion. The outline of the mechanism is presented here:

4-1. Up/Down & On/Off Device of IB Magnet

At conceiving a walking type locomotion, it is essential, in the leg-returning phase, to cancel the magnetic power of the sticking parts and simultaneously allow them float from the surface in some distance so that the part (foot) does not bump to unpredictable irregularities of the surface, if there are any. In MAVERICK, as shown in Fig. 7, the effect of IB magnet is created by moving the magnet up/down by a short-stroke air-cylinder. When the magnet is lifted up more than some distance, the frame is designed to be pulled up, too. This design simplifies the mechanism in which, single actuator can control such motions as decrease of the magnetic sticking force followed by release of the sticking parts, and inversely, contact of the sticking parts on surface followed by generation of magnetic sticking force. The IB magnet is of a sealed construction, as shown in Fig. 7, having strong magnetic matter called a middle yoke. Any ferro-dust attracted to the surface is easily released at the returning phase.

4-2. Driving Mechanism

Six units of IB magnet are used. Each three units are linked to make A & B groups. The two groups are designed to make forward/backward motions by the driving mechanism of staggerred phases. The principle of generating forward movement is as follows:

As shown in Fig. 9, the rotary motion of the forwarding motor located at the center is transmitted to the incomplete pinions, which, in their toothed periphery, mesh with racks. They advance the foot units mounted with IB magnets. When the rotation of the incomplete pinion reaches its toothless periphery, a restore spring quickly retreats the foot unit (magnet unit). In this way, forward/backward movements of foot are produced with quick returning charcteristic. The returning speed is adjusted with an air-damper to attenuate the impact. The A & B groups do not return simultaneously as their moving phases are staggerred each other. Besides, the IB magnet units are so adjusted as to stick (touch down) at the forwarding movement and to release (lift off) at the returning movement, in synchonization with such movements of the foot (magnet unit). Continuous advancing motion is made in this way.

Change of advancing direction is made by repeating the sticking/releasing motions in timing aligned with left/right turns of A or B group pivoting upon a same axis. As the mechanism of the driving-power-transmission is simplified, the incomplete pinions of Fig. 9 rotate against the driving gear and forward/backward motions are produced at the same time. At every rotation, inverse motion is also produced to cancel forward or backward movement so that pivoting motion upon a fixed point is realized. In summary, the present mechanism realizes both uni-directional advancing movement and pivoting movement by means of two actuators and one set of actuators which lift up and release down magnets.

Acknowledgement is due to Mr. Yukio Nishihama, of the Robot Systems Dept. of K.K. Daikin Industries, Mr. MinehisaImazato, Mr. Yoshiaki Kudo, and Mr. Naoki Sugiyama, for their cooperation to this study.

Reference

[1] Hirose,S., et al. Internally Balanced Magnetic Unit, J. Robotics Society of Japan, vol.3-1, pp 10-19 (1935)

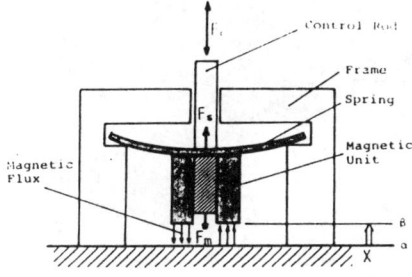

Fig.1 The basic structure of Proposing Internally Balanced (IB) Magnet

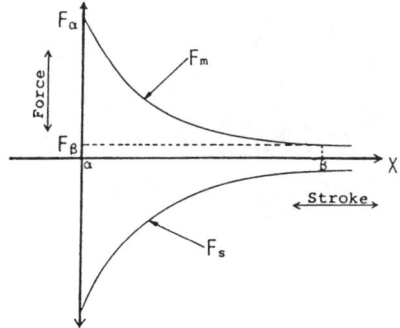

Fig.2 The kinematic relation between magnetic and spring force

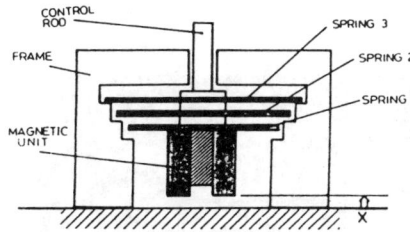

Fig.3 The basic structure of proposing stepped-multiple-spring system

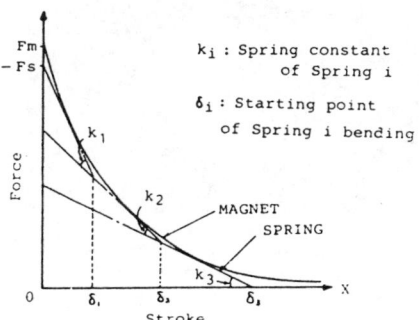

Fig.4 The relation between stroke and force of the proposing stepped-multiple-spring system

k_i : Spring constant of Spring i

δ_i : Starting point of Spring i bending

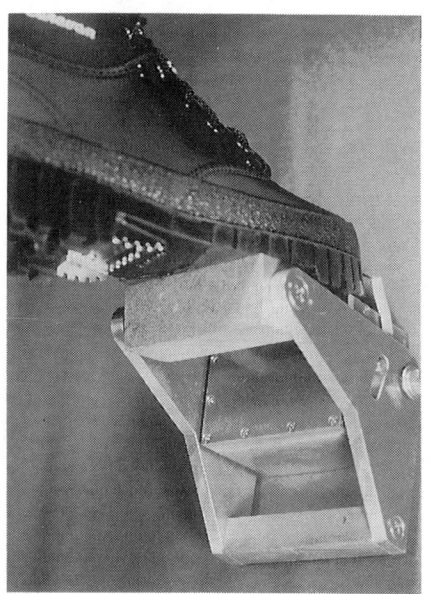

Fig.5 Man-wearing IB magnet with control lever

Fig.6 Man walking experiment on walls and ceilings using IB magnets

Fig.7 IB magnet with sealed construction and its driving pneumatic-cylinder

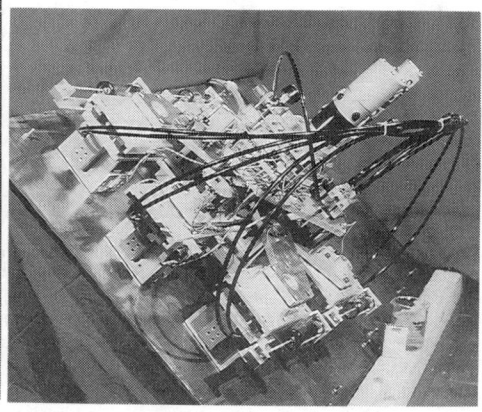

Fig.8 Walking experiment by the constructed Magnetic Wall Climbing Vehicle

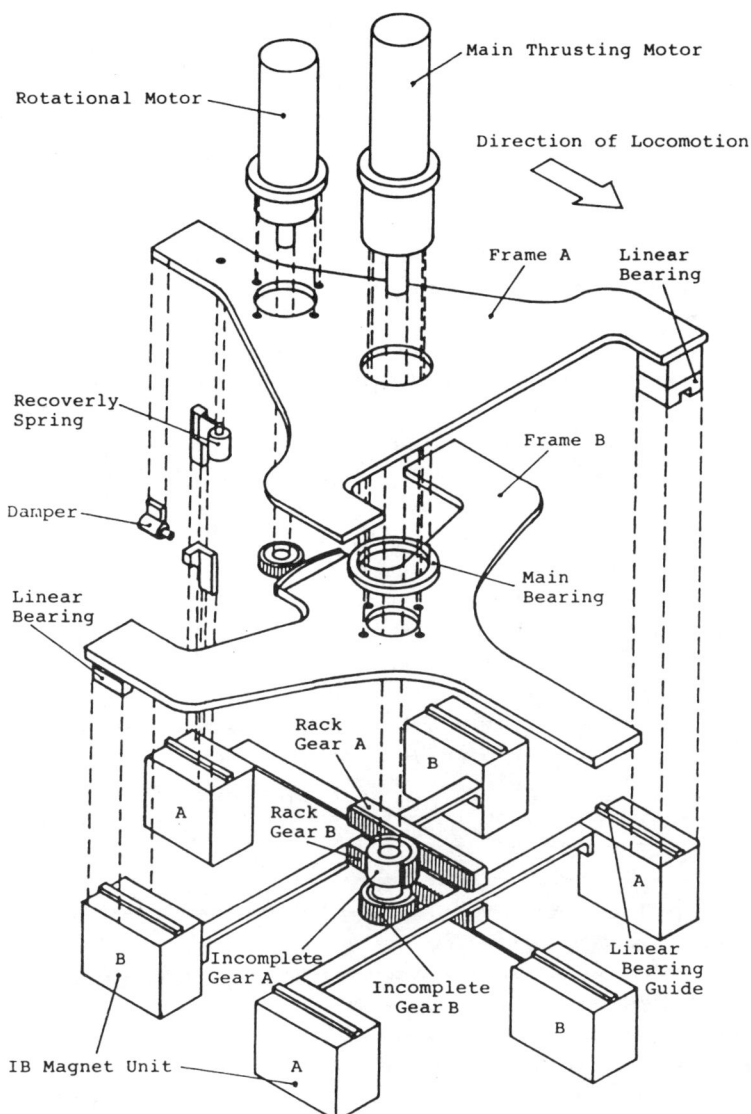

Fig.9 The driving mechanism of the constructed Magnetic Wall Climbing Vehicle

Experimental Development of a Walking Transport Robot

B.D. Petriashvili, M.A. Bilashvili and V.O. Margvelashvili

SUMMARY

The advisability of use of space pantographs with three-coordinate actuators as limbs for walking robots is substantiated on the basis of analysis of various kinematic schemes for walking propulsive devices. Analyzing the drive method and operation cycles of such limbs the control position is preferred. Considering walking robot motion on real soil, the criteria for unstable motion are determined. Reasons for unstable motion are given and the method for stabilization of motion is selected. The "ricksha" type two-legged walking robot model with plane pantograph limbs and hydraulic two-coordinate actuators, control of which is realized by position method, is described.

The development of robotechnics revealed the necessity for the creation and testing of walking robots of different kinds.

The range of possible application of walking robots defines a variety of uses and special problems and, consequently, variety in their construction: kinematics, types of drives and control systems.

The walking robot equipped with adaptive legs is the transport vehicle destined for motion on soil or off-road for travelling to places where the use of other transport is unsuitable or simply impossible. At the same time being provided with special suspension equipment, the walking robot will be effective in different spheres of human activity: agriculture, forestry, in carrying out geological, building and warehouse operations under extreme conditions.

Depending on the purpose of the walking robot, its development needs the solution of some problems, such as: kinematics, construction of limbs, control methods of its different systems and degree of automatization of their operation, and the criteria for the walking robot's stable motion.

Let us consider some of these problems, the solution of which in connection with a ground walking robot brought us to the creation of the "ricksha" type model.

In the literature many different constructions of walking propulsive devices are dealt with (1,2,3). Omitting pure mechanical walking propulsive devices being constructed by trajectory synthesis method, let us consider some prospective schemes supporting volumetric "service areas" of reference points.

Figure 1 and 2 show coupling versions of leverage and telescopic legs on the walking robot's frame. The degree of freedom of all of the given schemes is w = 3, but there are some differences in their kinematics.

These mechanisms do not have too many degrees of freedom or kinematic connections. Consequently, it is necessary to control all movable joints - i.e. joints must be constructed either in the form of active joints or provide separate drives between all adjacent units.

Simultaneous control of the three drives for the walking robot's legs to obtain the supporting units necessary trajectory is a rather difficult problem and cannot be solved without a vehicle-borne computer. However, as is well known from (4,5,6), realization of special control algorithms for different models of walking robots is possible.

The control problem for walking robot drives will be simplified incomparably if we use orthogonal two-coordinate (7) or three coordinate schemes of legs, though in the given cases, dimensions (strokes) of drives will be commensurable with the range of supporting units.

The necessary relationship between the range of legs supporting units on the sections of its trajectory and sizes of drives may be established if we use well-known mechanisms for converting curves, the so called pantograph mechanisms.

Plane pantograph walking propulsive devices are well-known (1,8). Their major defect is a small plane "service area" and necessity of using additional mechanisms for turning. These defects may be eliminated in spatial pantograph legs. Let us consider one of the possible versions. The leg shown in figure 3 is transformable into a spatial (8) plane pantograph leg. In comparison with the plane pantograph, the construction of spatial pantograph is complicated. In the kinematic chain of the mechanism one spherical three-motion, one prismatic one-motion and revolute one-motion kinematic pairs are added.

Three drives along coordinate axes control the three degrees of freedom of the motion of the mechanism. One of them provides a change of "gauge" within necessary limits, as well as permitting the walking robot to turn. The operational logic of this drive is determined by the coordination principle as well as the turning rate.

We shall consider the operation of two main drives on the example of plane pantograph legs. For simplicity in the walking robot's control system, let us consider simultaneous motion of two legs. Figure 4 illustrates the kinematic scheme of the pantograph legs placed on the frame of the walking robot. Kinematic pairs V_1, V_2, H_1, H_2 are constructed in the form of a power cylinder. Displacement of the pantograph driven link is controlled by transducers placed at points a,b,e,c,k,d.

Figure 5 illustrates the block diagram of the walking robot's control system. Changes of phases of the driven link's operation, represented by the separate leg motions, are supported by the corresponding connection order of the above-mentioned transducers and signal delay blocks 1,2,3,4,5.

Transducers that stop the lowering of the legs when stepping on obstacles may be threshold switches fixed in the supports, or force transducers that regulate the load of the lowered legs' vertical drive.

The parameters controlling the foot path are: the length of horizontal track of the driving member, its speed at different parts of the trajectory and the height of any surmountable obstacle. Theoretically, such control systems may ensure accident-free motion for a multilegged robot in off-road conditions even with unideal operation of the drives because of the strict succession of supporting polygons.

For testing the accepted concepts, walking robot models were created. In the first laboratory model the pantograph propulsive device was realized using two-coordinate drive in the form of an electromechanic differential block, figure 6. The model was tested under laboratory conditions on artifical obstacles.

To understand the real interaction processes of artificial legs with the ground, as well as to test the above-mentioned control system, the "ricksha" type model was created, figure 8. Model pantograph legs are propelled by the hydraulic drive using electric control, the operational logic of which is analagous to that mentioned above. The power unit of the model is mounted on the vehicle. It is an internal combustion engine. The model's characteristics are: power, 7.5 kW, pressure in the hydraulic system is $(10 \div 15)$ MPa, speed of the model is 1 m/s, net mass is 750 kg. Horizontal force capability of one leg is nearly 0.4 t. The scale of pantograph is $=:1$. The height of the cleared obstacles is 500 mm. Maximum range of legs is 1750 mm. Speed and direction of the model's motion is determined by the drive-operator. It also has the choice of method for clearing separate untypical obstacles by the way of immediate control of lef drives. The operator may regulate the legs' vertical adaptation range for measurement of maximum speed, to increase the robot's stability and for off-loading its drives.

During the model's operation on real soil, the criteria of stable motion were revealed and generalized. To enumerate these criteria: keeping motion speed and direction assigned by the operator - this depends on the operation model's drives as well as longitudinal and transverse slippages of the transport on the surface of motion; the keeping of the assigned robot frame's orientation in space or relative to surface of motion - this depends on the relief of the surface of the motion, deflection of legs and soil under legs; keeping of assigned thrust force - it changes depending on the vertical and horizontal supporting power of the soil.

For a real walking robot it is necessary to take into consideration all these phenomena to make corresponding corrections in the operation of the drives and to change the parameters of support during motion under changing conditions of the soil.

On the basis of results obtained from the experiment the operation of multileg walking robots with a pantograph scheme of legs, illustrated in figure 3, is accepted. With an increased "service area" and simplicity of its drive control, the given scheme obtains essential quality – during robot motion, energy is not spent on supporting its weight; locking the vertical drive is enough.

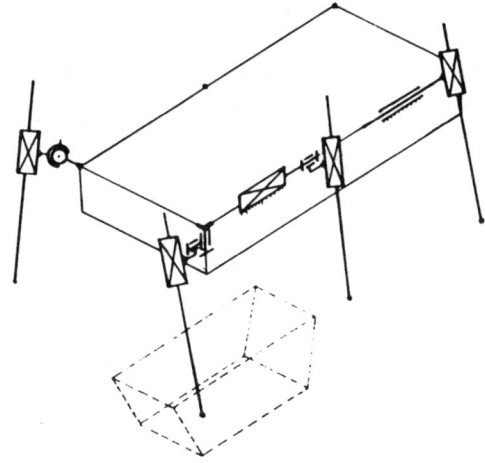

Fig. 1.

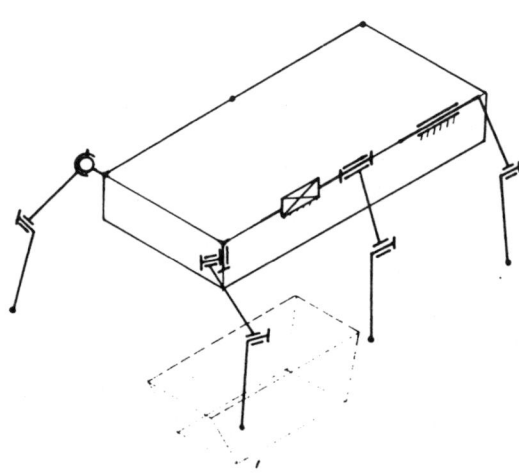

Fig. 2.

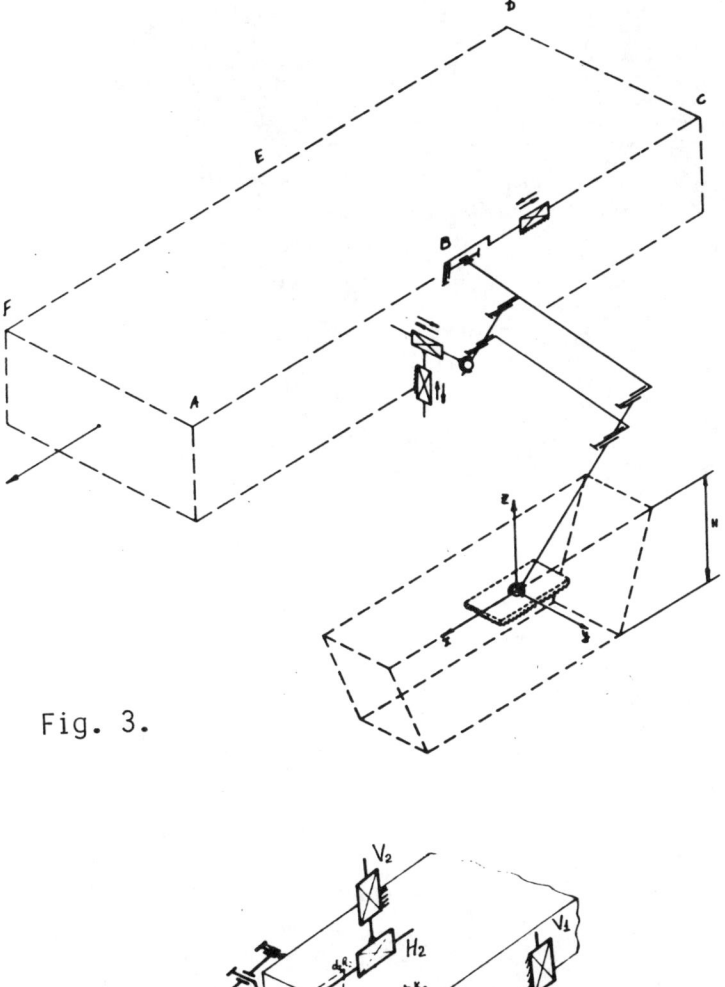

Fig. 3.

Fig. 4.

Fig. 6.

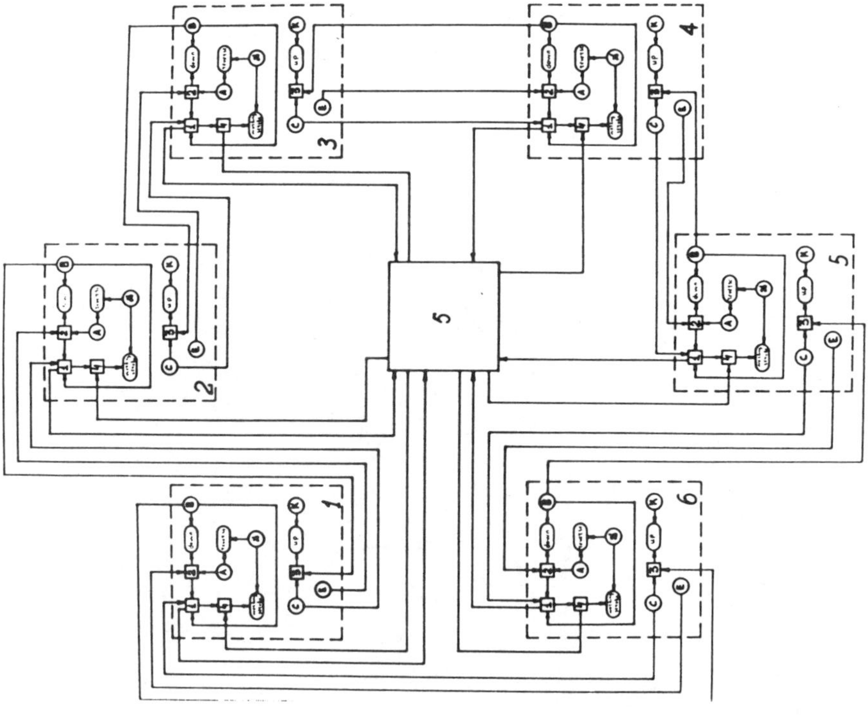

Fig. 5.

Fig. 7.

Fig. 8.

Legs that Deform Elastically

H. B. Brown Jr and M. H. Raibert

Carnegie-Mellon University, Pittsburgh, U.S.A.

Abstract

Central to the design of a legged system is the mechanical design of the leg itself. Legs are the elements that exert forces on the body to propel the body forward for transport, to counteract gravitational loading, and to keep the body in an upright posture. Most legs designed for legged machines are intended to be rigid, yet animals have legs that deform substantially under load. Compliance in legs can improve efficiency, reduce maximum loading, and simplify control. This paper compares various types of elastic storage mechanisms and considers the design of legs that use them.

Introduction

An intriguing characteristic of legs found in nature is their ability to deform elastically during running (Cavagna 1970). The elements primarily responsible for elastic deformation are the muscles and tendons. Elastic deformation is used by biological systems to recover a portion of the energy expended during a stride, and to return that energy on the next stride. This can reduce the total cost of transport (Dawson and Taylor 1973, Alexander and Vernon 1975, Cavagna et al. 1977, McMahon 1984). The net energy required to accelerate and decelerate the legs in their swinging motion may also be reduced through elastic storage in a flexible spine. Another function of compliance in biological legs is to reduce the impact forces and peak loads that are experienced by the leg, the body, and the support surface. Finally, there is the possibility that the compliant character of biological limbs can simplify the control task performed by the nervous system (Bizzi et al. 1982).

In contrast to the compliance of legs found in nature, most legged machines have legs that are designed to be stiff. Because legged machines typically walk rather than run, there may be less of a need for compliant legs. For instance, walking can be energy efficient if the legs move the body in a purely horizontal motion with no actuators absorbing energy. Several legs have been designed according to this principle (Lucas 1894, Hirose and Umetani 1980, Waldron and Kinzel 1983). Impact forces may be kept small during walking by bringing each foot into contact with the ground at low relative speed. Small leg mass and low speed reduce the cost of accelerating the swing motion of the leg.

In this paper we follow nature's lead by concentrating on legs that have elastic elements that deform during each stride. We have used such legs in machines that balance actively as they run. The purpose of the elastic element is to conserve energy associated with the bouncing motion, to reduce impact forces, and to simplify control (Raibert 1986). We restrict the discussion to legs that move in the sagittal plane, the plane determined by the gravitational vector and the heading. Motion of the foot in the plane can be obtained by various combinations of linear and rotary joints as shown in Fig. 1. We also ignore other possible uses for elastic energy storage, such as conserving energy associated with the swinging motion of the legs.

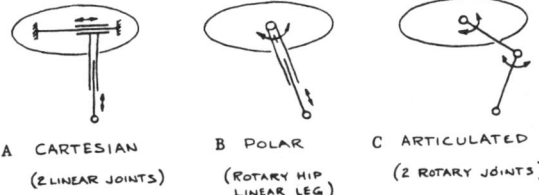

A CARTESIAN   B POLAR   C ARTICULATED

(2 LINEAR JOINTS)   (ROTARY HIP   (2 ROTARY JOINTS)
                     LINEAR LEG)

Figure 1: Three planar leg configurations: A) Cartesian: two orthogonal joints produce independent motion. The upper joint may be troublesome because it must tolerate substantial torques while providing smooth sliding motion, and the entire mass of the leg must be accelerated in the fore/aft direction to obtain horizontal foot motion. B) Polar: rotary hip and telescoping leg provide motion in polar coordinates. The inertia loads associated with the fore/aft motion of the foot are smaller than for a Cartesian leg, because only the end of the leg moves at full foot speed. Cartesian and polar designs both require linear sliding joints, which are difficult to build with precision and resistance to side load. C) Articulated: two rotary joints avoid the friction, wear, and size disadvantages of linear joints. The drawback is the kinematic coupling of the two joints—a purely vertical or radial motion requires movement of both joints. This coupling necessitates larger ranges of joint travel than the polar leg to achieve the same foot motion, and it is difficult to resolve the springiness into the vertical direction without degrading the speed and precision of control in the horizontal direction.

Before describing the specific designs we have studied, we turn to a brief discussion of energy storage in elastic materials. For discussions of other important issues in leg design see (Hirose and Umetani 1980, Vohnout et al. 1983, Waldron and Kinzel, 1983).

Mechanisms for Elastic Storage

Figure 2 shows the ratio of storable elastic energy to the mass of the material for several systems. Each system has properties that recommend it for use in leg springs, but each has drawbacks as well. Steel is an isotropic material which can easily be formed into shapes such as coils. However steel springs have a relatively poor energy to weight ratio, about 140 J/kg. Fiberglass has about six times the energy capacity of steel, but because of its directional properties is not so easily fashioned into a spring. Fiberglass is most readily used in bending as a leafspring or other beam shape.

Rubber and animal tendon have substantially higher energy capacities than steel, about 5000 J/kg. Because these materials can undergo large elastic strains they can provide usable deflections in pure tension. The 10% strain of the achilles tendon of an animal is compatible with the short lever arm to which it is attached behind the ankle joint. For example, Alexander's (1974) data for the dog indicate that its Achilles tendon acts through a 27 mm lever to move a 120 mm long link, resulting in a 1:4.4 ratio. Rubber, because it strains roughly 500%, is not equivalent to

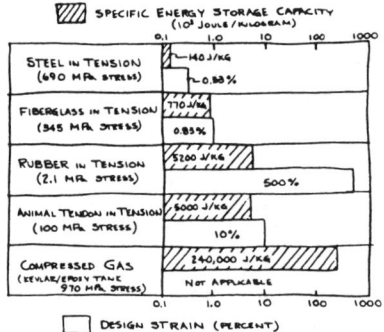

Figure 2: Strain energy per unit mass for various materials.

animal tendon in terms of its deformation, so it cannot be used directly in a leg design like the dog's. Rubber is often used in the form of torsion springs and bushings that shear tangentially when loaded. These might be usable as springs in rotary leg joints.

Gas compressed in a container with high specific strength has a very high energy capacity, 240,000 J/kg. A usable gas spring requires a cylinder and piston or comparable hardware, which increases the effective weight of the elastic material. Frictional and thermodynamic losses can be substantial. Still, gas springs may be used effectively if they are compatible with an overall design.

Telescoping Legs

After initial attempts to use steel springs in compression, we turned to compressed gas as a mechanism for elastic energy storage in legs for running machines. A standard pneumatic cylinder with low friction seals (ARO model 0418-2209-100) formed the main structure of the leg in our first successful running machine. The cylinder was mounted to the body of the machine by a simple hip joint, and a rubber bumper at the end of the piston rod served as a foot. The pneumatic circuit shown in Fig. 3 provided thrust and retraction, and controlled the springiness.

This approach to a springy leg was adequate for a series of experiments on the control of machines that balanced actively as they ran (Raibert and Brown 1984, Raibert et al. 1984). One function of the air spring used in these experiments was to recover part of the hopping energy during landing and return it during the subsequent upward acceleration. In an optimized design this could contribute to efficient locomotion. A second function was to provide a cushion for the upper leg and body. This cushion reduced the system's unsprung mass, the maximum loads produced by foot impact, and the peak forces transmitted to the sprung part of the system. A third function of the air spring was to simplify the control. The details of the vertical bouncing motion were determined largely by the passive oscillation of the body rebounding on the springy leg—the control system excited and modulated the oscillation but was not responsible for the details of the trajectory

Legs that Deform Elastically 439

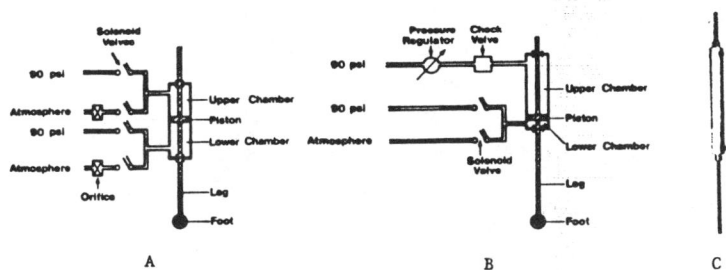

Figure 3: Pneumatic telescoping leg used in one-legged hopping machines. It consists of an air cylinder with a cushioned foot at one end of the rod. Electric solenoid valves control air flow to both chambers of the cylinder to extend and retract the leg, while trapped air makes the leg springy. Two pneumatic circuits were used: A) *Top Control*—Air enters the top chamber to provide downward thrust on the foot during the support portion of the running cycle. The thrust is controlled by adjusting the pressure of the air in the top of the cylinder when the foot touches the ground. This circuit was used in a planar one-legged hopping machine (Raibert and Brown 1984). B) *Bottom Control*—The control system adjusts thrust by regulating the pressure in the bottom chamber of the leg cylinder while a pressure regulator and check valve maintain a fixed charge of air in the top of the cylinder leg. The top chamber acts as a passive spring. To provide thrust, the control system exhausts air from the bottom chamber of the leg during support. Special quick-exhaust valves are used to dump air rapidly to permit maximum thrust. Bottom control is more efficient than top control because a smaller volume of air is exhausted on each cycle. This circuit was used in a three-dimensional one-legged hopping machine (Raibert et al. 1984). Sensors measured the length of the leg, the air pressure in both chambers of the air cylinder, the angle of the leg with respect to the body, and contact between the foot and the ground. The unsprung mass of the leg is 0.91 kg and moment of inertia about the hip is $0.11 \, kg \cdot m^2$.

(Raibert 1986).

To study running on several legs we developed a leg that could lengthen and shorten rapidly, and precisely control thrust. Rapid shortening was needed so that the recovery leg could have adequate ground clearance to swing forward while the stance leg was substantially compressed. The need to control thrust arose when coordinating the relative thrust delivered by a pair of legs that provided support at the same time.

To satisfy these requirements—rapid retraction during recovery and precise control of thrust—we designed the leg shown in Fig. 4. It has a long stroke hydraulic actuator that operates in series with a passive air spring. This design was used in a biped that runs and hops, and in a quadruped that runs with a trotting gait (Hodgins et al. 1986, Raibert et al. 1986). Although this design has been useful for experiments, it has several limitations. The air spring is relatively heavy and leakage degrades its resilience. The sliding joint is mechanically complex and bulky, and subject to wear and looseness. Measuring leg length requires a long, specially made sensor. Wires to the foot must go through slack cables that are vulnerable to a variety of hazards. These limitations have motivated us to explore legs that use only rotary joints.

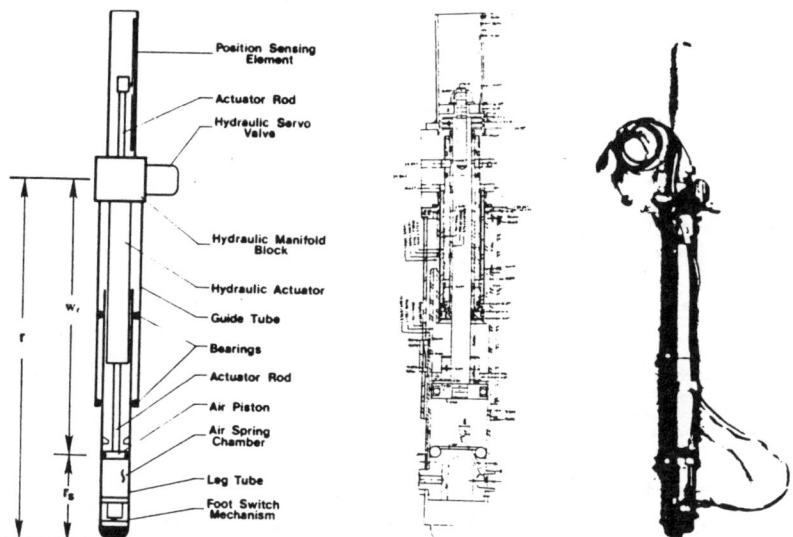

Figure 4: Hydraulic/pneumatic telescoping leg. A long stroke hydraulic actuator provides controlled axial thrust and rapid retraction. An air chamber near the foot provides the spring. To reduce friction in the hydraulic actuator, all high-pressure seals are clearance seals (.025 mm), with O-ring seals used to contain low-pressure leakage oil at the rods. Space between concentric cylinders provides paths for control and leakage flow to the lower end of the hydraulic actuator. The hydraulic actuator is servoed with a conventional high-bandwidth flow-control servo valve. The air cylinder forms the lower part of the leg and slides inside plastic guide buttons mounted in the upper leg tube. The foot includes a pneumatic check valve that allows makeup flow to the air spring, but prevents out-flow when the air spring is compressed. The hydraulic actuator has a 0.23 m travel and the air spring has a 0.01 m travel. At 17.5 MPa (2500 psi) hydraulic pressure, maximum thrust is about 950 N and maximum speed is about 2 m/s. The unsprung mass is 0.29 kg and moment of inertia about the hip is $0.14 \text{ kg} \cdot \text{m}^2$. This leg design was used in a planar biped (Hodgins et al. 1986), and in a quadruped running machine (Raibert et al. 1986).

Articulated Leg

The challenge in designing a leg with rotary joints is to include the compliance in a way that does not make the control too complicated. We are considering two candidates, shown in Fig. 5. The somewhat anthropomorphic design, shown in Fig. 5a, incorporates a rotary ankle joint connecting a rigid foot to the leg. An elastic tendon acts through a lever behind the ankle to provide the needed downward force at the toe. A suitable tendon material for such a design has not been found.

The design shown in Fig. 5b employs a noncompliant tendon and a leafspring foot. Energy can be stored in the bending of the foot if an elastic material such as fiberglass is used. This design

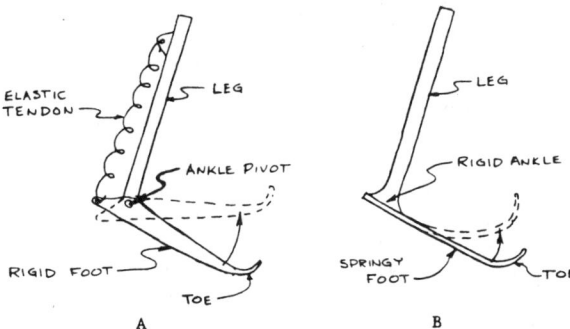

Figure 5: Articulated legs.

combines the structural and elastic functions into a single unit, minimizing mass. Because of the distributed nature of the spring, the effective unsprung mass of the leg is small. This should result in low impact forces during running.

A difficulty with both legs shown in Fig. 5 is that motion of the toe during deflection of the foot is not along the axis of the leg. Assuming the toe is rigidly fixed to the ground during stance, the ankle will move along a path that is approximately circular, centered at the toe (Fig. 6a). During vertical bouncing this causes the ankle and the lower part of the leg to move horizontally. This kinematically induced motion increases with the foot's angle α with respect to the support surface, and adds to the effective unsprung mass of the foot. Thus there is a tradeoff between a long foot that minimizes α and a short foot that minimizes mass. In principle one could compensate for this horizontal deflection by introducing an additional pair of leafsprings (Fig. 6b). Properly designed, such a mechanism could deflect with a nearly vertical motion of the toe, although it may be difficult to build.

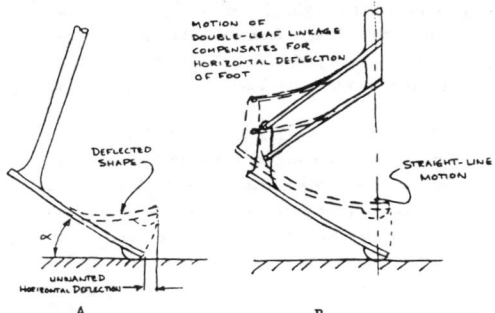

Figure 6: A) Deflection of leafspring foot introduces an undesirable horizontal motion of the toe.
B) Introduction of an additional pair of leafsprings compensates for the horizontal motion, yielding a nearly vertical motion of the toe.

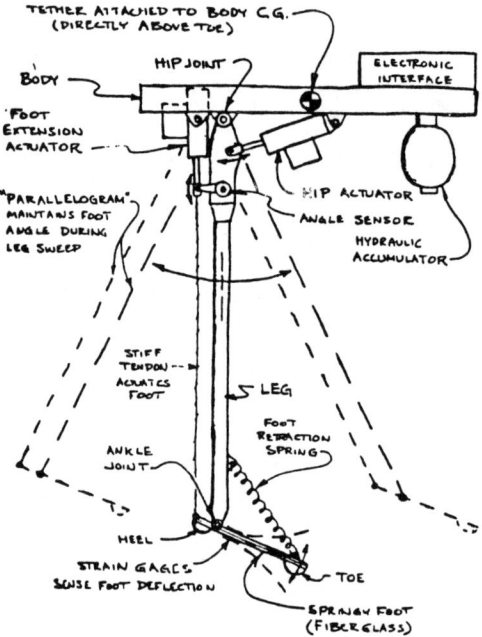

Figure 7: One-legged hopping machine using articulated leg. The foot is a leafspring that deflects during hopping. The ankle is actuated through an inelastic tendon and hydraulic actuator mounted at the hip. A retraction spring attached to the toe maintains tension in the tendon. The linkage makes the foot angle with respect to the body nearly independent of the hip angle. Potentiometers measure the two joint positions and a strain gauge measures foot deflection. The leg is intended for planar operation. The unsprung mass is 0.063 kg and moment of inertia about the hip is 0.056 kg · m².

To test the springy-foot concept we designed and built the planar one-legged hopping machine shown in Fig. 7. The leg has a rotary ankle joint so that the foot can be actuated through an inelastic tendon extending up to a linear hydraulic actuator mounted near the hip. This location minimizes the rotational inertia of the leg. The four-bar linkage formed by the leg and tendon keeps the foot angle nearly constant with respect to the body as the leg swings fore and aft. A significant shortcoming of this leg is its inability retract substantially. The ability of a leg to retract is important on a machine with several legs.

One way to get large retraction from a leg like the one shown in Fig. 7 is to add a kneelike joint. The ankle pivot and tendon could then be eliminated because vertical thrust could be obtained by a combination of knee and hip motions. However, the disadvantages of this coupling between actuators may ultimately dictate use of a third actuator at the ankle. This would be the primary thrust actuator, while the knee joint would be used mainly for gross changes in leg length during flight.

Summary

Elastic deformation can play an important role in the function of legs for machines that run. Elastic elements can store energy to improve machine efficiency, cushion impacts to reduce internal loading, and simplify control. We have built machines that run successfully with telescoping legs that use air springs for elastic storage. We are currently exploring a new leg design that incorporates a fiberglass leafspring foot and a pivot ankle driven through an inelastic tendon.

Acknowledgements

This research was supported by contract MDA903-81-C-0130 from the Engineering Applications Office of the Defense Advanced Research Projects Agency and by a grant from the System Development Foundation.

References

Alexander, R. McN. 1974. The mechanics of jumping by a dog. *J. Zoology (London)* 173:549-573.

Alexander, R. McN., Vernon, A. 1975. The mechanics of hopping by kangaroos (Macropodidas). *J. Zoology (London)* 177:265-303.

Bizzi, E., Chapple, W., Hogan, N. 1982. Mechanical properties of muscles: implications for motor control. *Trends in Neuroscience*, 5:395-398. 4:2738-2745.

Cavagna, G. A. 1970. Elastic bounce of the body. *J. Applied Physiology* 29:279-282.

Cavagna, G. A., Heglund, N. C., Taylor, C. R. 1977. Mechanical work in terrestrial locomotion: Two basic mechanisms for minimizing energy expenditure. *American J. Physiology* 233:R243-R261.

Dawson, T. J., Taylor, C. R. 1973. Energetic cost of locomotion in kangaroos. *Nature* 246:313-314.

Hirose, S., Umetani, Y. 1980. The basic motion regulation system for a quadruped walking vehicle. *ASME Conference on Mechanisms*.

Hodgins, J., Koechling, J., Raibert, M. H. 1986. Running experiments with a planar biped. *Third International Symposium on Robotics Research*, G. Giralt, M. Ghallab (eds.). Cambridge: MIT Press.

Lucas, E. 1894. Huitieme recreation—la machine a marcher. *Recreations Mathematiques* 4:198-204.

McMahon, T. A. 1984. *Muscles, Reflexes, and Locomotion*. Princeton: Princeton University Press.

Raibert, M. H. 1986. *Legged Robots That Balance* Cambridge: MIT Press.

Raibert, M. H., Chepponis, M., Brown, H. B. Jr. 1986. Running on four legs as though they were one. *IEEE J. Robotics and Automation*, 2.

Song, S. M., Vohnout, V. J., Waldron, K. H., Kinzel, G. L. 1981. Computer-aided design of a leg for an energy efficient walking machine. *Proceedings of 7th Applied Mechanisms Conference*, Kansas City, pp. VII-1-VII-7.

Vohnout, V. J., Alexander, K. S., Kinzel, G. L. 1983. The structural design of the legs for a walking vehicle. *Proceedings of 8th Applied Mechanisms Conference*, St. Louis, pp. 50-1-50-8.

Waldron, K. J., Kinzel, G. L. 1983. The relationship between actuator geometry and mechanical efficiency in robots. In *Theory and Practice of Robots and Manipulators, Proceedings of RoManSy'81*, A. Morecki, G. Bianchi, K. Kedzior (eds.). Warsaw: Polish Scientific Publishers, 305-316.

Waldron, K. J., Vohnout, V. J., Pery, A., McGhee, R. B. 1984. Configuration design of the adaptive suspension vehicle. *International J. Robotics Research* 3:37-48.

Features of Mechanisms Synthesis of Walking Robot Propelling Agents

V.V. Korenovski and A. Ja. Progrebnjak

Mechanical Engineering Research Institute
Academy of Sciences, Moscow, USSR

INTRODUCTION

Recently wide interest has been shown for the use of mechanisms as propelling agents for walking robots. It is anticipated that the use of such mechanisms will help to solve some problems of the walking propelling agents: realization of the cyclic work of the propelling agent itself when accumulating energy in the mechanism, possibility of balancing the workking propelling agent and possibility of increasing the speed of motion of the robot.

Some mechanisms for the walking propelling agents |2,3| have been proposed as ones reproducing them approximately with various degree of approximation.

There are mainly classic methods of synthesis for obtaining necessary mechanisms parameters in the works cited.

Unfortunately classic methods oriented mainly by optimization synthesis with one criterion are tightly fastened to the scheme of the mechanism. That is why the results of syntesis are compared against one criterion.

The constancy of speed or translated motion of the propelling agent of the supporting limb are not essential for the control of walking robots, except for straightforwardness. Therefore, the problem of optimization by two criteria: by position and by speed arises. If the dynamics of the mechanism is taken into account then the third criterion - by acceleration should be submitted for consideration. It follows that the full multicriteria synthesis of the mechanism will be hardly realized by classic methods.

Most probably these methods are required for determination of parameters of the mechanism in the first, rough approximation by one of the enumerated criteria. At further specifications of the parameters of the mechanism one has to use more universal numerical methods, based on multicriteria optimization of many parameters. In particular the numerical gradient methods may be such universal ones.

In this paper the results of using such approach are discussed - the first approximate values of the parameters are determined by common methods of classical synthesis and then they are specified with the help of numerical methods of optimization. The results of synthesis of mechanisms of various types both plane and spatial are given as well.

PUTTING A TASK

The four-bar mechanism, where the path of some point on a connecting rod is a straight line motion approximately along the length of trajectory at the full revolution of the crank (Fig.1), is known |1|. Angle ψ in this case may be within narrow range $43.5° < \psi < 45°$. The supporting point C moves along trajectory L touching the soil at forward and backward motions. In such case this mechanism should be equipped with an additional device for lifting the leg in transfer phase.

In seems advisable to change parameters of the mechanism in such a way that the shift of the leg will be carried out at some height. In this case when moving on an even surface the mechanism lifting the leg can be switched off and the frequency of the crank's revolution can also be increased because of small size of the leg's lifting and this increases the translational speed of the robot.

Thus, the initial requirements of the task of synthesis of a mechanism can be formulated i.e. to find the parameters of the mechanism the point of the connecting roc C which describes the trajectory with an approximately straight section. The remainder of the trajectory must all lie on one side of the straight section. In order for the crank OA to exist, it shouls be observed that the duration of passing the straight section of the trajectory should be not less than half cycle. The revolution of the driving link and the motion of the point C on the supporting section should be uniform.

METHOD OF SOLUTION

We failed to formulate mathematically the task of synthesis with such requirements, that is why we used synthesis by specifying precision points on the calculated trajectory through which the point C |4| should pass. It is known that five given positions simply determine one family of cognate mechanisms meeting the same requirements i.e. three mechanisms connected by the transformation of Roberts-Chebyshev.

For all this the resulting family of mechanisms will not necessarily meet the rest of the requirements. It is required to evaluate the mechanism's fitness as a propelling agent. It was supposed that the work of choosing parameters of the propelling agent would be realized in interactive mode when working with computer. With a view to widening the possibility of choosing the mechanism, four positions of the point C were used rather

than five. In this case it is possible to work out only eight
equations (because of the number of the coordinates of points)
for nine parameters of the mechanism. The redundant parameter
allows to get many solutions i.e. to determine geometric
places of positions of many joints and thus to get the family
of mechanisms satisfying the task and passing through four
points. Groups of mechanisms corresponding to additional requirements are chosen from this family.

The task is solved in two stages. In the first stage it is
necessary to calculate geometric places of the initial positions
of the joint A and the point C in the system of coordinates connected with the point of suspension bracket of the crank OA.
As a result of algebraic transformations of the initial system
of eight equations we get the expression for the length of the
crank

$$\cos \varphi_0 (Ar^2 + Cr \cos \varphi_0 + Er \sin \varphi_0 + G) +$$
$$+ (Br^2 + Dr \cos \varphi_0 + Fr \sin \varphi_0 + H) \sin \varphi_0 = 0$$

where A, B, C, D, E, F, G, H - coefficients depending on the length
m_1, m_2, m_3 (see Fig.2).

Therefore choosing the value of the angle φ_0 and knowning m_1,
m_2, m_3 it is possible to determine its length r, the coordinates
X_A, Y_A and X_C, Y_C. At this stage the programme automatically
choses solutions satisfying the conditions $Y_C > Y_A$, $Y_C > h+r$,
where h - height of the clearance.

At the second stage of choosing the geometric places of the
points A and D (points of the suspension bracket of the crank D
and connecting rod with the crank B) were calculated.

For determining the circular curve of points and the curve
of centres, four angles of the connecting rod's turn (corresponding to the points A_1-A_4, C_1-C_4) were defined.

As a result of the algebraic equation's transformations connecting the angles of connecting rod $\bar{d}_0,\ldots,\bar{d}_3$ and the length of the connecting rod, frame and rocker we get the equation connecting the length of the frame d with the parameters considered

$$(Ad^2 + Cdr\cos\alpha + Edr\sin\alpha + Gr^2)\cos\alpha +$$
$$+ (Bd^2 + Ddr\cos\alpha + Fdr\sin\alpha + Hr^2)\sin\alpha = 0$$

where A,B,C,D,E,F,G,H - coefficients.

Having determined d one can get geometric place for B. Changing the angle of inclination of the frame d one cam choose the required mechanism. Among natural limits it is possible to point out the necessary position of the geometric places B and D below the straight line $C_1 - C_4$ and also the observing of Grashoff condition i.e. the condition of existence of the crank OA.

The last test of the mechanism was the form of the trajectory C which was reached by turning the crank.

As a result of machine analysis the mechanism with dimensions OA = 0.49; AB = 1.3; BD = 1.0; AC = 1.59; OD = 1.49; BC = 1.015 has been chosen. The trajectory of this mechanism is at the distance 1.97 and below the point D and has the excess of length of shift over the surface in apogee $\approx$ 0.22 (see Fig. 3).

Unfortunately the supporting part od trajectory C is only a small part of the cycle, that is why there were attempts to improve the mechanism.

Further specification of parameters of the mechanism was conducted by numerical methods of optimization. Minimization of quality function was conducted by gradient method.

The quality function represented a function of mistake of reproducing a given trajectory by two criteria: position and speed.

$$F = \sum_{i=1}^{n} [\bar{Y}-Y(x_1,x_2,x_3,x_4)]^2 + \sum_{i=1}^{n} |\bar{X}-X(x_1,x_2,x_3,x_4)|^2$$

where x_1, x_2, x_3, x_4 - variable parameters a,b,c,f accordingly

$\bar{Y}$ - design coordinate, constant on the supporting length

$\bar{X}$ - design coordinate, changable with the constant speed on the supporting length.

The design coordinate values for the step were calculated by the known ratios

$$\alpha_i = \frac{-\sum_{i=1}^{4} \left(\frac{\delta F}{\delta x_i}\right)^2}{\sum_{i=1}^{4} \sum_{j=1}^{4} \frac{\delta F}{\delta x_i} \frac{\delta F}{\delta x_j} \frac{\delta^2 F}{\delta x_i \delta x_j}}$$

where $\delta F/\delta x_i$, $\delta^2 F/\delta x_i^2$, $\delta^2 F/\delta x_i \delta x_j$ - quotient derivatives, changed for the finite different ratios

$$\frac{\delta F}{\delta x_i} = \frac{F(x_1,\ldots,x_i+\Delta x_i,\ldots,x_4)-F(x_1,\ldots x_i-\Delta x_i,\ldots x_4)}{2\Delta x_i}$$

$$\frac{\delta^2 F}{\delta x_i^2} = \frac{F(x_1,\ldots x_i+\Delta x_i,\ldots x_4)-2F(x_1,\ldots x_i,\ldots x_4)+F(x_1,\ldots x_i-\Delta x_i\ldots x_4)}{\Delta x_i^2}$$

$$\frac{\delta^2 F}{\delta x_i \delta x_j} = \frac{F(x_1,\ldots x_i+\Delta x_i, x_j+\Delta x_j,\ldots x_4)-F(x_1,\ldots x_i-\Delta x_i, x_j+\Delta x_j,\ldots x_4)}{4\Delta x_i \Delta x_j} -$$

$$- \frac{F(x_1,\ldots x_i+\Delta x_i, x_j-\Delta x_j,\ldots x_4)-F(x_1,\ldots x_i-\Delta x_j, x_j-\Delta x_j, x_4)}{4\Delta x_i, \Delta x_j}$$

MAIN RESULTS

The work of specifying parameters of the mechanism war carried ozt in two stages: 1) the accuracy of reproducting the suuporting part of trajectory was improved and its duration was increased; 2) the task of approximation of the real trajectory to the calculated one for the whole turn of the crank was solved.

At the first stage the supporting trajectory was tested in twelve points and the following values of parameters have been obtained: a = 0.9932557; b = 1.2424399; c = 1.6564047; f = = 0.6968343 radian.

As a results of optimization the supporting part of the trajectory increased up to 1.2 mean quadratic deviation from the straight line decreased. Selfcrossing of the supporting part of the trajectory by the transfer phase path was a negative factor. Therefore at given parameters the mechanism can work only in the swing phase.

At the second stage there were attempts to approximate the real trajectory of the supporting point of the mechanism to the calculated one. The form of the calculated trajectory and the results of approximation are shown in Fig.4.

For choosing parameters the quality function with introduced various weight coefficients for the supporting phase and the phase of transfer was used.

$$F = \sum_{i=1}^{12} [\bar{Y}-Y(x_1,x_2,x_3,x_4)]^2 + \sum_{i=1}^{12} [\bar{X}-X(x_1,x_2,x_3,x_4)]^2 + \sum_{i=13}^{23} \{0.5[\bar{Y}-Y(x_1,x_2,x_3,x_4)]\}^2 + \sum_{i=13}^{23} \{0.1[\bar{X}-X(x_1,x_2,x_3,x_4)]\}^2$$

The introduction of the weight coefficients allows to choose the main lengths of trajectory for best approximation to them. Difference of the weight coefficients for X and Y in the transfer phase is explained by defferent requirements of the shape of the transfer path and the speed on it.

As a result of analysis on the computer the values of parameters meeting constraints imposed on the trajectory have been obtained a = 0.9592717; b = 1.2875596; c = 1.6217215; f = 0.6868657 rad, at $r = 0.49$; $d = 1.49$; $h_m = 1.926$ which did not change in the process of synthesis. Maximum exceeding of the lenght of transfer under the lenght of support was $\Delta h_{max} = 0.12$.

Analysis of the results of synthesis shows that among four bar mechanisms we can find one in which the supporting point will circumscribe the trajectory where the supporting length of it will be satisfactory, but the accuracy of following the trajectory along the axis Y here is not high.

The mechanism in the regime of rocking the driving link OA follows the straight line trajectory more accurately. Other mechanisms of rocking movement which can serve as propelling agents have been considered. The quality function also was worked out in two criteria: by position and constancy of speed on the supporting phase. The lever mechanism is shown in Fig.5 where the deviation of links B and C from the vertical connected by the ratio E. The parameters A,B,E at $D = 0.5$, $C = 1$ and the uniform rotation of the driving link B were determined. With the parameters $A = 0.8217004$; $B = 0.1790218$; $E = 0.4087071$; $\alpha_{max} = 50°$ relative accuracy of following the supporting trajectory x-x is $o = 0.17$ %.

In the spatial Watt mechanism it is necessary to carry out approximate spatial transfer of the supporting point E on the plane P along the line of parallel plane of the body of the walking robot with the constant speed at the uniform rocking of the driving link AD round the axis Z (Fig.6).

The quality function represented three criteria dependence on the position (Y,Z) and the speed (X(t))

$$F = \sum_{i=1}^{n} [\bar{Z}-Z(x_1,\ldots,x_4)]^2 + \sum_{i=1}^{n} [\bar{Y}-Y(x_1,\ldots x_4)]^2 + \sum_{i=1}^{n} [\bar{X}-X(x_1,\ldots x_4)]^2$$

The values of parameters were determined at which the obtained trajectory of the mechanism has been more close to the given one

$a = 0.5985 \quad B^* = 0.1842 \quad \varphi = \pm 30° \quad z_{min} = 0.6$

$b = 0.4825 \quad H = 0.5829 \quad \Delta H = 0.6 \quad z_{max} = 1.2$

$c = 0.6 \quad y_* = 0.5 \quad d = 1.0$

The results of analysis of the trajectory of the mechanism are given in Fig.7.

CONCLUSION

The interactive method of synthesis of the mechanisms when making special requirements of them has been considered. It is shown that the formulated problem can be solved by choosing simple mechanisms obtaining the highest accuracy of following the given trajectory on the basis of multicriteria choosing.

REFERENCES

1. Артоболевский И.И. и др. Научное наследие П.Л. Чебышева. Из-во АН СССР, М.-Л., 1945.
2. K.J. Waldron et al. Computer-aided Design of Leg for an Energy Efficient Walking Machine. Mechanism and Machine Theory. Vol. 19, N 1, 1984.
3. A.D. Ryan et al. Adjustable Straight-line Linkages Possible Legged-vehicle Applications. Transactions of the ASME. Vol. 107, June 1985.
4. Юкало П.А. Аналитический синтез плоского четырехзвенного механизма. Машиностроение, М., 1981.

Walking Robot Propelling Agents 453

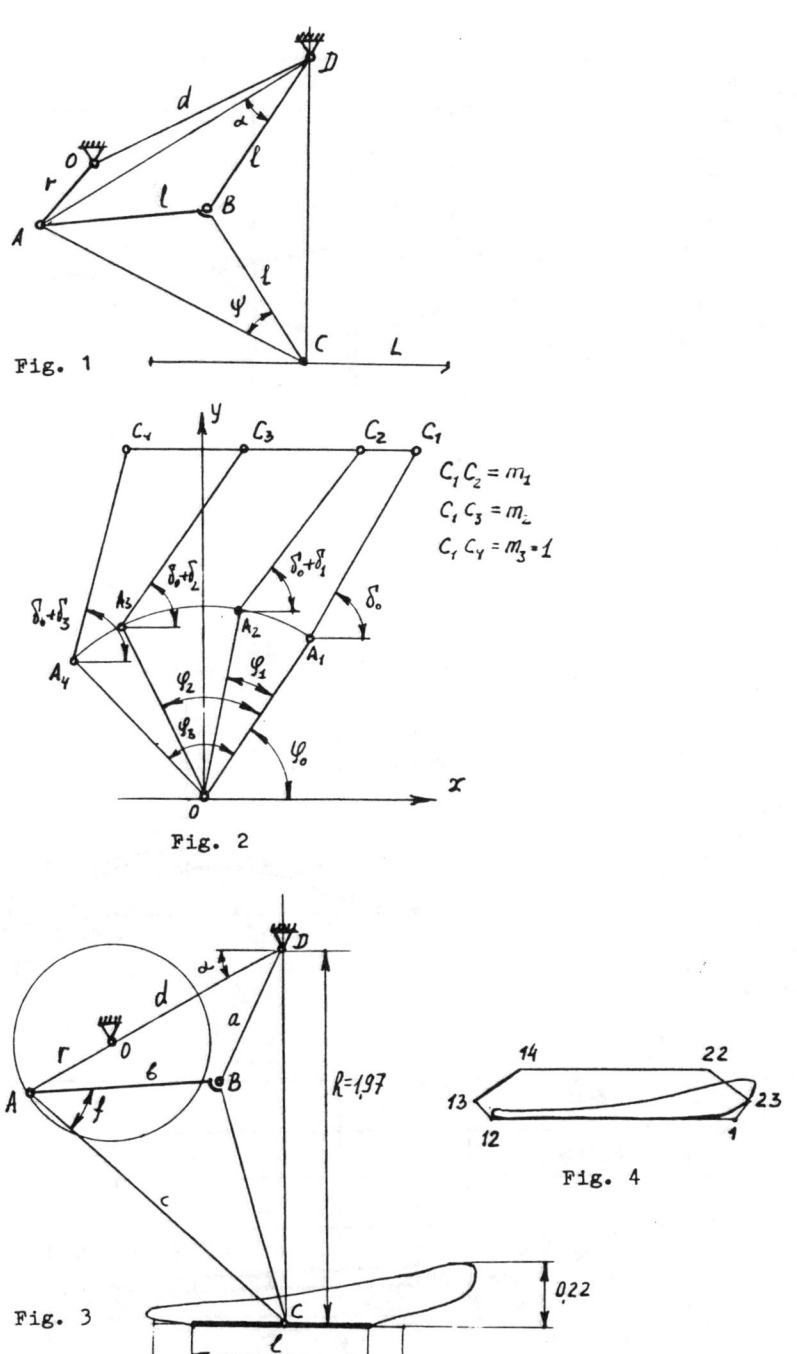

Fig. 1

Fig. 2

$C_1C_2 = m_1$
$C_1C_3 = m_2$
$C_1C_4 = m_3 = 1$

Fig. 3

Fig. 4

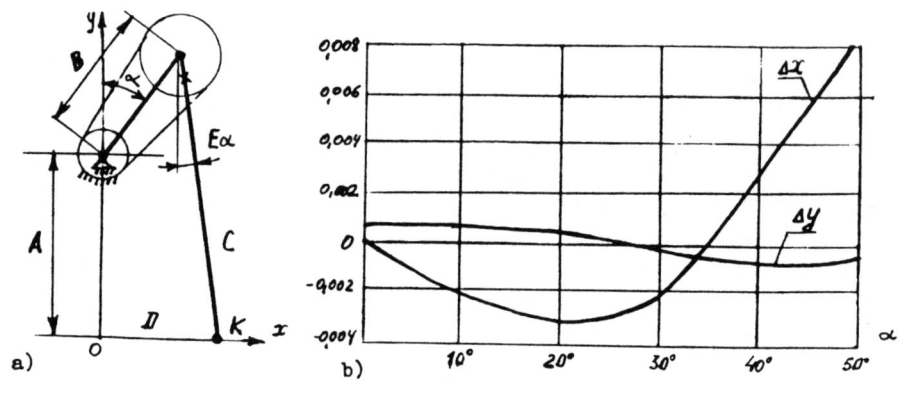

Fig. 5

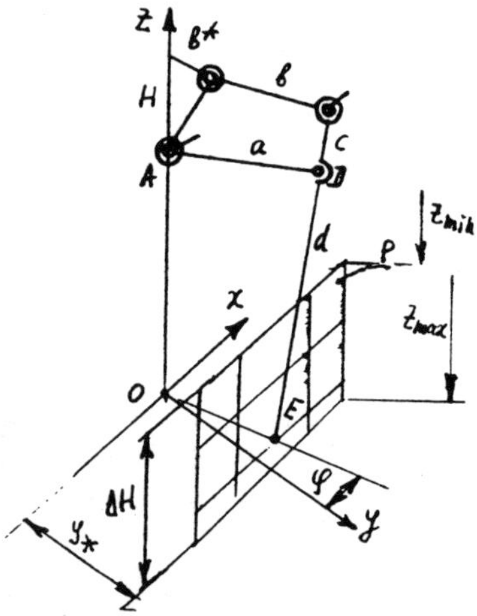

Fig. 6

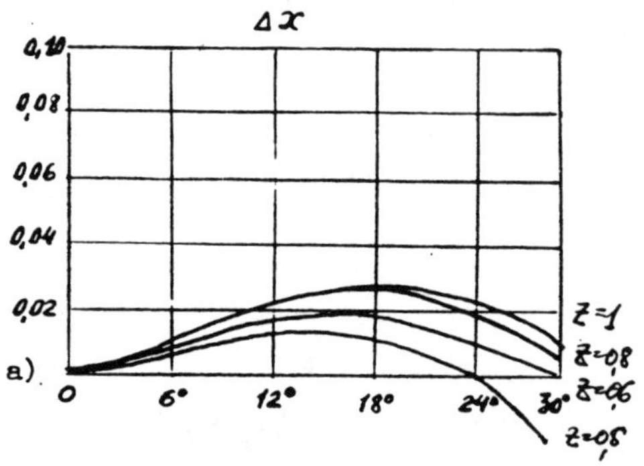

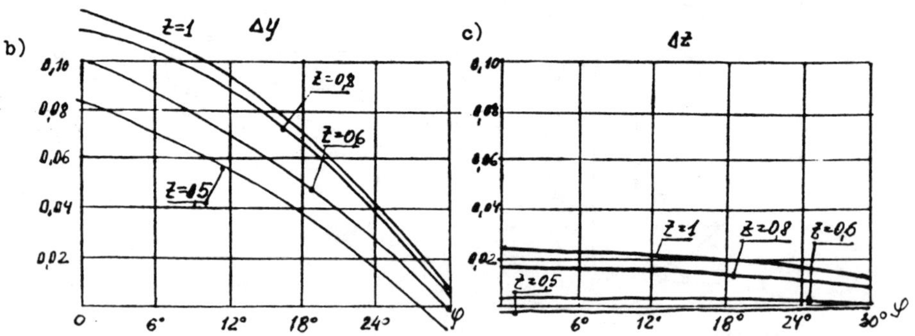

Fig. 7

Avoiding Obstacles by a Mobile Robotized Vehicle

M. Badida and J. Buda

Technical University, Kosice, Czechoslovakia

SUMMARY

The progress achieved in the FMS design (especially in the field of functional and spatial integration) leads to the drawing up of heteroarchical manufacturing systems' structures. Up-to-date design of such structures is considerably connected with conventional views on workshop layout. The new views on spatial layout will appear in near future. According to the possibilities of computational techniques certain explosion in the spatial structure anticipations will follow. Simulation technique will enable a fast comparison of more variants and their evaluation. Simulation of avoiding obstacles forms an important aspect of the development of anticipation. The authors present a strategy of avoiding obstacles for mobile robotized vehicle (MRV). This strategy is based on the points of intersection and angles. Selected simulation examples for concrete situations are included.

1. INTRODUCTION

The strategies of solving a given problem arise from geometrical interpretation of the description of the obstacle in the two-dimensional Euclidian space, i.e. on the ground-plane. The obstacles are characterized by polygons the tops of which are given.

For simplification of solving the task with graphical depiction of environment and its simulation the MS is substituted by a material point. It is supposed that especially on the basis of these parameters arises "enlarging" of the obstacle, respectively the obstacles will be enlarged by a value (the so called transformational parameter) which can be expressed as follows

$$\mu = 1.5\vartheta + \psi \qquad (1)$$

where: ϑ - is MRV diameter, the shape of which is circular or the width of MS the shape of which is oblong
ψ - the value eligible by the operator
1.5 - constant determined experimentally.

The statement of the path between two points in such environment is reduced to the search for a path on the graph created by the vertice of the "enlarged" obstacles (Fig. 1). Further, suppose that $B_v^{(o)}[\xi_v^{(o)}, \eta_v^{(o)}]$ is a point of MRV's path in which the obstacle situated on the reference path will be identified. The fact whether the obstacle will be identified in time depends on the perfection of applied technical device. At the given point of the path MRV will stop and it will start to create the strategy of avoiding the obstacle. The point $B_v^{(o)}$ is considered to be the starting-point for further solving.

The task of MRV is to realize the procedure from the initial point $B_v^{(o)}[\xi_v^{(o)}, \eta_v^{(o)}]$ to the final point $B_c[\xi_c, \eta_c]$. There are several possible paths from $B_v^{(o)}$ to B_c. The shortest of them is abscissa (reference path) connecting the two considered points. By the analysis of the obstacle and reference paths mutual position the four basic mutual positions can be stated (Fig.1):

1. Reference path and the obstacle have no moint in common.
2. There is just one common point of reference path and the obstacle (enlarged by the transformational parameter μ).
3. Reference path and obstacle (enlarged by the transformational parameter μ) have just one edge in common.
4. The obstacle lies on the reference path.

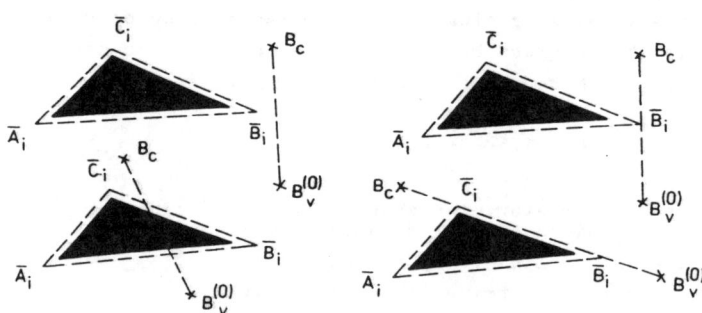

Fig. 1 The possibilities of mutual positions of the obstacle and reference path

The above mentioned possibilities of the mutual positions are exhausting for all the obstacles of arbitrary or n-angular shape.

The subject of our further solving is the fourth case in which it is necessary to realize obstacle avoiding, resp. avoiding of the varieties of obstacles.

2. CHARACTERISTICS OF INTERSECTIONAL-ANGULAR STRATEGY

Intersectional-angular strategy of avoiding obstacles has been proposed for the case of avoiding one n-angular obstacle variety of n-angular obstacles respectively.

Suppose that there is an n-angular obstacle on the reference path between the initial and final point (Fig. 2). The successful avoiding is possible supposing the following strategy will be applied.

Lead the straight line across the initial point $B_v^{(o)}$ and final point B_c. The straight line can be expressed analytically:

$$p_1^{(o)}: \quad y = k_{B_v^{(o)} B_c} x + q_{B_v^{(o)} B_c} \qquad (2)$$

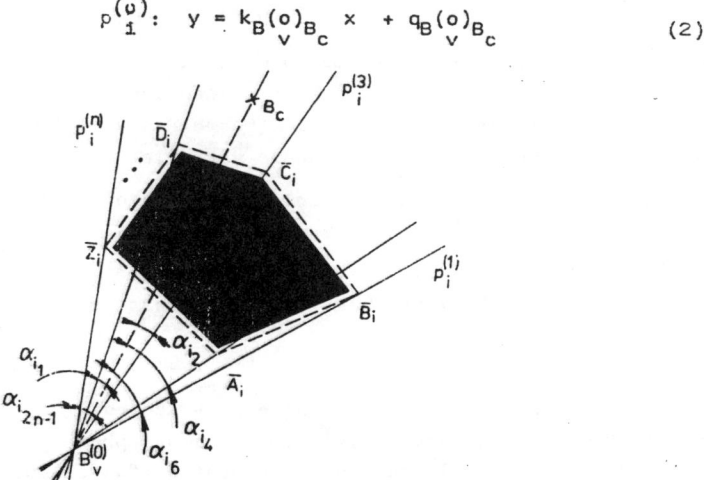

Fig. 2 Intersectional-angular strategy

where: k - straight line $p^{(o)}$ direction
   g - the section limited by the straight line on the coordinate y

An obstacle in the work plane is defined by its vertices. On the basis of this assumption it is possible to draw lines $p_z^1 \ldots p_z^n$ through the initial point B_v^o and the vertices $A_z \ldots Z_z$ and to determine the set of gradients for individual lines. Determination of angles formed by the line p_z^o and the set of lines p_z^1, $p_z^2 \ldots p_z^n$ presents the solution of the strategy for this particular task.

The angle between two straight lines, i.e. the straight line $p_i^{(o)}$ with one of the straight lines $\{p_i^{(1)}, p_i^{(2)}, p_i^{(3)}, \ldots, p_i^{(n)}\}$ can be determined from the direction of both straight lines.

Since the preceding supposition is fulfilled, the angles can be determined from the relations:

$$p_i^{(o)} \wedge p_i^{(4)} \Longrightarrow \alpha_{i_1} = \text{arctg} \frac{k_B(o)_{v B_C} - k_B(o)_{v \overline{D}_i}}{1 + k_B(o)_v k_B(o)_{v \overline{D}_i}} \qquad (3)$$

In an analogical way it is possible to determine all the other angles. These are considered to be the so called angles of "all the possibilities" of path's segments in the strategy step.

The examined angles create two basic varieties of angles:

a) variety of angles of "all the possible" upper path's segments

$$\{\alpha_{i_{2n-1}}\}_{i=1}^{\infty}$$

b) variety of angles of "all the possible" lower path's segments

$$\{\alpha_{i_{2n}}\}_{i=1}^{\infty}$$

In the following step of solving one angle is selected from both upper and lower MRV path's segments. In the case of upper path's segment selection the follwing condition must be fulfilled:

$$\alpha_{HDS} = \max \{\alpha_{i_1}, \alpha_{i_3}, \alpha_{i_5}, \ldots, \alpha_{i_{2n-1}}\} \qquad (4)$$

In the case of lower path's segment selection the fulfilment of analogical condition is required:

$$\alpha_{DDS} = \max\{|\alpha_{i_2}|, |\alpha_{i_4}|, |\alpha_{i_6}|, \ldots, |\alpha_{i_{2n}}|\} \quad (5)$$

By the above given procedure the angles of two paths segments are gained:

$$\{\alpha_{i_{HDS}}, \alpha_{i_{DDS}}\} \quad (6)$$

The optimizational criterion will decide which of the paths segments will be used for further processing

$$\alpha_{i_c} = \min\{\alpha_{i_{HDS}}, \alpha_{i_{DDS}}\} \quad (7)$$

where: α_{i_c} — angle of the path
$\alpha_{i_{HDS}}$ — angle of the upper path's segment
$\alpha_{i_{DDS}}$ — angle of lower path's segment

By the indicated path the new initial point $B_o^{(1)}$ will be reached: it is possible to lead a new straight line $p_i^{(o)}$ across it. In this case, too, it is inevitable to examine the mutual position of the straight line $p_i^{(o)}$ and remaining obstacle's edges. The procedure is analogical to the one used in preceding strategy.

The described strategy of avoiding n-angular obstacle can be successfully applied in the cases of more obstacles.

After leading the straight line $p_i^{(o)}$ across the initial point B_c it is essential to determine which of the obstacles is to be avoided as the first. Simultaneously with the determination of the obstacle's edge situated at the shortest distance from the initial point $B_v^{(o)}$ [$\xi_v^{(o)}, n_v^{(o)}$] the nearest obstacle is determined, i.e. this one which is to be avoided as the first. The application of intersectional-angular avoiding strategy causes the avoiding of n-angular obstacle and at the same time the determination of a new initial point $B_v^{(1)}$ [$\xi_v^{(1)}, n_v^{(1)}$]. From the point $B_v^{(1)}$ the above mentioned steps will be applied until the direct movement of MRV is possible; directly from the i-th initial point to the final point.

3. SIMULATION'S RESULTS COMPARISON

The strategy of avoiding a definite number of arbitrary n-angular shaped obstacles by means of MRV were verified on the personal computer OLIVETTI M 20. The mentioned computer has a 16-bit processor and the memory capacity of 156 KB. The simulaton programmes were written in BASIC. The simulation program was adjusted so that it enabled an interactive statement of parameters like: the number of the obstacles, starting-point, final point and a transformational parameter.
Simulation verification of the designed strategy of avoiding obstacles demonstrated its advantages for the proposed application. When applying the intersection - angular strategy, MRV could set the following trajectory in all simulated cases. The selected examples of simulation are given in the Fig. 3 and 4.

The optimal (the shortest) avoiding path is usually achieved by the application of intersectional-angular strategy.

By extending the mentioned strategy by the artificial intelligence methods and especially by the method of fuzzy sets for approximate defining of environmental carriage's characteristics, it is possible to achieve further improvement of the method. Nevertheless, it would require further research activity.

REFERENCES:

1. Buda, J. et al.: Artificial Intelligence in Machinery Manufacturing Processes. UTRIN/INPRO Praha, 1981

2. Buda, J., Badida, M.: Artificial Intelligence in Machinery Manufacturing Processes. ES VŠT, Košice, 1984

3. Buda, J., Badida, M.: Simulation of Mobile Systems for Avoiding Obstacles. In.: Proceedings 3rd International Conference on Automated Guided Vehicle Systems. Published by IFS (Publications) Ltd, UK, Stockholm, Sweden, 1985

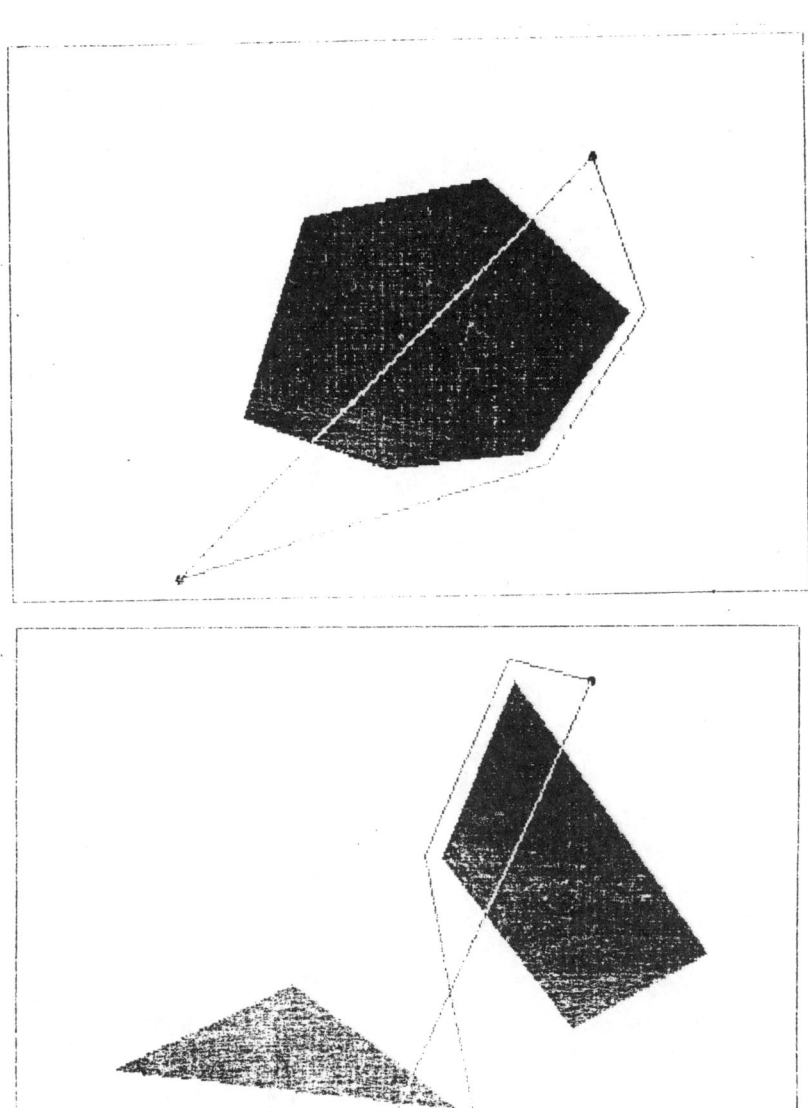

Fig. 3 The aplication of intersectional-angular strategy -

Fig. 4 The aplication of intersectional-angular strategy

Part 9
Application and Performance Evaluation

The Automation of the Mine Support Erection Technology with Remotely-Controlled Manipulators

Yu. A. Tzeitlin, V. Ya. Potyomkin and L.N. Prokopishin

Institute of Geotechnical Mechanics of the Ukrainian Academy of Sciences, USSR

Synopsis.
The system discussed in the context of our paper pertains to the mine support erection technology in the underground development workings as a subject for robotization. The proposed software and hardware arrangement ensures the automation of the mine support erection performed with the remotely-controlled manipulators.

Development working is one of the most labour-consuming processes in the underground mining of mineral deposits. It involves rock crushing, spoil removal and transportation and subsequent provision of mine supports ensuring an adequate cross-section of the heading as well as its stability. The mine support erection process is nowadays the least mechanized and the most laborious procedures which occupies about 30-35% of the total face time.

The automation of the mine support erection due to the use of remotely-controlled manipulators is now a viable proposition for better productivity and working safety for miners.

Fig. 1 shows one of the most promising technological flowsheet for mine development working with a selective boom-ripper. According to this flowsheet, the supports are being erected in sequence as the drifting advances. Each support represents a set of sections assembled of steel arches interconnected with intermediate steel strips and provided with a wire mesh on the top and either side. Each section is built up of 35 components, the major ones are shown in Fig.2.

A support erection procedure consists in supplying the components to the spot of installation along the monorail track erected inside the heading and subsequent assembling the above components into a single whole. Section assembling may be performed with a set of remotely-controlled manipulators fitted for the job and possessing appropriate load-carrying capacity and maneurability and provided with alignment control systems to adapt the assembling procedure to a variety of components and environmental parameters.

The main peculiarities of the assembling procedure under specific mine development conditions are as follows: constraint and variable working environment, a wide diversity in weight and overall dimensions of the components to be asaembled, probable deviations in the parameters of the components under assemblage from the standard ones, unavailability of a stable robust ground the components might be assembled on.

With the above in view, the mine support assembling procedure involves the following steps:

1. Instrumental or visual assessment of a free space available for the manipulation.

2. Comparison of the available working area with the actually required one and adjusting the first, accordingly, if necessary.

3. Preparation of the erection site on the working ground for the first unit of the section to be assembled.

4. Erection of a left-hand prop and fixing it in an appropriate position as per local coordinate system and with due regard to the location of the preceding support section.

5. Fixing the left-hand intermediate steel strip and assembling the joints 2 and 3 (fig. 2).

6-7. The above steps are repeated for the right-hand prop.

8. Arch capping mounting and fixing.

9. Assembling the joints 4 and 5.

10. Mounting the mesh 6.

Fig. 3 illustrates the software and hardware arrangement developed for the realization of the above described support assembling procedure with a set of remotely-controlled manipulators. Due to the implementation of a special algorithm for assemblage and also due to some structural features of the manipulators, the above system makes it possible to accomplish the support erection and assembling with visual or remote control only, excluding the necessity in provision of the informational set of environmental sensors.

To reduce the vibration set up by transporting the heavy components (e.g. props and arch cappings), their handling from the magazine to the erection site is performed with a couple of manipulators. Prop and arch capping joining is accomplished with a specially-designed arrangement which is retracted back into the magazine on completion of the section assembling.

The software and hardware implementation in the support section assembling procedure performed with conventional cyclic manipulators poses not only the kinematic and dynamic problems pertaining to the integrated mechanisms with multiple degrees of freedom which have been discussed in details in the papers (1,2), but also some specific mecanical problems stemming from the automation of the assembling procedure in a poorly organized environment. The above problems arise from the necessity to take into account the probable deviations in the locations and forms of the conventionally movable or fixed components under assemblage from those of the standard ones, in view of the fact that in the process of section assembling the above deviations may be considerable.

Under such circumstances, the considerations presented in the paper (3) may be successfully used for the elaboration of the effective algorithm for the assemblage.

Using a mechanical model for the component assembling procedure in the mine development environment, the condition for the software and hardware implementation in assembling may be represented by the following inequality:

$$\left(\frac{\partial H}{\partial q^{(1)}}\right)_{q=q(\kappa)} = \left[\frac{\partial U}{\partial q^{(1)}} - \frac{\partial U}{\partial q}\left(\frac{\partial F}{\partial q}\right)^T \psi \frac{\partial F}{\partial q^{(1)}}\right]_{q=q(\kappa)} > 0$$

where:
$$\psi = \left[\frac{\partial F}{\partial q^{(1)}} \left(\frac{\partial F}{\partial \dot{q}}\right)^T\right]^{-1},$$

H is Lagrangian function used for the minimization of the strain energy $U[q(k), S(k)]$ with due regards to the constraints imposed by the environmental conditions $F[q(k)]$;

$q = q(k)$ is the function of the integral argument "k" which characterizes the interrelationship of the components; the sequence $q_i(k)$, $k = 1, 2, ..., n$ is defined as matrix columns;

$q^{(1)}$ is a part of the matrix column $q(k)$ which includes the main coordinates significant for the given assembling procedure proper;

$S(k)$ is the matrix column of the integral argument "k" which characterizes the location of the manipulator grip pole in the local coordinate system.

In order to use the above unequality in a mathematical model for the software and hardware implementation for the support section assembling procedure, the component rigidity values should be determined under field conditions as well as the models of the manipulators' strain properties should be concretized. It is considered expedient to establish the same by the site experiments under actual geological conditions of the mine development.

The mathematical simulation results of the support assembling procedure give the possibility to conclude that in spite of considerable difficulties encountered in the software and hardware implementation under mine development conditions, the above technique may be successfully realized with conventional means excluding the necessity in provision of sophisticated metering systems for monitoring the environmental conditions.

References.

1. Popov E.P., Vereschagin A.F., Zenkevich S.L. "Robots-manipulators". Dynamics and algorithms". Moscow: "Nauka"Publishers, 1978, 398 pp.

2. "Dynamics in robot control", (E.I.Yurevitch, ed.), Moscow: "Nauka"Publishers, 1984, 334 pp.

3. "Robotization of assembling procedures", Okhotsimsky D.E., chiefed. Moscow, "Nauka"Publishers, 1985, 254 pp.

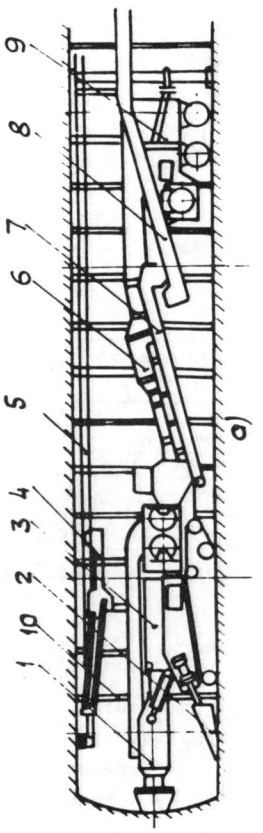

Fig. 1. Basic flowsheet of mine development working with a selective boom-ripper:
1 - the boom of the ripper; 2 - mucking pan; 3 - selective boom-ripper;
4 - arch support manipulators; 5 - monorail; 6 - local exhaust ventilation pipe; 7 - adjoined mucker; 8 - ore gantry; 9 - auxilliary equipment;
10 - steel arches.

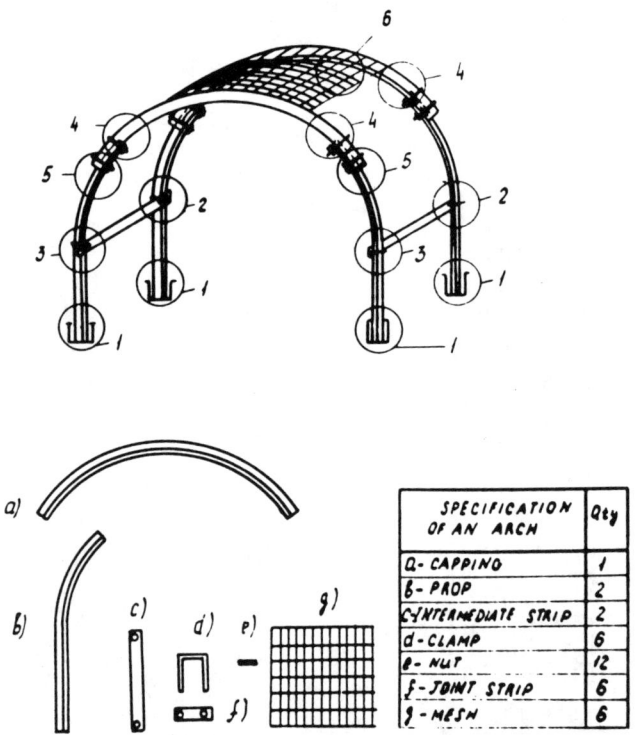

Fig. 2. Step-wise assembling procedure of the components:

1. Erection of the prop and fixing the bottom joint.
2. Fixing the rear end of the intermediate strip.
3. Fixing the front end of the intermediate strip.
4. Assembling the upper joint of the arch capping and the prop.
5. Assembling The lower joint of the arch capping and the prop.
6. Mounting the mesh 6.

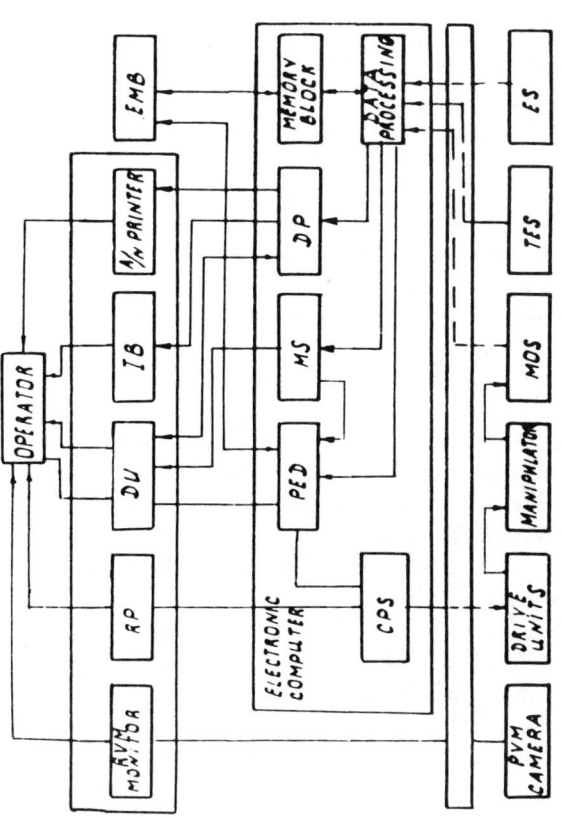

Fig. 3. Block diagram of the manipulator control:

PVM – process visual monitor; RP – reference program; DU – display unit; IB – information board; A/N printer – alphabetical-numerical printer; EMB – environment memory block; PED – program execution display; MS – mathematical model; DP – data presentation; CPS – control pulse shaping; MOS – sensors of manipulator operation; TES – technological equipment sensors; ES – environment sensors.

Experimental Investigations of Robots and Manipulators

B. Heimann and S. Tschakarow

Institute for Mechanics, Berlin

Summary: In the present paper it has been tried to give a systematic description of most important problems in the field of experimental investigations of robots and manipulators. The main topic is devoted to development of foundations for computer aided measurement.

1. Introduction

It is well-known that in dependence on main tasks of robots it is necessary to measure special characteristic geometric, kinematic and dynamic parameters describing
a) the accuracy of positioning of a rigid body with respect to a prescribed spatial position
b) the spatial position of robot links with respect to an inertial frame
c) the forces and torques acting in special parts of robots, reactions in joints, stresses in segments
d) the vibrational behaviour (eigenfrequencies, eigenmodes, response curves, damping parameters).

Taking into account that from the point of modelling the robot can be described as a system
- with variable structure
- whose equations of motion are nonlinear
- with closed-loop characteristics.

Because of high accuracy of needed on-line measurement computer-aided testing is necessary on the base of suitable mathematical models.

2. Computer-aided testing in the field of robotics

Applying computer-aided testing method it is useful to divide

the investigations into
- measurement task (strategy and means for receiving data)
- data processing (on-line processing in order to estimate characteristic properties of robots).

The block diagram containing the hardware equipment necessary for experiment automization is given in fig. 1.
Some additional blocks can simply be built into this diagram in order to calculate other characteristics we are interested in, for example software blocks for estimation of statistic values, dynamic characteristics, response curves.

3. Specific experimental investigations

3.1. Measurements of geometric perameters

Let us consider a manipulator consisting of a series of n links together by joints. Kinematic parameters which specify such a mechanism are illustrated in fig. 2.
Using homogeneous transformations the relations between coordinate frames are as follows:

$$(KS)_i = (KS)_o \cdot \underline{T}_i$$

where

$$\underline{T}_i = \left[\begin{array}{c|c} R & \begin{array}{c} r_x \\ r_y \\ r_z \end{array} \\ \hline 000 & 1 \end{array} \right]_i \quad (1)$$

$(KS)_o$ — inertial frame, $(KS)_i$ — body-fixed frame of i-th link, $\vec{r}_i = (r_x, r_y, r_z)_i^T$.
The rotation transformation has the form ($c_\theta := \cos\theta$, $s_\theta := \sin\theta$)

$$\underline{R}_i = \begin{bmatrix} c_\theta c_\psi & c_\theta s_\psi s_\phi - s_\theta c_\phi & c_\theta s_\psi c_\phi + s_\theta s_\phi \\ s_\theta c_\psi & s_\theta s_\psi s_\phi + c_\theta c_\phi & s_\theta s_\psi c_\phi - c_\theta s_\phi \\ -s_\psi & c_\psi s_\theta & c_\psi c_\phi \end{bmatrix}_i \quad (2)$$

From the point of experimental investigations the general task consists in measuring the values

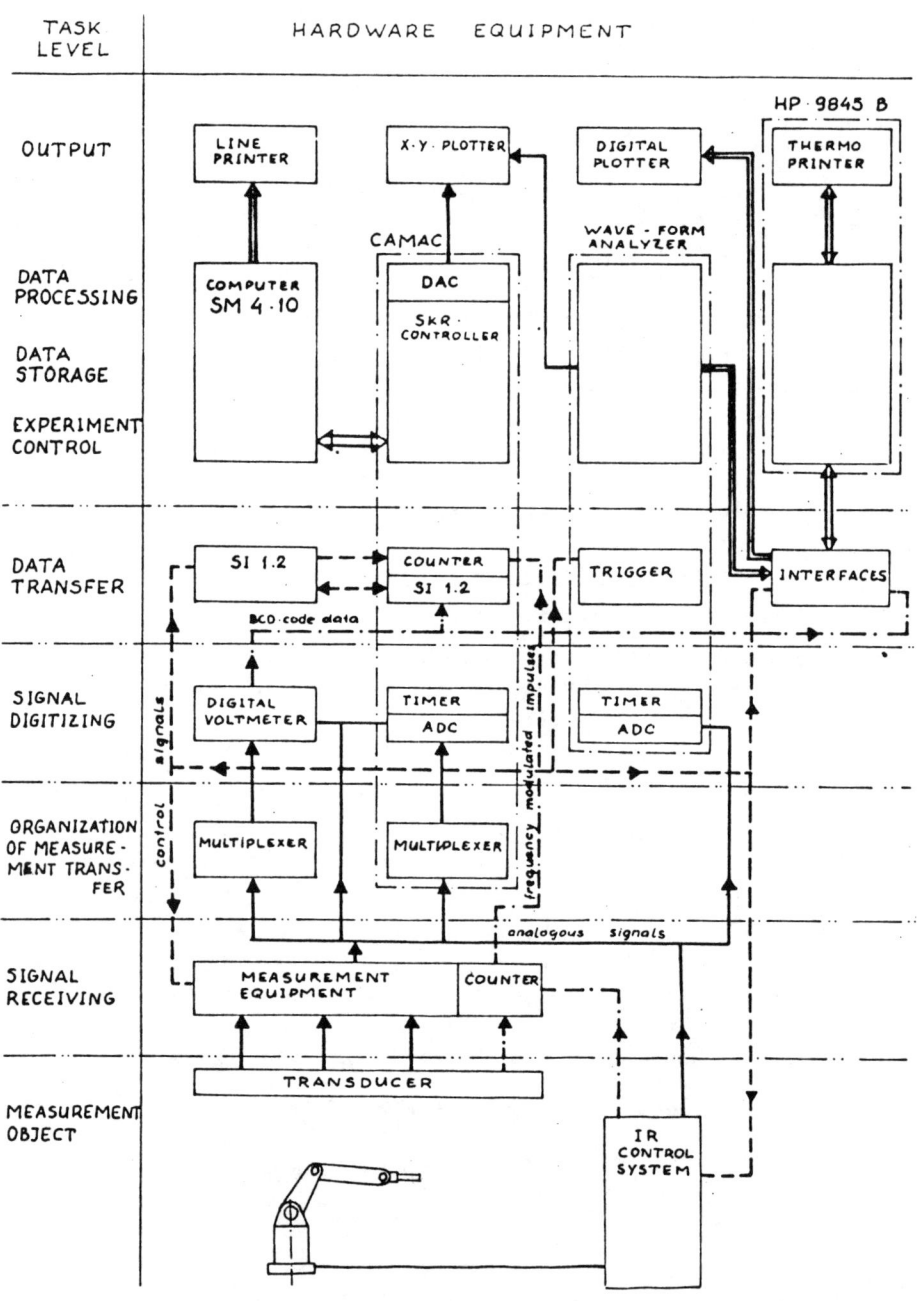

Fig.1: Hardware equipment

$$(r_x, r_y, r_z, \phi, \psi, \theta)_i$$

in order to estimate the position and orientation of robot links with respect to an inertial frame.

We regard the more simple case of measuring the derivation of i-th link with respect to a prescribed position (for instance, end position). Furthermore, assuming that the derivations are small we get from equ. (2)

$$\Delta \underline{T}_i = \begin{bmatrix} 1 & -\Delta\theta & \Delta\psi & | & \Delta r_x \\ \Delta\theta & 1 & -\Delta\phi & | & \Delta r_y \\ -\Delta\psi & \Delta\phi & 1 & | & \Delta r_z \\ \hline 0 & 0 & 0 & | & 1 \end{bmatrix}_i \quad (3)$$

The following remarks and results are based on these well-known geometric foundations (1) - (3).

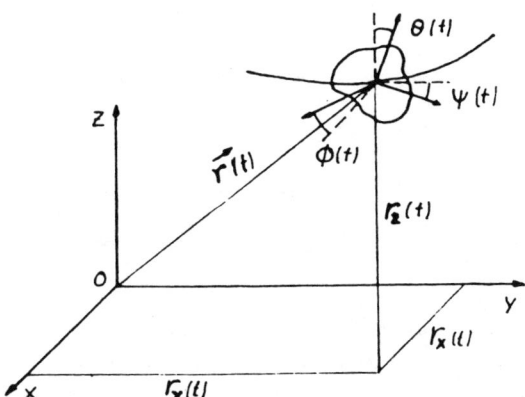

Fig.2: Coordinate frames

3.1.1. Measurement of derivations

If we restrict ourselves to measurements which represent only the derivation of a prescribed robot position, then we can express the derivation by using small translation $\overrightarrow{\Delta r} =$

$(\Delta r_x, \Delta r_y, \Delta r_z)^T$ and small rotation $(\Delta \phi, \Delta \psi, \Delta \theta)^T$, see fig.3.

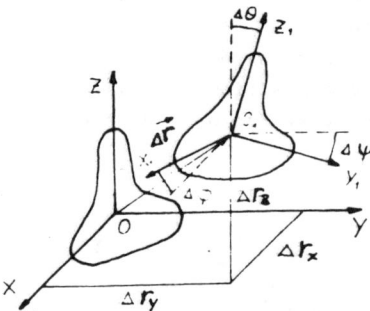

Fig.3: Small transformations

In this case the measurement task consists in the construction of a data block

$$(\Delta r_x, \Delta r_y, \Delta r_z, \Delta \phi, \Delta \psi, \Delta \theta)_N \quad . \tag{4}$$

Using the data processing block the estimation of following important characteristic values
- maximal derivation : $\max |\overrightarrow{\Delta r}|$
- mean value : $\overline{\Delta r} = \frac{1}{n} \sum_{n=1}^{N} |\overrightarrow{\Delta r}|$
- standard derivation : $s(\overrightarrow{\Delta r}) = \{\frac{1}{N-1} \sum_{n=1}^{N} (\overrightarrow{\Delta r}_n - \overline{\Delta r})^T (\overrightarrow{\Delta r}_n - \overline{\Delta r})\}^{\frac{1}{2}}$
- probabilistic distribution

can be performed.
In such cases where the orientation $(\Delta \phi, \Delta \psi, \Delta \theta)$ can be neglected, it is possible to reduce the measurement task to the estimation of a point position. From the practical point of view this measurement task is the most important one.

3.1.2. Measurement of global spatial motion

A parametric vector notation $\vec{r}(t)$, $(\phi(t), \psi(t), \theta(t))$ is being used for description of global motion of robot links. Additional, a time discretization with the time step Δt has been needed in comparison to measuring task explained in Chap.3.1.1.

Applying the same procedure we get the following data block

$$\begin{bmatrix} r_{x_1}, r_{y_1}, r_{z_1}, \phi_1, \psi_1, \theta_1 \\ \vdots \\ r_{x_n}, r_{y_n}, r_{z_n}, \phi_n, \psi_n, \theta_n \end{bmatrix}_N \quad ; r_{x_k} := r_x(k\Delta t), \quad k=1,2,\ldots,n \qquad (5)$$

where $T = t_1 - t_0 = n\Delta t$, N number of experiments.
By using the corresponding data processing blocks the calculation of
- kinematic determined metric (work space)
- relative geometric or time-dependent derivations
  (accuracy of position and orientation with respect to given trajectories)
- motion as function of time (path, velocity, acceleration).

Because of big work space, the high accuracy and the on-line measurements the technical realization is vers complicated. Therefore, a system for measuring only the position as a function of time has been realized as a first step.

3.2. Measurement of forces and torques

Forces and torques acting in special parts of robots can be described as functions of geometric parameters h, φ or as functions of time t. Fig.4 shows the schematic scheme of a robot axis ("active" d.o.f.)

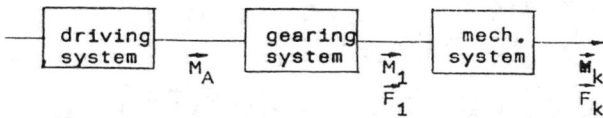

$\vec{M}_a$ driving torque
$\vec{M}_k, \vec{F}_k$ torque and force in k-th link

Fig.4: "Active" d.o.f.

By discretization of measured forces and torques with respect to $\Delta h, \Delta \varphi$ or Δt we get the data block

$$\begin{vmatrix} H_{A_1}, H_{11}, \ldots, H_{k1}, F_{11}, \ldots, F_{k1} \\ \vdots \\ H_{A_n}, M_{1n}, \ldots, M_{kn}, F_{1n}, \ldots, F_{kn} \end{vmatrix} \quad (6)$$

Applying the corresponding data processing block we find results concerning
- maximal values of forces and torques $\max|\vec{M}|$, $\max|\vec{F}|$,
- elasticity and backlash in robot's axis (by comparison of $\vec{M}_i$ with $\vec{M}_j$, $\vec{F}_i$ with $\vec{F}_j$, respectively,
- relationships $\vec{M}_k = \vec{M}_k(\vec{M}_A)$, $\vec{F}_k = \vec{F}_k(\vec{F}_A)$ as a base for control designing.

3.3. Vibration investigations

The vibrational investigations are based on multi-body systems. The equations of motion have the general form

$$\underline{A}(\vec{q})\ddot{\vec{q}} + \vec{B}(\vec{q},\dot{\vec{q}}) = \vec{Q}$$

where $\underline{A}(\vec{q})$ denotes the inertia matrix
$\vec{B}(\vec{q},\dot{\vec{q}})$ represents the CORIOLIS-forces, the centripetal forces, the bearing elasicity and damping,
$\vec{Q}$ the vector of generalized forces
$\vec{q}$ the vector of generalized coordinates.

Interpreting the vibration $\vec{x} = \vec{q} - \vec{q}^s$ as disturbances with respect to given trajectory $\vec{q}^s(t)$ then by assuming small vibrations we obtain after linearization

$$\underline{a}(\vec{q}^s)\ddot{\vec{x}} + \underline{b}(\vec{q}^s,\dot{\vec{q}}^s)\dot{\vec{x}} + \underline{c}(\vec{q}^s)\vec{x} = \Delta\vec{Q} \quad (7)$$

where
$$\underline{a}(\vec{q}^s) = \underline{A}(\vec{q}^s) \quad ; \quad \Delta\vec{Q} = \vec{Q} - \vec{Q}^s$$
$$\underline{b}(\vec{q}^s,\dot{\vec{q}}^s) = \frac{\partial}{\partial\dot{\vec{q}}}\left[\vec{B}(\vec{q},\dot{\vec{q}})\right]_{\vec{q}=\vec{q}^s}$$
$$\underline{c}(\vec{q}^s) = \left[\frac{\partial\underline{A}}{\partial q_1}\ddot{\vec{q}}\bigg|\ldots\bigg|\frac{\partial\underline{A}}{\partial q_n}\ddot{\vec{q}}\right]_{\vec{q}=\vec{q}^s} + \frac{\partial}{\partial\vec{q}}\left[\vec{B}(\vec{q},\dot{\vec{q}})\right]_{\vec{q}=\vec{q}^s}$$

Equ.(7) is the basic relation for designing and realization vibration experiments. From (7) follows that the vibration behaviour is depending on $\vec{q}$. Thus, the eigenfrequencies and eigenmodes are function of $\vec{q}$, for instance.

As an example the vibrational behaviour of an antropomorphic robot excited by a harmonic external force is given in a three-dimensional projection in fig.5.

For a given robot configuration (fixed $\bar{q}$) and in the case of harmonic excitation $\Delta\vec{Q} = \vec{F}\sin\Omega t$ the measurement task consists in registration of response curves for amplitudes $a_i(f)$ and phases $\psi_i(f)$, f frequency.
The same discretization procedure of $a_i(f)$ and $\psi_i(f)$ with the frequency step Δf yields

$$\begin{bmatrix} f_1, a_{11} \cdots a_{k1}, \psi_{11} \cdots \psi_{k1} \\ \vdots & \vdots & \vdots \\ F_n, a_{1n} \cdots a_{kn}, \psi_{1n} \cdots \psi_{kn} \end{bmatrix} \quad \text{where } f = f_{max} - f_o = n\Delta f.$$

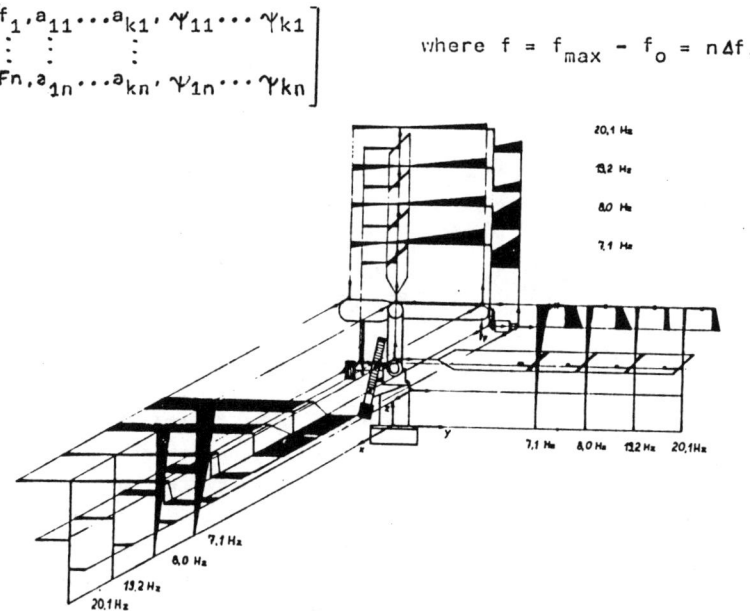

Fig.5: Results of vibration research in threedimensional projection

References:

/1/ Warnecke, H.J.; Schraft, R.D.: Industrieroboter, Krauskopfverlag, Mainz, 1982.
/2/ Tschakarow, S.: Verringerung der dynamischen Belastungen durch geeignete Geschwindigkeits/Weg-Steuerfunktionen, Tagungsmaterialien Symposium "Grundlagen der Dynamik und Steuerung von IR", Nov. 1984, Warnemünde.
/3/ Heimann, B.; Tschakarow, S.: Einige Grundlagen zur Systematik der experimentellen Untersuchungen an IR, Maschinenbautechnik, 1986.

Minimization of Vibrations of a Gantry Manipulator During Positioning

K. Tomaszewski and A. Golaś

Technical University of Mining and Metallurgy, Cracow, Poland

SYNOPSIS

   The paper is concerned with problems of minimization of manipulation head vibrations of a mobile manipulation machine under different conditions of motion of the gantry and head carriage. The vibro-acoustic parameters of the manipulation machine model were measured on a test rig, the experimental test results were analysed, and the effect of the model's basic design parameters on its vibro-acoustic properties and positioning accuracy was determined.

1. INTRODUCTION

   The dynamic analysis of manipulation machines, characterized particularly by high displacement velocities of the manipulation head, requires consideration of the effect of elasticity and attenuation of the kinematic chains of the mechanisms of these machines.
   Theoretical and experimental investigations of vibrations of the manipulation head and of the related vibrations of the robot hand constitute an important problem to be solved so that the positioning mechanisms shall be designed in a rational way, especially when high positioning accuracies are needed.

2. THE DESIGN AND OPERATION PRINCIPLE OF THE MANIPULATOR

   The gantry manipulator (Fig.1) consists of mechanical and electric assemblies. The mechanical part contains:
   - two rails placed on poles,
   - a bridge with manipulation head carriage, robot hand and gripper
   The electrical part comprises: A - motor of bridge travel along the x, B - motor of manipulation head carriage along y, C - motor of robot hand's travel along z, D - electromagnet of gripper closing and opening, E - motor of bridge positioning, F - motor of manipulation head carriage positioning, H - electromagnet of the mechanical brake, system controlling the A - H motors.

The task of the manipulator is to take a container by means of the robot hand from any stand and to carry it over to any other stand in the shop. This task can be performed automatically or using an individual program, if required by the operator.

The bridge's travel along the shop's direction, analogous to the travel of the gantry crane, is provided by the motor A. The bridge can be stopped at any one of the 63 stands, the situation of which is determined by the arrangement of 63 detecting elements: $P_{A1} - P_{A63}$ and $P_{E1} - P_{E63}$. At both sides of the P_{An} there are two additional detecting elements: P_{An-1} and P_{An+1} the task of which is to signal that the bridge is approaching the stand $P_{An} = P_{En}$ from one side or the other.

This causes braking of the linear motor A by current reversal. After deceleration of the bridge down to about 0,1m/s the braking by current reversal will be switched off and the bridge arrives at the predetermined stand P_{An}. At the moment when the P_{An} detecting element is reached, a mechanical brake is applied and the bridge stops at position P_{En}. The motor E blocks the bridge in this position. The drive A is equipped with a speed sensor allowing the measurement of the bridge's velocity. The bridge's velocity can be controlled by means of the motor A. A wedge introduced into the socket P_{En} shortens the detecting element which provides the signal of the completed bridge's travel along the x direction.

The movement of the head carriage with the robot's hand and gripper in the direction of the y axis is provided by motor B. The manipulation head carriage can assume nine positions (A - H) along the bridge (Fig.1) which are determined by the switches $P_{Ba} - P_{Bh}$. The number of possible positions where the head carriage can be stopped in the shop (the number of stands for manipulation) is 63 x 9 = 567.

The vertical motion of the robot hand is realized by means of the linear motor C. The robot hand's travel is limited by two switches P_{Cg} and P_{Cd} (the upper and lower position).

The movements of the gripper are realized by the electromagnet D. The gripper can assume two positions open and closed.

The bridge or manipulation head carriage can move only when the gripper is closed at the upper grab position.

The manipulation is controlled by push-buttons situated on the desk. There are four possible types of control: the automatic or a manual one, the control from the tester and the individual control.

3. EXPERIMENTAL INVESTIGATION OF THE MANIPULATOR

The range of investigations comprised vibrations of the bridge 3 with head carriage during positioning.

The vibration measurements were performed by means of the Brütel-Kjaer apparatus. The measuring points were arranged as shown in Fig.2. Figure 3 shows the graphical interpretation of the results of the investigation.

Oscillograms show that control of and truck 1 velocity has an important influence on displacement x_1, acceleration $\ddot{x}_1$, amplitudes and time of their decay. Vibration parameters of head carriage 2 x_1, $\ddot{x}_1$ are much less when velocity $\dot{x}_1 \leq 0{,}1$ m/s than when $\dot{x}_1$ is about $0{,}3$ m/s. It proves the importance of optimal choice of end truck 1 motion parameters during the positioning process.

4. POSITIONING PROCESS CONTROL

Let us assume that the set of differential motion equations of an electromechanical system under consideration has the form

$$\frac{dx}{d\tau} = \omega$$
$$I_r \frac{d\omega}{d\tau} = M_e + M_m \, \text{sgn}(-\omega) \quad (1)$$

where: α - angle displacement of the end truck 1, ω - angular velocity, I_r - equivalent moment of inertia of the system, M_m - load torque equivalent to mechanical loss, M_e - driving torque.

At a first approximation it can be assumed that $M_e = c\tilde{i}$ where: c - electromechanical coefficient, $\tilde{i}$ - current intensity. Motion equation (1) in normalized coordinates takes the form:

$$\frac{d\ell}{dt} = \nu \quad , \quad \frac{d\nu}{dt} = i + \mu \, \text{sgn}(-\nu) \quad (2)$$

where:

$$i = \frac{\tilde{i}}{\tilde{i}_N} \quad , \quad \mu = \frac{M}{M_N} \quad , \quad \nu = \frac{\omega}{\omega_N} \quad , \quad \ell = \frac{\alpha}{\alpha_N} \quad , \quad \alpha_N = \omega_N T_M$$

Index N indicates nominal value.

$$T_M = I_r \frac{\omega_N}{M_N}, \quad t = \frac{\tau}{T_M}$$

Let us make an assumption:

$$i = \chi U \quad (3)$$

where: χ - constant, U - control voltage.

Let us note

$$x_1 = \ell \quad, \quad x_2 = v \quad, \quad x_3 = 1 \qquad (4)$$

Using equation (4), motion equation (1) can be written in the form:

$$\begin{bmatrix} \dot{x}_1 \\ \dot{x}_2 \end{bmatrix} = \begin{bmatrix} 1 & 0 \\ 0 & 1 \end{bmatrix} \begin{bmatrix} x_2 \\ x_3 \end{bmatrix} + \begin{bmatrix} 0 \\ 1 \end{bmatrix} \text{sgn}(-x_2) \qquad (5)$$

where: $x_3 = \chi U$

Let us discuss limits, where state variable must be satisfied during the control process. The most dangerous position of head carriage 2 is its position on the end of the bridge crane 3.

In this position vibration amplitudes are the most dangerous during positioning motion of end truck 1.

Motion equation of an end truck 2 is as follows:

$$(m_2 + \tfrac{33}{140} m_b) x_d + \tfrac{3EI}{l^3}(x_1 - \tilde{x}_1) + h(\dot{x}_1 - \dot{\tilde{x}}_1) = 0 \qquad (6)$$

where: m_2 - mass of head carriage 2, m_3 - mass of crane bridge beam 3, l - longitude of a crane bridge beam, h - damping coefficient, E - Young's modulus, I - moment of inertia of the beam $\tilde{x}_1 = \alpha r$, r - radius of end truck's roll, x_1 - displacement of the head carriage 2. Natural frequency of the system is described by equation 7

$$\omega_o = \sqrt{\frac{3EI}{\left(m_2 + \tfrac{33}{140} m_b\right) l^3}} \qquad (7)$$

Displacement amplitude x_d is linear function of input force which equals

$$F(t) = \tfrac{3EI}{l^3} \tilde{x}_1(t) + h \dot{\tilde{x}}_1(t) \qquad (8)$$

Input force F(t) will be limited if velocity of end truck 1 $x_1(t)$ is limited.

In state variable notation it is expressed

$$|x_2| \leq \nu_{max} \qquad (9)$$

where: ν_{max} - maximum value of normalizated angular velocity which generates maximum permissible vibration displacement x_d. The second very important state variable limit ought to limit dynamic forces. It is obtained by limitation of acceleration of end truck 1

$$\dot{x}_2 \leq \varepsilon_{max} \qquad (10)$$

Using equation 2,

$$|i + sgn(-\nu)| \leq \varepsilon_{max} \qquad (11)$$

The condition will be satisfied, if a stronger condition will be satisfied

$$i \leq \varepsilon_{max} - \nu \qquad (12)$$

Then x_3 limit is as follows:

$$|x_3| \leq i_{max} \qquad (13)$$

where: $i_{max} = U_{max} \chi = \varepsilon_{max} - \mu$

The problem of good positioning in the sense of dynamical forces and, consequently, vibration amplitude limitation can be solved by finding an optimal control. The control will transport end truck 1 from a point $x_1(0)$, $x_2(0)$ to target point $x_1(t_r) = 0$, $x_2(t_r) = 0$, where t_r is a normalized control time satisfying limits (9) and (13). Transportation process must be realized in the shortest possible time to satisfy maximum productivity.

The quality coefficient for the time optimal problem has the form:

$$Q = \int_{t_o}^{t_r} dt \qquad 14)$$

Introducing $x_4 = Q$ we obtain state equations in the form

$$\begin{bmatrix} \dot{x}_1 \\ \dot{x}_2 \\ \dot{x}_4 \end{bmatrix} = \begin{bmatrix} 1 & 0 & 0 \\ 0 & 1 & 0 \\ 0 & 0 & 1 \end{bmatrix} \begin{bmatrix} x_2 \\ x_3 \\ 1 \end{bmatrix} + \begin{bmatrix} 0 \\ 1\ sgn(-x_2) \\ 0 \end{bmatrix} \mu \qquad (15)$$

Such defined optimization problem was solved using Pontriagin's principle. The Hamiltonian form for the problem is as follows:

$$H(x,U,\Psi) = -\Psi_0 + \Psi_1 x_2 + \Psi_2[\varkappa U + \text{sgn}(-x_2)] \quad (16)$$

The optimal condition is

$$\Psi = \frac{\partial H}{\partial U} = 0 \quad (17)$$

Hence

$$U = \frac{\mu}{\varkappa} \text{sgn } x_2 \quad (18)$$

The solution of the equation 17 has the form:

$$\Psi_1 = \Psi_{10}, \quad \Psi_2 = \Psi_{20} - \Psi_{10} \cdot t \quad (19)$$

It indicates the change of the control's sign and causes

$$U = \pm \frac{i_{max}}{\varkappa}$$

The solution is true when the strong limit condition is satisfied

$$|x_2| < \mathcal{V}_{max} \quad (20)$$

To find control in case when x_2 obtains value $\mathcal{V}_{max}$ it is necessary to solve the set of differential equations in the form:

$$\sum_{i=1}^{2} \frac{\partial g(x)}{\partial x_i} \dot{x}_i = 0 \quad (21)$$

where: $g(x) = x_2 - \mathcal{V}_{max}$ $\quad (22)$

Solution of equation (21) gives time periods where equation (22) is satisfied

$$U = \frac{\mu}{\varkappa} \text{sgn } x_2$$

The set of motion trajectory equations has the form

$$\dot{x}_1 = x_2$$
$$\dot{x}_2 = U\varkappa \text{ sgn}(\Psi_{20} - \Psi_{10}t) + \mu \text{sgn}(-x_2) \quad (23)$$

The solution of equation (23) gives us motion trajectories on phase plane $x_1 x_2$.

If $U(t) = \pm \frac{i_{max}}{\varkappa}$ then $x_1 - x_1(o) = \frac{x_2^2 - x_2^2(o)}{2[\varkappa U - \mu \text{ sgn}(-x_2)]} \quad (24)$

If $U(t) = \pm \frac{\mu}{\varkappa}$ then $x_2 = \pm \mathcal{V}_{max} \quad (25)$

Switchover curves are described by the equation

$$x_1 = \pm \frac{1}{2} \frac{x_2^2}{i_{max} + \mu} \quad (26)$$

Equations (24) - (26) allow control function in feedback loop to be found.

The proposed control system, which is planned to be realized assures limitation of bridge acceleration according to equation (10) and vibration minimization depends of assumed ε_{max}. Up to the present in the control solution there was no possibility of the bridge acceleration's being limited. It was reasoned by mechanical blockade where action caused an impulse change of velocity which generated vibration Figure 3.

5. CONCLUSIONS

The performed theoretical analysis of the manipulation machine's dynamic model confirmed the essential effect of the design parameters (mass, elasticity and attenuation) on the vibrations of the manipulation head. The dynamic calculations gave a basis for developing design directions for minimization of head vibrations, leading to increased positional accuracy.

REFERENCES

1 KUMAR A. and PRAKASH S.: Analysis of Mechanical Errors in Manipulators. Proceedings of the 6th IFToMM World Congress. New Delhi 1983.

2 VOLMER J. und Autorenkollektiv: Industrieroboter - Entwicklung. Berlin. VEB Verlag Technik 1983.

Minimization of Vibrations of a Gantry Manipulator 487

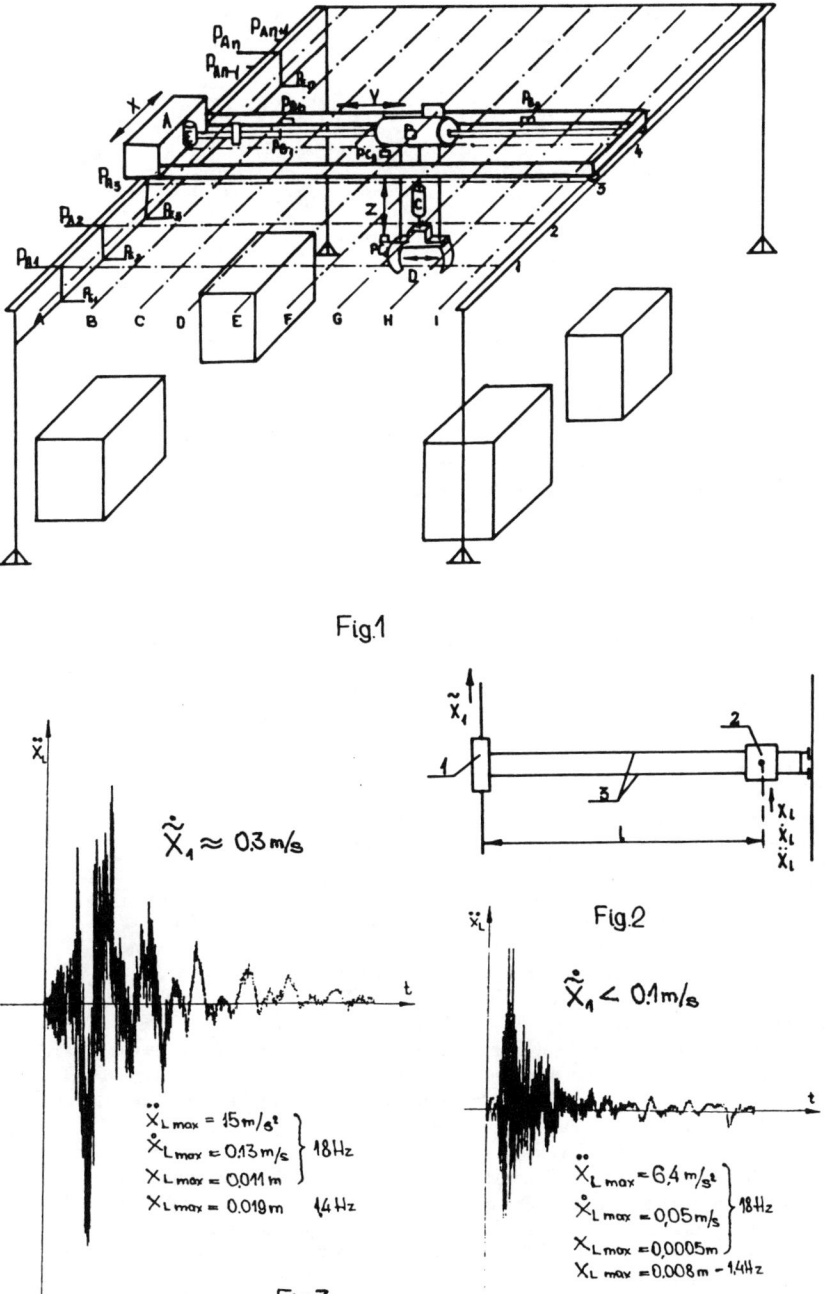

Fig.1

Fig.2

Fig.3

Experimental Evaluation of Feedforward and Computed Torque Control

C.H. An, C.G. Atkeson, J.D. Griffiths and J.M. Hollerbach

MIT Artificial Intelligence Laboratory, Cambridge, U.S.A.

Abstract. Trajectory tracking errors resulting from the application of various controllers have been experimentally determined on the MIT Serial Link Direct Drive Arm. The controllers range from simple analog PD control applied independently at each joint to feedforward and computed torque methods incorporating full dynamics. It was found that trajectory tracking errors decreased as more dynamic compensation terms were incorporated. There was no significant difference in trajectory tracking performace between the feedforward controller using independent digital servos and the full computed torque controller.

1 Introduction

Despite voluminous publications on the control of robot arms, there have been few experimental evaluations of the performance of proposed controllers. A major reason for this is the lack of a suitable manipulator for testing that fits these publications' modeling assumptions. Commercial robots are characterized by high gear ratios, substantial joint friction, and slow movement. As a result, their dynamics are approximated well by single-joint dynamics (Goor, 1985; Good, Sweet, and Strobel, 1985). Moreover, hardly any commercial robots allow the control of joint torque, which is required in many of the proposed controllers.

Direct drive arms are increasingly overcoming some of the performance limitations of highly geared robots (Asada and Youcef-Toumi, 1984; Curran and Mayer, 1985; Kuwahara et al., 1985). The manipulator dynamics are made more ideal by the reduction of joint friction and backlash effects, and the control of joint torques becomes more feasible. Hence direct drive arms have the potential for serving as good experimental devices for testing advanced arm control strategies. When gearing is eliminated, however, the full nonlinear dynamic interactions between moving links are manifested.

This paper reports on two sets of experiments with the MIT Serial Link Direct Drive Arm (SLDDA) involving a subset of proposed control strategies. The first set of experiments is based on a hybrid control system. There is an independent analog servo for each joint with the position and velocity references, and feedforward commands generated by a microprocessor. Since most commercial arms are controlled by simple independent PID controller for each joint, an independent PD controller was tested on this arm to provide a baseline for comparison. The PD controller was augmented by feeding forward first gravity compensation and then the complete rigid body dynamics to ascertain any trajectory following improvements attained by taking the nonlinear dynamics more fully into account. The second set of experiments shows the preliminary results of digital servo implementation, using one Motorola 68000 based microprocessor to control all the joints of the SLDDA. The on-line computed torque approach is compared to the PD and to the feedforward approaches using the digital servo.

The accuracy of the manipulator dynamic model impinges on the performance of feedforward and computed torque control. Since friction is negligible for direct drive arms, and presuming that one has good control of joint torques, the issue of accuracy reduces to how well the inertial parameters of the rigid links are known. In our previous work, we developed an algorithm for estimating these inertial parameters for any multi-link robot as a result of movement, and applied it to the SLDDA (An, Atkeson, and Hollerbach, 1985). The present paper presents results of utilizing the estimated model to control the robot by both off-line (feedforward) and on-line (computed torque) computation of the joint torques.

1.1 Control Algorithms

The full dynamics of an n degree-of-freedom manipulator are described by

$$n = J(q)\ddot{q} + V(q,\dot{q}) + G(q) + F(q,\dot{q}) \tag{1}$$

where n is the vector of joint torques (for rotational joints), q is the vector of joint angles, J is the inertia matrix, V is the vector of coriolis and centripetal terms, G is the gravity vector, and F is a vector of friction terms. The simplest and most common form of robot control is independent joint PD control, described by

$$n = K_v(\dot{q}_d - \dot{q}) + K_p(q_d - q) \tag{2}$$

where $\dot{q}_d$ and q_d are the desired joint velocities and positions, and K_p and K_v are $n \times n$ diagonal matrices of position and velocity gains.

Feedforward control augments the basic PD controller by providing a set of nominal torques n_{ff}:

$$n_{ff}(q_d, \dot{q}_d, \ddot{q}_d) = \hat{J}(q_d)\ddot{q}_d + \hat{V}(q_d, \dot{q}_d) + \hat{G}(q_d) + \hat{F}(q_d, \dot{q}_d) \tag{3}$$

where the hat (^) refers to the modelled values. When this equation is combined with (2), the feedforward controller results:

$$n = n_{ff}(q_d, \dot{q}_d, \ddot{q}_d) + K_v(\dot{q}_d - \dot{q}) + K_p(q_d - q) \tag{4}$$

The feedforward term n_{ff} can be thought of as a set of nominal torques which linearize the dynamics (1) about the operating points $q_d, \dot{q}_d$, and $\ddot{q}_d$. Therefore, it is reasonable to add the linear feedback terms $K_v(\dot{q}_d - \dot{q}) + K_p(q_d - q)$ as the control for the linearized system. These feedforward terms can be computed off-line, since they are function of the parameters of the desired trajectory only.

On the other hand, the computed torque controller computes the dynamics on-line, using the sampled joint position and velocity data. The control equation is:

$$n_{ct}(q_d, q, \dot{q}_d, \dot{q}, \ddot{q}_d) = \hat{J}(q)\ddot{q}^* + \hat{V}(q,\dot{q}) + \hat{G}(q) + \hat{F}(q,\dot{q}) \tag{5}$$

where $\bar{\mathbf{q}}^*$ is given by,
$$\bar{\mathbf{q}}^* = \ddot{\mathbf{q}} + \mathbf{K}_v(\dot{\mathbf{q}}_d - \dot{\mathbf{q}}) + \mathbf{K}_p(\mathbf{q}_d - \mathbf{q}). \tag{6}$$

If the robot model is exact, then each link of the robot is decoupled, and the trajectory error goes to zero. Gilbert and Ha (1984) have shown that the computed torque control method is robust to small modelling errors.

Previously, Liegeois, Fournier, and Aldon (1980) suggested feedforward control as an alternative to the on-line computation requirements of computed torque control, although they did not present any experimental results. Golla, Garg, and Hughes (1981) discussed different linear state-feedback controller using a linearized model of a manipulator. Asada, Kanade, and Takeyama (1983) presented some results of applying a feedforward control to the early version of a direct drive arm at the Robotics Institute of CMU, though for quite slow movements and for inertial parameters derived by CAD modeling. The computed torque method has been considered by several other investigators (Paul, 1972; Markiewicz, 1973; Bejczy, 1974; Luh, Walker, and Paul, 1980). Although simulation results have been presented, there has been no published report on the actual implementation of this controller, mainly due to the lack of an appropriate manipulator or on-line computational facility.

In this paper, we first use the feedforward controller to evaluate the accuracy of our estimates of the link inertial parameters, and to compare its performance against several other simpler control methods for high speed movements. Then, we present some preliminary results on the implementation of a computed torque controller, again using the estimated inertial parameters of the links.

1.2 Estimates of inertial parameters

The inertial parameters in the feedforward computation for the SLDDA were estimated by an algorithm developed previously (An, Atkeson, and Hollerbach, 1985). It was shown that the unknown inertial parameters of each link (mass, center of mass, and moments of inertia) appear linearly in the rigid body dynamics of a manipulator. Then a least squares algorithm was used to compute the estimates of these parameters.

The accuracy of the inertial parameters was verified initially by comparing the measured joint torques to the torques computed from the estimated parameters. This comparison, together with the torques computed from parameters derived by CAD modelling, is shown in Figure 1 for a 1.3s trajectory of all the joints moving 250 degrees. The results were actually superior for the dynamically estimated parameters than for the CAD-modelled parameters. A more practical verification of the estimated parameters is in generating feedforward torques as part of a control algorithm. The results of such experiments are presented in the next section.

2 Robot Controller Experiments

In this section, performances of several different controllers for full motion of the SLDDA are evaluated using the hybrid controller. The reference positions and velocities, and the

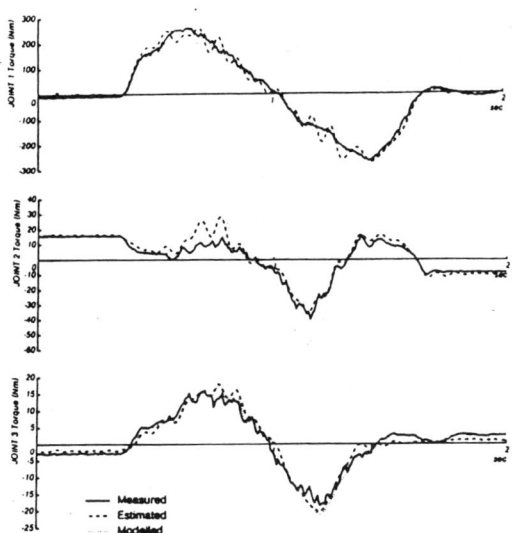

Figure 1: The measured, CAD-modelled, and estimated joint torques.

feedforward torques are generated by a microprocessor and input to three independent joint analog servos. We present evaluations of the following five control methods used for high speed movements of all three joints of the manipulator:

1. PD controller with position reference only:

$$\mathbf{n} = -\mathbf{K}_v \dot{\mathbf{q}} + \mathbf{K}_p(\mathbf{q}_d - \mathbf{q})$$

2. PD controller with position reference and feedforward of gravity torques:

$$\mathbf{n} = \hat{\mathbf{G}}(\mathbf{q}_d) - \mathbf{K}_v \dot{\mathbf{q}} + \mathbf{K}_p(\mathbf{q}_d - \mathbf{q})$$

3. PD controller with position and velocity references:

$$\mathbf{n} = \mathbf{K}_v(\dot{\mathbf{q}}_d - \dot{\mathbf{q}}) + \mathbf{K}_p(\mathbf{q}_d - \mathbf{q})$$

4. PD controller with position and velocity references plus feedforward of gravity torques:

$$\mathbf{n} = \hat{\mathbf{G}}(\mathbf{q}_d) + \mathbf{K}_v(\dot{\mathbf{q}}_d - \dot{\mathbf{q}}) + \mathbf{K}_p(\mathbf{q}_d - \mathbf{q})$$

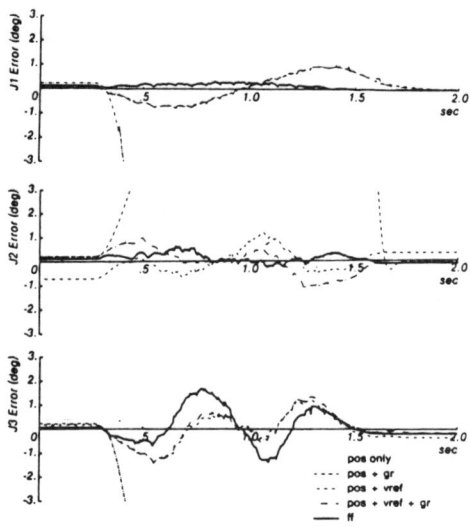

Figure 2: Trajectory errors of the 5 controllers for full 1.3s motion.

5. PD controller with position and velocity references plus feedforward of full dynamics:

$$n = \hat{J}(q_d)\ddot{q}_d + \hat{V}(q_d, \dot{q}_d) + \hat{G}(q_d) + K_v(\dot{q}_d - \dot{q}) + K_p(q_d - q)$$

In these experiments, friction was neglected. The nominal position and velocity gains were adjusted experimentally to achieve high stiffness and overdamped charateristics without the feedforward terms.

A fifth order polynomial in joint space was used to generate the reference trajectory. The joints moved from $(80°, 269.1°, -30°)$ to $(330°, 19.1°, 220°)$ in 1.3s, with peak velocities of 360 deg/sec and the peak accelerations of 854 deg/sec^2 for each joint. For control methods (2), (4), and (5), the estimates of the link inertial parameters given by An, Atkeson, and Hollerbach (1985) were used to compute the feedforward torques.

The trajectory errors for the above 5 controllers are shown on Figure 2. The errors for the first controller are very large and are out of the graph range. Adding a gravity feedforward term does not help very much, and the trajectory errors for Controller 2 are also very large. This was expected since gravity feedforward is a static correction to Controller 1, and the dynamic effects dominate the response for high speed movements. Modifying the first controller by adding a velocity reference signal improved the response greatly. As with Controller 2, adding a gravity feedforward term did not reduce the trajectory errors very much, and influenced mainly the steady state errors for joints 2 and 3.

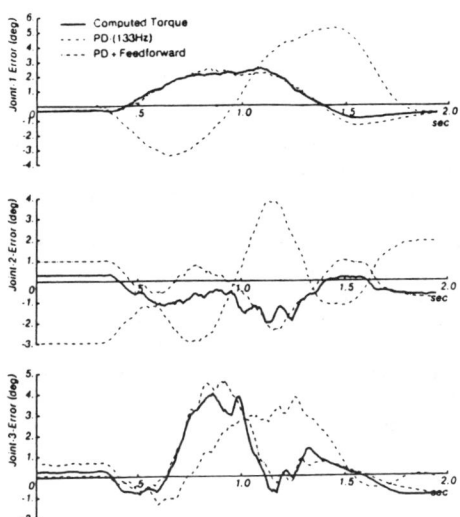

Figure 3: Trajectory errors of the three digital controllers for full 1.3s motion.

The full feedforward controller reduced the trajectory errors significantly for joints 1 and 2, with peak errors of only 0.33° and 0.64°, respectively. For joint 3, the feedforward torques did not help because of the light inertia and the dominance of unmodelled dynamics in the motor and in bearing friction. The high feedback gains make this joint somewhat robust to these unmodelled dynamics; yet, the trajectory errors could not be reduced below 1.4° with the feedforward torques based on the ideal rigid body model of the link.

2.1 Computed Torque Controller Experiment

In this section, some preliminary results are presented for the computed torque method implemented on the SLDDA. In this implementation, the analog servos are disabled, and the feedback computation is done digitally by one Motorola 68000 based microprocessor using scaled fixed-point arithmetic. Written in the C language, the controller, including the full computation of the robot dyamics, runs at a 133 Hz sampling frequency. Although further improvements in computation time are possible, this speed was adequate in demonstrating the efficacy of dynamic compensation. The details of this implementation are discussed in (Griffiths, 1986).

A similar fifth order polynomial trajectory as in the previous section was used for this experiment. Figure 3 shows the trajectory errors for three controllers: the digital PD controller, the feedforward controller using a digital servo, and the computed torque controller. The computed torque and the feedforward controllers both show a significant reduction in tracking errors for joints 1 and 2 compared with the PD control, with no clear distinction between feedforward and computed torque. The tracking errors for joint 1 range from 4.4° to 2.2° and for joint 2 go from 3.5° to 2.0° with the addition of dynamic component. As before, the trajectory errors for joint 3 were not reduced by the computed torque or the feedforward controller. Again, this seems to indicate that our model for the third link may not be very good.

The trajectory tracking performance of the computed torque controller is not as good as that of the analog feedforward controller of the previous section. The main reason for this is the slow sampling frequency (133 Hz) of the digital controller, as compared to the 1 KHz sampling frequency at which the reference inputs were given to the analog servos. Improvements in the computation time should also improve the tracking performance of the computed torque controller.

3 Conclusions

We have presented experimental results of using an estimated dynamic model of the manipulator for dynamic compensation via feedforward and computed torque control methods. The results indicate that dynamic compensation can improve trajectory accuracy significantly and that the estimated rigid body model of the manipulator is quite accurate and adequate for control purposes for joints 1 and 2. The unmodelled dynamics of the light third link, including the motor dynamics and friction, are dominant and yield larger trajectory errors than at the other two joints. Therefore, for joint 3, it may be necessary to use a more complete model to improve trajectory following.

The results of the digital implementation of the feedforward and computed torque controllers were not as good as the hybrid feedforward controller. This indicates that if a robot was being used solely for free space movements without significant variation of its loads, then a hybrid controller using an independent analog servo for each joint may be quite adequate. A hybrid controller, however, is not flexible, and cannot handle varying loads or interactions with the environment. Future experiments with the MIT Serial Link Direct Drive Arm will concentrate on improving the computation time for the digital control system and on issues of force control.

Acknowledgments

This paper describes research done at the Artificial Intelligence Laboratory of the Massachusetts Institute of Technology. Support for the laboratory's artificial intelligence research is provided in part by the Defense Advanced Research Projects Agency under Office of Naval Research contracts N00014-80-C-050 and N00014-82-K-0334 and by the Systems Development Foundation. Partial support for CGA was provided by a Whitaker Fund Graduate Fellowship and for JMH by an NSF Presidential Young Investigator Award.

References

An, C.H., Atkeson, C.G, and Hollerbach, J.M., 1985, "Estimation of inertial parameters of rigid body links of manipulators," *Proc. 24th Conf. Decision and Control*, Fort Lauderdale, Florida, Dec. 11-13.

Asada, H., Kanade, K., and Takeyama, I., 1983, "Control of a direct-drive arm," *Trans. of ASME*, 105, pp. 136-142.

Asada, H., Youcef-Toumi, K., and Lim, S.K., 1984, "Joint torque measurement of a direct-drive arm," *Proc. 23rd Conf. Decision and Control*, Las Vegas, Dec. 12-14, pp. 1332-1337.

Bejczy, A.K., Tarn, T.J., and Chen, Y.L., 1985, "Robot arm dynamic control by computer," *Proc. IEEE Conf. Robotics and Automation*, St. Louis, Mar. 25-28, pp. 960-970.

Curran, R., and Mayer, G., 1985, "The architecture of the AdeptOne direct-drive robot," *Proc. American Control Conf.*, Boston, June 19-21, pp. 716-721.

Gilbert, E.G., and Ha, I.J., 1984, "An Approach to Nonlinear Feedback Control with Applications to Robotics," *IEEE Trans. Systems, Man, Cybern.*, SMC-14, pp. 879-884.

Golla, D.F., Garg, S.C., and Hughes, P.C., 1981, "Linear-state feedback control of manipulators," *Mech. Machine Theory*, 16, pp. 93-103.

Good, M.C., Sweet, L.M., and Strobel, K.L., 1985, "Dynamic models for control system design of integrated robot and drive systems," *ASME J. Dynamic Systems, Meas., Control*, 107, pp. 53-59.

Goor, R.M., 1985, "A new approach to robot control," *Proc. American Control Conf*, Boston, June 19-21, pp. 385-389.

Griffiths, John D., 1986, Experimental Evaluation of Computed Torque Control, B.S. Thesis, MIT, Mechanical Eng..

Kuwahara, H., Ono, Y., Nikaido, M., and Matsumoto, T., 1985, "A precision direct-drive robot arm," *Proc. American Control Conf.*, Boston, June 19-21, pp. 722-727.

Liégeois, A., Fournier, A. and Aldon, M., 1980, "Model Reference Control of High-Velocity Industrial Robots," *Proc. Joint Automatic Control Conf.*, San Francisco, CA, August 13-15.

Luh, J.Y.S., Walker, M.W., and Paul, R.P.C., 1980, "Resolved-Acceleration Control of Mechanical Manipulators," *IEEE Trans. Auto. Contr.*, AC-25(3), pp. 468-474.

Markiewicz, B., 1973, "Analysis of the Computed Torque Drive Method and Comparison With Conventional Position Servo for a Computer-Controlled Manipulator," Jet Propulsion Laboratory, Pasadena, CA, March 15.

Paul, R. C., Sept. 1972, "Modeling, trajectory calculation and servoing of a computer controlled arm," AIM-177, Stanford University Artificial Intelligence Laboratory.

Experimental Research and Development of Methods for Improving Kinematic and Dynamic Robot Characteristics

A.N. Ananjev, E.G. Ananjeva and E.G. Nakhapetjan

Institute for the Study of Machines, Academy of Sciences,
Moscow, USSR

The purpose of this investigation is the development of methods aimed at increasing reliability and rapid functioning of industrial robots. The particular case of research of a robot with hydraulic drive helped to solve the problem of improving its kinematic and dynamic characteristics due to: 1) rational choice of executing mechanism's parameters and drive's parameters; 2) braking regime control. The questions of working out and identification of a mathematical model describing control system, drive and executing mechanisms were considered.

The analysis of a great number of industrial robot experimental investigations has shown that the basic reserve for improvement of rapid functioning and limiting of dynamic loadings on robot's hand links is the choice of rational braking laws. Such laws diminish the time of braking and oscillation damping time at the end of the motion. This time may range from 30 to 70% of overall movement time. The analysis of factors defining the duration of transition processes and dynamic loadings was carried out. A complicated dependence of kinematic and dynamic robot characteristics on a great number of mechanism's parameters, drive and control system's parameters was revealed. Therefore a complex analysis which combined experimental investigations and mathematical modelling was implemented. The increase of the volume and extension of experimental investigation problems required the creation of a multipurpose test stand. The experimental static, kinematic and accuracy characteristics were carried out by means of this stand.

Characteristics of rigidity, motion laws, values of maximum acceleration, amplitude-frequency characteristics, forces of friction, displacements of robot hand's links, etc. were found (fig.1)

Experimental data analysis has shown; 1) The time of braking and oscillation damping time at the end of the motion composes a major part of overall movement time and depends on a combination of robot mechanism's own properties and a choice of braking laws.

2) Mass load capacity, values of gain coefficients in feed-back loops, mechanism's rigidity, etc. mainly influence the kinematic and dynamic robot characteristics. 3) Simultaneously taking into account the execute mechanism, drive and control system own properties is the peculiarity of the developed robot's mathematical model [1].

For robots intended for transport operations the complex estimation of kinematic and dynamic characteristics was suggested:

$$\varphi = 1 - \frac{1}{\sqrt{3}}\left[\sum_{i=1}^{3} P_i B_i^2\right]^{1/2}; \quad P_i = 1; \quad B_1 = \frac{\omega_{cp}}{\omega_H}; \quad B_2 = 1 - \frac{\varepsilon_{max}}{\varepsilon_H}; \quad B_3 = 1 - \frac{L}{L_H}; \quad L = \frac{t_o}{\Delta x};$$

where ω_{cp} - average velocity;

ε_{max} - maximum acceleration;

Δx - error of positioning;

t_o - time of braking and oscillation damping time;

$\omega_H, L_H, \varepsilon_H$ - standardised coefficients which were obtained during the experimental investigation of different constructions of Industrial robots.

The smaller estimation φ is the better kinematic and dynamic characteristics the robot. Quality of robot mechanism's control influences greatly the estimation φ.

Methods of experimental research permit to control mechanisms using kinematic and dynamic characteristics [2].

The aim of such control is to achieve rapid functioning and to diminish dynamic loadings on robot hand's links. Examples of control of the mechanisms of robots are shown in Table 1.

Table 1

The type of drive		ω_{cp}	$\dfrac{t_o}{T_y}$	ε_{max}	φ
		s^{-1}	%	s^{-2}	-
Robot with hydraulic drive	1	0.80	46	17.0	0.27
	2	1.00	29	10.1	0.15
Robot with electric drive	1	0.59	54	8.8	0.41
	2	0.72	36	15.0	0.28
Robot with pneumatic drive	1	0.72	48	52.2	0.55
	2	1.10	41	42.5	0.42

Note: 1 - robot mechanisms' characteristics without control

2 - robot mechanisms' characteristics with control

T_y - overall movement time.

The next step after the cycle of experimental investigations was mathematical modelling aimed at improving robot's characteristics. A mathematical hydraulic drive robot model was worked out. The system of equations is divided into 3 blocks:

Block 1 - describes the work of control valves;

Block 2 - describes the work of hydraulic drives;

Block 3 - describes the work of executing mechanisms.

$$A\ddot{f} = K_{oc}(x_o - x_2) - B\dot{f} - Cf, \tag{1}$$

$$A\ddot{f}_1 = K_{oc1}(y_o - x_4) - B\dot{f}_1 - Cf_1,$$

$$\dot{p}_1 = \delta'(f_o + f)\sqrt{\tfrac{2}{\rho}}(p_n - p_1 - \varphi_1\dot{x}_1 - \varphi_2\dot{x}_1^2 \operatorname{sign}\dot{x}_1) -$$
$$- \delta'(f_o - f)\sqrt{\tfrac{2}{\rho}}(p_1 + \varphi_1\dot{x}_1 + \varphi_4\dot{x}_1^2 \operatorname{sign}\dot{x}_1) - b_o\dot{x}_1,$$

$$\dot{p}_2 = \delta'(f_o - f)\sqrt{\tfrac{2}{\rho}}(p_n - p_2 + \varphi_1\dot{x}_1 + \varphi_2\dot{x}_1^2 \operatorname{sign}\dot{x}_1) +$$
$$+ \delta'(f_o + f)\sqrt{\tfrac{2}{\rho}}(p_2 - \varphi_1\dot{x}_1 - \varphi_4\dot{x}_1^2 \operatorname{sign}\dot{x}_1) + b_o\dot{x}_1, \tag{2}$$

$$\dot{p}_3 = \delta_1''(f_o + f_1)\sqrt{\tfrac{2}{\rho}}(p_n - p_3 - \varphi_{11}\dot{x}_4 - \varphi_{21}\dot{x}_4^2 \operatorname{sign}\dot{x}_4) -$$
$$- \delta_1''(f_o - f_1)\sqrt{\tfrac{2}{\rho}}(p_3 + \varphi_{11}\dot{x}_4 + \varphi_{41}\dot{x}_4^2 \operatorname{sign}\dot{x}_4) - b_{o1}\dot{x}_4,$$

$$\dot{p}_4 = \delta_1''(f_o - f_1)\sqrt{\tfrac{2}{\rho}}(p_n - p_4 + \varphi_{11}\dot{x}_4 + \varphi_{21}\dot{x}_4^2 \operatorname{sign}\dot{x}_4) +$$
$$+ \delta_1''(f_o + f_1)\sqrt{\tfrac{2}{\rho}}(p_4 - \varphi_{11}\dot{x}_4 - \varphi_{41}\dot{x}_4^2 \operatorname{sign}\dot{x}_4) + b_{o1}\dot{x}_4,$$

$$m_1\ddot{x}_1 = F(p_1 - p_2) + C_1 i(x_2 - i x_1) + b_1 i(\dot{x}_2 - i\dot{x}_1) - W_{TP1}\operatorname{sign}\dot{x}_1,$$

$$J_1\ddot{x}_2 = -C_1(x_2 - i x_1) - b_1(\dot{x}_2 - i\dot{x}_1) + C_2(x_3 - x_2) + b_2(\dot{x}_3 - \dot{x}_2),$$

$$J_2\ddot{x}_3 = -C_2(x_3 - x_2) - b_2(\dot{x}_3 - \dot{x}_2) - 2m_3(x_5 + d)\dot{x}_3\dot{x}_5, \tag{3}$$

$$m_2\ddot{x}_4 = \tfrac{1}{2}[M_1\dot{x}_5^2 - M_2\dot{x}_4^2 - M_3(x_5 \Pi(x_4))^2] + C_3(x_5 \Pi(x_4))\Pi'(x_4) +$$
$$+ b_3\Pi'(x_4)(\dot{x}_5 - \Pi'(x_4)\dot{x}_4) + F_1(p_3 - \alpha p_4) - W_{TP2}\operatorname{sign}\dot{x}_4 - W_n(x_4),$$

$$m_3\ddot{x}_5 = m_3(x_5 + d)\dot{x}_3^2 - M_1\dot{x}_4\dot{x}_5 - C_3(x_5 - \Pi(x_4)) -$$
$$- \tfrac{1}{2}(M_3(x_3 - x_2)^2 - b_3(\dot{x}_5 - \Pi'(x_4)\dot{x}_4)),$$

where x_0, y_0 -coordinates of the given point;

$x_1 \div x_5$ - generalised coordinates;

$C_1 \div C_3$ - coefficients of rigidity;

$b_1 \div b_3$ - scattering coefficients;

f, f_1 - control valve's cross sections;

P_n - input pressure;

$P_1 \div P_4$ - pressure in the forcing and discharge chambers of the hydraulic cylinders;

K_{oc}, K_{oc_1} - gain coefficients of feed-back loops;

$m_1 \div m_3$ - reduced masses;

J_1, J_2 - reduced moments of inertia;

$\Pi(x_4)$ - variable transmission relation of the linear motion mechanism;

W_{TP_1}, W_{TP_2} - reduced forces of friction;

$W_{np}(x_4)$ - reduced force of robot hand's weight;

$A, B, C, f_0, \rho, a, d, b_0, b_{o1}, \varphi_1, \varphi_2, \varphi_4, \varphi_{11}, \varphi_{21}, \varphi_{41}, F, F_1, i$ - constant coefficients;

$$M_1 = \frac{dm_3}{dx_4} \; ; \quad M_2 = \frac{dm_2}{dx_4} \; ; \quad M_3 = \frac{dc_3}{dx_4} \; .$$

The identification of the mathematical model was carried out by using experimental planning methods separately for each mechanism. Velocities and pressure in forcing and discharge chambers of the hydraulic cylinders were taken as control parameters. Mean square root deviations between teoretical and experimental data were chosen as the measure of proximity. For example, mean square root deviations for the rotation mechanism were equal to:

$$K_1(\bar{\alpha}) = \frac{1}{n}\sum_{i=1}^{n}\left(\frac{p_{1e}-p_{1t}}{p_{1e}}\right)_i^2 \; ; \quad K_2(\bar{\alpha}) = \frac{1}{n}\sum_{i=1}^{n}\left(\frac{p_{2e}-p_{2t}}{p_{2e}}\right)_i^2 \; ; \quad K_3(\bar{\alpha}) = \frac{1}{n}\sum_{i=1}^{n}\left(\frac{\dot{x}_{3e}-\dot{x}_{3t}}{\dot{x}_{3e}}\right)_i^2 ;$$

where $p_{1e}, p_{2e}, \dot{x}_{3e}, p_{1t}, p_{2t}, \dot{x}_{3t}$ - experimental (e) and

teoretic (that is obtained as the result of mathematical modelling) values (t) of pressure (p) and velocity ($\dot{x}_3$); $\bar{\alpha}$ - vector of robot mechanism's parameters.
The sum of those deviations was concidered:

$$Q(\bar{\alpha}) = \frac{1}{m} \sum_{R=1}^{m} K_R(\bar{\alpha}),$$

where $\frac{1}{m}$ - weight coefficients of the R-th deviation.
The aim of identification was to find such region of robot parameters' alteration $G_o(\bar{\alpha})$ where $Q(\bar{\alpha}) \to 0$.
The result of identification was as follows: the region of robot parameters' alteration $G_o(\bar{\alpha})$ where estimation $0 < Q < 0.075$ was found.

Two methods of improving and maintaining constant robot characteristics in a given mathematical model were considered. The first consists in rational choice of constructive mechanism parameters' combination: masses, moments of inertia, gain coefficients of feed back loops, feeding pressure, rigidity, etc. The influence exerted by those parameters' combinations upon kinematic and dynamic characteristics was evaluated with the help of estimation $\Delta\Phi = \frac{\Phi}{\Phi_H}$ and its components $\Delta L = \frac{L}{L_H}$ and $\Delta \mathcal{E} = \frac{\mathcal{E}}{\mathcal{E}_H}$; ΔL - characterizes robot rapid functioning properties, $\Delta \mathcal{E}$ - characterizes dynamic loadings of robot hand's links, Φ_H, L_H, $\mathcal{E}_H$ - standardised values of corresponding estimations taken as initial. Values $\Delta\Phi < 1$, $\Delta L < 1$, $\Delta \mathcal{E} < 1$ correspond to the improvement of robot's characteristics. In the region G_o such vector of parameters $\bar{\alpha}$ was found, for which estimations $\Delta\Phi$, ΔL, $\Delta \mathcal{E}$ improved by 21, 25 and 19%
(Fig. 2). It means that on the robot design stage it is possible to solve the task of rational parameters' combination

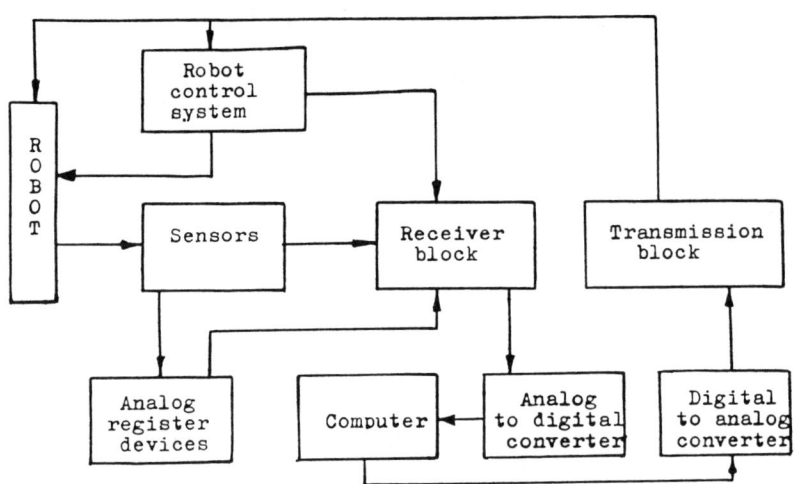

Fig. 1

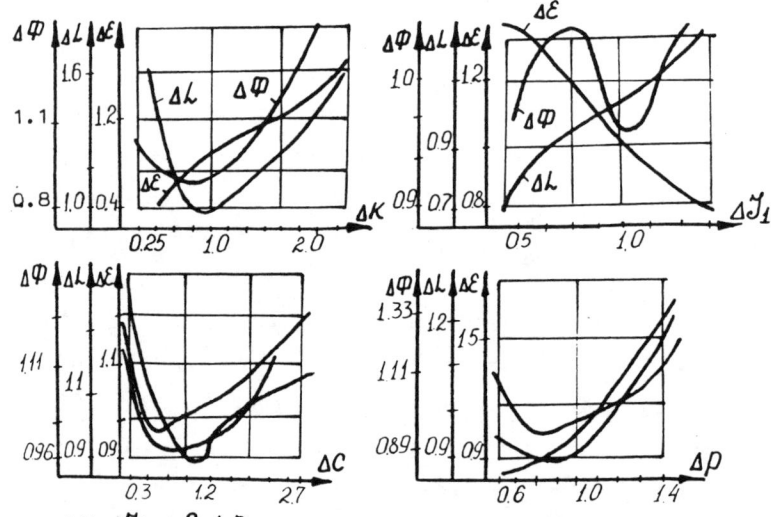

$\Delta K, \Delta J_1, \Delta C, \Delta p$ — standardised values of gain coefficient of feed back loop, moment of inertia, coefficient of rigidity and feeding pressure.

Fig. 2

choice by using mathematical modelling methods.

The second method of improving robot characteristics includes braking regime control which is aimed at: 1) reduction of time and lenght of braking process; 2) exclusion of hand's oscillations while positioning. Braking laws with linear and linear decreasing acceleration were chosen. This laws were realised by putting corresponding signals from a computer to the robot control system. By using proposed braking laws estimation φ was improved by 27%, the time of braking and oscillation dumping time was diminished by 15% and dynamic loadings - by 28%.

The results obtained during the mathematical modelling were verified during experimental test of robot controlled by a computer.

The combination of the experimental research methods and the mathematical modelling methods permits to solve complex problems connected with the improvement of industrial robot characteristics.

References

1. Ananjev A.N., Nakhapetjan E.G. " Experimental tests of Industrial robots mechanisms", Proceedings 4-th CISM-IFToMM Symposium on theory and practice of robots and manipulators, ROMANSY-81 Warsawa, p. 630-641.

2. Нахапетян Е.Г. "Диагностирование оборудования гибкого автоматизированного производства", М., Наука, 1985г., 225с.

Part 10
Synthesis and Design 2

Kinematics and Torque Control of Multi-Fingered Articulated Robot Hand

A.E. Samuel and P. Ridley

Engineering Design Group Department of Mechanical
and Industrial Engineering
The University of Melbourne, Melbourne, Australia

SYNOPSIS
The paper reviews some of the special problems related to synthesis of multi-fingered robot hands, where co-operated motion must be achieved under some "best" control strategy. The concepts of finite screws are applied to the motion of the hand and workpiece in the working volume of the hand to achieve some simple but non-trivial target tasks. It is demonstrated that combined torque and displacement control is the most appropriate mode of co- operated motion. A method of determining the "best" joints to control in either mode is discussed. An algorithm for computer control of the Melbourne hand is presented.

1. INTRODUCTION

The versatile robot gripper has received considerable attention in the recent past (see for example Rovetta,1977;Kato,1982; Wright and Cutkosky, 1984;-passive grip; Skinner,1975;Okada,1982;Salisbury and Craig,1982; Jacobsen at al.,1984; Cutkosky,1985 - active grip). Yet in spite of the all the available information on grippers, several problems remain unresolved. In fact the designer of the gripper has yet to obtain proper guidance for synthesis of the robot hand in the following three areas:
        (i)   workspace topology,
        (ii)   hand dynamics and servo design and
        (iii) computer algorithms for real time control of the hand.

Workspace topology will eventually decide the most appropriate layout of the fingers on the palm of the hand. The concept of workspace for a single chain manipulator is difficult to define and workers generally agree that this problem is also unresolved (see for example Tsai and Soni,1984; Kumar and Waldron,1981). For the multiloop chain of the hand-object in co-operated motion the concept is even more elusive.

Hand dynamics will critically effect the response characteristics of the hand-object system.

Computer control algorithms for real time control of the hand must employ the best strategy for moving the object. This strategy must be based on the best description of the motion and must include the proper balance between finger tip forces displacements and velocities.

The remainder of this paper examines the motion and some possible control strategies for the Melbourne hand, a three fingered hand with nine actuated joints. For a description of the design of the hand refer to Samuel and Ridley(1985a,1985b).

2. KINEMATICS

2.1 Location of a Rigid Body in Space

In general it is necessary to specify only six independent constraints to fix a rigid body in 3D space. One way of applying six independent constraints is to specify the location of three points on the grip triangle (fig. 1). Unfortunately this provides nine co-ordinates [$P(x_i, y_i, z_i)$; i=1,3], with three quadratic conditions relating three "superabundant" coordinates to the six independent constraints:

$$(x_1-x_2)^2 + (y_1-y_2)^2 + (z_1-z_2)^2 = d_{12}^2, \text{ and similarly for}$$
$$d_{23}^2 \text{ and } d_{31}^2.$$

Hunt(1978) demonstrated that three geometric elements could be used to fix a rigid body in space with six independent constraints. These elements are the point line and plane:
        point   - P[x,y,z],
        line   - l[L,M,N;P,Q,R],
        plane - [t,u,v,s].
This approach has the considerable advantage that superabundant coordinates can be related back to six constraints by solution of linear equations.

For cooperated motion to be achieved by three fingers manipulating a workpiece, it is necessary that the following conditions be met throughout the manipulation:
        - the grip triangle shall remain invariant and
        - the the workpiece shall be held in a stable grasp.
Stable grasp may be defined in terms of finger tip forces and the coefficient of friction at the points of contact (Macarthy,1984;Ridley,1986). Manipulation of the workpiece involves the movement of the grip triangle from an initial location (A_1,O_1,B_1) to some other predetermined location (A_2,O_2,B_2). Complete trajectories, even those passing around obstacles, may be formed from a series of such "finite screw" motions.

2.2 Point/Line/Plane approach (ref. fig.2)

(1) Define the initial and final locations of the body in space using point, line and plane (s_i, l_i, π_i, i=1,2) such that π_i contains l_i and l_i contains s_i. In the terminology of the grip it is convenient to make the plane π_i the plane and the line l_i one side of the grip triangle, and the point s_i one fingertip contact.
(2) Obtain six independent equations whose simultaneous solution will provide the Plucker coordinates of the finite screw axis. These equations are found from the geometric conditions:

#1: plane π_i intersects the screw axis at a constant angle (ie. angle α between the outward pointing normal n_i and L_{12} is constant) -- this condition provides three independent equations for the direction cosines of the finite screw axis, or Plucker coordinates L, M, N.

#2: perpendicular distance r_1 between l_i and L_{12} is constant
#3: perpendicular distance r_2 between s_i and L_{12} is constant.
-- these conditions provide another three independent equations for the position coordinates of the finite screw axis, or Plucker coordinates P,Q,R.

(3) The finite screw angle ϕ_{12} is subtended by the two planes α and β at the screw axis L_{12}. Plane α_{12} contains S_1 and the line L_{12} and plane β contains the point S_{12} and line L_{12}.

(4) The cardinal screw pitch h_{12} can be determined from the points of intersection, O_i, of the planes π_i with the screw axis L_{12}. Thus:

$$h_{12} = |O_1 O_2| / o_{12}$$

2.3 Trajectory

The finite screwing motion is such that every point on the body traces a portion of a helical path. Equations describing the helical trajectory of each point in the workpiece can be most simply expressed if one of the axes of the coordinate system used is aligned with the screw axis (fig. 3).

2.4 Closure Equations

Given that we can describe the path each fingertip will trace, the next problem is to determine the finger joint angles consistent with each point on the path. The forward and inverse closure equations for a 3R finger are shown on figs. 4 and 5. These equations are in their simples form when related to finger root coordinates (x_r, y_r, z_r).

2.5 Instantaneous Velocity of the Workpiece

Consider the instantaneous motion of a rigid body. Providing that the velocities of three points in the body are known, then:

- the instantaneous screw axis (ISA) of the motion can be found,
- the velocity of any other point in the body may be found and
- if the body is undergoing finite screw displacement, then the finite screw axis and the ISA will coincide.

Any infinitesimal displacement of a rigid body can be described as an infinitesimal screwing motion about the ISA. Instantaneously every point in the body is moving along a helical trajectory. There are a total of ∞^2 such helices, all coaxial with the ISA.

Phillips(1984) demonstrated the procedure for finding the screw coordinates of instantaneous motion (ref. fig. 6).
- plane α is defined by three points A,O and B,
- plane β is defined by taking the vectors v_A and v_B and relocating their origins to point O. The plane defined by the three points a,o,b (the tips of the velocity vectors) is plane β.
- point O' is determined by drawing a line through O parallel to the unit normal of the plane β. point O' is the point of intersection of this line with plane β. Points A' and B' are found similarly.
- A'A'' and B'B'' are the components of v_A and v_B in plane
- the ISA is the intersection of two planes, namely:
the plane π_A with normal A'A'', containing A and the plane π_B with normal B'B'' containing B.

Thus the Plucker line coordinates [L,M,N;P,Q,R] may be computed.
- the screw pitch h = QO'
- screw coordinates P^*, Q^*, R^* are given by:

$$P^* = P+hL; \quad Q^* = Q+hM \text{ and } R^* = R+hN.$$

The velocity of any point in the body may be calculated (say the centre of mass $G(x_G, y_G, z_G)$):

$$\dot{x}_G = P^* - \begin{vmatrix} y_G & z_G \\ M & N \end{vmatrix} \qquad \dot{y}_G = Q^* - \begin{vmatrix} z_G & x_G \\ N & L \end{vmatrix} \qquad \dot{z}_G = R^* - \begin{vmatrix} x_G & y_G \\ L & M \end{vmatrix}$$

3. GENERATION OF MOTION

Consider a hand comprised of three fingers, each of the type shown in fig.7. In order to achieve a desired instantaneous motion, with twist coordinates [L,M,N;P^*,Q^*,R^*], it will be necessary to drive a linearly independent set of six joint axes from the nine available actuated joints. The six velocities required to achieve the motion are calculated at any instant from :

$$\begin{vmatrix} \omega_1 \\ \omega_2 \\ \omega_3 \\ \omega_4 \\ \omega_5 \\ \omega_6 \end{vmatrix} = [\$_{indep}]^{-1} \begin{vmatrix} L \\ M \\ N \\ P^* \\ Q^* \\ R^* \end{vmatrix}$$

where: [$\$_{indep}$] is a matrix (6×6) of six linearly independent screws.

ω_i - required joint velocity of the ith independent axis.

Let the set of six driven joints be called 'active' joints and the remaining three be called 'passive' joints. In the following it will be shown how the active joint velocities are determined from the required motion of the workpiece.

3.1 Reciprocal Connections

Each finger in the hand may be considered as a six axis manipulator with three revolute joints and a spherical joint at the tip. Comparisons can be made between the Stewart platform (fig. 8) and a general 6R serial manipulator and the following may be deduced (Hunt,1985):

- identical motions could be generated transitorily in the end effector of each mechanism;
- the screw coordinates [L_i,M_i,N_i;P^*_i,Q^*_i,R^*_i] of the ith. joint axis of the 6R manipulator can be written down as the ith. column of a [6x6] Jacobian matrix [J].
- if [J] is inverted the rows of [J]$^{-1}$ yield the screw coordinates of the linear actuators of the transitorily equivalent Stewart platform.
- the screws coordinates [L'_i...R'_i] of the six linear actuators of the transitorily equivalent Stewart platform are reciprocal to the screw axes of the original 6R manipulator, such as one finger say [L_i...R_i].
- the inverse Jacobian may be found by traditional means,

OR

- written down by inspection once the reciprocal connections are identified. The reciprocal connection is the line which is reciprocal, simultaneously, to five of the finger joint axes. Since the finger joint axes are of zero pitch, the reciprocal connections either intersect 'locally' or at infinity (parallel axes). Hence the six legs of the Stewart platform, transitorily equivalent to a finger, may be found from lines which intersect five of the six revolute axes at a time.

In summary: the physical significance of reciprocity is that if five of the six degrees of freedom of a six axis manipulator are locked, then the single remaining degree of freedom is along the line which intersects the locked axes. This line is the reciprocal screw to the five locked revolute axes, and forms one leg of the transitorily equivalent Stewart platform.

In order to achieve full mobility of the workpiece the hand only requires six active joints. Selection of any two of the three joints per finger leads to the creation of two reciprocal connections passing through the ball joint at the tip. Fig. 10 shows the simplest selection of six reciprocal connections on the workpiece, if the two proximal joints of the finger are chosen as active joints. If the line coordinates of the reciprocal connections are written in a common coordinate frame, such as (x_w, y_w, z_w), attached to the finite screw axis shown in fig. 3, the matrix equation can be written :

$$\begin{vmatrix} V_A \\ V_B \\ V_C \\ V_D \\ V_E \\ V_F \end{vmatrix} = \begin{vmatrix} L_A' & M_A' & N_A' & P_A' & Q_A' & R_A' \\ L_B' & M_B' & N_B' & P_B' & Q_B' & R_B' \\ L_C' & M_C' & N_C' & P_C' & Q_C' & R_C' \\ L_D' & M_D' & N_D' & P_D' & Q_D' & R_D' \\ L_E' & M_E' & N_E' & P_E' & Q_E' & R_E' \\ L_F' & M_F' & N_F' & P_F' & Q_F' & R_F' \end{vmatrix} \begin{vmatrix} P^* \\ Q^* \\ R^* \\ L \\ M \\ N \end{vmatrix}$$

where : $[L, M, N; P^*, Q^*, R^*]$ are the screw co-ordinates of a desired twist
$V_A \ldots\ldots\ldots V_F$ are the velocities of the equivalent linear actuators.

Ridley(1986) develops the theoretical procedure for selecting active joints in a general way.

The velocities of the equivalent linear actuators can be simply related to the finger parameters as for example for finger #1:

$$V_A = (l_1 + l_2 c_2 + l_3 c_{23}) \dot{\theta}_1$$
$$V_B = l_2 s_3 \dot{\theta}_2 \quad , \text{ where}$$

and are equal to the active joint velocities needed to generate the motion. The other active joint velocities may be calculated similarly.

4. CONTROL STRATEGY

The control objective is to obtain cooperated motion between three fingers as they manipulate the workpiece from initial to goal position and orientation within the dextrous workspace of the hand. The workpiece will be under the influence of a general system of external forces and moments.

4.1 Coordinate Frames.

During the computation it will be necessary to work between four coordinate frames, namely base frame (located at the base of the macro manipulator), palm frame (located in the palm), finger root frame and work frame (located in the workpiece).

Palm coordinates are attached to the tip of the macro manipulator. In the case of the equispaced fingers of fig. 11, palm coordinates are defined by the intersections of the root joint axes. Finger tip velocities and forces are related to joint torques and velocities by the Jacobian matrix $[J]_r$ denoted here in root coordinates. The simple relationships between fingertip position and joint angles are shown in figs. 4 and 5.

Workpiece coordinates are located along the finite screw axis of the motion (fig. 3). Trajectory equations are most simply written in workpiece

4.2 Schematic Control Algorithm

Cooperative control can be achieved by position servoing six of the nine actuated joints and torque servoing the remaining three. Prior to manipulation taking place the following input data is needed:

(a) Physical properties -
- mass properties of the workpiece; mass (m), principal mass moments of inertia (I_1,I_2,I_3) and location of centre of mass $(G(x_w,y_w,z_w))$.
- hand geometry ; finger link lengths (l_{ij}), joint motion limits $(\theta_{ij} max)$, location of finger root axes $(x_j,y_j,z_j)_w$ relative to palm frame origin 0_{pw}. (the w subscript denotes workpiece coordinates).
- equivalent lumped masses of the finger links (m_{ij})

(b) Trajectory plan -
- current and goal locations of the workpiece in palm coordinates; either as the grasp triangle (x_{Ti},y_{Ti},z_{Ti}) or point\line\plane $(S_i,l_i,_i)$.
- cycle time.

(c) External loads -
- screw coordinates of the external wrench $(\$_E)_p$ at the palm.

(1) Finite screw axis (L_{12}) is calculated in base coordinates. Next create a set of workpiece coordinates with z_w axis along the screw axis(fig. 3).

(2) Check that the motion is physically possible. If the joint angle limits are violated, the macro manipulator will need relocation.

(3) Specify the helical trajectory for each fingertip in workpiece coordinates. These trajectories can be transformed bac to finger root coordinates via the palm coordinates.

(4) Generate unique set of time series of joint angles, velocities and accelerations consistent with cooperative motion along each trajectory.

(5) Transform external wrench into workpiece coordinates and calculate fingertip forces (F_A,F_O,F_B) necessary for stable grasping. Calculate joint torques.

(6) Position servo six active joints to set points generated in (4) above, and torque servo remaining joints to set points generated in (5).

Generally it will be satisfactory to torque servo the three distal joints and position servo the remaining six joints. However another selection of joints will be necessary if the mechanism reaches a configuration where active joints lose their linear independence or the 'passive', torque servoed, joints can not bring about force closure.

5. REFERENCES

Cutkosky M.R.(1985) Robotic Grasping and Fine Manipulation, Kluwer Acad.
Hunt K.H.(1978) Kinematic Geometry of Mechanisms, Clarendon, Oxford.
Hunt K.H.(1985) Robot Kinematics- a Compact Analytic Inverse Solution for Velocities, ASME (in press)
Jacobsen S., Wood J, Knutti D. and Biggers K.(1984) The Utah/MIT Dextrous Hand; Work in Progress, Int. J. Robot. Res. 3, 4, pp21-50.
Okada T.(1982) Computer Control of Multijointed Finger System for Precise Handling, IEEE Trans. on Sys. Man and Cybern. SMC-12, 3, pp289-299.

Kato I.(1982) Mechanical Hands Illustrated, Hemisphere, N.Y.
Kumar A. and Waldron K.(1981) The Workspaces of a Mechanical Manipulator, ASME Journal of Mech. Design, 103, July, pp.665-672.
Phillips J.(1984) Freedom in Machinery Vol.1, Introducing Screw Theory, Cambridge University Press.

Ridley P.R.(1986) Mechanics and Control of an Articulated Multi Fingered
    Robot Gripper, M. Eng. Sci. Design Group University Of Melbourne.
Rovetta A.(1977) On Specific Problems of Design of Multipurpose Mechanical
    Hands in Industrial Robots, 7th. ISIR, Tokyo, pp337-343.
Salisbury J. and Craig J.(1982) Articulated Hands: Force Control and
    Kinematic Issues. Int. J. Rob. Res. $\underline{1}$, 1, pp4-17.
Samuel A.E. and Ridley P.R.(1985a) Design and Development of a Flexible
    Robot Hand, Trans. I.E. Aust. Mech. Eng., $\underline{ME10}$, 3, Sept.pp208-212.
Samuel A.E. and Ridley P.R.(1985b) Motion and Control of a Flexible Robot
    Gripper ,Int. Conf. on Adv. Rob'tics., Tokyo, pp295-302.
Tsai Y and Soni A.(1984) The Effect of Link Parameter on the Working Space
    of General 3R Robot Arms, Mechanism and Machine Th., $\underline{19}$, 1, pp9-16.
Wright P.K. and Cutkosky M.R.(1984) Design of Grippers, in Handbook of
    Industrial Robotics, S. Nof.,editors. Wiley,N.Y. ch.2.4.

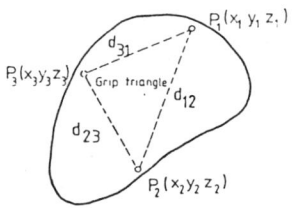

Fig. 1.   Location of a rigid body in space using three points

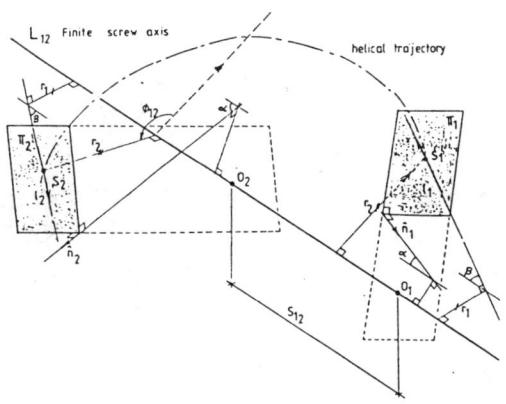

Fig. 2.   Point/line/plane approach to a spatial displacement

Trajectory Equations

$T = r_T c(\phi+\theta_T)$ $x_T = -\phi r_T s(\phi+\theta_T)$ $\ddot{x}_T = -(\dot\phi r_T s(\phi+\theta_T) + \phi^2 r_T c(\phi+\theta_T))$

$T = r_T s(\phi+\theta_T)$ $y_T = -\phi r_T c(\phi+\theta_T)$ $\ddot{y}_T = (\dot\phi r_T c(\phi+\theta_T) - \phi^2 r_T s(\phi+\theta_T))$

$T = h_{12}\phi + z_T$ $z_T = h_{12}\phi$ $z_T = h_{12}\phi$

here: (x_T, y_T, z_T) Cartesian co-ords of the fingertip T (ie. A,O or B)

r_T finite screw radius

ϕ finite screw angle

θ_T initial angle of fingertip T relative to x_T axis

$s(\phi+\theta_T) = \sin(\phi + \theta_T)$

$c(\phi+\theta_T) = \cos(\phi + \theta_T)$

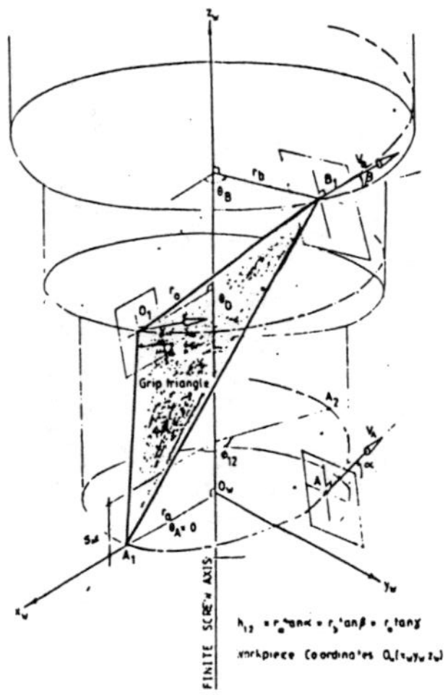

Fig. 3. Helical trajectory

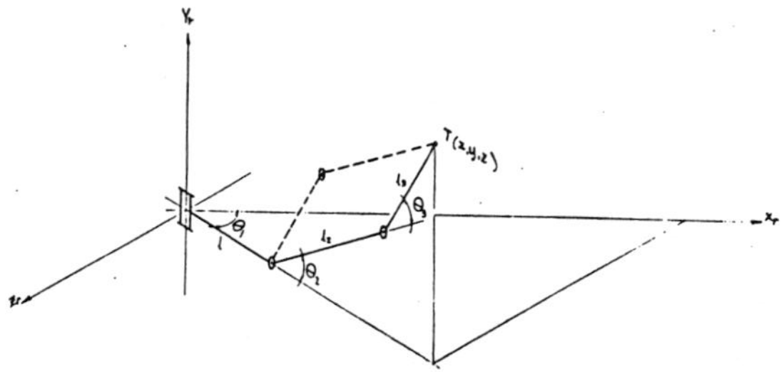

Fig. 4. 3R Manipulator schematic

i) Forward Solution

$$x_T = (l_1 + l_2c_2 + l_3c_{23})c_1$$

$$y_T = (\quad l_2s_2 + l_3s_{23})$$

$$z_T = (l_1 + l_2c_2 + l_3c_{23})s_1$$

ii) Inverse Solution

$$\theta_1 = \tan^{-1}(z_T/x_T)$$

$$\theta_2 = \tan^{-1}\left|\frac{y_T}{(x_T^2+z_T^2)^{0.5}-l_1}\right| - \tan^{-1}\left|\frac{l_3s_3}{l_2 + l_3c_3}\right|$$

$$\theta_3 = \cos^{-1}\left|\frac{[y_T^2 + ((x_T^2+z_T^2)^{0.5} - l_1)^2 - l_2^2 - l_3^2]}{2l_2l_3}\right|$$

where: $c1 = \cos\theta_1$, $s1 = \sin\theta_1$, $c23 = \cos(\theta_2+\theta_3)$, $s23 = \sin(\theta_2+\theta_3)$

Fig. 5. 3R Manipulator closure equations

Torque Control of Multi-Fingered Articulated Robot Hand 515

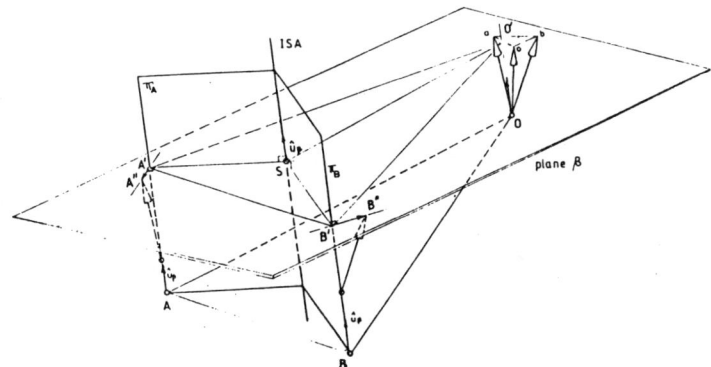

Fig. 6. Determination of ISA screw co-ordinates (Phillips, 1984)

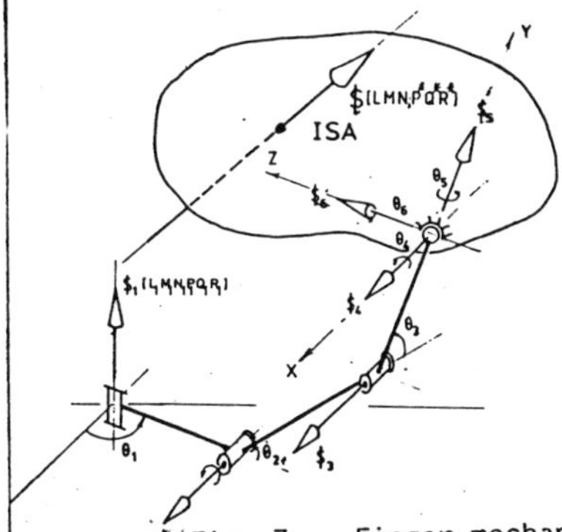

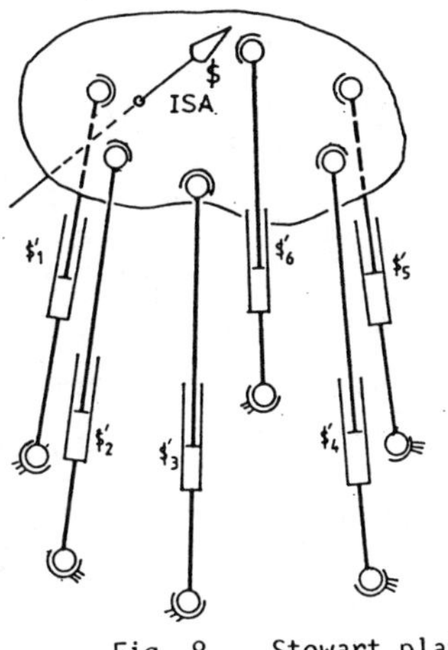

Point	x	y	z
		Co-ordinates in Frame (x_t, y_t, z_t)	
1	0	$-(l_3 + l_2 c_3 + l_1 c_{23})$	$l_2 s_3 + l_1 s_{23}$
2	0	$-(l_3 + l_2 c_3)$	$l_2 s_3$
3	0	$-l_3$	0
4	0	0	0
5	0	0	$\dfrac{l_1 + l_2 c_2 + l_3 c_{23}}{s_{23}}$
6	0	$-\dfrac{(l_1 + l_2 c_2 + l_3 c_{23})}{c_{23}}$	0

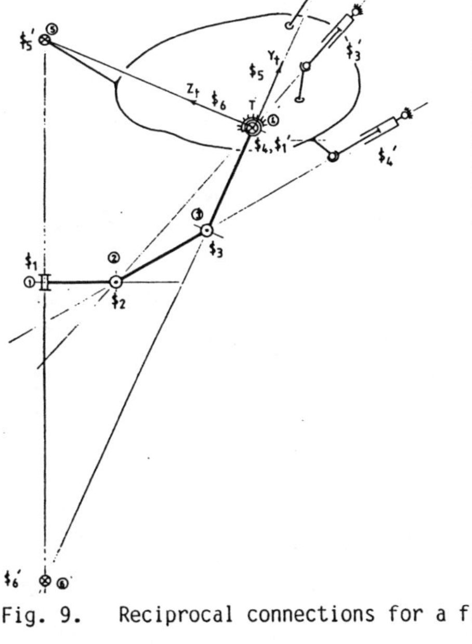

Fig. 7. Finger mechanism

Fig. 8. Stewart platform

Fig. 9. Reciprocal connections for a finger

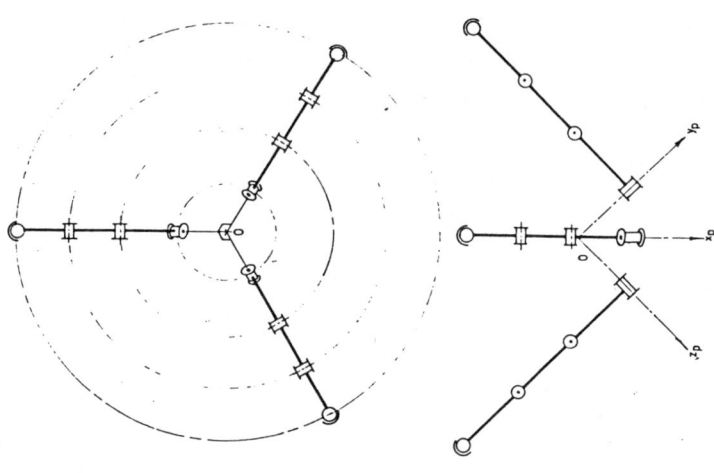

Fig. 11. Three fingered hand - palm layout

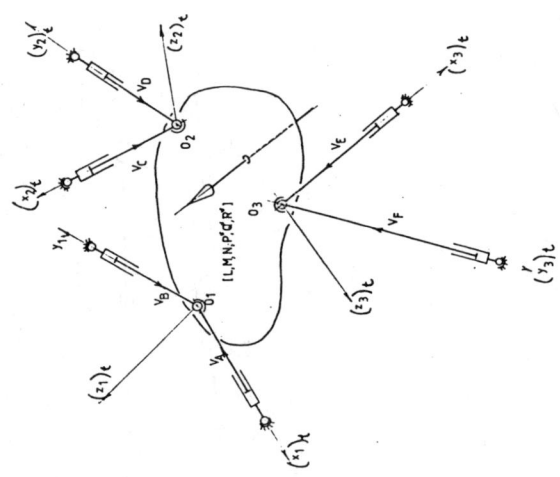

Fig. 10. Six reciprocal connections using active proximal joints

Progress towards a Robotic Aid for the severely Disabled

S. Michalowski

Department of Mechanical Engineering
Stanford Univerisity, Stanford, U.S.A.

I. INTRODUCTION

During the past several years a number of projects have studied the feasibility of using robots to serve the needs of the severely disabled (1-4). The goal has been to build a device for use in the performance of activities of everyday living, as well as in recreational and vocational tasks, by persons whose upper and lower limb functions are impaired due to disease or injury.

Work in this area differs from that in mainstream robotics, where the emphasis is on completely autonomous systems, and the anticipated benefits (so far unrealized) of using robots on a large scale are expected to occur as humans are displaced from tedious or hazardous workplaces. The complete replacement of humans by machines in all but the simplest industrial tasks, however, requires a degree of sensation and intelligence that is beyond the present state of the art. By contrast, developers of assistive devices for the disabled can take advantage of the assumed presence of an expert user, whose cognitive and perceptual faculties are, in most cases, intact. Thus the field of rehabilitative robotics is related to efforts to develop other "intelligent," interactive tools: expert systems, office and communication equipment, personal computers, etc.

The introduction of "high-tech" devices and methods into an everyday human environment poses some special challenges to the designers of rehabilitative aids. A robot that does not pass a certain threshold of utility (defined as a combination of capability and ease of use) will be abandoned if it produces frustration, boredom or embarrassment. A successful manipulation aid should be a practical alternative to human assistance, special-purpose mechanical or electronic devices or animal helpers. It should enhance the disabled person's sense of independence and privacy. As a device on the cutting edge of technology, it should be an object of admiration, thus helping to overcome the social stigma of disability.

This paper reviews the status of the Robotic Aid project, an ongoing effort funded by the United States Veterans Administration and carried out at Stanford University and the Rehabilitation Research and Development Center of the VA Medical Center in Palo Alto, California.

II. BACKGROUND

A prototype Robotic Aid has been undergoing evaluation trials at the VA Medical Center in Palo Alto (5-7). Its major components are: a PUMA 260 robotic arm and gripper, a Z-80 based supervisory computer, a speech recognition unit with a vocabulary of 70 words, a speech output device, and a small flat panel display. The arm is clamped to a table and can be "piloted" by means of explicit position and velocity commands.

Motions consist of rotations and translations in one of two coordinate systems - one fixed with respect to the base of the robot, the other embedded in the gripper. Any configuration of the arm can be assigned a name and the arm may be returned to that position at a later time with a single command.

Over thirty disabled veterans, and more than one hundred able-bodied individuals, have been trained in the use of the Robotic Aid, learning simple tasks such as picking up a cup of water and taking a drink. A few disabled persons have spent a considerably longer time learning how to program and execute more complex motion sequences such as preparing a meal or performing basic personal hygiene and grooming tasks (8,9).

Based on clinical experience, a major effort is under way to enhance the capabilities of the Robotic Aid and to make it easier to use. The important elements of this upgrade are: increased participation of the user in the operation of the device, incorporation of sensors into the motions of the robot, enlarged physical range of manipulation, and enhanced autonomy of the robot for selected generic tasks.

III. REAL-TIME HARDWARE IMPLEMENTATION

The new version of the Stanford/VA Robotic Aid has two parts: a mobile manipulator and a stationary user console (10,11). The manipulator arm is a commercial model - the Unimation PUMA 260. Each of its six revolute joints is actuated by a DC servo motor and is equipped with an optical encoder. A dedicated microprocessor (supplied by the manufacturer) performs a closed-loop servo algorithm to move the joint to a commanded position. For smooth operation, joint positions must be supplied at a rate of 35 Hz.

The PUMA arm is mounted on a three-wheeled omnidirectional mobile base, as shown in the figure. The wheels are arranged along the sides of an equilateral triangle. The circumference of each wheel consists of a set of twenty free-wheeling rollers so that the wheel can move parallel to the axis of rotation. By controlling the three wheel velocities, true freedom of motion in the plane has been achieved, in contrast to vehicles whose wheels must be physically reoriented before the direction of motion can change. For simplicity, the motors and servo controllers of the vehicle are of the same type as those of the arm. A fourteen-segment bumper (not shown in the photograph) surrounds the robot. A small CCD TV camera is mounted on the arm to provide a close-up view of the remote manipulation process.

The arm is equipped with a two-fingered gripper. Each finger has six photoelectric proximity sensors that detect the reflection of an infared light source off the surface of a nearby object. The strength of the reflection is digitized and multiplexed on the hand itself. The resulting stream of data is sent to a dedicated microprocessor. A force-sensing wrist, instrumented with eight pairs of strain gauges, is used to measure the six forces and torques in the wrist joint.

The mobile computer system is based on the 22-bit QBUS which is compatible with a set of products manufactured by Digital Equipment Corporation: the LSI11 family of 16-bit microprocessors (including the KXT11 single board processor) and the new MicroVAX 32-bit machines. The QBUS was chosen because of the great variety of available peripheral devices and, most importantly, to take advantage of a high-level real-time programming language - MicroPower Pascal.

The main processor is a DEC LSI11/73. Five kinds of peripherals also reside on the bus: 512 kB of solid state memory, a sixteen-channel analog-to-digital convertor, a four-channel serial interface, four bi-directional 16-bit parallel interface modules, and a DEC KXT11-C single board processor. The timing of events on the bus is controlled by an internal 10 MHz. oscillator. An external 120 Hz. clock interrupt synchronizes the execution of applications programs with the robot's servo controllers.

One of the serial lines is used to provide bidirectional communication over a 4800 baud FM radio link to the stationary console. Another serial line provides a connection to an on-board terminal which features a 40-character, 16-line LCD display and keyboard. The terminal displays status and error information and also can be used by the development staff to enter commands to the computer.

The parallel I/O modules link the LSI11 with the servo controllers, the bumper system and a variety of control and diagnostic functions. The KXT11 microprocessor controls the hand and wrist of the robot. It records readings from the twelve photoelectric proximity sensors that are mounted on the fingers of the gripper and the eight pairs of strain gauges that record forces and torques in the wrist joint. The processed data is transferred to the main system memory through a DMA controller.

IV. REAL-TIME CONTROL SOFTWARE

The computer programs that run on the mobile robot are written in a dialect of Pascal. The complete memory image consists of a kernel and a set of user-written processes that execute concurrently. Through a system of semaphores, priority tables and inter-process communication facilities, the processes gain temporary control of the CPU in an organized way. At this time, the applications package consists of some two dozen modules, linked together through a variety of synchronization flags and shared data structures. Some of the modules perform housekeeping functions such as maintaining the communication link with the user console or servicing interrupts for the joint/wheel controllers.

At the heart of the control structure are the modules that compute the real-time motion of the arm and the vehicle, each corresponding to a special-purpose function of the robot. At this time, seven of these subprocesses are implemented: piloting and navigation functions of the arm and base, and simple control of the hand, the wrist and the bumpers. At any instant, some combination of these modules is activated, usually on operator command. Results of the separate computations are reported in a standard format as velocities in a labeled coordinate system. A central module performs the coordinate transformations and kinematic computations that result in a set of joint/wheel angles. These are passed on to the driver processes. The whole structure is highly modular, so that future motion subprocesses will be easily "plugged in". By design, there is no built-in mechanism for interpreting the combined effect of the operation of the subprocesses although, in principle, the resulting motion may be chaotic (for example, the robot may try to use force data to turn a knob while simultaneously using the same data to avoid collisions between the gripper and the environment). Furthermore, certain

combinations of motion-computation routines can overtax the resouces of the
CPU, producing jerky motion of the robot. The role of coordinating the
behavior of the system is assigned to the operator who may, for example,
disable a certain function based on his or her evaluation of the external
situation and an understanding of the high-level goals that are being
pursued. This results in a very flexible control system, although it places an
extra burden on the designers to create an efficient human/machine interface.

V. THE USER CONSOLE

The stationary component of the Robotic Aid's computer system is an IBM
PC/AT, equipped with two high-resolution color graphics controllers. The PC/AT
is connected to a DECTALK speech synthesis unit, a Kurzweil KVS1.1 speech
recognition system, a 4800 baud radio link module and an ultrasonic head
position detector. All console functions are enabled by voice command. The
console is designed for use by a wheelchair-bound person and includes a working
area for the manipulator.

During operation, the user monitors a variety of text and graphic
displays. Displayed information includes: the position of the robot within the
environment (based on an on-board dead-reckoning algorithm), defined vehicle
and arm configurations and trajectories, engineering status information
(battery voltages, for instance) and, during piloting, the nonzero velocity
components of the arm and/or vehicle.

The position of the operator's head is monitored by a pair of ultrasonic
ranging units. A dedicated Z-80 processor performs a simple triangulation
algorithm, providing the PC/AT with (x,y) data at a rate of 3 Hz. Head motions
can be used to steer the arm or vehicle in two selected directions, to select
menu items on a screen or to define labeled positions that can be assembled
into trajectories.

VI. FUTURE DESIGN CONSIDERATIONS

The Robotic Aid is intended to evolve into a practical manipulation device
for quadriplegics and other severely disabled persons. At the same time, the
laboratory prototypes serve as test-beds for the exploration of fundamental
issues in human-robot interactions. If the human and the robot are to be
successfully integrated into one efficient manipulation system, the interface
cannot be designed as a stand-alone entity that links a naive user to a
standard commercial robot: the entire system must be designed with human
participation in mind.

The architecture of the robot control software has been designed
specifically to accommodate the unique characteristics of human interaction.
The selection of sensors and the basic elements of the robot's data structures
and motion-control algorithms are defined to optimize functions and tasks that
humans will want to perform. Similarly, the human users of the system have to
have a good understanding of how the robot can be used, based on training,
testing and evaluation.

The potential applications of interactive human/robot systems extend
beyond rehabilitation. If a system of this type is designed correctly, the
entities (human and robot) on both sides of the interface can be AUGMENTED:

the robot, for instance, can receive information and guidance from the human in areas where autonomous functioning is either very difficult (such as vision) or poorly defined (such as goal-directed behavior). The human, on the other hand, can use the robot to establish control over special environments that are remote (undersea, for example), inaccessible due to disability, or subject to peculiar constraints. Two examples of this last case may be helpful: when performing remote manipulation in outer space, the robot should be able to compensate for unfamiliar effects such as a non-inertial coordinate frame. On the opposite end of the size spectrum, manipulations on a microscopic level might require special assistance by the robot in correcting for scale-dependent phenomena such as surface tension.

It is usually assumed that when people avail themselves of robots, their role is limited to the highest level functions - those pertaining to goal identification and task supervision. We take the view that human interaction can occur at any level of robot operation, provided that the interface is properly designed. Here are the various modes of human participation:

- PILOTING. The operator can give explicit motion commands such as "go left", "turn clockwise", etc. Piloting also occurs if the robot mimicks the motions of a human arm in a master-slave telemanipulation scheme.

- PERCEPTION. A human can provide objective sense data that the robot cannot acquire on its own. For example, if the robot has a black-and-white vision system, relevant color information can be supplied by the operator.

- ADVICE. Parameters of a robotic algorithm can be adjusted in real-time based on global knowledge or reasoning. For example, if a mobile robot is trying to navigate around an obstacle, the operator can offer a suggestion as to the most fruitful direction to explore.

- FINE TUNING. During the execution of complex, preprogrammed routines, the user can modify the motions in real time to account for small discrepancies between the robot's model of the world and the actual locations of objects.

- ERROR CORRECTION. The operator may conclude that some part of the robot is failing and may suggest corrective action, for example to ignore the readings from a particular sensor.

- HAZARD IDENTIFICATION. The user may foresee impending trouble and take steps to avoid it.

- GOAL SELECTION. This is probably the most significant human contribution to the robot's behavior. It involves using human judgement to select a task from the robot's repertory and, most importantly, to set up the initial conditions to maximize the chances for successful completion. For example, when using a robotic hand to grasp an elongated object, the user can establish the initial orientation of the gripper (by using explicit motion commands such as "left", "turn", etc.) so that the hand will encounter the small dimension of the object as soon as possible.

- TASK ORGANIZATION. If a complex goal is selected, a robot could be instructed on the best method to use. Instructions may include such information as the order in which procedures are to be invoked, what to do if certain procedures fail, etc.

- CONFIGURATION. For a given task, the choice of the correct end effector and/or tool, is of great importance. In certain cases, the choice may have to be made by the human operator.

- TERMINATION: In many instances it may be difficult to determine when a task has been completed, especially in cases when the system is failing to achieve the desired goal. For example, a robot navigating through an impossible maze can probably go on trying forever - once the user is convinced that all attempts to reach the destination will fail, he may simply instruct the robot to stop trying.

- CONFLICT RESOLUTION. In a complex robotic system, situations can arise in which different parts of the control structure produce conflicting or incompatible motion requests. For example, one algorithm may try to exert a force on an object while another, operating independently, may be trying to avoid contact with the external world, for reasons of safety. Similarly, a robot's sensing or computational resources may become overloaded due to the complexity of the physical situation. In these cases, the operator can evaluate the relative importance of the various functions and deactivate selected routines.

- OBJECT IDENTIFICATION: This form of interaction may be very primitive - for example, labeling of a manipulator configuration. A more interesting situation arises if the robot has partial knowledge about an object. By supplying an identification, the operator allows the robot to consult its database, which may contain additional information about the object.

- TEACHING: Complex types of robot behavior may be built up from simpler ones based on user instructions. In the future, the robot may be taught how to set a table (for example), using existing primitive capabilities such as "go to," "grasp," "place" and the appropriate elements of an existing object database - "glass," "plate," etc.

- ACTION: In certain situations, the manipulative capabilities of the human may be available. Simple examples: the operator can disentangle a sensor cable that has become wrapped around the arm or remove a physical obstacle in the path of a mobile robot.

The Robotic Aid research group is currently involved in the design of a next-generation device which will incorporate the above considerations explicitly. In particular, a natural language interface will provide a way to teach the robot about objects and operations in an everyday, homelike environment.

VII. ACKNOWLEDGMENTS

The author is the project manager and primary software designer of the Robotic Aid. He wishes to thank the other members of the research team, past and present, for their valuable contributions to the project: Larry Leifer, Urs Elsasser, Machiel van der Loos, Lin Liang, Larry Edwards, Hisup Park, David Jaffe, Wade Henessey, James Miles, Scott Walter, Charles Wampler, John Jameson, Charles Buckley, John Walecka and Walter Conti.

VIII. REFERENCES

(1) Funakubo H. A newly created electrical total arm prosthesis of module type controlled by microcomputer and voice command system. Acta Medicotechnica 29(2), 45-48, 1981.

(2) Scheider W., Scmeisser G, Seamone W. A computer-aided robotic arm/-worktable system for the high-level quadriplegic. IEEE Computer, 41-47, January 1981.

(3) Guittet J, Kwee HH, Quetin N, Yolon J. The Spartacus Telethesis: Manipulator control studies. Bulletin of Prosthetics Research BPR-32, 69-105, Fall 1979.

(4) Leifer L. Rehabilitative Robotics. Robotics Age, 4-15, May 1981.

(5) Leifer L. Restoration of motor function - a robotics approach. Uses of Computers in Aiding the Disabled, Josef Raviv, ed. North-Holland Publishing Co., New York, 1982.

(6) Buckley C. Organization of a control system for a manipulative aid. Proc. 6th Annual Conf. on Rehabilitation Engineering, San Diego, CA., 350-352, 1983.

(7) Buckley C, Leifer L. Fast interactive control of a manipulator for the severely disabled. Proc. American Control Conference, Arlington, VA, 1195-1198, 1982.

(8) Angius B, Engelhardt KG, Leifer L. Recreational applications of a robotic aid. Proc. 6th Annual Conf. on Rehabilitation Engineering, San Diego, CA., 366-368, 1983.

(9) Awad R, Engelhardt KG, Leifer L. Development of training procedures for an interactive voive-controlled robotic aid. Proc. 6th Annual Conf. on Rehabilitation Engineering, San Diego, CA., 276-278, 1983.

(10) Michalowski S, Leifer L, Van der Loos M. Computer system for an advanced manipulation aid. Proc. 9th Annual Conference on Rehabilitation Engineering, Minneapolis, MN, 1986.

(11) Van der Loos M, Michalowski S, Leifer L. A mobile manipulation aid for the severely disabled. Proc. 9th Annual Conference on Rehabilitation Engineering, Minneapolis, MN, 1986.

Logical Structures for Collision Avoidance in Assembly with Robots

A. Rovetta

Department of Mechanics, Politecnico di Milano, Milan, Italy

ABSTRACT

This paper deals with the realization of a logical structure for the use of robots in assembly.
The development of algorithms of obstacle avoidance, by means of analysis of trajectories, is performed.
The strategy for the obstacles avoidance is the first step in the construction of a supervisory structure for controlling the physical interactions between the environment and the robot.

1. INTRODUCTION

This paper deals with the development of a logical and applicative structure for assembly robots.
Use of algorithms for collision avoidance is described.
They permit the motion from prefixed initial positions to final fixed or variable goal positions of the robot.
By means of the determination of visibility, with the evaluation of risk potential function, possible trajectories are determined.
The main elements of the research have been translated into two programs, the first on a Personal Computer in Basic language, the second on the operative computer Series 1, in AML language, for the motion of the robot 7565 IBM.
The software of the logical structure is flexible, and may be adapted to every robot, both for obstacle avoidance and for collision avoidance among several arms of a robot.

2. ANALYSIS OF THE ASSEMBLY PROCESS

According to the definition of assembly, the structure "environment + objects + man" in the final condition is different from the initial configuration.
The process of assembly is a functional variation of objects.
The logical sequence of the alterations of the state functions (partial assemblies, couplings, blockings, etc.) can be analyzed with the operative logic, following the expression:

$$(F1, F2) \cup (L1, L2) = E \qquad (1.1)$$

where:
Fi is a particular fact,
Li is a general law,
E is the event, namely the final assembly operation.

The relation (1.1) is consequent to the inference of the presence of facts and the existance of laws.
The E event is definitive.
In this scheme the feedback, that is the immediate influence of the event E on the F source of the phenomenon, is not evidenced.
The modern engineering uses process schemes where feedback is fundamental.
Therefore the scheme has to assume the following form:

$$\begin{array}{c} \uparrow F\ U\ L\ E \\ \llcorner \text{--L R---} \end{array}$$

This conception implies and expresses the closing of the logical consequence between cause and effect.
The principle of operative feedback is applied to the assembly theory, without modifying the logical terms.
In variable terms, if we consider the variations of facts/elements $\Delta E = E' - E$, we obtain:

$$\Delta F\ U\ L = \Delta E$$
$$\leftarrow \text{--L R---}$$

The meaning of the closing of the logical loop, in a continuous interactive process, is tied up with the actual developments of the technique, of technology and of science.
Protagonist are CAD, CAM, the computers and the man-machine systems.
In the event "assembly", from the contact between the bodies, variations of initial values and of physical state of components occur.
The assembly modifies itself during the process, as the components do.

3. GENERAL SCHEME OF THE LOGICAL METHOD

The scheme of the logical process is constituted by the necessity to define the principal and fundamental elements of the recognition process of the assembly.
In this structure, may be evidenced connections between:
- robot and object,
- robot and environment,
- object and environment.

3.1 THE CONNECTIONS BETWEEN ROBOT AND OBJECT

They may be expressed by:
a) robot's mobility and robot's base mobility;
b) sensors of various type;
c) mechanical actions;
d) relations of recognition between robot and object;
e) object form: geometry in plane, in space (CAD) ;
f) grasping between extremity and object.

3.2 CONNECTIONS BETWEEN ROBOT AND ENVIRONMENT

They present fundamental elements like:
a) sensors of every type;
b) supercontrol influence;
c) internal obstacle;
d) obstacle of reduced reliability, environment;
e) diagnostics and autodiagnostics.

3.3 CONNECTIONS BETWEEN OBJECT AND ENVIRONMENT

They present fundamental elements like:
a) sensors (influence by the sensorial signals, which define the relation between object and environment);
b) direct actions of the environment on the object (gravity's actions; inertial actions, etc.)
c) direct action of the object on the environment.

4. FUNCTIONAL ANALYSIS OF THE ASSEMBLY PROCESS

The different logical structures in assembly are pointed out by the following relations:

$$F \cup L \longrightarrow\!\!\!> E \qquad (4.1)$$

$$F \cup L \longrightarrow\!\!\!> E$$

$$L'r <\!\!\!\longleftarrow\!\!- \qquad (4.2)$$

$$Fx \supset (Qx == Ex) \qquad (4.3)$$

These series of logical relations allow to interpret the assembly, realized with the robot.

The relation (4.2) considers a closed loop of logical type, to reproduce the operative structure of the robot.

It is positive to consider that in the case of the automatic rigid (and not flexible) machine, the logical scheme has only one possible direction. It must necessarily become a " loop " in a system "robot/ environment/object", where the total structure varies during the process.

The feedback logical structure must be practically applied to assembly.

5. SIMULATION MODEL DESCRIPTION

By means of the program a mobile or a fixed obstacle may be avoided.

The first phase of the simulation model represents the configuration of the obstacle; the second phase performs robot and obstacle motions, by applying the collision avoidance strategies.

5.1 The scenary where robot is running is represented by means of geometric elements. They may be polygons of regular or irregular

shape, with data obtained from sensors or from internal files.
The obstacle is characterized by shape, dimensions, security parameters (distance, tolerances), which influence the robot's motion strategies.

A main parameter is also the "risk potential" used in the operative phase of the program.
It defines the effective presence of the obstacle with reference to the interaction with the robot.
The limit risk potential defines the boundary condition for the robot, in order to avoid dangerous possible collisions with the obstacle.

5.2 In the operative phase, the robot initial and final positions are represented.
After examining the conditions of obstacles, the visibility and duration evaluation procedures are applied.
The strategy for the choice of the motion is pointed out, in order to move the robot it self. The possible considerated strategies are (fig. 1):

Ea: calculation of a complete trajectory which is fulfilled until the risk potential overgoes the prefixed limit; after that, a new trajectory is calculated;

Eb: the trajectory is calculated at every step, and is continuously adjourned.

Ec: the trajectory is chosen at the beginning of the motion with a principle of optimization and is prosecuted until the
risk potential is too high. A completely new trajectory is calculated, with the criterium of optimization.

Ed: the trajectory is determined, in consequence of the complete knowledge (also statistical) of the obstacle's motion.

The strategy Eb has been followed in a particular form in this research.
At every step, the following computations are executed:
- analysis of relative position of obstacle and robot (initial / final places);
- determination of visibility of the relative positions;
- computation of trajectories, with control of feasibility;
- determination of optimal trajectory, by following prefixed optimization criteria.

5.3 MODELLING OF ROBOT AND OF OBSTACLE

The modelling has considered a bidimensional analysis of robot and of obstacle.
The same strategies may be applied to threedimensional cases, as developed in the prosecution of the research.
The robot is represented by the point which is closest to the obstacle.
The obstacle is simulated by a series of rectilinear sides, which are connected to constitute a polygon.

The polygonal shape reproduces objects, with a digital representation that is transferable to the results of a vision system.

Fig.2 reports an example of robot motion, following strategy Ec.

Line 14 reports the followed trajectory.

6. ALGORITHMS FOR THE COLLISION AVOIDANCE

6.1 Analysis of interference.

For defining the obstacle's position with reference to the robot, every apex has been considered as origin of a relative reference system X,Y, in the counter/clock wise.

With reference to the position R of the robot (see Fig.3), a parameter $RI(i)$ is calculated, which is "1" if the robot and obstacle are occupying the same semiplane, with respect to X.

Otherwise, RI is "0".

The product $DR = P < RI(i) >$ for all the sides ($i=1,N$) has value =1 if there is interference.

If DR=0, it means that at least for one side there was no interference of robot and obstacle.

The algorithm is simple and may be applied in a very quick way.

6.2 Criterium for the visibility.

The visibility is determined by examining the relative position of the edges of the obstacle from the initial point A to the final point B of the robot's motion.

Two cones of visibility are calculated and, depending on their amplitude, the connection from A to B may be direct (visibility=1) or indirect (visibility=0) and necessity of obstacle avoidance.

The criterium is efficient and permits high precision, as verified with the simulation program and with the experimental tests on robot IBM 7565.

7. ANALYSIS OF THE ALGORITHM FOR OBSTACLE AVOIDANCE

The avoidance algorithm is composed of a flexible sequence of operations, made of subsequent steps.

Therefore the global avoidance collision procedure is represented from:

$$Ei = A_{i,1} \wedge A_{i,2} \wedge A_{i,3} \wedge \ldots \wedge A_{i,n}$$

where $A_{i,j}$ are consequential operations.

Ei may be assumed as rigid operation.

The optimal choice of avoidance occurs among different series of sequences, following:

$$OP = E_1 \vee E_2 \vee \ldots \vee E_m$$

Because every Ei is a rigid operation, the flexibility of

obstacle avoidance is due to the flexibility of every operation $A_{i,j}$ and to the choice of the E sequences.
The optimization is connected to the parameter of functionality.
When time duration is chosen, then

$$\frac{\partial OP}{\partial t} = 0$$

is the minimum condition.

When trajectory length is chosen, then

$$\frac{\partial OP}{\partial l} = 0$$

is the minimum condition.

8. DESCRIPTION OF THE LOGICAL STRUCTURE

The single A_i phases may be summarized:

A_1 = obstacle shape analysis
A_2 = definition of robot B final point
A_3 = scene analysis from robot A initial point
A_4 = scene analysis from robot B final point
A_5 = visibility determination
A_6 = trajectories proposal
A_7 = trajectories feasibility analysis
A_8 = choice of optimal trajectory.

In a global expression:

$E_i = A_1 \wedge A_2 \wedge A_3 \wedge A_4 \wedge A_5 \wedge A_6 \wedge A_7 \wedge A_8$

where a binary logical structure may be ensured.

In order to obtain a real flexibility, the optimization is obtained from the relation

$OP = E_a \vee E_b \vee E_c \vee E_d$

where every E_j operation is valid.
Only supervisory control makes the choice of the most suitable strategy E_j.

It follows that:

1) $A_{i,j}$ operations are fundamental elements of the program.
In the simulation program they are represented by subroutines in Basic language.
In AML they may constitute " aggregates " able to contain

elementary operations.
2) A high number of different strategies may be developed, using the elementary operations A_i in different connections in the software.
3) The operating strategy may be modified with some elementary substitution in the process.
4) It is possible to modify, during the motion, the operating strategy, by using different E_i stretegy with analogous A_i, j operations.

9. CONCLUSIONS

A preliminary simulation program evaluates the effective significance of the theoretical analysis. The preliminary simulation program effects the following operations:

- it models the obstacle with a plane polygonal form;
- it models an arm as a point in the plane;
- it utilizes dimensions of the plane corresponding to the robot 7565 IBM, on which the research work is developed;
- it utilizes the graphical characteristics of the colour for the simulation;
- it utilizes the BASIC language with an open formulation, to facilitate the implementation on the computer Series 1 of the robot 7565, the AML language;
- it simulates the obstacle avoidance in the plane, following easy recognitions procedures, which use the matrix analysis of plane figures, through reference system, with the sides of the objects;
- it proposes optimization criteria: minimum trajectory lenght, minimum motion duration, etc.;
- it develops an interactive dialog between the program's user and the computer for the future simplicity, also off-line, of the program.

The theoretical analysis can point out after the simulation, the characteristics of the risk potential, in relation to the robot and the obstacle, having the operative and dynamical conditions of the robot's arm.

The program in AML language is running on a robot IBM 7565 with the computers Serie 1, The program is structured in hierarchical form, and is developed with macroinstructions.

The structure of the program has indicated that the phase representation permits the off-line programming.

The optimization's choice of the robot's motion may be programmed off-line by using more strategies, with supervisory procedures.

REFERENCES

(1) Thomas Lozano Perez, Automatic Planning of manipulator transfer movements, Massachussets Institute of Technology, December 1980
(2) Thomas Lozano Perez, M.T.Mason, Russel H. Taylor, Automatic Synthesis of Fine-motion Strategies for Robots, The International Journal of Robotics Research, Spring 1984;
(3) I. Del Gaudio, S.Del Sarto, Motor Fan Flexible Automatic Manufacturing System, 4-th International Conference on Assembly Automation, 1980;
(4) B.Dufay, Apprentissage par induction en robotique, application a la synthese de programme de montage, Institut National Polytechnique de Grenoble, Juin 1983;
(5) Y.Descotte, J.C.Latombe, Compromising among antagonist constraints in a planner, Institut National Polytechnique de Grenoble, January 1984
(6) J.L.Nevins, D.E. Whitney, Computer-controlled Assembly, Scientific American, February 1978;
(7) T.Lozano Perez,he esign of a Mechanical Assembly System, Artificial Intelligence Laboratory, Massachussets Institute of Technology, December 1976;
(8) M.Myrup Andreasen, S.Kahler, T.Lund, Design for Assembly, Springer-Verlag, 1983;
(9) T.Sata, F.Kimura, H.Hiraoka, M.Enomoto, An Approach to Model Based Robot Software for Industrial Application, IFIP Working Conference, Como, Italy, June 1984;
(10) A.Rovetta, Logical Structure of Assembly with Robots , IEEE Robotics and Automation Conference, S.Francisco, April 1986.

ACKNOWLEDGEMENT

This work was developed in cooperation with dr. eng. Franco Zecchini and dr. eng. Giuseppe Frosi, IBM Italia.

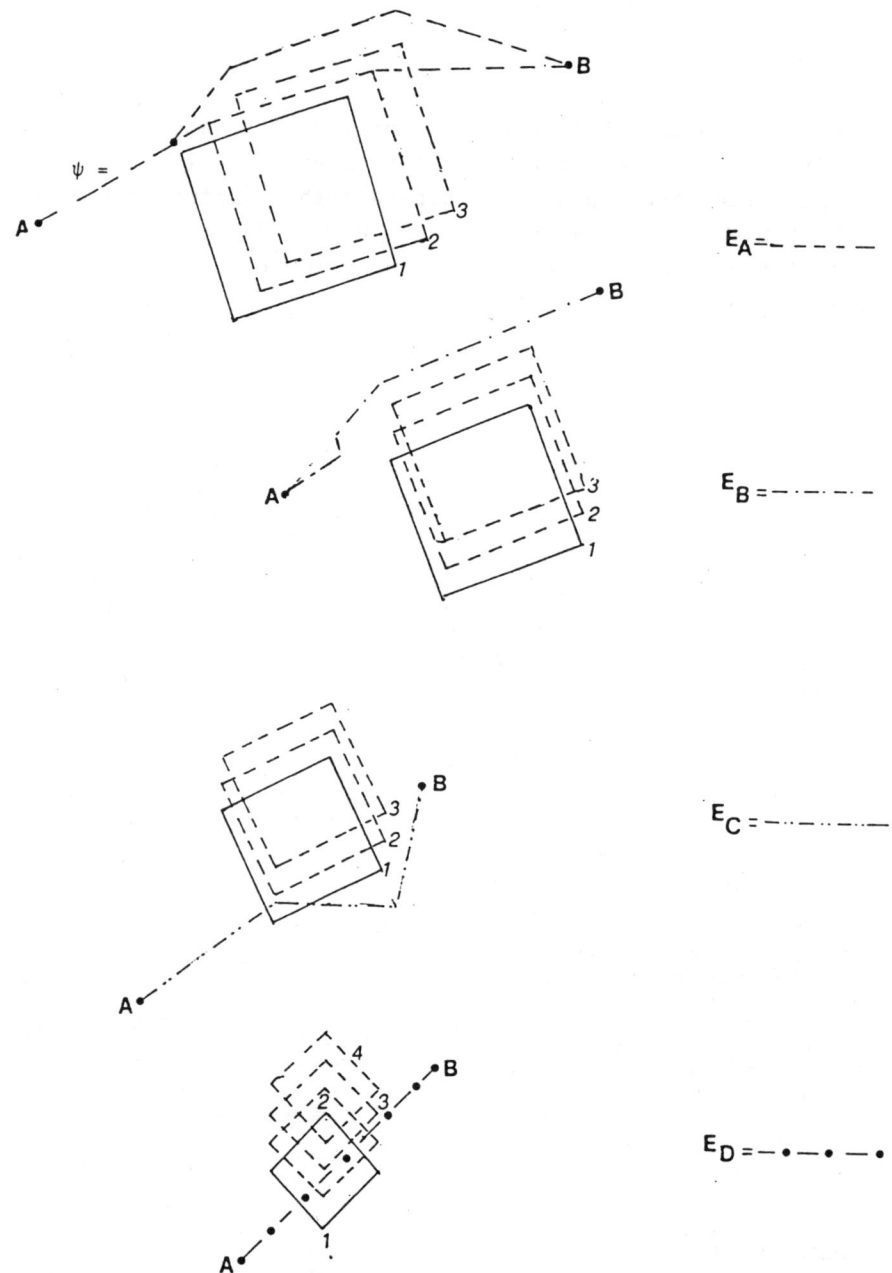

Fig. 1

Logical Structures for Collision Avoidance in Assembly with Robots 533

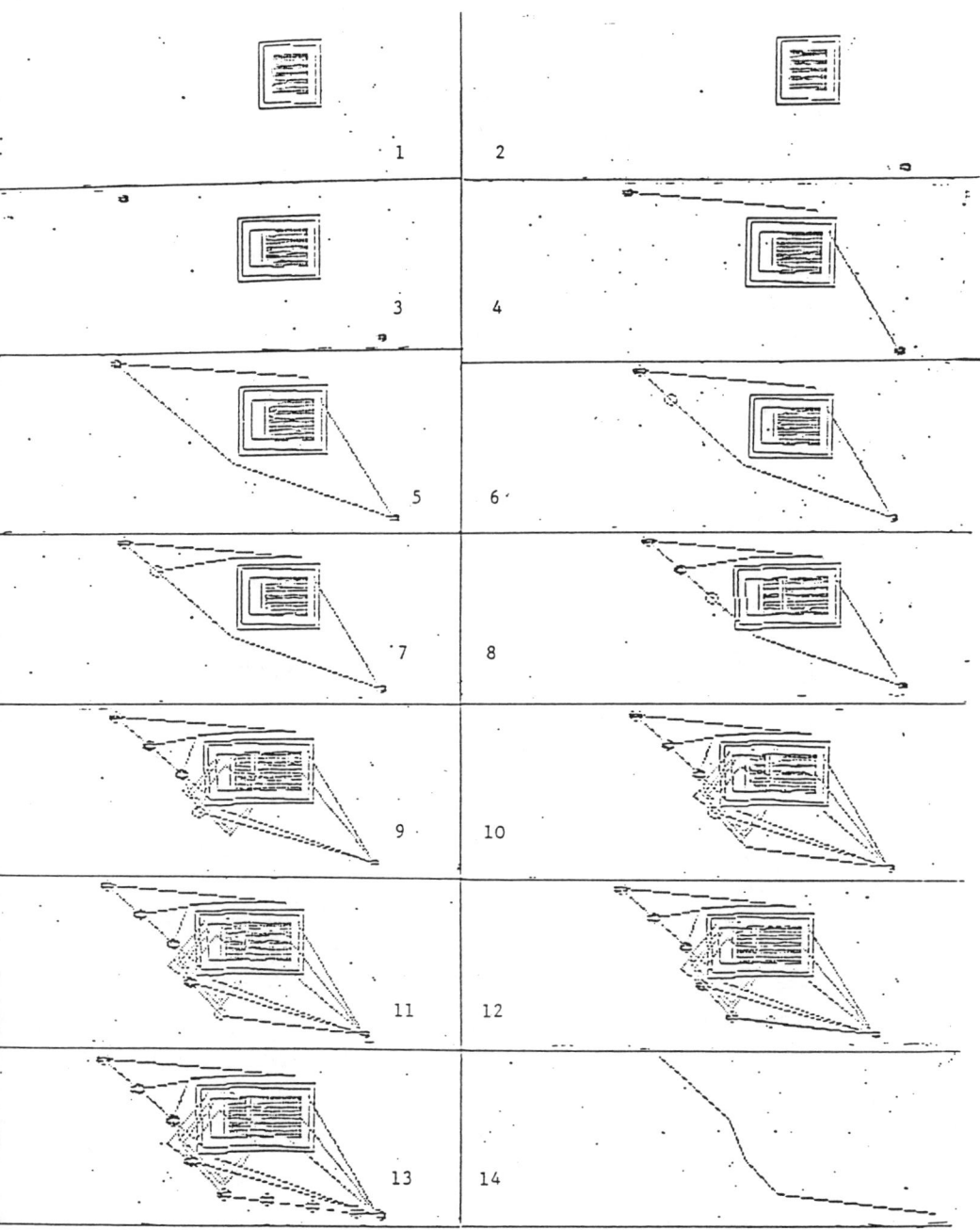

Fig. 2

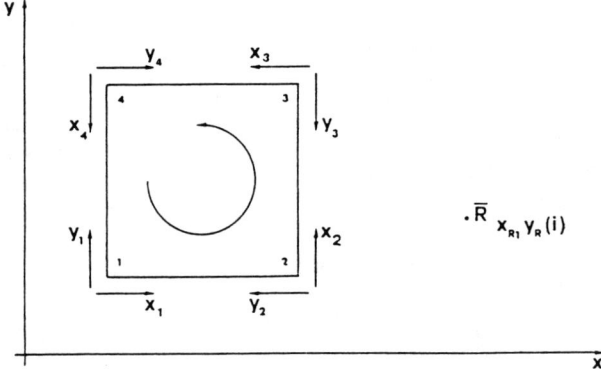

Fig. 3

Repositioning-Unit for very Fine and Accurate Displacements. Analysis and Design

F. Artigue, C. François and J.G. Pontnau

LIMRO-I.U.T. de Cachan, Université de Paris-Sud, Cachan, France

INTRODUCTION

Displacement and orientation of work pieces in space are the typical tasks for which industrial robots are well suited. Nevertheless these robots, which consist generally of several rigid bodies connected sequentially by revolute or prismatic points, are mainly designed to perform large scale displacements. In the same way as they are built according to different coordinate structures (i.e. cartesian, cylindrical, spherical or articulated) the analysis of the task planning may be a criterion for the choice of the suitable robot architecture. The SCARA robot is a typical example of this choice. On the contrary, when task planning needs eds small amplitude displacements, high positioning accuracy and high resolution, commercially available robots with classical architecture have not the required performance. For instance, the superposition of moving joints leads to several drawbacks such as cumulation of backlash, too much bulk and robot control tricky to carry out. For these reasons, in the last few years, several laboratories have designed three dimensional displacement units which are better suited to perform fine and very accurate displacements (figure 1). Such units with various geometrical architectures have been designed and built at LIMRO laboratory (1).

The aim of this paper is to describe these units and their control in a synthetic way.

BACKGROUND

A three dimensional displacement unit with co-inserting axis presents some similarities with the configuration of prehensors with three, four, five or six fingers (2). The mobile parts of the unit are driven by linear actuators. This type of actuator links with the static base and also characterizes the involved number of degrees of freedom of the unit. So, knowledge of both the allowed motions and the number of degrees of freedom of the joints leads to a complete description of the possible displacements of the mobile part.

The up to date modules are equipped with six linear actuators. The main differences between them arise from the disposition and the type of the actuators. Two configurations are commonly used:

- the first one (2x3R) is represented in figure 2a. The linear actuators include a swivel at each end. The five degrees of freedom of each link are obtained through the three rotating degrees of freedom of each end of the actuators (one of them being common for both ends).
- the second one (3R + 2T) is represented in figure 2b.

The linear actuators define always a five degree of freedom link, but these degrees of freedom are arranged together at one end through a swivel and sliding plate. Such a disposition allows two crossed translations perpendicular to the controlled displacements of the linear actuator.

To describe complex motions requiring both translations and rotations, these modules are controlled through the computed translation values applied to each linear actuator.

FEEDBACK CONTROL OF THE LINEAR ACTUATORS

A planned motion is described by tensor operator including in an Euclidean frame a translation vector T and a rotation vector R. The motion of a point M tied to the mobile part due to a rotation defined by R is calculated by a classical rotation matrix. Nevertheless, in the typical applications of these modules, i.e. small amplitude displacements, it is possible to give simplified expressions. For instance, a V' vector derived from a V vector through a R rotation may be calculated by the CAMPBELL HANDSDORFF formula, as follows (3)

$$\vec{V}' = \vec{V} + \vec{R} \wedge \vec{V} + 1/2 \vec{R} \wedge (\vec{R} \wedge \vec{V}) + 1/6 (\vec{R} \wedge (\vec{R} \wedge (\vec{R} \wedge \vec{V})) + \cdot \cdot 1/n! (\vec{R} \wedge (\vec{R} \wedge \vec{V}) + \vec{\epsilon}$$

The amplitude displacement (given by $\|R\|$) and the required accuracy allow limitation of the expansion in power series. For a few degrees rotation, a first order approximation is enough. This over-simplification enables either to save time of computer calculation and so to carry out a kinematic control of the module.

For each linear actuator, the following notations are used:
- $\vec{D}$ is the displacement vector of a point tied to the link between the linear actuator and the mobile part of the module.
- $\vec{p}$ is the required lengthening of the linear actuator
- $\vec{n}$ is the unit vector along the initial direction of the linear actuator.
- l the initial length of the linear actuator. It is important to point out that, for the (3R + 2T) modules, only the dispositions with the linear actuators arranged perpendicularly to the moving plates are considered.

Referring to the figures 3 and 4, the lengthening values of the linear actuators are fairly well calculated as follows:

$$2 \times 3R \text{ Module} \quad p = \|\overrightarrow{M_1 M_2} + \vec{D}\| - \|\overrightarrow{M_1 M_2}\|$$

$$\text{with } \overrightarrow{M_1 M_2} = l\,\vec{n}$$
$$3R + 2T \text{ module } p = \overline{D\vec{n}}/\vec{n}.\vec{n}$$

For small amplitude displacements and after simplification the p expressions are similar for the two types of module i.e.:

$$p = \vec{D}\,\vec{n}$$

FEEDBACK CONTROL FOR SMALL AMPLITUDE DISPLACEMENTS

When small amplitude displacements are required, the following conditions are implemented:
- the displacement due to the rotation is calculated with only the first order term of the CAMPBELL HANSDORFF formula.
- the lengthening value is calculated from the simplified expression $p = \vec{D}\,\vec{n}$.
Under these conditions the lengthening values may be explicitly written. After generalization for the six linear actuators, we have (4):

$$\{P\} = \{M\}\,\{Tx, Ty, Tz, Rx, Ry, Rz\}$$

$\{M\}$ is the control matrix of the module. We are now in position to analyse the configurations of the linear actuators which are associated with the most simple control law to implement.

Such conditions may be carried out through a limitation of the number of linear actuators driving an elementary displacement (translation or rotation on a single axis), like in figure 5. The performance related to the position accuracies may also be optimized, nevertheless the analysis shows that the geometrical configurations obtained are more or less identical to the preceding ones (5) (figure 6).

STRESS ANALYSIS

Prior to the discussion, it must be pointed out that for conventional robot arm, the calculation of resultant stress acting on each joint from forces and torques experienced by the module is very tedious work. For a module, if F and M designate respectively the force and the angular momentum experienced, $f_1, f_2 \ldots f_6$ the stress on each linear actuator, we have:

$$\vec{F} = f_1 \vec{n}_1 + f_2 \vec{n}_2 + \ldots f_6 \vec{n}_6$$
$$M = \vec{a}_1 f_1 + \vec{a}_2 f_2 + \ldots \vec{a}_6 f_6 \text{ with } \vec{a} = \overrightarrow{OM} \wedge \vec{n}i$$

In matrix form:

$$\{F, M\} = \{P\}\,\{f_1, f_2 \ldots f_6\}$$

In fact, the $\{P\}$ matrix which is the transpose of the $\{M\}$ matrix is the characteristic matrix of the system.

From these expressions the stress acting on the linear actuators may be calculated; this knowledge enables one to outline a dynamic control of the module which is very useful in automatic assembly operations. Another attractive way is to carry out a theoretical stress model which is effectively applied by the actuators to conclude that there is no jamming or collision during an assembly operation.

MECHANICAL CONSTRUCTION

Until now five repositioning units have been built and tested at LIMRO laboratory (two 3R + 2T type and three 2 x 3R type). Among them, two are powered by hydraulic actuators and three are powered by electric actuators. A comparison between the performances of the two types of repositioning units allows listing of both the drawbacks and the advantages inherent in each module as follows:
- firstly, the advantages of the 3R + 2T module. When the simplified equations are used, positional accuracies are better than with the 2 x 3R module. Also, when the complete equations are used, time computation may be saved.
- secondly, the drawbacks of the 3R + 2T module. Mechanical construction is more complex, more costly than with the 2 x 3R module. Also, move identical work pieces, the 3R + 2T module is more heavy and bulky than the 2 x 3R module.
 This is a great disadvantage if the module is the end effector of an arm robot.
 To conclude this analysis and to take advantage of the one or the other module, it must be pointed out that the decoupling between the mobilities is easier with the 3R + 2T module, thus enabling a possible modular architecture.
 Figure 7 shows the modules which have been built at the LIMRO laboratory.

CONCLUSION

Assembly automation has lead several laboratories to study and develop specific products which are well suited to the requirements of fine amplitude displacement and high positioning accuracies. These repositioning units which are built according to different architecture are analysed in this paper.
 To be operational, these modules have to be transported in the working zone by a conventional robot arm or simply by a pick and place manipulator. In the first case there is a strong redundancy between the robot arm and the repositioning unit. The control strategy has to take into account this redundancy.
 At LIMRO laboratory we have used sequentially controlled arms with measurement of the displacement.

REFERENCES :

[1] C.FRANCOIS et F.ARTIGUE: "Module de repositionnement et capteur de déplacement tridimensionnels pour des robots d'assemblage dans l'hypothèse de petits déplacements". Actes du 7 ème Congrès Français de Mécanique.Bordeaux 1985.

[2] J.C GUINOT et BIDAULT : "Etude d'un préhenseur articulé muni de sens tactile" Rapport CNRS Gréco A.R.A .Pole Mécanique et Technologique.Besançon.1983.

[3] J.F PRICE :"LIE GROUPS and COMPACT GROUPS".London Mathematical Society.Lecture Note,Serie 25.Cambridge University Press.

[4] F.ARTIGUE et C.FRANCOIS :"Assemblage par prise de référence" Journées A.R.A.Toulouse 1984.

[5] F.ARTIGUE : "Analyse cinématique de système de repositionnement et de mesure tridimensionnels pour l'assemblage automatique".Thèse d'Etat,Université Paris VI.1984.

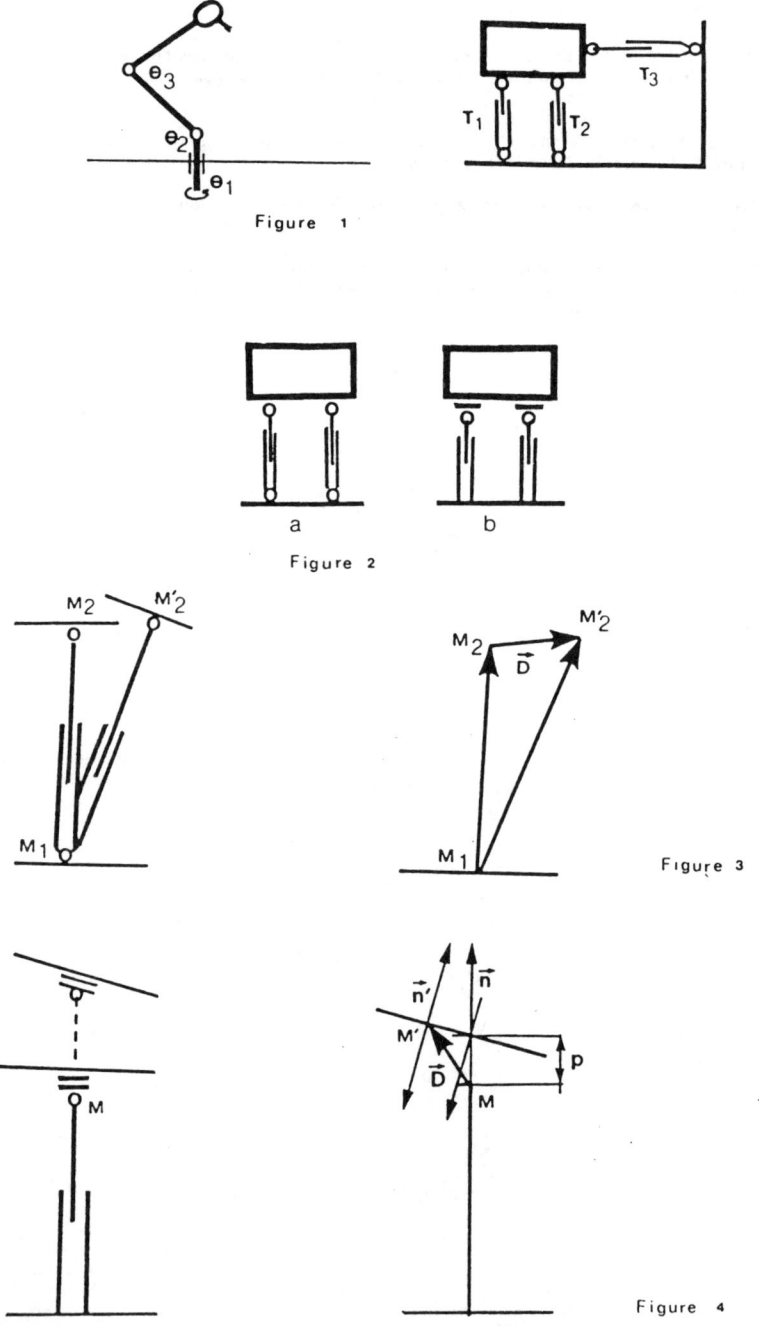

Figure 1

Figure 2

Figure 3

Figure 4

Very Fine and Accurate Displacement 541

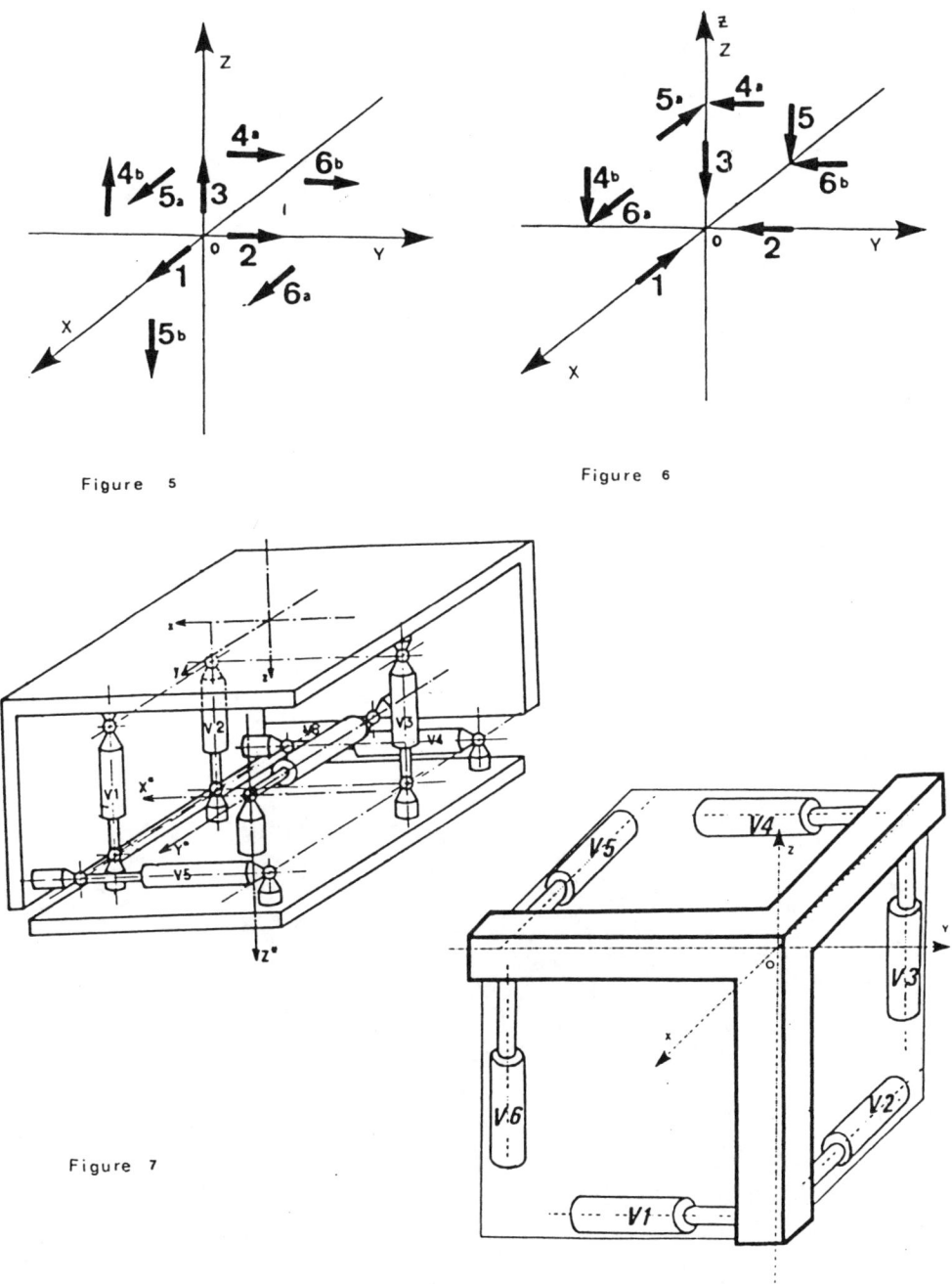

Figure 5

Figure 6

Figure 7

Part 11
Synthesis and Design 3

Polyarticulated Mechanical Structure for Decoupling the Position and Orientation of a Robot

M. Fayet and A. Jutard

INSA, Lyon, France

SUMMARY

In this paper we present a mechanical structure with revolute joints which can be used eventually on a robot and which allows the independent functioning of the positioning and the orientation of the terminal organ. It generalises the principle of the pantograph. First a three-bar system is proposed followed by the structure of the robot itself, replacing one of these bars, so we can consider just a two-bar system. An original guiding system is envisaged for these bars in the arm and we show that no guiding is necessary in the forearm. The problem of singular configurations is approached and some solutions are found for those which cause most problems.

1. A MECHANICAL STRUCTURE ALLOWING THE REPRODUCTION OF A MOTION AT A FIXED POINT AROUND ANY OTHER POINT

The design of a great number of handling machines seeks to carry out the decoupling of the functions of positioning and orientation in a structure which only has revolute articulations. This is the case in the simple pantographs used in public work machines, in mechanical shovels etc. and in a more sophisticated way on the five axes ASEA robot which has two pantographs and a differential mechanism (1). Regarding this robot the aim is to make the control models as simple as possible.

However, in all of these examples there is only partial decoupling. On the contrary, the generalized pantography, which is proposed here, fulfills this aim completely, namely that three degrees of mobility (D.O.M.) are used in the positioning of a terminal organ point and three others are used separately in its orientation.

2. GENERALISED 3-BAR PANTOGRAPH

The first idea consists of placing three bars (Fig. 2a) inside

the arm and the forearm; a guiding system allows them to stay
parallel to the respective axes of these two organs. Thus, the bars
$a_1 b_1$, $a_2 b_2$, $a_3 b_3$ only need to be of the same length OO' as that
of the arm and the bars $b_1\ c_1$, $b_2\ c_2$, $b_3\ c_3$ all of the same length
so that the aims stated in 1 are achieved: when the triangle $a_1\ a_2\ a_3$
is moved around O, the triangle $c_1\ c_2\ c_3$ follows from the former
through translation, whatever the position of the arm and the forearm.

Note that all the joints at the $a_1\ b_1\ c_1$, $a_2\ b_2\ c_2$, $a_3\ b_3\ c_3$
points are spherical. In order to ensure large movements they will
be made by means of three revolute joints. One of these will be
changed using a self-aligning bearing to allow the passage of the
singular configuration defined by two aligned axes. (Fig. 2b).

3. CRITICAL REMARKS

The three bar solution presents many problems, cost, bulkiness
as well as problems of rigidity.

This is the reason why we suggest a second design in which the
carrier part is classical. We shall also show that only two
bars are required to direct the terminal organ.

4. GENERALIZED 2-BAR PANTOGRAPH
4.1. Presentation

A drawing of the mock-up is presented here (Fig. 4.1a) and a
sketch (Fig. 4.1b). The system appears to be very simple. This is
due to the fact that the terminal organ is articulated on the
structure itself by means of three revolute joints (Fig. 4. 1a)
between the solids 2, 13, 14 and 3. These three joints together
are symbolised by a spherical joint between 2 and 3 in Fig. 4.1b.

With this device it is the frame itself which plays the part of
carrier and at the same time takes the place of the bars $a_1\ b_1$ and
$b_1\ c_1$.

Thus two bars only can be used. They are guided in a very simple
way in 1 and two others no longer need to be guided in 2.

4.2. Guiding System Inside 1

Bar 8 is held parallel to 5 by the pliable parallelogram $g_2\ g'_2$
$f_2\ f'_2$. The joint 5-4 is a revolute pair and 4 can turn with regard
to arm 1 around $\underline{z_1}$.

On the other hand in Fig. 4.1b, 8' is held parallel to 8 by the parallelogram g_3 g'_3 a_2 a_3.

But if we come back to guiding the 8, it must be noted that a_2 point of 10 must move on a sphere Sa_2 (Fig. 4.2) with the centre a_1. So the device formed must allow this movement on Sa_2.

The reasons for this are the following. Due to the two revolute joints of the parallel axes in g_2, the point g'_2 moves, with regard to 4 on a spherical surface of centre g_2. But since 4 turns around 1, this spherical surface generates a toric volume in $\ell g'_2$. The point g'_2 can then move in this torus in which, it can be noted that, the central part (not hatched) cannot be reaches by g'_2. Besides a_2 can be in a similar toric volume ℓa_2 following from the former through translation of vector g_2 a_2. So that the movement of a_2 is possible, Sa_2 must be interior to ℓa_2. The constructive dimensions which allow the greatest possibility of movement for g_2 should answer to the relation: $a_1 \, a_2^2 = R^2 - \frac{d^2}{4}$.

4.3. Absence of guiding inside 2

Now let us show that no guiding is necessary for the bars 9, 9', in 2. It only needs link 11 to articulate on 9' and connected to 9 as shown in Fig. 4.1b so that these two bars are coplanar. Thus the quadrilateral b_2 b_3 c_2 c_3 is a parallelogram.

Indeed if 1 and 2 as well as the points b_2 and b_3 have determined positions, the points c_2 and c_3 are on the spheres Σ_2 and Σ_3 of centres b_2 and b_3 (Fig. 4.3). On the other hand, b_2 b_3 c_2 c_3 being a parallelogram c_2 c_3 is parallel to b_2 b_3 and since c_2 and c_3 are linked to solid 3, in spherical articulation in c_1, these points must be on circles δ_2 and δ_3 which would describe these points if 3 turns around the axis (c_1, δ) parallel to b_2 b_3.

Finally c_2 and c_3 are on the respective intersections of Σ_2 and δ_2 and of Σ_3 and δ_3. Or in other words, if Σ is the cylinder generated by the right c_2 c_3 turning around (c_1, δ) and if Γ_2 and Γ_3 are the intersections of Σ with Σ_2 and Σ_3, c_2 and c_3 are on the intersections of Γ_2 and δ_2 and of Γ_3 and δ_3.

4.4. Application of the mechanism theory

The previous geometrical demonstration can be proved by the mechanism theory showing that the D.O.M. of all the mechanism of the Fig. 4.1b is indeed 6 as we wish.

The graph associated with Fig. 4.1b is represented in Fig. 4.4. Six finite faces appear on the graph 4. They define six independent elementary circuits. The unit screws of each of these circuits are of rank 6 except those of the circuit (5,6,8,7) which are of rank 3. According to the formulae proposed by Hunt and other authors 5,6,8, which can sometimes be faulty but which is quite appropriate here, the D.O.M. is then as expected in general: $d = 39-5 \times 6 - 3 = 6$. 39 is the total number of degrees of freedom (D.O.F.) in all the joints.

5. SINGULAR CONFIGURATIONS (S.C.)
5.1. Geometrical approach

The construction seen on §4.2 shows two solutions c_2 c_3 and c_2' c_3'. But the intersections $\Gamma_2 \cap \delta_2$ and $\Gamma_3 \cap \delta_3$ can be double or indeterminate. It is then a question of singular configurations.

The double solutions are obtained when the right b_2 b_3 cuts $b_1 c_1$ (Fig. 5.1a). This position gives c_2 and c_3 the possibility to move following the arrows f or f'. So it must be avoided by a thrust which prevents the plane c_1 c_2 c_3 from containing b_1 c_1.

The solution of Fig. 4.1a would require a thrust situated in B, in the middle of the arc where 14 moves.

Thus we prefer the Fig. 5.1 solution where two thrusts are placed on the extremities of the arc where 14 moves.

The indeterminate intersections appear when b_2 b_3 and c_1 are aligned, so $\Gamma_2 \equiv \delta_2$ and $\Gamma_3 \equiv \delta_3$ (Fig. 5.1). Triangle 3 can then turn freely around b_2 c_1. But this position is not admitted by the preceding thrusts since this configuration also assumes that b_1 c_1 is in the plane c_1 c_2 c_3.

5.2. Approach using the mechanism theory

We shall not examine the S.C. here due to the very classical external structure and the 3 S.C. being very well known (7). We shall thus consider 0, 1, 2 as forming one single solid I.

In order to explore the S.C. due to the internal mechanism which is the multi-loop type, we shall on the other hand vary its three D.O.M. separately. We shall make the triangle of control a_1 a_2 a_3 turn one after another around the axes $(a_1, \underline{z}_1)$, $(a_1, \underline{z}_2)$ and $(a_1, \underline{u})$ ($\underline{u} // \underline{a}_2 \underline{a}_3$).
a) Rotation around $(a_1, \underline{z}_1)$

In this case the mechanism is equivalent to the simplified (Fig. 5.2a). All the part interior to I act as one single solid 8. This is a plane mechanism. The graph which is associated is given in (Fig. 5.2a). The matrix of the unit screws can be:

$$[M] = \begin{bmatrix} \hat{S}_{a_1}, \hat{S}_{c_1}, \hat{S}_{b_3}, \hat{S}_{c_3}, \hat{0}, \hat{0} \\ \hat{S}_{a_1}, \hat{S}_{c_1}, \hat{0}, \hat{0}, \hat{S}_{b_2}, \hat{S}_{c_2} \end{bmatrix}$$

Its rank is inevitably maximum 5 since we have, in fact, here the double-parallelogram mechanism examined in (8). The D.O.M. is thus indeed 1. Note that each independent circuit of the graph presents screws whose rank is generally 3. But it can go down to 2 when a_1 c_1 b_3 c_3 for example are in the same plane. However, this cannot occur simultaneously in general in the two circuits (similar to the ASEA transmission (1)). It is like this because of the location of the points a_1 a_2 a_3 at the vertices of a rectangular isosceles trangle. However, the points $a, b, b_2, b_3, c_1, c_2, c_3$ can, if in a very particular position, be in the same plane. We meet the S.C. again mentioned in 5.1 which is avoided owing to the thrusts.

b) Rotation around (a_1, z_2)

The study of the mechanism during this movement is symmetrical to that shown in a): the solids 9-9' interior to 2 have now become one piece. However, the S.C. which would bring the triangle a_1 a_2 a_3 to contain (a_1, z_1) is no longer really an S.C. In fact if we take the geometrical point of view, the guiding system of 8 and 8' does not allow the evolution from this S.C. towards two possible solutions.

c) Rotation around $(a_1, \underline{u})$ $(\underline{u} // \underline{a_2 - a_3})$

In this case the solids 8 and 8', 9 and 9' behave as two solids 8 and 9 and the equivalent mechanism appears (Fig. 5.2). The graph 5.2c' presents three finite faces but if the link (5,6,8,7) is replaced by one single liaison between 5 and 8 (89) of the screw $\hat{S}_{g'_2} = (\underline{0}, \underline{g'_2 g_2} \times \underline{v})^T$ ($\underline{v}$ unit vector of the liaison 5-6 in g_2), the graph 5.2c" is given with two finite faces only.

The matrix of the unit screws can take the form of:

$$[M] = \begin{bmatrix} \hat{S}_{a_1}, \hat{S}_{a_2}, \hat{S}_{b_2}, \hat{S}_{c_2}, \hat{S}_{q_1}, \hat{0}, \hat{0}, \hat{0}, \hat{0} \\ \hat{S}_{a_1}, \hat{S}_{a_2}, \hat{0}, \hat{0}, \hat{0}, \hat{S}_{c_2}, \hat{S}_{g_2}, \hat{S}_{g_2}, \hat{S}_{a_1}^2 \end{bmatrix}$$

The first line of screws corresponds to the circuit c_1 = (I,10, 8,9,3). All the axes of the liaisons are parallel to $\underline{u} = \lambda \underline{a_2 a_3}$. The rank equal to 2 if a_1 a_3 b_2 b_3 c_1 c_2 c_3 are in a same plane. But this eventuality is not admitted by the thrusts seen in a).

The last line of the screws corresponds to the circuit c_2=(I,10,8,5,4)

The rank of the screws involved is generally 5. In fact if a vectorial shape is given to the screws expressed at point a_1, for this second line we have:

$$\left[\begin{array}{c}\underline{u}\\0\end{array}\right]\left[\begin{array}{c}u\\a_1\vec{a}_2\times u\end{array}\right]\left[\begin{array}{c}z_1\\a_1\vec{a}_2\times z_1\end{array}\right]\left[\begin{array}{c}0\\q'\vec{g}\times v\end{array}\right]\left[\begin{array}{c}z_1\\a_1\vec{g}\times z_1\end{array}\right]\left[\begin{array}{c}z_1\\0\end{array}\right]$$

where the first line of vectors has only two elements.

But, if a_2 is on the axis (a_1, z_1), $a_1\ a_2\ z_1 = 0$, two screws are the same and the whole thing (6,5,4) can turn around (a_1, z_1). The rank of the M matrix stays 8. The D.O.M. of the equivalent mechanism stays 1, but it is used for this movement around (a_1, z_1). Then, the usual movement cannot arrive to the terminal organ. However, if $(a_1 g_2, z_1, u) = 0$, the rank of the previous six screws is 2 and the rank of M becomes 7. Then the equivalent mechanism has 2 D.O.M. and the usual movement can go on.

In order to overcome this, we propose adding between 4 and 5 an elastic liaison presented in Fig. 5.2d which allows us to obtain an eventual degree of freedom to get out of this S.C.

Note that this solution allows a_2 to sweep across at least a part of the non-hatched area in Fig. 4.2. The movement of the control triangle 10 thus becomes more convenient in a larger area.

We have considered here the greatest number of S.C., without being able to confirm that we have seen them all. In particular, we have considered the linear dependencies only between screws belonging to one circuit. Nevertheless we think that the cases which have been studied are the ones which present the most problems. We would like to add that we are preparing a paper for the M.M.T. magazine on the subject of the systematic research of the S.C. of a multi-loop mechanism.

CONCLUSIONS

The mechanism presented works very well in the complete decoupling of the functions of positioning and orientation. It does not present any great problems in its construction. If there are any S.C. which do present problems, we have presented some quite simple solutions. These solutions allow movements of greater magnitude.

REFERENCES

1. ASEA Documentation CK091502E/S 72184 VASTERAS SUEDE 1983
2. KOENIG Leçons de cinématique PARIS HERMANN 1895
3. PRUDHOMME Cinématique - Théorie - Applications PARIS DUNOD 1954
4. BERGE Graphes et hypergraphes PARIS DUNOD 1970
5. HUNT Kinematic, Geometry of Mechanisms OXFORD CLARENDON PRESS 1978
6. MOROSHKIN Seminara pro Teori i Masch. i Mekh 1954
7. WITNEY The Mathematics of coordinated Control Prosthetic Arm and Manipulators Transactions of ASME SERIE F, Décember 1972
8. DAVIES Mechanical Networks II. Mech. and Mach. Theory - vol. 18 - n°8 PERGAMON PRESS 1983
9. BAKER On relative Freedom between Links in Kinematic Chains with Cross Jointing Mech. and Mach. Theory - vol. 15 - PERGAMON PRESS 1980

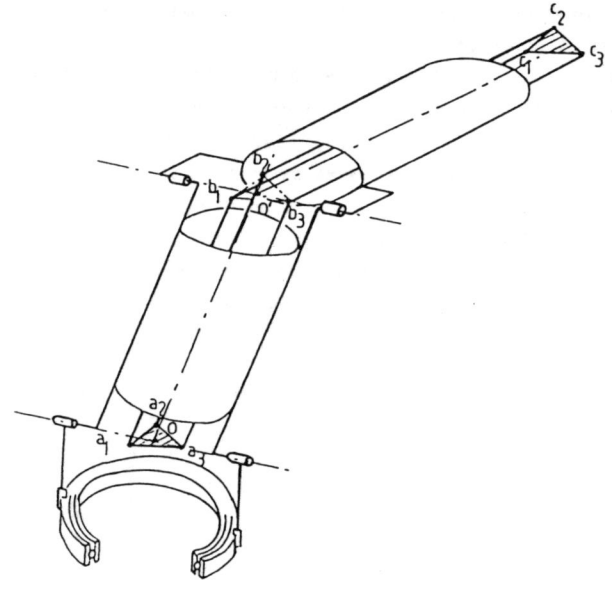

Fig. 2a

Fig. 2b

Polyarticulated Mechanical Structure 553

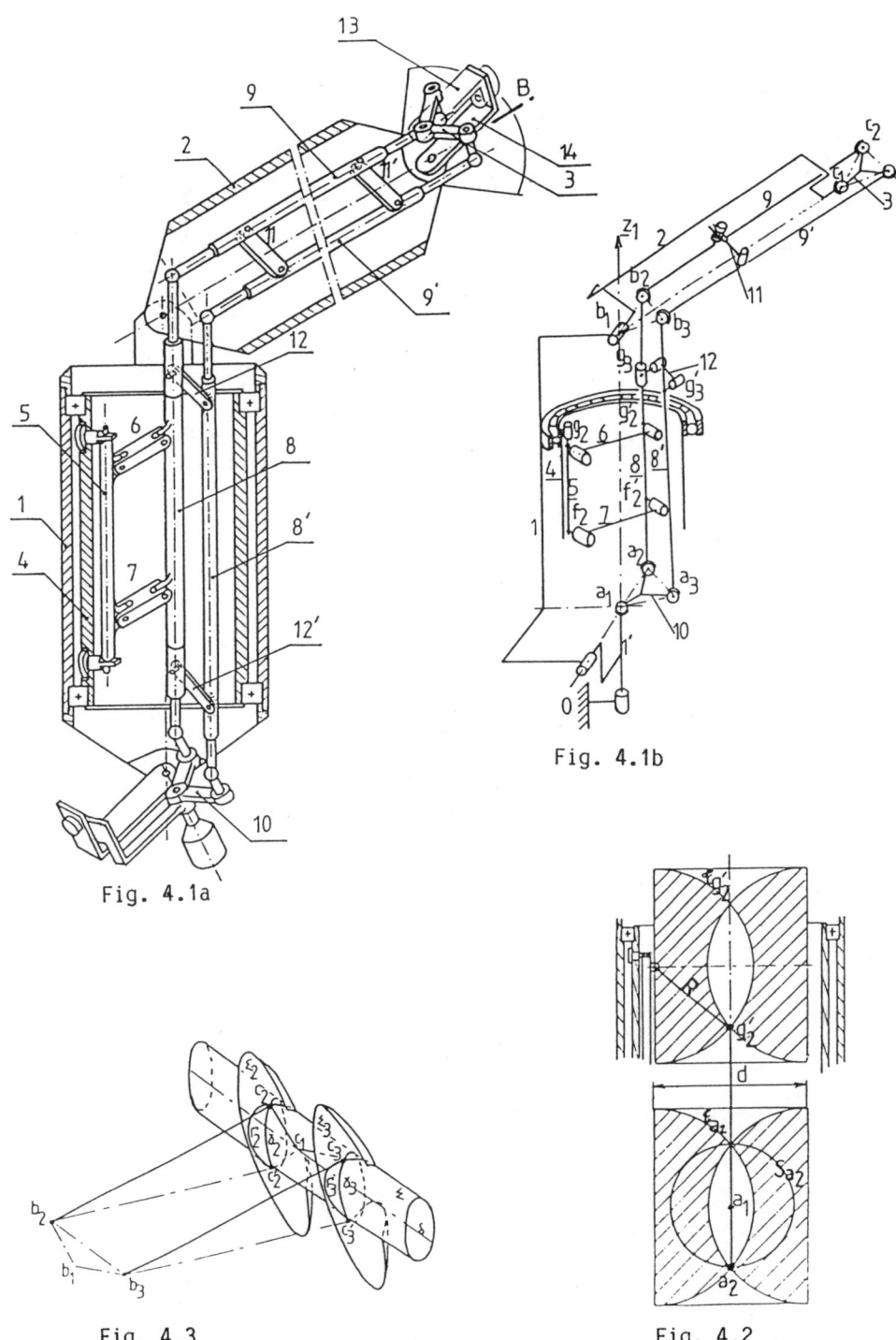

Fig. 4.1a

Fig. 4.1b

Fig. 4.3

Fig. 4.2

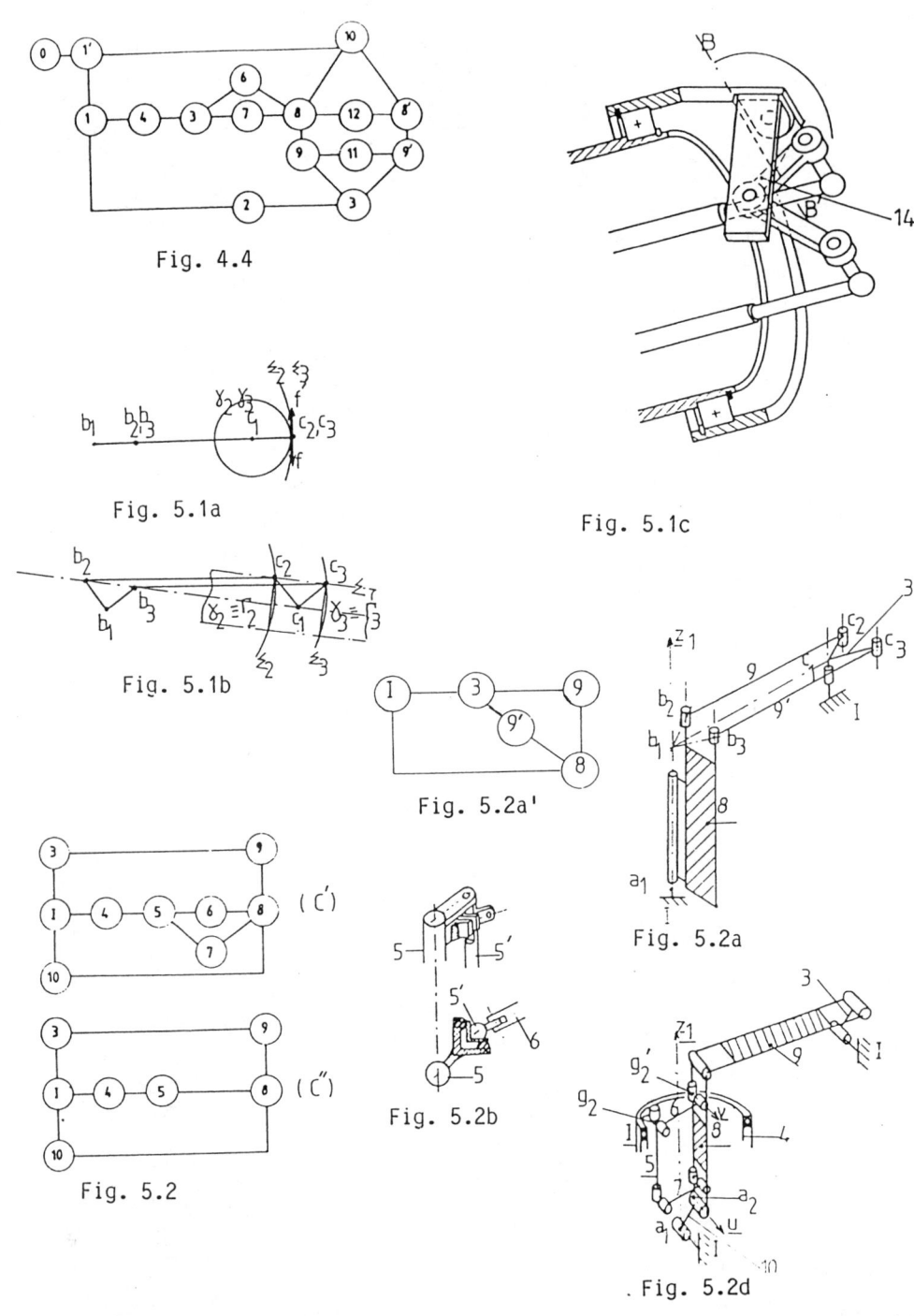

Fig. 4.4

Fig. 5.1a

Fig. 5.1c

Fig. 5.1b

Fig. 5.2a'

Fig. 5.2a

Fig. 5.2b

Fig. 5.2

Fig. 5.2d

Application of ℓ-Coordinates in Robotics

A. Sh. Koliskor, V.I. Sergeyev

Mechanical Engineering Research Institute
Academy of Sciences, Moscow, USSR

SUMMARY

    The keypoint of ℓ-coordinates is the specification of a free rigid body's position and motion in space by 6 distances between the points of the body and the points of the fixed base. The motion equations in the ℓ-coordinates are uniform and expose the six corresponding segment lengths' time dependence.

    This paper covers the ℓ-coordinate design method in the field of robotics mechanical design, controls, sensors and inspection as well as patented technical solutions. The developments are illustrated by an industrial robot with 30 degrees of freedom, a navigational robot with end-effector, six absolute frame coordinates feedback, six component force sensor and a navigational gripper's path recording system for assessment and diagnostics.

    The application of the ℓ-coordinate developments in robotics are capable of further advances.

ℓ-COORDINATES

    The keypoint of ℓ-coordinates is the specification of a free rigid body's position and motion in space using distance only. In ℓ-coordinates a free rigid body's location and orientation in space are determined by six linear arguments - the lengths of six segments $\ell_1, \ell_2, \ldots, \ell_6$ connecting the points of the body and the points of the fixed base, so that the geometrically constrained structure is provided as shown on Fig. 1. These two examples of ℓ-coordinate structures show their common feature to look like a spatial hinged truss where bars correspond to segments and connect spherical joints, located in the points on base and body. If two or three bars meet in one point on the base or on the body, each of them is considered connected with the base or the body with its own spherical joint. The necessary condition is that there is no line which intersects all six segments simultaneously.

The graph theory and combinatory application made it possible to find and classify all the structures regarding the number of points and the interconnections. The ℓ-coordinate structures are versatile enough to meet the needs of robotics. Each of them has its particular features which predict structure applicability for various tasks.

The ℓ-coordinate free-rigid-body motion equations have the form:

$$\ell_1 = F_i(t) \; ; \; i = 1, 2, \ldots, 6 \tag{1}$$

They differ from the commonly used Eulerian-Cartesian equations in using linear parameters only.

APPLICATION ADVANCES

In industrial robot mechanics the equations (1) may be used for designing motive, sensory and navigational systems.

MOTIVE SYSTEMS

Applicability of ℓ-coordinates in robotic manipulation systems may be used for designing mechanisms with six or more degrees of freedom provided by linear translations. Designs of this kind are known as "platform type" or "in-parallel actuated" (Stewart, 1965; Danilevsky, 1977; Hunt, 1978, 1982; McCallion, 1979; Fichter, 1980; Yang and Lee, 1984), but they have no more than six degrees of freedom. The ℓ-coordinate motive systems may be based on any structure, so that different kinematical and dynamical parameters such as workspace, speed, acceleration, power consumption, rigidity and etc. may be obtained.

The robotomodule in Fig. 2 designed in the Mechanical Engineering Research Institute is based on the structure shown in Figure 1. The design has two plates connected with six adjustable legs by ball joints so the controlled performance of prescribed relative dislocations with six degrees of freedom is obtained. The workspace of the module is determined by the extremes of pneumocylinders and ball joints' location on the plates. This module may be used either independently or as a part of a design including similar modules and spacers of different size and shape.

Another development of the Mechanical Engineering Research Institute is the industrial robot with the arm including five robotomodules described above. The arm has 30 degrees of freedom so it is highly manoeuverable and fits different technological tasks in cluttered workspaces. The ability of quick design rearrangement provided by inserting spacers and varying the number of modules makes it extremely flexible in changing its kinematics and dynamics to meet production problems. Such an arm has no gearing and its elements are not exposed to high surface stresses, torsional, bending and cyclic stresses. So it is very still, accurate and has long-term reliability. The elements of the arm are unified and interchangable (Fig. 3 and 4).

The length of the legs may be controlled by servo control or by relay control. The servo drives enable the output plates of each module to perform continuous path motion in the workspace frame. The relay control system changes the extremes of a leg length. The total amount of the extremes of six legs combinations - the number of relative positions of the upper plate is 2^6. Thus if five modules are connected, the total number of the output plate positions in space rises to 2^{30}. Such an arm is a spatial stepping system with nearly continuous path feature due to the extremely high density of the end effector positions in the workspace frame. The control guidance in this case has the combination of ON and OFF switches of the head and rod ends of the drive cylinders as set variables.

THE ℓ-COORDINATE CONTROL SYSTEMS

In the field of control ℓ-coordinates provide the opportunity to obtain current information about a free body's location and orientation in space by measuring distances only. What is of special importance is that the information about the body's (gripper) position and motion is obtained directly in the absolute fixed coordinate frame, while industrial robot control is based on the relative coordinate frames transformations of each link.

Fixed coordinate frame control (robotic navigation) of the industrial robot provides direct measurement of current values of six ℓ-coordinates of the gripper and the description of its position and motion in space. The arm of the industrial robot with the six fixed coordinate frame control - navigational robot - is shown in Fig. 5. The arm consists of links 1 with joints 2. The drives provide relative revolutions in each joint. The current location and orientation of the end-effector (gripper) 3 in space is described according to equations (1) by six ℓ-coordinates:

$\ell_1 = aA$, $\ell_2 = aB$, $\ell_3 = aD$, $\ell_4 = dB$, $\ell_5 = dD$, $\ell_6 = bD$

The points a, b, d belong to the gripper 3, and the points A, B, D belong to the base 4. The feedback system includes six devices for measuring linear translations to obtain ℓ_1, ..., ℓ_6 values. While the industrial robot's arm is in operation the current values of ℓ-coordinates are used as feedback data for minimizing error.

The fixed coordinate frame control (robotic navigation) of the end-effector enables improvement of robot's performance and proper dynamics due to the exclusion of link and joint elastic deformations, backlash and inaccuracies. So the navigational robot is expected to be one more step towards the software-hardware system integrity using computing to overcome limits of workmanship.

THE ℓ-COORDINATE SENSOR SUPPLY

The six component force sensory supply of industrial robots on the basis of ℓ-coordinates needs special load sensors. Such a sensor is shown in Figure 6. The flanges 1 and 2 are supplied with ball-and-socket joints and connected by six bars to realize one of the ℓ-coordinates structures. Each bar has an axial force transducer 3. Six components may be computed into the load in any application point on the flange, any arm link, end-effector or remote centre.

THE ℓ-COORDINATE SYSTEM FOR INDUSTRIAL ROBOTS INSPECTION AND DIAGNOSTICS

A sketch of the system is shown in Figure 7. The real gripper's motion information in form (1) makes it possible to inspect the industrial robot's performance from the view of accuracy and dynamics. The differences between performed gripper motion and the predicted one may be revealed. So automatic inspection trajectory tracking features are readily obtained.

What is of special importance is that there is no need to measure all the parameters of six motion laws simultaneously, because the robot can perform any desired number of cycles for a proper statistic sample. In each cycle one may record some of the motion law's parameters, one or two for example, and others in another cycle. It reduces the number of measuring devices, two to one for example. In the last case the measuring device 2 is to perform recording in each sample of cycles located in one point on the base and it should be located in the other point of the base for another cycle's sample recording. So the current distance measurements of the point on the body and the point on the base is the keypoint to the system performance.

CONCLUSION

The illustrated developments and applications of ℓ-coordinates in mechanics, control, sensors, inspection and diagnostics of the industrial robots do not cover the range of application of this analytical tool in science and manufacture.

The proper way to use this tool is the integrated - system approach.

REFERENCES

1. Koliskor A.Sh. The ℓ-coordinates approach to the industrial robots design and assesment. Machines & Tooling, 1982, 12.
2. Arzumanjan K.S.,Koliskor A.Sh.,Sergeev V.I. FMS robotomodules structural synthesis. The Design and Operation Problems of FMS in Machinery. Thesisies, V1, Vilnus, 1984
3. Danylevsky V.N. Manipulator. USSR patent 558788, Bul.19,1977
4. Warnecke H.I.,Brohdeck B.,Schiele G. Results of the examination of industrial robots on a test stand. Proceedings of the Fifth World Congress on Theory of Machines and Mechanisms, 1979, p.816-820
5. Koliskor A.Sh.The industrial robot control system. USSR patent 112767, Bul. 45, 1984.
6. Koliskor A.Sh. Force and moment sensor. USSR patent 974155, Bul.42, 1982.
7. Daich D.M., Koliskor A.Sh., Sergeyev V.I. Spatial dislocations measuring system. USSR patent 1040318, Bul.33, 1983.

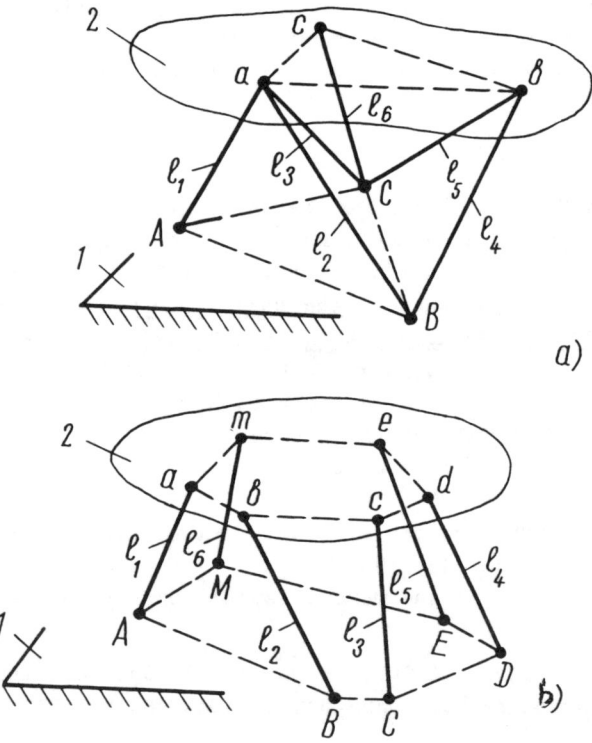

Fig. 1

Fig. 2

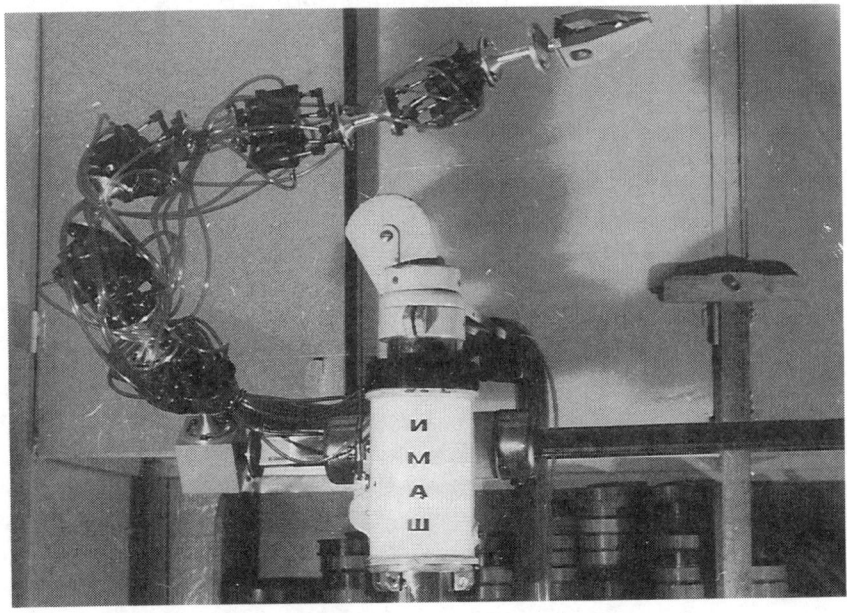

Fig. 3

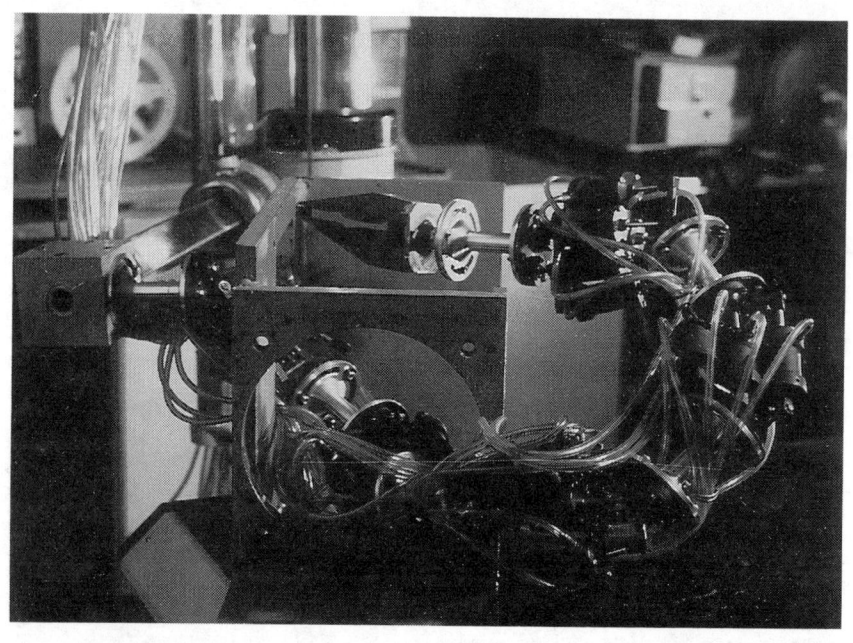

Fig. 4

Application of *l*-Coordinates in Robotics 563

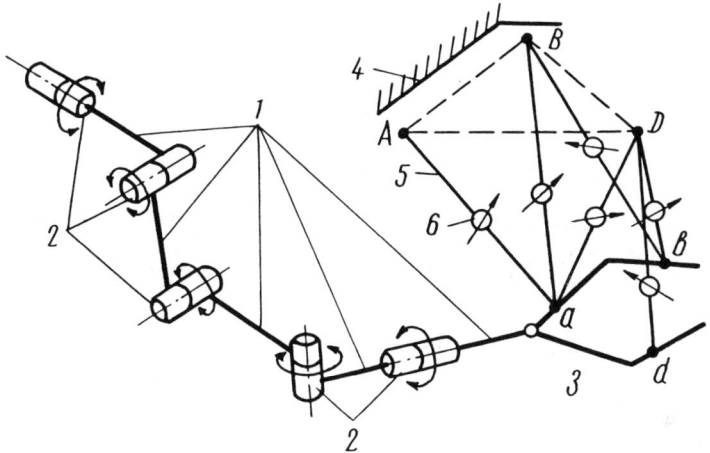

Fig. 5

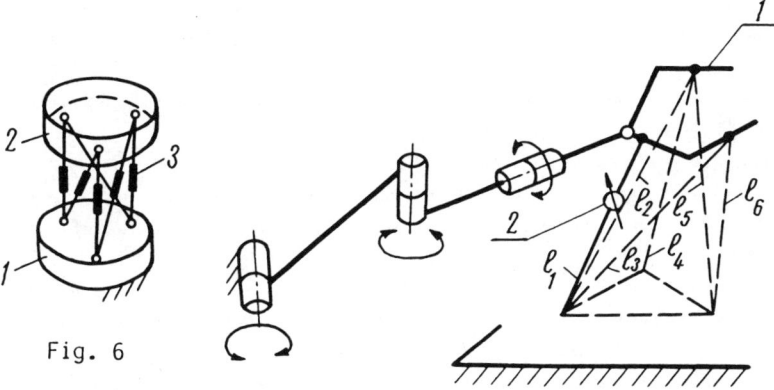

Fig. 6

Fig. 7

Design of Spring Mechanisms for Balancing the Weight of Robots

J.M. Hervé

Ecole Centrale de Paris, Châtenay-Malabry, France

1. INTRODUCTION

The dynamic analysis of robots highlights the dominating role that gravity has in the balance of energies needed to power them. Currently, most robots are made up of heavy elements rotating around horizontal axes. In this case it seems essential to find the best solutions for balancing gravitational forces, without increasing the inertia of the robot. Some elastic spring mechanisms have already enabled this aim to be achieved, (see bibliography). Some general principles are presented here which allow a more systematic and scientific design of spring balancing mechanisms.

2. THEORY OF THE SPRING MECHANISMS

Consider a rigid body of weight P and with centre of gravity G which rotates around a horizontal axis (see figure 1). Let A be the point of intersection of the perpendicular from G to the axis of rotation, R the distance AG and θ the angle subtended between AG and the upward vertical. The torque which tends to make the rigid body rotate under the influence of gravity will be: $C = P.R.\sin\theta$.

Let us now assume that a mechanism of negligible mass deforms an elastic spring (e.g. helical spring) also of negligible mass and of constant stiffness k, as the robotic arm leans. Ignoring the exact nature of this mechanism, some useful deductions are possible.

Denoting the magnitude of the spring's deformation by x, the force exerted by the spring is: $F = -k.x$.

Let p be the moment arm or "transmission radius" of this force; p is more precisely the ratio of the restoring torque M acting on the robotic arm and, arising from the spring deformation due to force F, i.e. $p = M/F$, p is a function of the configuration of the the balancing mechanism.

As the mechanism conserves energy, then $M.d\theta = F.dx$ or $p = dx/d\theta$, which provides a geometric interpretation of the transmission radius. The desired equilibrium is expressed by the equation $M = -C$ i.e. $F.p = -P.R.\sin\theta$. It is now proposed to find practical solutions for the variables x and p which satisfy this equilibrium equation.

When the arm is in an upward position, the equilibrium does not require any

restoring force from the spring and it is therefore natural to choose x = 0 for
θ = 0.

The principle of power conservation allows us to write:

C.dθ = F.dx   i.e. P.R.sinθ.dθ = k.x.dx

By integrating we obtain an expression linking the gravitational and spring potential energies:

P.R.(1-cosθ) = k.x^2/2

After using a trigonometric relation, we obtain:

2.P.R.sin^{2}θ/2 = k.x^2/2

A necessary relation between x and θ is therefore

$x = \pm 2.\sqrt{\frac{P.R}{k}}$ .sinθ/2   i.e. x must be proportional to sinθ/2 .

$p = \pm 2.\sqrt{\frac{P.R}{k}}$ .cosθ/2   i.e. p must vary as cosθ/2

Reciprocally, the relation F.p = -2.P.R.sinθ/2.cosθ/2 = -P.R.sinθ shows that the preceding equations are sufficient.

3. A NEW PATENTED DEVICE

The counter-balancing device incorporates a rigid auxiliary component (B) hinged about a second horizontal axis B, shown by its projection on the plane of the figure 2. This axis is parallel to the first axis A and is held fixed here thus.

The auxiliary component (B) incorporates two equal transverse arms of lengths b, extending to each side to the second axis B.

Means (M) are provided for driving the auxiliary component for rotation about the second axis B at an angular velocity w equal to half of the angular velocity W of the robot arm about axis A. This will thus provide the relation w = 1/2 W, as an absolute value. Such mechanical means (M) may consist of a mechanism involving a chain, belt or other inextensible flexible device, or of a mechanism with gear wheels or of an articulated mechanism.

The distant ends of the two transverse arms of the auxiliary component (B) are capable of bearing against a straight portion of a component (T) which is movable in translation. The element (T) is subjected to a biasing force returning the component (T) to an initial position. This force is provided by a system of one or several springs.

The operation of the counter-balancing device is as follows. When the robotic arm is inclined at an angle θ relative to the upward vertical, the auxiliary component (B) is inclined relative to its initial position at an angle θ/2, either in the same direction or in opposite direction, depending upon the transmission mechanism

(M) adopted. The movable component (T) is displaced by a distance d relative to the initial position so that $d = b.\sin\theta/2$. The elastic spring or springs generate a return force N upon the movable component (T) equal to $N = k.d = k.b.\sin\theta/2$, k being the stiffness constant of the spring. This return force is transmitted by the straight portion of (T) to one of the ends of the component (B). This force N is exerted at a distance h from the axis B such that $h = b.\cos\theta/2$. The moment of this force about the axis B has the value $N.h = k.b.\sin\theta/2.b.\cos\theta/2 = 1/2.k.b^2.\sin\theta$. Owing to the fact that the velocity of rotation of the robot arm is twice that of the component (B), the torque which is transmitted to the robot arm from (B) by driving means (M) is divided by 2.

This torque will balance that generated by the gravitational force if the following equation is true

$1/4.k.b^2 = P.R$

The counter-balancing device according to this invention enables correct balancing for full revolutions of a robotic arm.

4. Elements of bibliography

*Journées ARA, besançon, 16-17nov.1983, rapport du LMI-ECP, p84,FRANCE.
*Brevet français n°82 15039, société AKR, FRANCE.
*Brevet européen n°85 400938.8, Ecole Centrale, FRANCE.

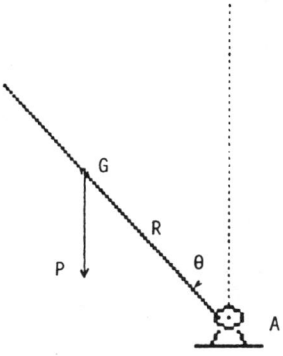

Figure 1

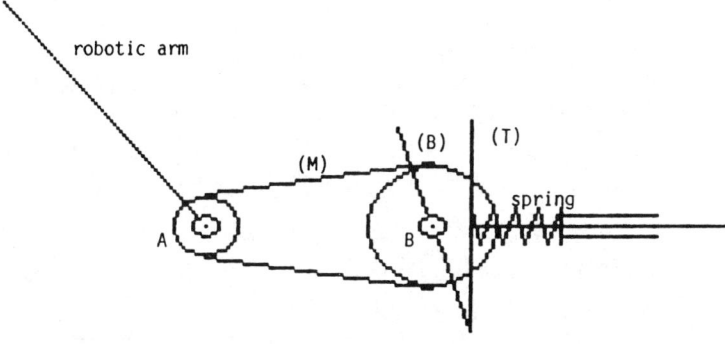

Figure 2

Structural and Geometrical Systematization of Spatial Positioning Kinematic Chains Employed in Industrial Robots

Fl. Duditza, D. Diaconescu and Gr. Gogu

University of Braşov, Romania

SUMMARY

It is considered that the open kinematic chain of a manipulator type industrial robot (IR) is formed of an orienting open kinematic chain (OC), whose main role is the angular displacement of a body around a characteristic point M, and of a positioning open kinematic chain (PC) having as the main role the linear displacement of the point M in space and, implicitly, the generation of a geometric locus, which forms the working space of IR.

In this paper, which is an expansion of (4), a structural systematization of trimobile PC, which describes non-degenerated workspaces, and two geometrical-kinematic systematization meant for the optimization of the PC "architecture" are proposed. These systematizations are useful in the optimization of IR design.

1. NOTATIONS

R, T -notations for revolute pair and prismatic pair respectively,

$\perp, \parallel$ -notations for perpendicular and parallel position respectively, of a pair of an open kinematic chain relative to the preceding pair.

$\perp^{\perp}, \perp^{\parallel}$ -notations are used in PC of the type-$X \perp T \perp^{\perp} Y$ and of the type-$X \perp T \perp^{\parallel} Y$ $(X, Y \in \{R, T\})$ respectively, in order to designate the perpendicular position of the pair Y with reference to the pair T, concomitant with the perpendicular position ($\perp^{\perp}$), respectively parellel position ($\perp^{\parallel}$), of the pair Y relative to the pair X.

$R \perp R(a_1) \perp R(b, a_2) - M(r, d)$ -an example of structural-geometric notation of a PC; according to the notation, the axis of the second pair (R) is situated at the distance a_1 perpendicular to the axis of the initial pair (R), the axis of the third pair (R) is situated at the distances b, a_2 perpendicular to the axis of the second pair (R), and the characteristic point M is situated at the distances r and d to the last axis (see Tab.1 and 2).

$\propto_m^n = m^n$ -variation with repetition of m different "elements" grouped by n; examples:

$\alpha^3_{(R,T)} \Rightarrow \alpha^3_2$, $\alpha^2_{(\perp,\parallel)} \Rightarrow \alpha^2_2$; $\alpha^{(a1,b,a2)}_{(=0,\neq 0)} \Rightarrow \alpha^3_2$ - variations with repetition: of two pairs (R,T) grouped by threes, of two relative positions ($\perp$, $\parallel$) grouped by twos and of two values (=0,$\neq$0) assigned to three quantities (a_1, b, a_2) respectively.

IR, OC, PC, M -notations for industrial robot (IR), elementary open orienting (OC) and positioning (PC) kinematic chain respectively and the notation of characteristic point (M).

2. PREMISES

A necessary, but not sufficient condition, for the work space of a PC to be non-degenerated (into surface or line) is that it be at least trimobile; by taking this condition into account, in the structural systematization the following premises are adopted:

    I. The PC is considered trimobile;

    II. The PC is considered to be formed only of pairs R and/or T;

    III. A pair of PC is either perpendicular ($\perp$) or parallel ($\parallel$) to the preceding pair.

Under these premises, PC of type-X$\perp$T$\perp$Y has two variants, X$\perp$T$\perp$Y and X$\perp$T$\parallel$Y, whereas the other types of PC have only one variant.

In the geometric systematization meant for the optimization of the workspace the premises IV and V are adopted:

    IV. The pair R has axis, and the pair T has direction (and not axis);

    V. Geometric coordinates associated with the PC include also the coordinates of the characteristic point M.

In the geometric systematization meant for the optimization of PC "architecture", the premises VI and VII are considered;

    VI. Axis is associated to pair T too; as a result, in this systematization both pairs R and pairs T have axes.

    VII. The characteristic point M is not considered.

3. SYSTEMATIZATION

    a. In keeping with Table 1, col. 1 and 2, $\alpha^3_{(R,T)} \Rightarrow \alpha^3_2 = 8$ distinct groupings of pairs R, T taken by threes can be conceived; for groupings of type-XRY and four of type-XTY, $X,Y \in \{R,T\}$; each XRY grouping can be associated with $\alpha^2_2 = 4$ distinct relative positions and 4+1=5 relative positions may be associated with each XTY grouping (see Tab. 1, col. 2). As a result 4x4+4x5=16+20=36 trimobile chains are obtained, among which only the 20 systematized in Tab. 1, col. 3 and represented in Tab. 2, col. 1 can have non-degenerated workspaces and, implicitly, can be employed as PC in IR.

From the analysis of the existing and applied variants it results that among the 20 structural variants represented in Tab. 2 the following four have the widest practical application; 17(R∥T⊥T) 3(R⊥R∥R), 4(R⊥R⊥T) and 20(T⊥T⊥T).

In literature, the problem of the structural systematization of PC is tackled in some papers, but the results established do not coincide with one another (1), ..., (3).

TABLE 1

SYNTHESIS OF TRIMOBILE POSITIONING KINEMATIC CHAINS OF INDUSTRIAL ROBOTS

Distinct arrangements of kinematic pairs $\alpha_2^3 = 8$	Configurations of relative positions of kinematic pairs from arrangement-type XRY and XTY ; $x\in\{R,T\}, y\in\{R,T\}$			General geometrical variants of trimobile positioning kinematic chains (M-characteristic point positionated in space by the chain) - table 2	Number of particular configurations of axes in which the chain can positionate the point M in space	Numbers of particular the positions of point M in the positioning chain	Numbers of particular chains
Nr. crt.	$\alpha^3_{(R,T)}$	$\alpha^2_{(\bot,\|)}$	XRY	XTY				
1	R∣R∣R	⊥,⊥	⊥,⊥	⊥,⊥	1 R⊥R(a₁)⊥R(b,a₂)-M(r≠0,d)	$\alpha^{(a_1,a_2,b)}_{(\neq 0,\neq 0)}$(0,0,0)=> $\alpha^3_2 - 1 = 2^3-1 = 7$	$\alpha^{(d)}_{(\neq 0,\neq 0)}$=> $\alpha^1_2 = 2^1 = 2$	7×2= 14
				⊥,⊥"	2 R∥R(a₁≠0)⊥R(a₂)-M(r≠0,d)	$\alpha^1_2 = 2^1 = 2$	$\alpha^1_2 = 2^1 = 2$	2×2=4
2	R∣R∣T			⊥,⊥"	3 R⊥R(a₁)∥R(a₂≠0)-M(r≠0,d)	$\alpha^1_2 = 2^1 = 2$	$\alpha^1_2 = 2^1 = 2$	2×2=4
					4 R⊥R(a) ⊥T - M(r,d)	$\alpha^1_2 = 2^1 = 2$	$\alpha^2_2 = 2^2 = 4$	2×4=8
3	T∣R∣R	⊥,∥	⊥,∥	⊥,∥	5 T⊥R⊥R(a) - M(r≠0,d)	$\alpha^1_2 = 2^1 = 2$	$\alpha^1_2 = 2^1 = 2$	2×2=4
4	T∣R∣T	∥,∥	∥,∥	∥,∥	6 R⊥R(a) ∥T - M (r≠0)	$\alpha^1_2 = 2^1 = 2$	$\alpha^0_2 = 2^0 = 1$	2×1=2
					7 T∥R⊥R(a) - M (r≠0,d)	$\alpha^1_2 = 2^1 = 2$	$\alpha^1_2 = 2^1 = 2$	2×2=4
5	R∣T∣R	∥,⊥	∥,⊥	∥,⊥	8 R∥R(a≠0)∥T - M (r≠0)	$\alpha^0_2 = 2^0 = 1$	$\alpha^0_2 = 2^0 = 1$	1×1=1
6	R∣T∣T				9 T∥R∥R(a≠0)-M(r≠0)	$\alpha^0_2 = 2^0 = 1$	$\alpha^0_2 = 2^0 = 1$	1×1=1
					10 R⊥T⊥⊥ R - M (r≠0,d)	$\alpha^1_2 = 2^1 = 2$	$\alpha^1_2 = 2^1 = 2$	1×2=2
7	T∣T∣R	Number of distinct configurations $\alpha^2_2 = 2^2 = 4$	$\alpha^2_2 = \alpha^2_2+1=$ =5		11 R⊥T∥R(a) - M(r≠0)	$\alpha^1_2 = 2^1 = 2$	$\alpha^0_2 = 2^0 = 1$	2×1=2
8	T∣T∣T				12 R∥T⊥R(a) - M(r≠0,d)	$\alpha^1_2 = 2^1 = 2$	$\alpha^1_2 = 2^1 = 2$	2×2=4
					13 R∥T∥R(a≠0)-M (r≠0)	$\alpha^0_2 = 2^0 = 1$	$\alpha^0_2 = 2^0 = 1$	1×1=1
Thus can be conceived : 4×4 (XRY) + 5×4 (XTY) = 36 structural schemes of trimobile chains, only 20 from these schemes can positionate in space a characteristic point M and, implicitly, can be used as positioning kinematic chains					14 T⊥R∥T -M (r≠0)	$\alpha^0_2 = 2^0 = 1$	$\alpha^0_2 = 2^0 = 1$	1×1=1
					15 T∥R⊥T -M (r)	$\alpha^0_2 = 2^0 = 1$	$\alpha^1_2 = 2^1 = 2$	1×2=2
					16 T⊥T∥R -M (r≠0)	$\alpha^0_2 = 2^0 = 1$	$\alpha^0_2 = 2^0 = 1$	1×1=1
					17 R∥T⊥T -M (r)	$\alpha^0_2 = 2^0 = 1$	$\alpha^1_2 = 2^1 = 2$	1×2=2
					18 T⊥T⊥R -M (r≠0)	$\alpha^0_2 = 2^0 = 1$	$\alpha^0_2 = 2^0 = 1$	1×1=1
					19 R⊥T⊥T -M (r)	$\alpha^0_2 = 2^0 = 1$	$\alpha^1_2 = 2^1 = 2$	1×2=2
					20 T⊥T⊥T -M	$\alpha^0_2 = 2^0 = 1$	$\alpha^0_2 = 2^0 = 1$	1×1=1
					TOTAL = 20	TOTAL = 34	TOTAL = 61	

b. The geometric systematization, meant for the optimization of the workspace, yields 20 general geometric variants enumerated in Tab. 1, col. 3 and represented in Tab. 2; these general variants were obtained by associating with the 20 structural schemes the distances specific to the configuration of axes of rotation and of the characteristic point M.

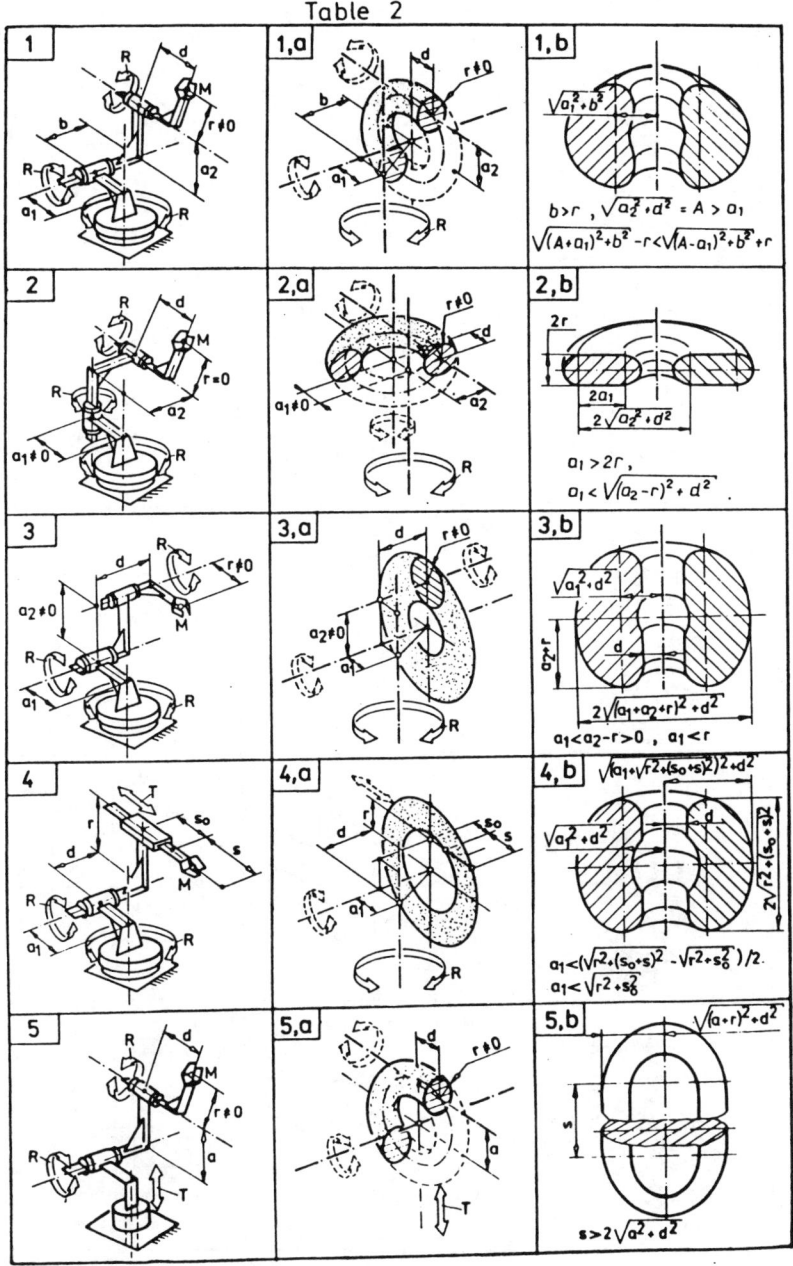

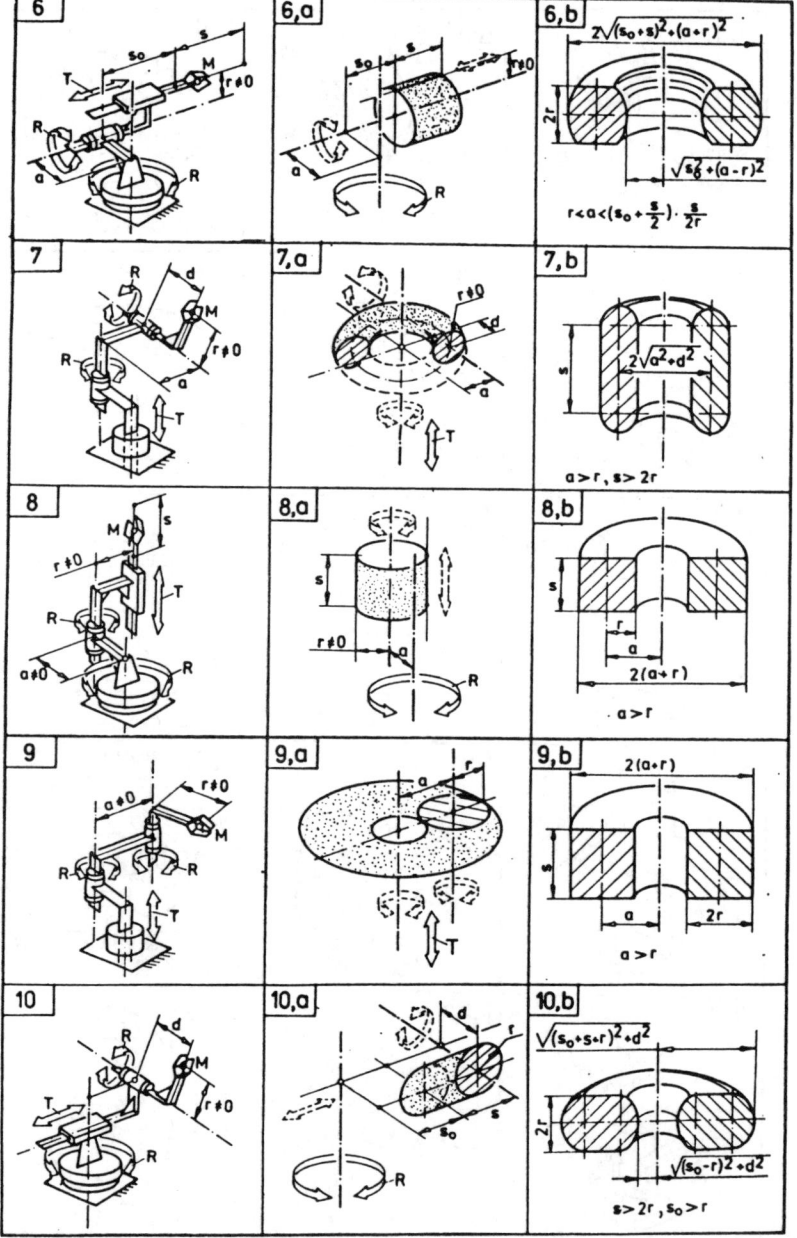

Systematization of Spatial Positioning Kinematic Chains

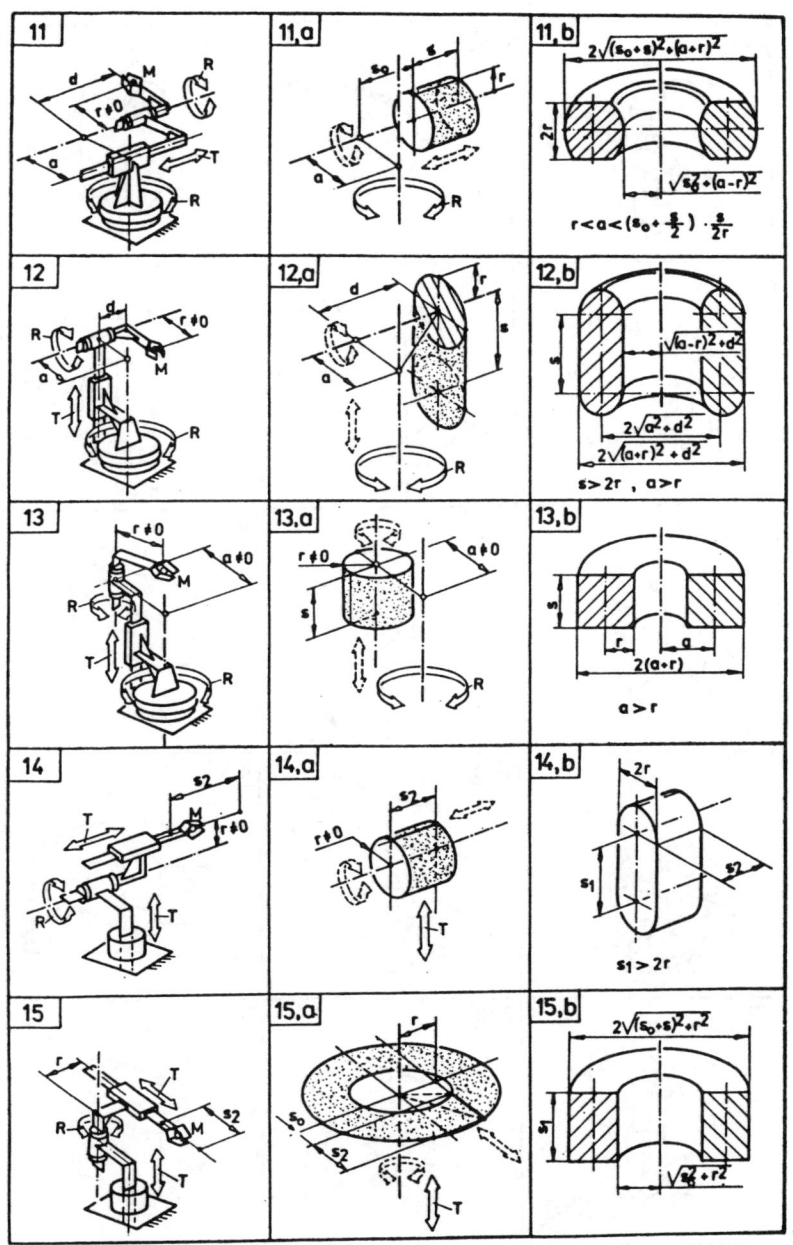

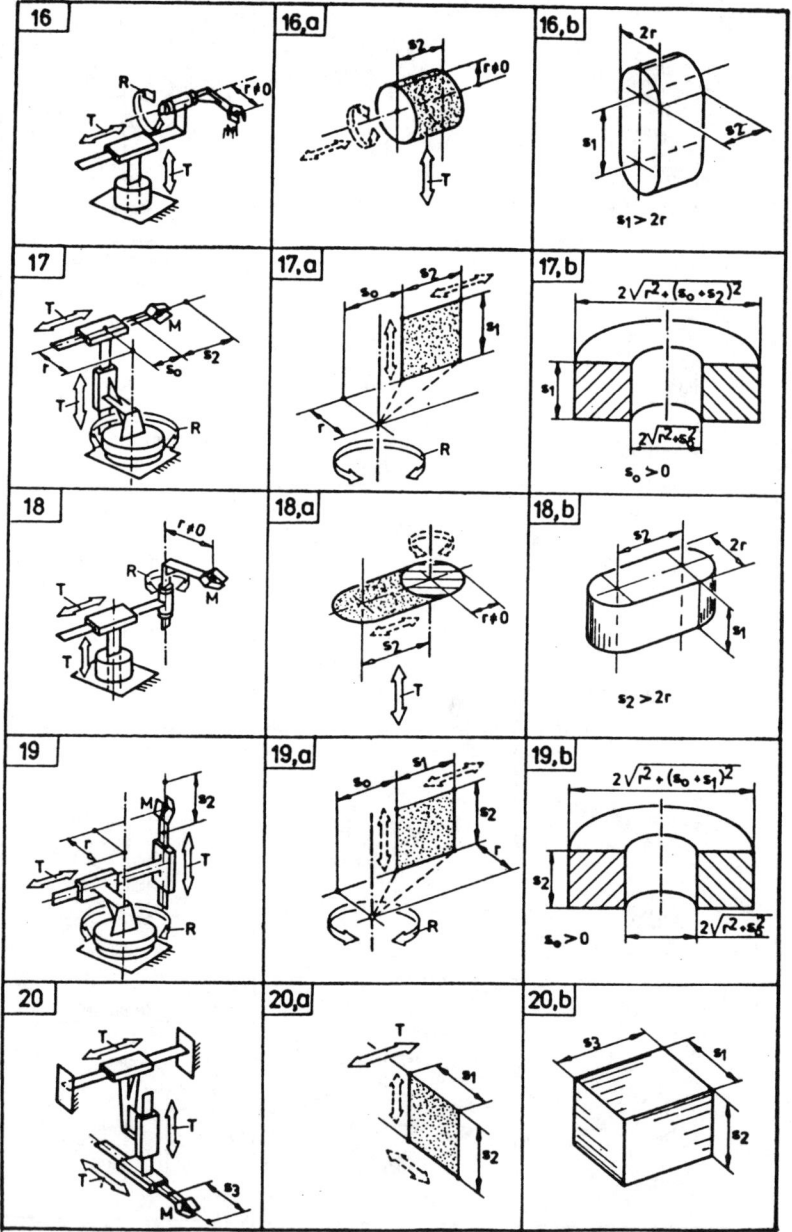

By taking into account the destination of systematization, in Tab. 2, col. 2, the line described by the point M during the relative displacement in the final pair, and the surface described as null by this line, during the relative displacement in the medium pair (the second) is represented qualitatively; according to Tab. 2, col. 2 and 3, during the relative displacement in the first pair (the pair adjacent to base), this surfaces generates the workspace of IR. The form of the workspace, illustrated by examples in Tab. 2, col. 3, is conditioned by observing the restrictions noted below the drawing.

By successively assigning null and/or non-null values to the distances specific to the rotation axes' configuration (see Tab. 1 and 2), according to Tab. 1, col. 4, 34 particular configurations are obtained; by taking the number of particular positions of the point M into account too (see Tab. 1, col. 5), finally 61 variants of distinct particular PC are obtained (see Tab. 1, col. 6). These results constitute the starting base for the study of the geometric optimization of workspace.

c. The geometric systematization meant for the optimization of the PC "architecture", results in 20 general variants too, in which, in contrast with the variants shown in Tab. 1, col. 3, there are in addition the distances between prismatic pair's axes to be considered, where as the point M is omitted.

By successively assigning null and/or non-null values to the coordinates associated to the rotation and translation axes, 95 variants of distinct particular PC are obtained; these variants are useful in the study of optimization of PC "architecture" and, implicitly of IR "architecture".

4. REFERENCES

1. WARNECKE,H.J.,SCHRAFT,R.D. A computer-aided method to design an industrial robot.Sec.Int.CISM-IFToMM Symp.Warsaw,Poland,1976.
2. PAVLENKO,I.I. Konstruktivnye i kinematičeskye varianty promyšlennyh robotov. Vestnik mašinostr.,nr.11,198o,s.3-5.
3. BOLOTIN,L.M. Analiz kinematičeskih struktur promišlennyh robotov. Mašinovedenie, nr.2,1984,s.33-39.
4. DUDITZA,FL.,DIACONESCU,D. Sinteza structurală a mecanismelor de transport din R.I. Symp.MTM,Timişoara, 1984,p.39-48.

Tasks and Methods of Constructing Mechanical Facilities and Control Systems of Industrial Robots Taking into Account their Force Interaction with the Equipment

S.N. Kolpashnikov and I.B. Tchelpanov

The control laws as well as the functional charts and the block diagrams of the industrial robot control circuits intended for realizing those laws are created on the basis of meeting the requirements of fast action and the present precision or the optimization conditions for the respective criteria of motion in the course of the free transfer of the objects being manipulated. The overwhelming majority of works dedicated to the analysis of the geometry, the kinematics and the dynamics of robot motion also refers to the tasks of free transfer mechanics.

Meanwhile practice shows that the most important and complicated stages of manipulation are the phases of hindered state or movement when the object which is in the gripping mechanism of the robot or the end-effector of the robot interacts, through direct physical contact with other objects, devices or units of the equipment. In this case, side by side with the necessity in the highly reliable performance of the programmable movements, there are set and solved new problems of limiting, stabilization at the preset level or minimization of component forces and/or torques of the forces of interaction of the object being manipulated with external objects.

Below is given the description of some typical problems.

1. The robot withdraws the object (the part, the billet) out of the container recess transferring it rectilinearly on the guides along the longitudinal axis of the recess (Fig.1). Due to the linear and the angular errors of the robot's and container's relative position, the program processing errors, the errors of location and those of the shape of the object surface etc, one may realize that, when the object is on the guides and in the gripping mechanism simultaneously, the strain of the robot manipulator mechanism and that of the guides do occur. That strain results from the effect of the contact interaction forces. At considerable interaction forces, the object jamming, the gripping mechanism's abnormal working conditions (for example, partial opening of the gripper), the container's displacement and other things may happen.

2. The robot passes the object to the chucking machine's jig or to the gripping mechanism of another robot (Fig.2). There is a phase in the process in concern when the object must be gripped, simultaneously, by two mechanisms that belong to different systems. The linear and the angular errors depending upon the same factors that were listed in Item 1 result in the strain of the mechanism. The effect of the above errors may be the conditions under which both the grippers would not be able to grip the object simultaneously, and in this case the transfer process is in progress while one of the mechanisms is opened and the other closed. The mechanics of this interaction is described in the works written by the authors together with Trubin I.A.

3. The robot holds the object which is being machined or is an element in the assembly process, when the realization of the technological operation is accompanied by force interaction. Here the robot may hold the object during grinding, pressing-in (Fig.3), cutting, etc. The requirement of decreasing the elastic displacements of the object being held as the effect of the forces depending on the technological equipment and the conditions of its operation is typical for those tasks.

4. The robot performs operations of parts assembling with small clearances. The simplest task of that type is the problem of assembling the shaft-and-bush pair (Fig.4). The methods of assembling such a pair were described in detail in many works. The specific feature of the task is in the fact that jamming is possible when the principal assembly motion along the pair's axis is being accomplished. On the basis of the analysis of the respective conditions, one may make a conclusion that the jamming occurs as the consequence of rather a small minimum positive clearance in the contact, the initial angles of skewing and the discrepancies between the actual elasticity characteristics and those specified.

5. The robot carries out technological operations, its end-effector interacts directly and in contact with the piece being machined. The typical example of the robot is the one intended for grinding curvilinear surfaces: its end-effector is a unit incorporating its own motor and abrasive wheel (Fig.5). The program of motion along the coordinates is prescribed by the contour of the surface being machined. However, in the course of processing, it is necessary to limit or to stabilize the normal and the tangential components of the forces of interaction of the abrasive wheel with the surface being machined.

The methods of solving the problems of mechanics of the mechanisms with closed kinematic loops comprehensively developed are a considerable part of modern theory of mechanisms. As to the robots operated under the conditions described above, the tasks for them are rather specific. Let us couch the task of controlling the movement in general sense in conformity with the robots under the duties of force interaction with external objects.

The position of the operating manipulator is preset by six generalized coordinates q_i. Let us assume that Q_i is the respective generalized forces stipulated by the pull of gravity and the contact interactions of the end-effector with the external objects. The program of movements is prescribed in accordance with all or only some generalized coordinates. At the force interaction, additional requirements may get superimposed directly onto the generalized forces or onto their linear combinations. Depending on the actual aim of the tasks, the following versions of requirements may be encountered:

- limitation: $/C_j^T Q/ \; a_j$ (1)
- minimization: $/C_j^T Q/ = \min$ (2)
- stabilization in respect of the programmed values: $/C_j^T Q - b_j / = \min$ (3)

Everywhere above Q is the six-dimensional vector of the generalized forces while C is the vector of coefficients preset. As a rule, by properly selecting the coordinate system and the generalized coordinates we manage to reduce these requirements to the appropriate requirements and, directly, to the generalized forces individually. This means that for task 1 the withdrawal of the object out of the recess requires the limiting or the minimization of components F_y and F_z of the forces along axes Y and Z and components M_y and M_z (possibly, M_x too) of the torque around axes Y, Z and X. As for the transfer of the object to the gripping mechanism according to task 2, the most rational method to be applied here is the limitation of all the component forces and torques. The assembling in compliance with task 4 requires the minimization of the side component forces and the component torques around axes Y and Z. In task 5 where grinding is carried out the component forces should get stabilized along axes X and Z.

The complexity of the problem of controlling the motion of the robot lies in the fact that at the force interaction with the equipment the program of movements along the coordinates should be executed and the above requirements to the components of forces and torques ought to be met at the same time. All said above may be divided rationally into two cases depending upon how many and which exactly generalized coordinates and generalized forces are used for presetting the programs and the requirements.

1. As to those generalized coordinates that are used for presetting the programs, limitations are not applied to the respective generalized forces, and vice versa. In this case there can exist alternatives of complete definability when either the generalized coordinate program or the generalized force requirement is preset for each of the six degrees of freedom, and alternatives of incomplete definability when neither programs nor requirements are preset for some degrees of freedom.

2. Both the programs of variation of the generalized coordinates and the requirements of generalized forces are preset for some degrees of freedom simultaneously. As in case 1, alternatives of complete and incomplete definability are possible.

As was mentioned above, it is often convenient to preset the generalized coordinates by applying the rectangular coordinate system. The typical variants for case 1 of task 2 are shown in Table 1. The programs of free transfer according to alternative 1 are preset for all the six generalized coordinates (three displacements and three angles of swing) while alternative 6 is carried out with presetting six requirements for the six generalized forces (three force and three torque components). The other alternatives occupy intermediate position. The alternatives are given in the order of decreasing of the number of the degrees of freedom. When one degree of freedom is cancelled, the requirement for the respective generalized force appears instead of the generalized coordinate variation program.

At the step of the preliminary analysis of the tasks, it is convenient, first, to assume that the motion along each of the generalized coordinates is preset individually (each drive presets the movement along the very coordinate along which this motion is required). As to those degrees of freedom for which programs are preset, the drives realize the processing of the programs. But there are several solutions for those degrees of freedom for which the requirements of the generalized forces are applied.

(a) As to the coordinates whom the requirements to the generalized forces correspond, the programs are preset in the same manner as they are for the other coordinates in such a way that for the ideal case (when neither error is available) the motion compatible with the couplings applied could be obtained. If there are errors of performing the movements but at the same time if there are large enough clearances that are intact in the course of the motion, then the respective reactions are not available, and the requirements of minimization of the respective generalized forces are met automatically.

For example, for the withdrawal of the object out of the recess according to task 1, the motion program is set along axis X, instead of the minimization requirements for all the component forces, with the exception of F_x, and all the component torques, there should be preset the programs of stabilizing coordinates Y and Z and the angles of swing around all the three axes. If, however, the clearances are small and exceeded by the stabilization errors, then contact will appear inevitably, and the interaction forces would not get controlled.

(b) Along the coordinates along which the requirements to the generalized forces are prescribed, the drives get cancelled, therefore the movements become free and uncontrolled. In this manner, the force contact is excluded along the corresponding coordinates. In this case the contact interaction is not cancelled, naturally, and at free movements shocks are possible. In addition to that, phenomena of the kind of jamming may occur at an unfavourable arrangement of the possible points of contact

(c) Elastic characteristics of the elements of the robot mechanism's kinematic chain are selected specially in such a manner that in the course of possible errors of the program executions the contact interaction could cause the limitation of the forces and the torques at the levels preset. The force restraint requirements cause the limitation from above of the stiffness of the manipulator gear. If the natural flexible compliance of the mechanism's links and of the joint assemblies is not sufficient, then a special elastic component with its characteristics selected specially for that purpose should be introduced into the mechanism. As it is applied to the problem of automatic assembling the shaft-bush pair, the task of selection of the characteristics of the elastic element inserted into the mechanism between the last link and the basement of the gripper was analyzed in detail in the works of the authors according to references [1, 2]. By means of a proper selection of the elasticity characteristics, the undesirable consequences of shocks can be reduced and the jamming can be averted at slow programmable movements.

(d) As to the coordinates along which the generalized forces requirements are preset, the method of control changes: instead of the programmed control along the coordinates the servo drives are switched to the duty of minimization or stabilization of the generalized forces. In this case the sensing elements are the transducers realizing the force and torque sensitization that are built in the gripper or mounted between the last link and the base of the gripper. This method as applied to the automatic assembling problems is described in detail in many works.

(e) As to specified coordinates, the programs are executed in the shape of combined programs, they involve simultaneous performance of several movements – for example, the main smooth movements and the auxiliary oscillatory motions. The oscillatory movements may play the role of the search (during the automatic assembling), or they may be designated for the oscillation linearization of the dry friction forces in the contact points as well as for eliminating the possibilities of seizure and jamming when the main movements are carried out.

At incomplete definability, when neither programs nor limitations for the appropriate generalized forces are preset for certain coordinates – for example, when the object held in the robot's gripper has got the axis of symmetry and rotation around it is of no significance – the task becomes simplified: unspecified movements may be brought about along those coordinates.

In case 2, when the programs of varying the coordinates and the requirements of the generalized forces are preset simultaneously for some degrees of freedom, considerable difficulties may appear in the course of creating the control programs. If the actual environment differs considerably from the specified, the mutual realization of the programs and the requirements to the forces may appear to be impossible. For example, while performing the operations of grinding, if there are great allowances, the execution of the program of motion corresponding to the contour in concern under the duties selected (at the preset speed of displacement) requires the increasing of the forces. In this case it may appear that this required increasing may appear to be impossible for the reasons of the robot's strength, averting of overheating and obtaining the proper quality of the surface being treated. Two methods are possible to overcome these contradictions.

First, the program of movements may be adapted correspondingly for realizing the motion along the other coordinates. For example, at great allowances in the course of the grinding processes, the force limitation requirements may be met due to decreasing of the feed speed. Second, the programs priority may be selected. As a rule, the limiting of the generalized forces is the most significant thing because it ensures the reliability and the safekeeping of the equipment and the units. Therefore, in case the incompatibility of the movement programs and the force limitation are available the priority is given to the latter. To attain the final end, the program should be also modified – for example, sequential passes ought to be carried out, that is, the movements must be able to be repeated.

Two principal ways of meeting the requirements to the generalized forces are available whose description follows.

1. Only purely mechanical facilities are used that make it possible to obtain the proper mechanical characteristics of the manipulator in respect of the acting forces. Preliminarily, the requirements to the forces should be converted into limitations as was said in Equation (1). This is possible for most of the tasks. Then one should take into consideration the fact that those limitations for the forces must be always realized at all possible errors. This imposes limitations from above on the stiffness characteristics.

The extreme cases are the elasticity characteristics (the dependence of the generalized force upon the respective generalized coordinate) shown in Fig.6. Characteristic 1 is linear, characteristic 2 is relay, it is obtained due to the initial tension, characteristic 3 is real for the case of the initial tension. Let us analyze first the case of the linear characteristic without tension.

If the mechanism of the manipulator is stiff enough, then, as was mentioned above, a special elastic element should be inserted into the kinematic chain. If the F_o is the permissible value of the generalized force and the δ_o is the possible displacement, then the stiffness coefficient for that coordinate can be defined in the following way: $c < F_o / \delta_o$.

In the elastic elements, blade springs, torsion bars, spiral and cylindrical springs etc may be used. Fig.7 shows the variants of the simpliest circuits where elastic motion occurs along a single coordinate only. If the elastic movements are to be obtained along several generalized coordinates, then two solutions are possible. First, two circuits may be combined together each of which will allow to have the movement along only a single coordinate (this example is illustrated in Fig.8). Second, one may construct the flexible elements that are not divided into the simpliest ones, and in such elements the same springs operate at the variation of different coordinates (the example is shown in Fig.9).

The first method is characterized by a simpler adjustment (this is accomplished individually for each of the degrees of freedom) while more compact arrangements are possible for the second method. The layout of the elastic element with initial tension that ensures the movement along only a single coordinate is illustrated in Fig.10. Similar circuits may be combined into a single element. It is not difficult to construct such elastic elements in which the exceeding of the threshold value by a single force gives rise to the possibility of free movements along several generalized coordinates. For example, Fig.11 shows the element that ensures the possibility of random plane movements when the force along axis X exceeds the initial tension force.

2.The appropriate characteristics are obtained by selecting the structure and the values of the parameters of the servo drive circuits. In this case both the main drives that create the transfer movements of the manipulator and the special drives inserted at the end of the kinematic chain – for example, the gripper shaft drives – may be used. The force-and-torque sensitization transducers whose operating principles and circuits are set forth in many works are usually applied in the capacity of the sensors, i. e. the instrument transducers.

Application of additional drives makes the design more complicated but allows to distribute the functions of transfer and limitation of the forces among the drives, which is important at the performance of such precise operations as assembling of pairs with small clearances. In addition to this, fixing of the displacement axes to the axes of the end-effector makes it possible to directly and independently vary the properties of elasticity along those axes along which the force limitation was preset.

But if the main drives are applied the control channels appear to be restricted in their operation generally, and this causes the necessity of simulation of the conversion of the coordinates in the control units. The method being described provides us, in principle, with ample opportunities of varying the coefficients of quasi-elasticity within broad limits, improving the dynamics of the processes, etc. In this case, however, the reliability may become worse because of the insufficient perfectness of the force--and-torque sensitization transducers.

In all the cases the elastic mechanical compliance of the manipulator may appear to be considerable. All the calculation equations involve the total elastic mechanical compliance of the elastic element and the mechanism, and the elastic mechanical compliance matrix elements are added to each other in the same axes.

If the interaction between the object being manipulated and the external objects takes place on a rather definite position, then the appropriate elasticity characteristics of the manipulator, as a whole, may be obtained by adjusting the elastic element correspondingly. In those cases when the elastic properties of the manipulator itself possess considerable asymmetry, the elastic element must compensate for this asymmetry.

In the course of carrying out the operations of automatic assembling with small clearances, the requirements for the stiffness coefficients relation and for the symmetry of the elasticity characteristics are stringent enough, and the meeting of those requirements at different positions may call for a rearrangement of the elastic elements since the manipulator elasticity characteristics usually vary considerably within the limits of the work area.

Under those conditions the circuits incorporating a feedback coming from the force-and-torque sensitization transducers are more flexible, their gain factor should be varied depending upon the position, that is, upon the values of the generalized coordinates.

When dealing with the automated assembly problems, one should also take into account the fact that the requirements to the coefficients of stiffness may vary considerably for different phases of the assembling process (according to the tradition, these requirements are reduced to the limitations of the position of the centre of the manipulator mechanism elasticity mechanical compliance together with the elastic element).

The new task is the problem of holding the object while it is being machined when considerable forces may get applied to it. So far as this task is concerned, the rational solution of the problem consists in using the feedback circuits coming from the force-and-torque sensitization transducers to the drives due to whose action an equivalent improvement of the mechanism's stiffness can be obtained.

REFERENCES

1. Kolpashnikov S.N., Polevoy A.I., Stoyanov V.M., Tchelpanov I.B. Problems of Mechanics at Automatic Assembling with Application of Industrial Robots. In book: Robotization of Assembling Processes, Moscow, NAUKA, 1985.
2. Kolpashnikov S.N., Tchelpanov I.B. Optimizing the Structure of Elastic Elements of Robots Applied for Assembling Processes. In book: Industrial Robots, Leningrad, MASHINOSTROYENIYE, 1986.
3. Kolpashnikov S.N., Tchelpanov I.B. Formalization of Requirements Imposed on Elastic Characteristics and Control Laws of Manipulators of Assembly Robots. "Proc. 1st ICAA, Brighton, UK, 1980" IFS, 1980, p.325-330.

Table 1

Alternative	Final Condition Alternatives	Figure	Final Parameters Preset	Parameters to Be Determined
1	Free motion		$P_x, P_y, P_z,$ M_x, M_y, M_z	$\delta_x, \delta_y, \delta_z,$ $\theta_x, \theta_y, \theta_z$
2	Part is gripped rigidly		$\delta_x, \delta_y, \delta_z,$ $\theta_x, \theta_y, \theta_x$	$P_x, P_y, P_z,$ M_x, M_y, M_z
3	Part is gripped in centres		$\delta_x, \delta_y, \delta_z,$ θ_x, M_y, θ_z	$P_x, P_y, P_z,$ M_x, θ_y, M_z
4	Part moves on guides		$P_x, \delta_y, \delta_z,$ $\theta_x, \theta_y, \theta_z$	$\delta_x, P_y, P_z,$ M_x, M_y, M_z
5	Setting of bush on the shaft		$P_x, \delta_y, \delta_z,$ M_x, θ_y, θ_z	$\delta_x, P_y, P_z,$ θ_x, M_y, M_z
6	Part moves on the plane		$\delta_x, P_y, P_z,$ M_x, θ_y, θ_z	$P_x, \delta_y, \delta_z,$ θ_x, M_y, M_z
7	Part point rests against the plane		$\delta_x, \delta_y, \delta_z,$ M_x, M_y, M_z	$P_x, P_y, P_z,$ $\theta_x, \theta_y, \theta_z$
8	Screwing of the nut		$\delta_x, \delta_y, \delta_z,$ M_x, θ_y, θ_z	$P_x, P_y, P_z,$ θ_x, M_y, M_z

ABSTRACT. There are set and solved new problems of limiting, stabilization at preset level or minimization of component forces and torques of the forces of interaction of the object being manipulated with external objects. The description of some typical problems is given together with illustrative examples.

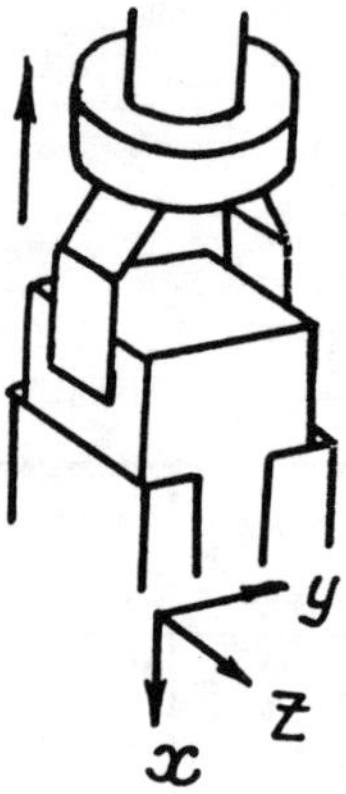

Fig. 1

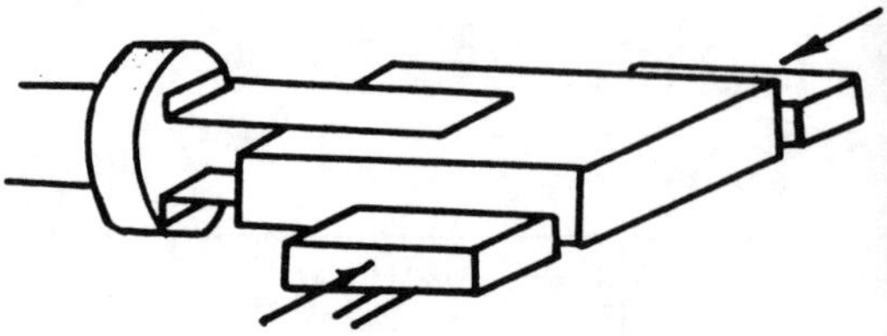

Fig. 2

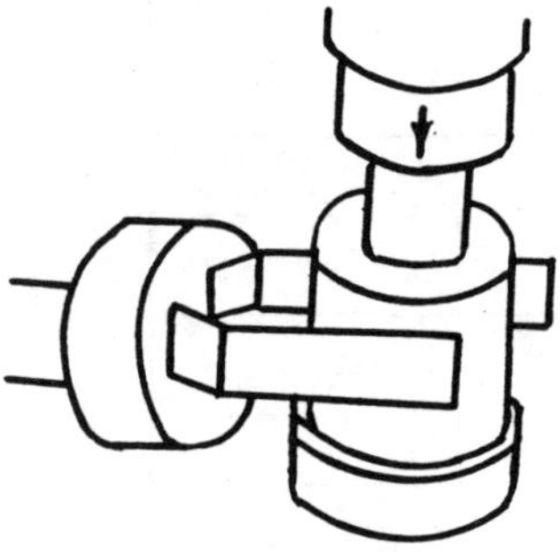

Fig. 3

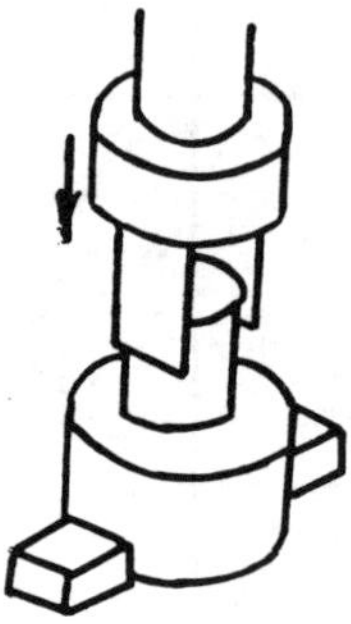

Fig. 4

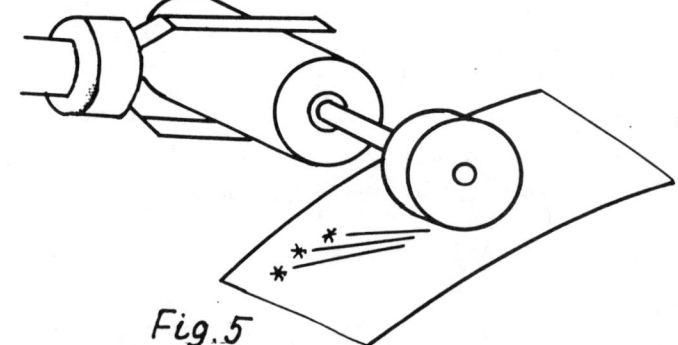

Fig. 5

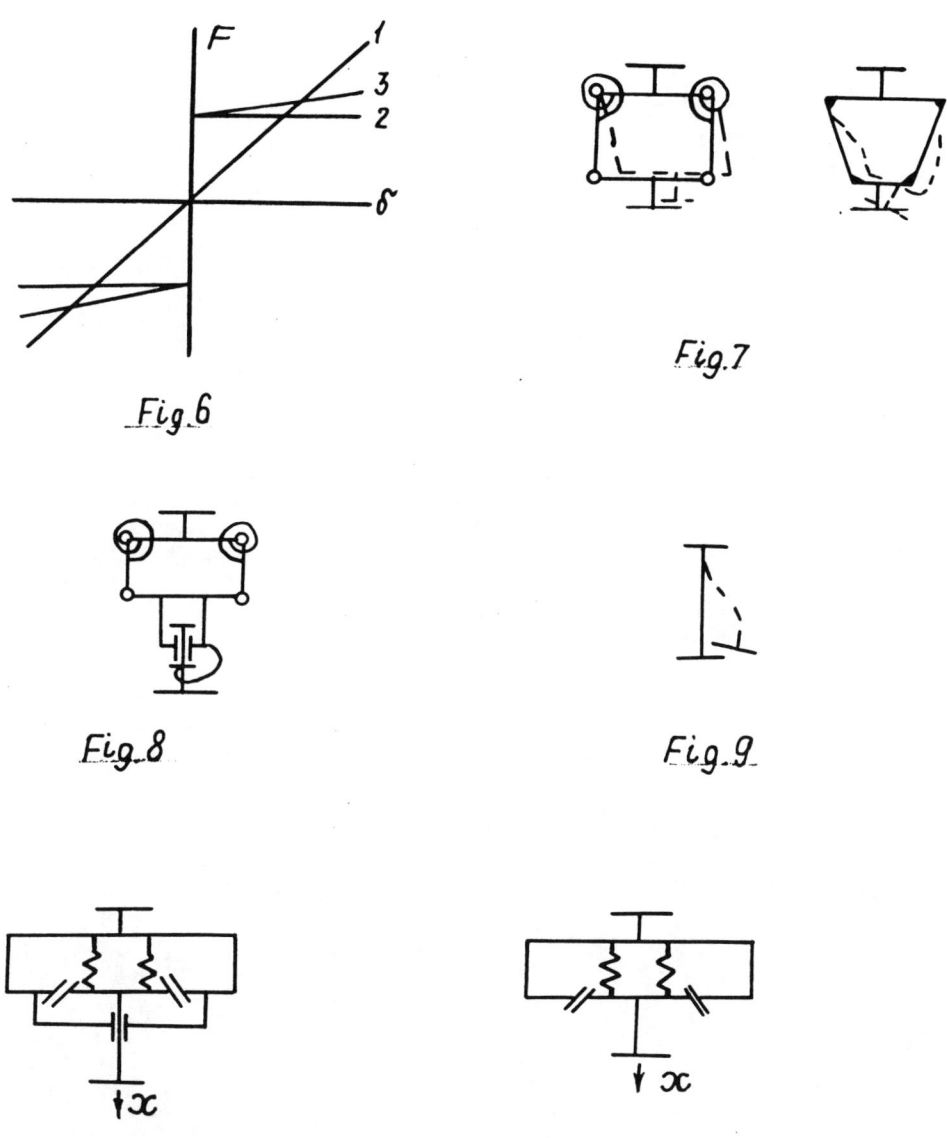

Fig. 6

Fig. 7

Fig. 8

Fig. 9

Fig. 10

Fig. 11

Part 12
Control of Motion 2

Contribution to Solving Dynamic Robot Control in Machining Process

*D. Vujić and **M. Vukobratović

*Mihailo Pupin Institute, **Goša Institute, Beograd, Yugoslavia

SUMMARY

In this paper the possibility for dynamic control of robots in machining process is presented. On the basis of closed chain theory the nominal dynamics is calculated and the dynamic control ensuring desired reaction force during machining process is synthesized. The numerical example of grinding process with a six-degree-of-freedom manipulator is presented.

INTRODUCTION

As it is well known from industrial practice, the constraints on the manipulator gripper's motion appear in many manipulation control tasks. In these situations the models of manipulation robots change from open to closed kinematic chains. Using the general dynamics theorems, i.e. the theorem about the center of gravity motion and the theorem about the change of kinetic moment, the dynamic model of closed kinematic chain is established. By addition of the models of actuators' dynamics we get the complete model of the dynamics of manipulation robot with constrained gripper motion. The simulation of the robot's dynamics as the closed kinematic chain is achieved. Further, the dynamic control ensuring desired reaction force during machining process using force feedback is syntesized. The numerical example of grinding process with a six-degree-of-freedom manipulator is presented. Using an interactive procedure for computer-aided design, all relevant dynamic characteristics are calculated and the control parameters are selected for the particular manipulation robot.

MATHEMATICAL MODEL OF MANIPULATOR WITH
CONSTRAINTS OF GRIPPER'S MOTION

The manipulator's position can be defined by the generalized position vector X_g which consists of n independent parameters [1]. For a manipulator with six degrees of freedom (n=6) we adopt the position vector

$$X_g = [x_A \ y_A \ z_A \ \theta \ \rho \ \psi]^T \tag{1}$$

where x_A, y_A, z_A are Cartesian coordinates of the gripper point which determine its position and θ, ρ, ψ, are Euler's angles which determine its orientation. Imposed gripper's motion reduces the number of degrees of freedom (d.o.f). Let n_r be this reduced number of d.o.f. It holds that $n_r \leq n$ (the equality holds when there is no constraint). We now introduce n_r free and independent parameters $u_1, \ldots, u_{n_r}$ which define the constrained position of the gripper. The reduced position vector X_r is introduced

$$X_r = [u_1 \ \ldots \ u_{n_r}]^T \tag{2}$$

We express the constrained motion by the second-order Jacobian form connecting the position vector (1) and the reduced position vector (2).

$$\ddot{X}_g = J_r \ddot{X}_r + A_r \tag{3}$$

where J_r is the reduced Jacobian and A_r is the associated reduced vector the dimensions of which are $J_r(n \times n_r)$, $A_r(n \times 1)$. For motion without constraints:

$$\ddot{X}_g = J\ddot{q} + A \tag{4}$$

Combining (3) and (4) we obtain

$$\ddot{q} = J^{-1} J_r \ddot{X}_r + J^{-1}(A_r - A) \tag{5}$$

The constraints produce reactions, forces and moments which are introduced into the dynamic model developed by general dynamics theorems, which can be presented in the following form

$$H\ddot{q} + h = P + D_1 F_A + D_2 M_A \tag{6}$$

or

$$H\ddot{q} + h = P + DR_A \tag{7}$$

where the matrix D and reaction vector R_A are

$$D_{(n \times 6)} = \left[D_1 \ (n \times 3) \ \vdots \ D_2 \ (n \times 3) \right], \quad R_A \ (6 \times 1) = \begin{bmatrix} F_A \ (3 \times 1) \\ \hline M_A \ (3 \times 1) \end{bmatrix}$$

In (7) $H=H(q):(n \times n)$ is the inertial matrix, $h=h(q,\dot{q}):(n \times 1)$ the vector consisting of gravity, centrifugal and Coriolis effects, $P=P(t):(n \times 1)$ the vector of driving torques (forces) in joints, $\ddot{q}:(n \times 1)$ the acceleration vector in internal coordinates, $D:(n \times 6)$ the matrix associated to the reaction vector and $R_A:(n \times 1)$ the reaction vector.

Depending on the constraint imposed and on manipulator's configuration, there are some condidions which should be satisfied by the six-component reaction R_A. Namely, there are $6-(n-n_r)$ scalar conditions which can be expressed in matrix form:

$$ER_A = 0 \tag{8}$$

where E is a matrix of dimensions $(6-n+n_r) \times 6$. Now, equations (5), (7) and (8) define the complete mathematical model of closed chain configuration.

We will first discuss the problem of calculating the nominal dynamics. The nominal dynamics (prescribed manipulation task) assumes that the forces, which we want to realize during motion, are given. Thus F_A, M_A and accordingly R_A are known. These values must be prescribed so that they satisfy (8). Now, the necessary driving torques (forces) can be solved from (7). However, if we want to calculate the unknown motion and reactions, then substituting (5) into (7), we obtain

$$HJ^{-1} J_r \ddot{x}_r + h - DR_A = P - HJ^{-1}(A_r - A) \tag{9}$$

Combining (9) and (8), the following matrix equation can be reached:

$$\begin{bmatrix} HJ^{-1} J_r & \vdots & -D \\ \hline 0 & \vdots & E \end{bmatrix} \begin{bmatrix} \ddot{x}_r \\ \hline R_A \end{bmatrix} + \begin{bmatrix} h \\ \hline 0 \end{bmatrix} = \begin{bmatrix} P \\ \hline 0 \end{bmatrix} + \begin{bmatrix} -HJ^{-1}(A_r - A) \\ \hline 0 \end{bmatrix} \tag{10}$$

where the dimensions are:

$$\begin{bmatrix} (n \times n_r) & \vdots & (n \times 6) \\ \hline ((6-n+n_r) \times n_r) & \vdots & ((6-n+n_r) \times 6) \end{bmatrix} \begin{bmatrix} (n_r \times 1) \\ \hline (6 \times 1) \end{bmatrix} + \begin{bmatrix} (n \times 1) \\ \hline ((6-n+n_r) \times 1) \end{bmatrix} =$$

$$= \begin{bmatrix} (n \times 1) \\ \hline ((6-n+n_r) \times 1) \end{bmatrix} + \begin{bmatrix} (n \times 1) \\ \hline ((6-n+n_r) \times 1) \end{bmatrix}$$

Equation (10) represents a system of n_r+6 equations which can be solved for n_r+6 unknowns $\ddot{X}_r$ and R_A. Now, by using (5) we may obtain $\ddot{q}$, i.e. we may calculate the manipulator's motion.

The dimensionality of system (10) can be reduced if we note that the six-component reaction R_A depends on $n-n_r$ independent parameters. Let us introduce the reduced reaction vector R_{Ar} of dimension $(n-n_r) \times 1$ consisting of these independent parameters. Now, R_A can be expressed as

$$R_A = GR_{Ar} \qquad (11)$$

where G is a matrix of dimensions $(6 \times (n-n_r))$.

Substituting (11) into (9) it follows that

$$HJ^{-1}J_r \ddot{X}_r + h - DGR_{Ar} = P - HJ^{-1}(A_r - A) \qquad (12)$$

or

$$\left[HJ^{-1}J_r \;\middle|\; -DG \right] \begin{bmatrix} \ddot{X}_r \\ \hline R_{Ar} \end{bmatrix} = P - h - HJ^{-1}(A_r - A) \qquad (13)$$

where the dimensions are

$$\left[(n \times n_r) \;\middle|\; (n \times (n-n_r)) \right] \begin{bmatrix} (n_r \times 1) \\ \hline ((n-n_r) \times 1) \end{bmatrix} = (n \times 1)$$

Thus, (13) represents a system of n equations which can be solved for n unknowns $\ddot{X}_r$ and R_{Ar}.

GRIPPER MOVING ALONG A SURFACE

We consider a manipulator with the gripper which cannot move freely but its point A is forced to move along a given surface (Fig.1). This time, the independent parameters approach is applied.

Let us define the relative position of the gripper's point A with respect to a surface by means of two parameters u_1 and u_2. This leads us to the parametric form of the moving surface (nonstationary constraint):

$$x = f_x(u_1, u_2, t), \quad y = f_y(u_1, u_2, t), \quad z = f_z(u_1, u_2, t) \qquad (14)$$

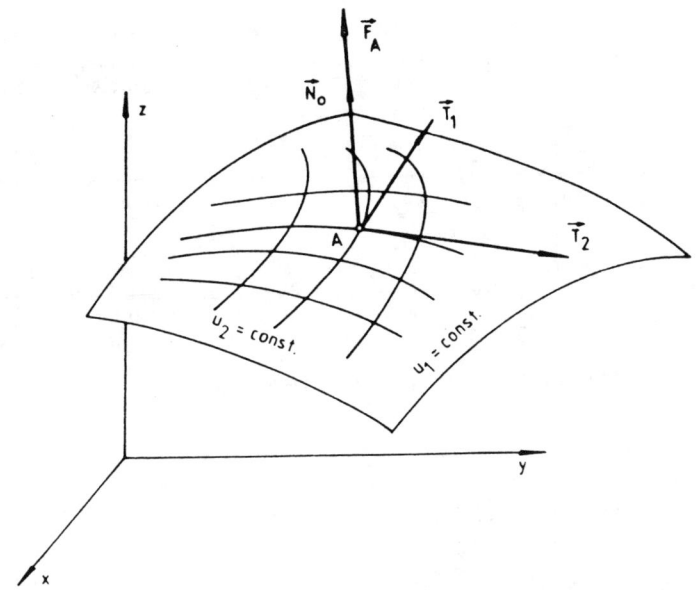

Fig. 1. Point A on a moving surface

Now, $n_r = n-1$ and therefore the reduced position vector, for a manipulator with six degrees of freedom, is

$$X_r = [u_1 \ u_2 \ \theta \ \rho \ \psi]^T \qquad (15)$$

i.e., $u_3 = \theta$, $u_4 = \rho$, $u_5 = \psi$

Now, the Jacobian form (3) can be obtained. The reduced Jacobian and the reduced associated vector are:

$$J_r = \begin{bmatrix} \frac{\partial f_x}{\partial u_1} & \frac{\partial f_x}{\partial u_2} & \\ \frac{\partial f_y}{\partial u_1} & \frac{\partial f_y}{\partial u_2} & 0(3 \times (n-3)) \\ \frac{\partial f_z}{\partial u_1} & \frac{\partial f_z}{\partial u_2} & \\ \hline 0((n-3) \times 2) & I((n-3) \times (n-3)) \end{bmatrix}, \ A_r = \begin{bmatrix} \alpha_x \\ \alpha_y \\ \alpha_z \\ \hline 0((n-3) \times 1) \end{bmatrix} \qquad (16)$$

where α_x, α_y, α_z can be written in the form (17)

$$\alpha_x = \frac{\partial^2 f_x}{\partial u_1^2}\dot{u}_1^2 + 2\frac{\partial^2 f_x}{\partial u_1 \partial u_2}\dot{u}_1\dot{u}_2 + \frac{\partial^2 f_x}{\partial u_2^2}\dot{u}_2^2 + 2\frac{\partial^2 f_x}{\partial u_1 \partial t}\dot{u}_1 + 2\frac{\partial^2 f_x}{\partial u_2 \partial t}\dot{u}_2 + \frac{\partial^2 f_x}{\partial t^2}$$

$$\alpha_y = \frac{\partial^2 f_y}{\partial u_1^2}\dot{u}_1^2 + 2\frac{\partial^2 f_y}{\partial u_1 \partial u_2}\dot{u}_1\dot{u}_2 + \frac{\partial^2 f_y}{\partial u_2^2}\dot{u}_2^2 + 2\frac{\partial^2 f_y}{\partial u_1 \partial t}\dot{u}_1 + 2\frac{\partial^2 f_y}{\partial u_2 \partial t}\dot{u}_2 + \frac{\partial^2 f_y}{\partial t^2} \quad (17)$$

$$\alpha_z = \frac{\partial^2 f_z}{\partial u_1^2}\dot{u}_1^2 + 2\frac{\partial^2 f_z}{\partial u_1 \partial u_2}\dot{u}_1\dot{u}_2 + \frac{\partial^2 f_z}{\partial u_2^2}\dot{u}_2^2 + 2\frac{\partial^2 f_z}{\partial u_1 \partial t}\dot{u}_1 + 2\frac{\partial^2 f_z}{\partial u_2 \partial t}\dot{u}_2 + \frac{\partial^2 f_z}{\partial t^2}$$

and I is a unit matrix of the corresponding dimension.

Finally the dynamic model (13) becomes

$$\left[HJ^{-1}J_r \;\vdots\; -D(G+G') \right] \left[\begin{array}{c} \ddot{X}_r \\ \hline S \end{array} \right] = P - h - HJ^{-1}(A_r - A) \quad (18)$$

and it can be solved for n unknown $(\ddot{X}_r, S)$. In (18)

$$G = [N_o^T \;\vdots\; 0_{(1\times 3)}]^T \qquad G' = [-\mu V_{Ar_o}^T \;\vdots\; 0_{(1\times 3)}]^T \quad (19)$$

where μ is the friction coefficient and $\vec{V}_{Ar_o}$ is the unit vector

$$\vec{V}_{Ar_o} = \frac{\vec{V}_{Ar}}{|\vec{V}_{Ar}|} \quad (20)$$

and $\vec{V}_{Ar}$ is the relative velocity of gripper's point A with respect to the surface.

NUMERICAL EXAMPLE

Let us consider the manipulation robot during grinding of the working object which is moved with constant velocity of 0.1m/s (Fig.2). During the period T_1 the manipulator moves freely, i.e. the manipulator with a cutting tool moves towards the working object. Let us assume that the contact between the gripper (cutting tool) and the working object is impactless and that the end of period T_1 is, at the same time that the start of the interval T_2 (point A_1 in Fig. 2) takes place. The motion from A_1 to A_2 is imposed, i.e. the manipulator is considered as a closed chain.

The relative manipulator's motion with respect to the surface, as mentioned before, is defined by means of two parameters u_1 and u_2. The parameter u_1 is constant during the grinding process and the parameter u_2 can be defined in the form:

$$u_2(t) = \frac{L}{T} t = \frac{0.4}{2} t = 0,2t \tag{21}$$

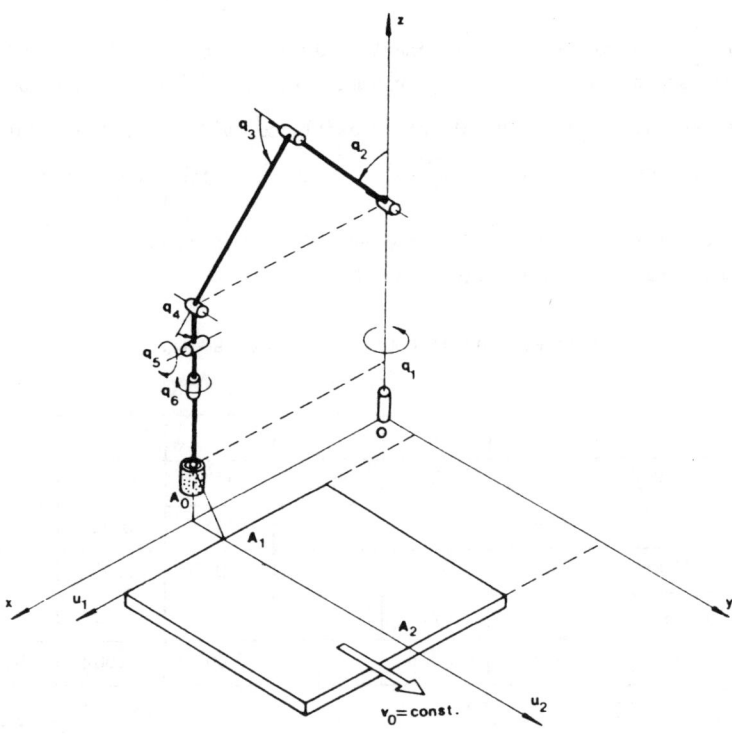

Fig.2. Scheme of manipulation task

If we introduce actuators parameters into (7), and use the relations between $\ddot{q}$ and $\ddot{X}_g$, i.e. $\ddot{X}_r$, we obtain the matrix equation which describes the dynamics of the total system in the form

$$\left[H^* J^{-1} J_r \;\vdots\; -D_1(G+G') \right] \begin{bmatrix} \ddot{X}_r \\ --- \\ S \end{bmatrix} = U_c - U^* - H^* J^{-1} (A_r - A) \tag{22}$$

where

$$H^*_{ik} = \left[H_{ik} + (J^*_r J_n N^2 n)_{ik} \right]$$

$$U_{ci} = (\frac{C_m}{R_r} N\eta u)_i \qquad (23)$$

$$U_i^* = \left[(B_c + \frac{C_e C_m}{R_r})N^2 n\dot{q} + h\right]_i$$

and $J_r^* \left[kgm^2\right]$ - the moment of motor inertia, J_n-a unit matrix of n-th order, $N_1 [-]$ - gear speed ratio, $N_2 [-]$ - gear moment ratio, $\eta[-]$-efficiency coefficient, $C_m\left[\frac{Nm}{A}\right]$-mechanical constant, $R_r[\Omega]$-rotor resistance, $u[V]$ - control signal, $C_e|\frac{V}{rad/s}|$ electical constant, $B_c\left[\frac{Nm}{rad/s}\right]$ - viscous friction coefficient.

The mechanical confiquration's parameters of the manipulator are given in tab. 1, and actuator's parameters are given in tab. 2.

TABLE 1: Mechanical confiquration's parameters

LINKS	1	2	3	4	5	6
mass $[kg]$	0.	5.	5.	1.	1.	2.
length $[m]$	0.8	0.8	0.8	0.15	0.15	0.3
$J_{x_i} [kgm^2]$	0.	0.25	0.25	0.002	0.002	0.01
$J_{y_i} [kgm^2]$	0.	0.01	0.25	0.002	0.002	0.002
$J_{z_i} [kgm^2]$	0.2	0.25	0.01	0.002	0.002	0.01

TABLE 2: Actuator's parameters

Actuator	C_M $\left[\frac{Nm}{A}\right]$	C_E $\left[\frac{V}{rad/s}\right]$	R_r $[\Omega]$	B_C $\left[\frac{Nm}{rad/s}\right]$	N_1 $[-]$	N_2 $[-]$	η $[-]$	J_r^* $[kgm^2]$
1	1.5	1.43	1.6	0.0058	31.17	31.17	0.8	0.00003
2	22.32	27.90	1.8	3.15	150	120	0.8	0.00079
3	14.88	18.6	1.8	1.4	100	80	0.8	0.00079
4	3.52	4.4	0.85	0.24	100	80	0.8	0.00001
5	3.52	4.4	0.85	0.24	100	80	0.8	0.00001
6	3.52	4.4	0.85	0.24	100	80	0.8	0.00001

Applied control is synthesized in the following form

$$u^i(t) = u^{io}(t) + \Delta u_i^\ell(t) + \Delta u_i^G(t) \qquad (24)$$

where $u^{io}(t)$ is the nominal (programmed) control, $\Delta u_i^\ell(t)$ is the local and $\Delta u_i^G(t)$ is the global control. The local control can be presented in the form

$$\Delta u_i^\ell(t) = -C\Delta x^i(t) = -C(x^i - x^{io}) \qquad (25)$$

where C is the matrix of constant position and velocity feedback gains, x^i is the perturbed state vector, x^{io} is the nominal state vector. The global control which ensures desired reaction force during grinding can be presented in the form

$$\Delta u_i^G(t) = K_i^G D_1 (\sum_{\ell=1}^{L} R_\ell - S_0) = K_i^G e_i \left[r_i^A \times (\sum_{\ell=1}^{L} R_\ell - S_0) \right] \qquad (26)$$

where K_i^G is the global gains vector, e_i is the unit vector of the i-th joint, r_i^A is the vector from force acting point to the i-th joint, $\sum_{\ell=1}^{L} R_\ell$ is the set of measured forces given by tranducers on the shaft of cutting tool, S_0 is desired (prescribed) reaction force.

The global gains vector and the matrix of constant feedback gains are of the form, respectively

$$K_i^G = (0.05, 0.05, 0.05, 1.0, 1.0, 1.0) \qquad (27)$$

CONCLUSION

In this paper the possibility for dynamic control of robots in machining process is presented. One preliminary analysis has shown that, using force feedback gain, not only is it possible to control the state coordinates but also to achieve the desired force during the process. In this paper the stability analysis and the choice of global feedback gains are not presented. It will be the subject of further work on these problems and the topic of the next papers.

The presented numerical example is based on one case of constraints on gripper's motion which are of interest to practice. However the algorithm for control synthesis and simulation of the robot dynamics as the closed kinematic chain can easily be applied to other types of constraints upon the gripper's motion.

$$C'' = \begin{bmatrix} 4086.77051 & 194.20882 & 0 & 0 & 0 & 0 & 0 & 0 \\ 0 & 650.66315 & 41.83725 & 0 & 0 & 0 & 0 & 0 \\ 0 & 0 & 577.94476 & 42.05196 & 0 & 0 & 0 & 0 \\ 0 & 0 & 0 & 95.27557 & 2.98742 & 0 & 0 & 0 \\ 0 & 0 & 0 & 0 & 94.26759 & 1.59622 & 0 & 0 \\ 0 & 0 & 0 & 0 & 0 & 0 & 93.73153 & 0.62463 \end{bmatrix} \quad (23)$$

The simulation results are shown in Figs. 3. and 4.

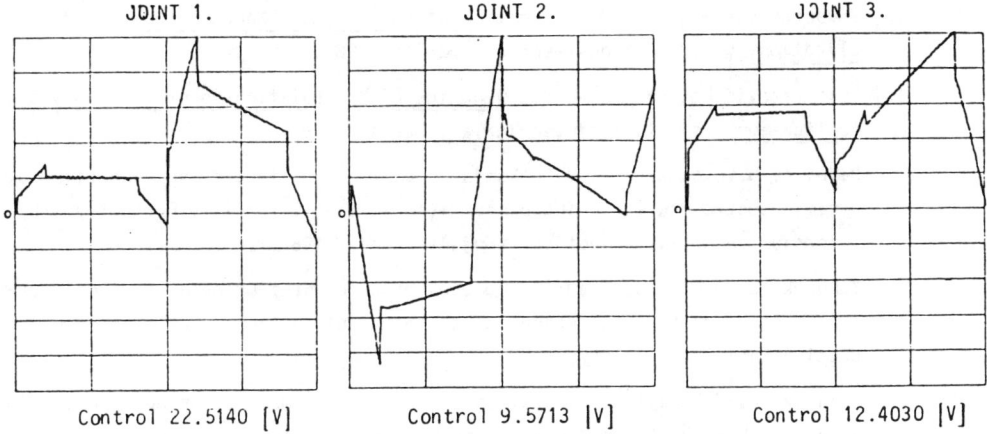

Fig.3. Control for joints 1, 2, 3

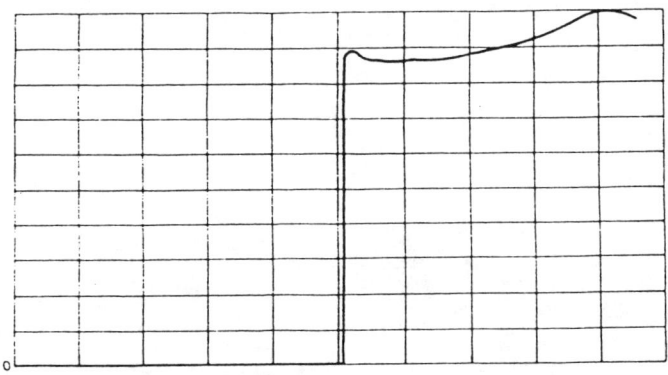

Cutting force F_r 28.0 [N]

Fig.4. External force obtained from (22)

Presented diagrams are plotted by TEKTRONIX plother and they are normalised on the max. value.

REFERENCES

[1] Vukobratović M., Potkonjak V., Applied Dynamics and CAD of Manipulation Robots, Vol. 6, Springer-Verlag, Berlin, 1985.

[2] Vukobratović M., Stokić D., Control of Manipulation Robots: Theory and Application, Vol.2, Springer-Verlag, Berlin, 1982.

[3] Vukobratović M., Potkonjak V., Dynamics of Manipulation Robots: Theory and Application, Vol. 1, Springer-Verlag, Berlin, 1982.

[4] D.Vujić, M.Vukobratović, V.Potkonjak, Contribution to the solving of robot dynamic control in machining process (in serbian), 132-141, Forth Yugoslav symposium on applied robotics, Vrnjačka Banja, 1985.

[5] Y.Furukawa, S.Ohishi, Adaptive Control of Creep Feed Grinding to Avoid Workpiece Burn, 64-69, Proceedings of the 5th, International Conference on Production Engineering, Tokyo, 1984.

[6] K.Nakayama, J.Takagi, T.Fukuda, In-process Measurement of Grinding Wheel Sharpness, 294-299, Proceedings of the 5th International Conference on Production Engeneering, Tokyo, 1984.

An Approach to Development of Real-Time Robot Models

N. Kirćanski, M. Kirćanski, M. Vukobratović and O. Timćenko

Mihailo Pupin Institute, Beograd, Yugoslavia

SUMMARY

A general organization of a new software-development system (SDS) for robotic controllers is described. The 3 main parts of SDS are shortly presented: 1. The program-system for the generation of symbolic kinematic models, 2) The program-system for the generation of dynamic robot models in a numeric-symbolic form, and 3. The time distribution analysis module. The first two parts are designed to produce computationally very efficient forms of models. The third module is intended for obtaining the maximal time intervals allowable for the computation of robot kinematics and dynamics. This module is based on an automated FFT algorithm.

INTRODUCTION

Beside the designers' experience, the development of modern robots and their control units requires special software development systems to facilitate the design of such complex systems. Having in mind that contemporary robots have to move along complicated continuous paths with more than 1 m/s tips speed and less than 0.3 mm contouring error, and that robots with high torque brushless motors (with about 10 m/s top speed and 5 G acceleration) are already being developed [1], one can expect that the exclusive use of designers' experience does not always yield satisfactory results. For example, while 20 ms sampling period is quite satisfactory for robots with up to 1 m/s speed, the question arises what sampling periods should be used in the case of high speed robots. Then, the ratio between the kinematic and dynamic (servosystem) sampling periods should not be chosen exclusively according to the designer's experience. How should the kinematic and dynamic models be evaluated with the minimal number of floating-point (F.P) operations? These questions have been separately discussed in the literature. For example, robot kinematics was considered in [2-7], while real-time dynamic modelling was also considered in [8-11]. However, the problem of determining the kinematic model's evaluation frequency, the sampling frequency of the joint coordinates and the dynamic module's evaluation period, has not been systematically elaborated in the literature.

In this paper a software development system (SDS) will be described, as an instrument for generating the software to be directly implemented on robot controllers. The basic elements of the SDS are: 1. the module for generating the symbolic kinematic model with the minimal number of F.P. operations [7], 2. the module for generating the nonlinear, linearized and sensitivity dynamic models with nearly minimal number of F.P. operations [11], and 3. the time-distribution analysis module. The last module makes use of an FFT based algorithm. Given the robot's parameters and a representative manipulation task it determines the maximal kinematic model's evaluation period and the dynamic servoregulator sampling period.

The SDS will be illustrated on robot PUMA-600 of Unimation Inc. Special attention will be paid to the frequency characteristics of the variables figuring in the dynamic robot control algorithms.

THE SOFTWARE DEVELOPMENT SYSTEM FOR ROBOTIC CONTROLLERS (SDS-RC)

The software development system for robotic controllers is primarily aimed at generating robot models appropriate for implementation on microprocessors. The SDS is usually located on the host computer, although it might also be put directly to the controller itself. In the last case, upon imposing robot's parameters, the robot programs would become "self-generated", implying that such a robot controller may be called "universal".

The global SDS organization is shown in Fig. 1. Upon imposing initial requirements, such as kinematic configuration and dynamic parameters, the model's "design" begins. It involves the generation of symbolic or numeric--symbolic expressions describing the kinematic and dynamic models of the robot. Besides, the model's "design" includes the optimization of the trigonometric expressions in order to evaluate them with the minimum number of numeric operations. The optimization algorithms need not be deterministic in nature, especially if extremely complex expressions are to be optimally evaluated. The last part of the SDS is the model "coding" module. It produces an output source program appropriate for compilation,

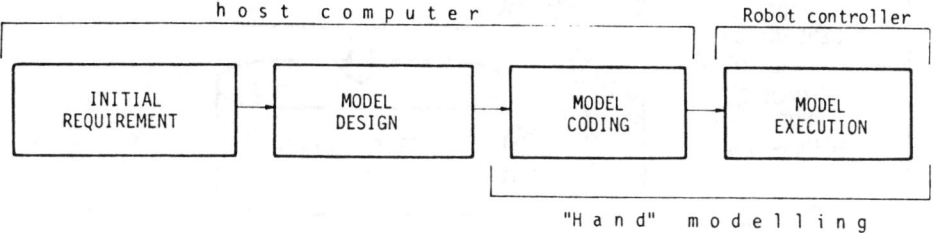

Fig. 1. The global SDS organization

A more detailed organization of the system SDS - robotic controller is shown in Fig. 2. It is evident that the SDS provides for the generation of the linearized dynamic model and the sensitivity model, beside the kinematic and the nonlinear dynamic model. Linearized models are important in the application of the extremly developed linear control theory, while the models of sensitivity with respect to the variation of robot's parameters are directly applicable in adaptive control algorithms [12] and examination of the control robustness. As we can see, almost all the models generated may be used at the strategic control level of the RC. This facilitates suboptimal or even optimal trajectory planning at this control level, as well as the preparation of the parameters for the lower control levels (such as feedback loop gains).

The tactical control level involves the transformation of external (world) coordinates into the joint coordinates using the kinematic model. The execution level usually performs the function of digital servosystems enabling tracking of the desired trajectories. This level, however, usually includes the compensation of the robot model's nonlinearities, and therefore requires the approximate or even exact dynamic model.

The time-distribution analysis (TDA) module determines the kinematic and dynamic models' evaluation periods, taking into account parameters of the mechanism and the actuators, as well as representative manipulation tasks (with given payload's mass, velocities and accelerations).

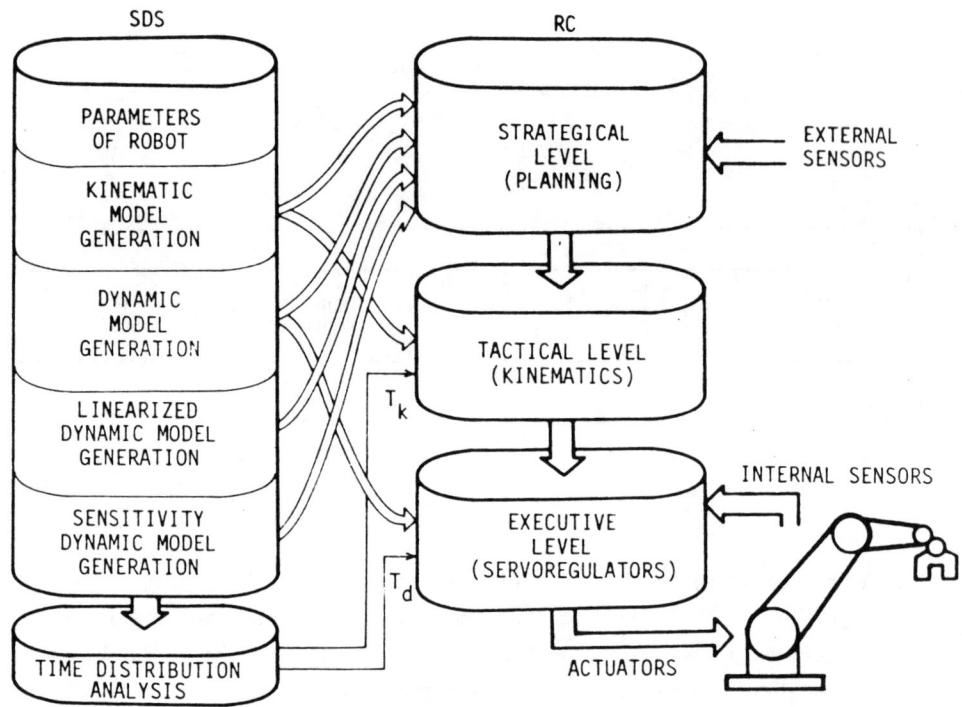

Fig. 2. Software development system interconnected with an advanced robotic controller

COMPUTER-AIDED GENERATION OF SYMBOLIC
KINEMATIC MODELS

This program module enables automatic generation of the manipulator's kinematic model in the symbolic form given an arbitrary serial-link manipulator with n degrees of freedom [7]. Input of the module are Denavit-Hartenberg's parameters of the mechanisms. On output it yields a computer program for evaluating manipulator hand position, rotation matrix and the Jacobian matrix using the minimal number of multiplications and additions. It makes use of either backward or forward recursive symbolic relations in computing the elements of the homogeneous transformations. For example, the elements $Tn(i-1)1\ell$, $\ell=1,2,3$ in the first row of the transformation matrix $^{i-1}T_n$ between the last and the $(i-1)$-st coordinate frame, is evaluated recursively from

$$Tn(i-1)1\ell = \cos q_i \; Tni1\ell - \sin q_i \; TRni\ell, \quad \ell=1,2,3$$

where

$$TRni\ell = \cos\alpha_i \; Tni2\ell - \sin\alpha_i \; Tni3\ell, \qquad \ell=1,2,3,4$$

Here q_i denotes the joint angle (the i-th joint is a revolute one in this example), α_i - the twist angle, $Tnij\ell$, $j=1,2,3$, $\ell=1,2,3,4$ the elements of matrix ^iT_n. Similar expressions were derived for prismatic joints, for revolute joints with parallel joint axes, for the elements of Jacobian matrices with respect to the base frame or the end-effector's coordinate frame. The problem of compressing the trigonometric expressions is avoided either by combining backward and forward relations, or carried out analytically on the general symbolic expressions. Thus the computer program obtained contains the minimal number of numeric operations.

This program package, however, does not resolve the inverse kinematic problem. It has to be elaborated manually for robots where the analytical solution exists, or the numerical solution or combined numerical-analytical solutions [13] have to be applied.

PROGRAM MODULE FOR COMPUTER-AIDED GENERATION OF DYNAMIC ROBOT MODELS

This program system enables automatic generation of analytical forms of the elements of dynamic model matrices [11]:

$$P = H(q, d)\ddot{q} + \dot{q}^T C(q, d)\dot{q} + g(q, d) \qquad (1)$$

where $H(q, d)$ is an $n \times n$ inertia matrix of robot arm, $C(q, d)$ is $n \times n \times n$ matrix corresponding to centrifugal and Coriolis' effects, and $g(q, d)$ is n-dimensional vector originating from gravity forces. The parameters (vector $d \in R^{n_p}$) are treated numerically, while the joint coordinates (vector $q \in R^n$) are treated symbolically. An arbitrary dynamic variable, as well as any term in H, C or g matrix, may be represented as a polynomial

$$v = \sum_{k=1}^{m} v_k x_1^{\epsilon_{1k}} \cdots x_N^{\epsilon_{Nk}} \qquad (2)$$

where v_k are numerical coefficients, and

$$x_i = \begin{cases} \cos q_i & i=1,\ldots,n \\ \sin q_i & i=n+1,\ldots,2n \\ q_i & i=2n+1,\ldots,3n=N \end{cases} \quad (3)$$

Suppose that the coefficients v_k form the column vector V and that the exponents ε_{ik} form the m×N matrix E. The ordered pair (V, E) will be called "the polynomial matrix" of variable v. It was proved [11] that the exponents ε_{ik} may take the values 0, 1 or 2. In this program package we implemented various algebraic operations in the space of polynomial matrices (like dot and cross products) and others, which are useful for obtaining the elements of dynamic model's matrices. Further, we implemented the algorithms for the computation of polynomials (2) reducing the computational burden as much as possible. The output of the package is the source program of the robot's model. This program, automatically generated by the computer, consists of simple statements leading to the computation of H, C and g matrix element. Each statement includes one F.P. multiplication and one or more F.P. additions.

TIME DISTRIBUTION ANALYSIS MODULE

The input of this module are the generated kinematic and dynamic models, the manipulation task (describing the most difficult dynamic actions to be performed by the robot) and the actuators' parameters.

At first, the task given in world coordinates is transformed into joint coordinates, using the symbolic kinematic model. Then, the frequency analysis is performed both in world and joint space, using a corresponding FFT module. Since these functions are not periodical, they are first multiplied by the factor $(1-\cos^2(t))$, thus removing the undesired harmonics. Accordingly, this module generates amplitude and phase power spectra, both in the joint and world coordinates space. It should be noted that the spectrum in joint space may have considerably wider bandwidth than the spectrum in world coordinates. This occurs if the manipulator is passing through the zones close to manipulator singularities.

The spectra corresponding to joint positions, velocities and accelerations are generated separately. Their bandwidths are considerably different, since the acceleration bandwidth is wider than the velocity one, while the velocity bandwidth is considerably wider then of the position one.

Further, this module analyzes the spectra corresponding to dynamic variables, which are useful in the dynamic control laws. Therefore, one obtains the spectra corresponding to the elements of the inertial matrix, matrices of centrifugal and Coriolis' effects (optionally) and the gravity vector. At the end, a table of bandwidths is formed. Analyzing this table one can see that the elements of the dynamic model matrices have considerably narrower bandwidths than the corresponding joint accelerations, as will be shown in the Example. If this ratio equals k (e.g. k=5), then the elements of the dynamic model matrices can be evaluated during k sampling periods. Since the evaluation of the dynamic matrices takes most of the floating-point multiplications and additions required for the evaluation of torques, owing to this analysis, the number of operations is, in fact, reduced k times.

At the end, let us emphasize that this SDS module is completely automatized, so that the user, upon specifying the representative task, obtains the required ratios between the sampling periods of various modules at the tactical control level, as well as the executive dynamic control level of the robot controller.

EXAMPLE

In this section we will point out the useful results which can be obtained using the developed SDS. We will consider the commercially available manipulator PUMA 600. Upon specifying the proper parameters of the mechanical arm of the PUMA 600 robot, one may generate the purely symbolic, computationally optimal kinematic model. Fig. 3 shows the segment of the obtained source code corresponding to the computation of the elements of the transformation matrix $^{0}T_3$. The entire program for computing the direct kinematic problem contains n_M=48 F.P. multiplications and n_A=22 F.P. additions.

The source program of the dynamic module requires n_M=744 F.P. multiplications and n_A=234 additions for the computation of H, C and g matrices of PUMA 600 robot. On the other hand, the source program for the model of the same robot with the first 3 degrees of freedom contains n_M=179 and n_A=150 multiplications and additions, respectively. The last model is indeed quite satisfactory for controller implementation. Fig. 4 shows the part of the generated source program - H_{11} element in numeric - symbolic form.

Upon specifying a characteristic test-task in world coordinates (1m/s average tip-speed), the time evaluations of joint coordinates are computed.

```
T3011=C23*C1
T3021=C23*S1
T3031=-S23                    S1=SIN(Q1)
T3012=-S1                     C1=COS(Q1)
T3022=C1                      Q23=Q2+Q3
T3032=0.                      S23=SIN(Q23)
T3013=S23*C1                  C23=COS(Q23)
T3023=S23*S1                  S2=SIN(Q2)
T3033=C23                     C2=COS(Q2)
```

Fig. 3. Source program segment corresponding to $^{0}T_3$ generated by kinematic module of SDS

```
F9=C3* 0.14178E+01
F8=C3*(-0.49072E+00)
F7=S3* 0.24789E+01
F6=C3*(-0.49072E+00)
F5=C3* 0.24789E+01
F4=C3* 0.21221E+01
F3=S3* 0.49072E+00
F2=C3* 0.49072E+00
F1=S3* 0.14178E+01
Q6=C3*F9
Q5=S3*(F7+F8)
Q4=C2*(F6 -0.11877E+01)
Q3=S2*(F5+ 0.66985E+01)
Q2=C2*(F3+F4 0.66985E+01)
Q1=S2*(F1+F2+ 0.11877E+01)
F4=C2*(Q5+Q6+ 0.29100E+01)
F3=S2* 0.12146E+02
F2=C3*(Q3+Q4)
F1=S3*(Q1+Q2)
Q2=C2*F4
Q1=S2*(F1+F2+F3)
HH(1,1)=Q1+Q2+ 0.35537E+00
```

Fig. 4. Source program segment corresponding to H_{11} generated by dynamic module of SDS

The joint coordinate, velocity and acceleration for the first degree of freedom is shown in Fig. 5. The similar functions are obtained for other joints.

The time-distribution analysis module generates the frequency spectra of the joint angles, velocities and accelerations. Fig. 6 shows the output of the FFT algorithm (amplitude spectra, 0÷25Hz) applied with 100Hz sampling frequency. We obtain the amplitude bandwidths of 2.13Hz for the joint coordinate, 2.96Hz for the joint velocity and 9.37Hz for the joint acceleration. The similar results are obtained for the other degrees of freedom.

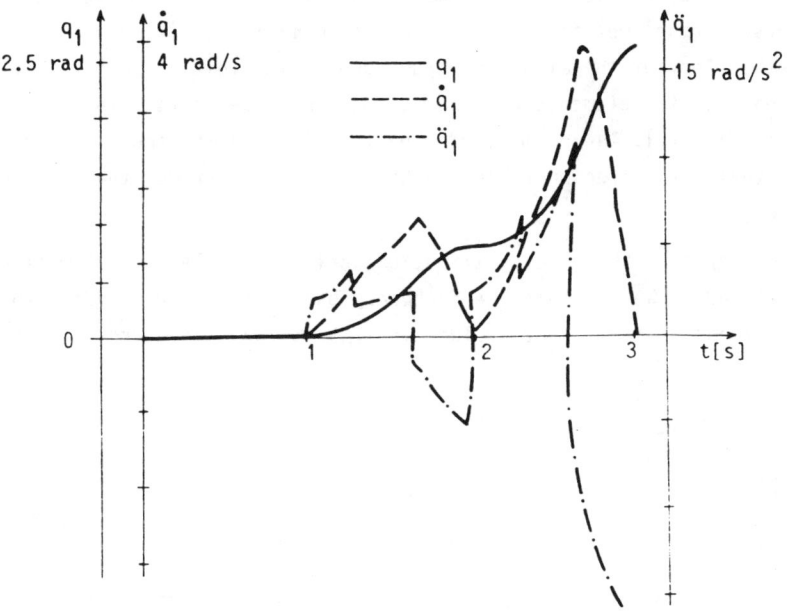

Fig. 5. Joint coordinate, velocity and acceleration of the first degree of freedom

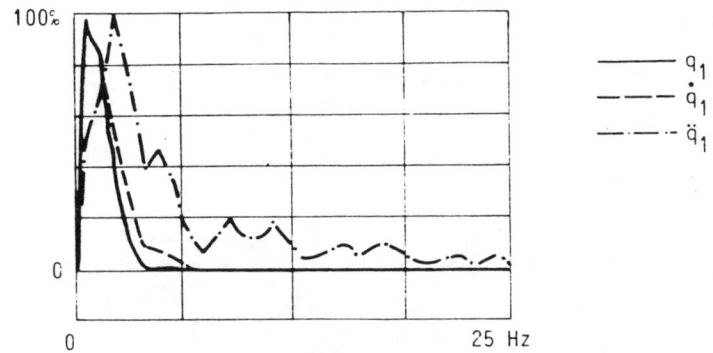

Fig. 6. Amplitude spectra of the first joint coordinate velocity and acceleration

Now, we come to the executive control level of a robot controller. We will suppose that at this level the following dynamic compensation is introduced:

$$P = H(q, d)\ddot{q} + G(q, d) \tag{4}$$

Depending on the controller's structure, this module may be located either in the direct branch (feedforward compensation [11]) or in the feedback branch ("inverse-torque" method [4]). The lack of C matrix in (4) does not influence the methodology of the algorithm described in the text to follow. The time dependence of the elements of dynamic model matrices during the movement is shown in Fig. [7]. Their amplitude spectra (0÷12Hz) are shown in Fig. [8]. We see that their frequency bandwidths are rather nerrower than that of q, $\dot{q}$ and $\ddot{q}$.

The output of the time distribution analysis module is systematized in Tables 1 and 2. We see that the frequency bandwidths corresponding to joints accelerations is about 10Hz, velocities 2.5 to 3Hz, and coordinates about

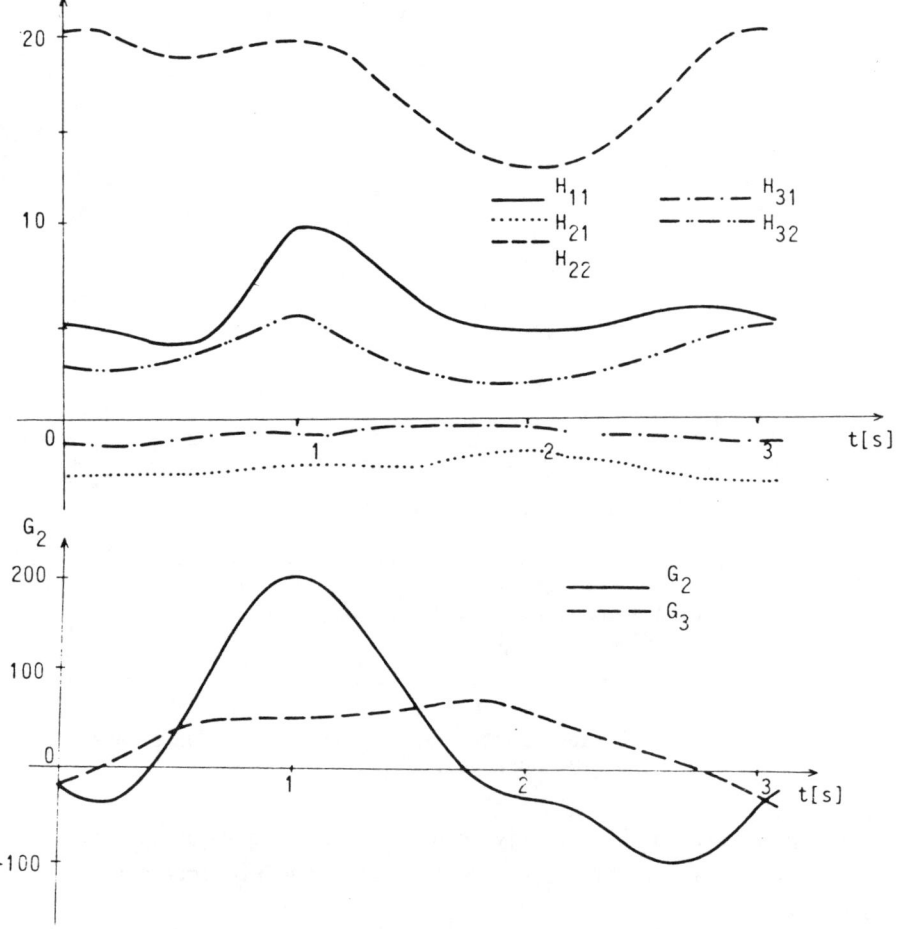

Fig. 7. Elements of dynamic model matrices time domain

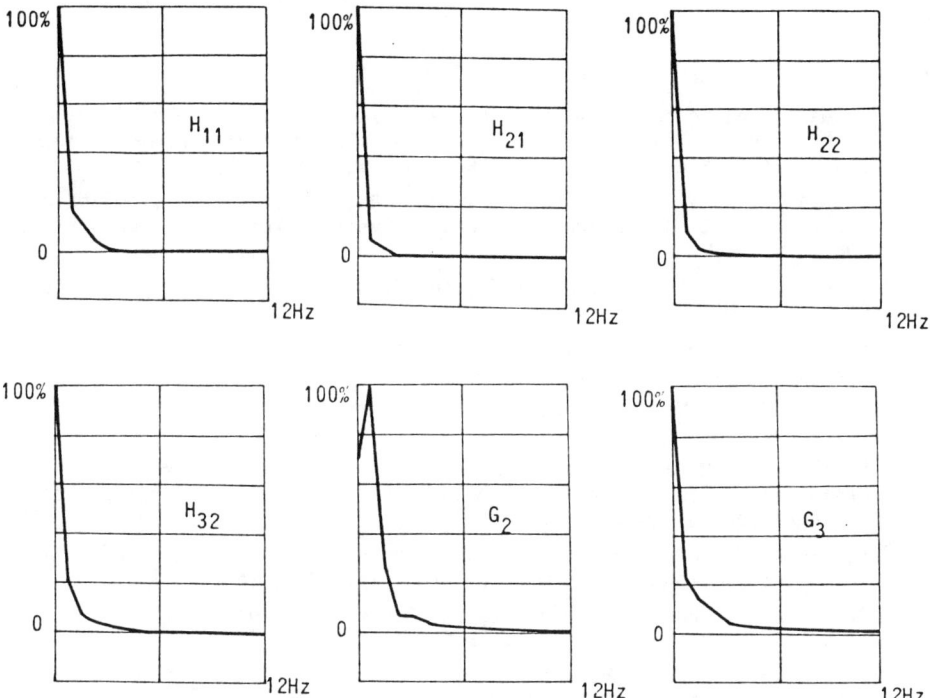

Fig. 8. Amplitude spectra of the elements of dynamic model matrices

Joint \ Boundary frequency Hz	Position amplitude spectra	Velocity amplitude spectra	Acceleration amplitude spectra
1	2.13	2.96	9.37
2	2.30	2.63	6.25
3	1.64	2.46	8.72

Table 1. Boundary frequences of position, velocity and acceleration amplitude spectra for joints of manipulator

Matrix element	H_{11}	H_{12}	H_{22}	H_{13}	H_{23}	G_2	G_3
Boundary frequency	1.15	0.27	0.27	0.21	1.15	1.64	1.48

Table 2. Boundary frequences of H and G matrix elements amplitude spectra

2Hz, while the bandwidth corresponding to the elements of dynamic model matrices is between 0.2 and 1.5Hz. Thus, the computation of dynamic model's matrices may be performed at least 5 times more-rarely than that of acceleration.

REFERENCES

[1] Asada H. and Youcef-Toumi K., "Analysis and Design of a Direct-drive arm with a Five-bar-link Parallel Mechanism", Journal of Dynamic Systems, Measurement, and Control, Vol. 106, pp 225, Sept. 1984.

[2] Khosla P. and Neuman C., "Computational Requirements of Customized Newton--Euler Algorithms", Journal of Robotic Systems, Vol. 2 (3), pp 309-327, 1985.

[3] Brady M. et al. (eds.), Robot Motion: Planning and Control, MIT Press, Cambridge, MA, 1982.

[4] Paul R.P., Robot Manipulators: Mathematics, Programming and Control, MIT Press, Cambridge, MA, 1981.

[5] Lee C.S.G., "Robot arm Kinematics, Dynamics and Control", Computer, 15 (12), pp 62-80, 1982.

[6] Vukobratović M. and Kirćanski M., Kinematics and Trajectory Synthesis of Manipulation Robots, Scientific Fundamentals of Robotics series, Vol. 3, Springer-Verlag, 1985.

[7] Kirćanski M. and Vukobratović M., "Computer-aided Generation of Manipulator Kinematic Models in Symbolic Form", 15. ISIR, Tokio, 1985.

[8] Bejczy A.K., "Robot Arm Dynamics and Control", Technical mem. 33-669, Jet Propulsion Lab., Pasadena, CA, 1974.

[9] Luh J.Y.S., Walker M.W. and Paul R.P.C., "On-Line Computational Scheme for Mechanical Manipulators", Journal of Dynamic Systems, Measurement, and Control, Vol. 102, pp 69-76, 1980.

[10] Hollerbach J.M., "A Recursive Lagrangian Formulation of Manipulator Dynamics and a Comparative Study of Dynamics Formulation Complexity, "IEEE Transaction on SMC, Vol. 10, pp 730-736, 1980.

[11] Vukobratović M. and Kirćanski N., "Real-Time Dynamics of Manipulation Robots", Scientific Fundamentals of Robotics, Vol. 4, Springer-Verlag, 1985.

[12] Vukobratović M. and Kirćanski N., "An Approach to Adaptive Control of Robotic Manipulators", Automatica, 1986.

[13] Takano M., Yashima K. and Toyama S., "A New Method of Solution of Synthesis Problem of a Robot and its Application to Computer Simulation System", 14. ISIR, Sweden, 1984.

Time-Optimal Robotic Manipulator Task Planning

S. Dubowsky, M.A. Norris and Z. Shiller

Department of Mechanical Engineering, MIT, Cambridge, U.S.A.

SUMMARY

A method is presented which finds the minimum time motions for manipulators. The method considers the full nonlinear manipulator dynamics, actuator characteristics, and accounts for obstacles in the work space and restrictions on the motions of the manipulator's joints. It is computationally practical and has been implemented in a Computer Aided Design (CAD) software package, OPTARM II, which facilitates its use. Examples of its application are presented and show that the technique yields substantial improvements in system performance.

INTRODUCTION

To achieve high productivity, manipulators must be capable of performing complex tasks quickly, in minimum time. Current industrial manipulator tasks are not planned to achieve minimum time motions. A number of approaches have been considered by researchers to find a method for the time optimal control of robot motions [1-3]. An effective algorithm for finding a manipulator's minimum time velocity profile along a given path has been developed [4], proven rigorously optimal [5], and shown to be a computationally efficient design tool[6]. It considers the full nonlinear dynamics of the manipulator, actuator capabilities that are functions of the system state, and other factors such as the limits on gripper grasping forces [7].

The algorithm can be extended to find the minimum time manipulator path [8,9]. This paper shows that this be done practically, for realistic problems, and that the method can be further extended to consider the constraints imposed by work space obstacles and manipulator joint motion limitations. These constraints are handled in a computationally efficient and flexible manner. The method has been implemented in a CAD package, called OPTARM II, with interactive graphics capabilities which enhance its practical use. Results show that substantial improvement in system performance can be achieved with this method.

THE ANALYTICAL METHOD

The basic time optimal control algorithm is presented and proven rigorously optimal in References [4] and [5]. The algorithm obtains the open loop torques/forces for the time optimal motion of a general rigid six degree-of-freedom manipulator along a prescribed path subject to actuator constraints (see Figure 1). The algorithm also provides optimal joint positions and velocities for closed loop control. It is shown that the time to traverse the path from the given initial condition, at S_0, to the required final condition, at S_f, will be minimal if the manipulator's acceleration $\ddot{S}$ is equal to either its maximum permissible value, $\ddot{S}_a$, or its minimum permissible value, $\ddot{S}_d$, at each point along the path. It is also shown that at each point on the path there exists a velocity limit, $\dot{S}_m(S)$, that if exceeded, the system constraints will be violated. Plotting $\dot{S}_m(S)$ versus S in the phase plane yields a velocity limit curve. The algorithm solves the minimum time problem by finding the switching points between $\ddot{S}_a(S,\dot{S})$ and $\ddot{S}_d(S,\dot{S})$ such that the velocity $\dot{S}$ is always maximum, but does not cross the limit curve, $\dot{S}_m$, in the phase plane.

Path Optimization without Constraints

With the ability to find the minimum time for a given path, t_p, it is possible to find the optimum path (the one with the smallest minimum time, t_m). First, the path is represented by a finite set of n scalar parameters, a_1 to a_n, which can be written as a vector: $[a_1, a_2, a_3, \ldots a_n]^T$. For each path, or set of path parameters there will be a value of t_p. To find the optimal path, $t_p(\underline{a})$ is defined as the cost function $J(\underline{a})$, which is to be minimized in the form of the unconstrained minimization problem:

$$\text{Minimize } J(\underline{a}) = t_p(\underline{a}) \qquad (1)$$
$$\underline{a}$$

Three path representations that proved most effective were: Straight lines connected by circular arcs; perturbations about a straight line by a Fourier series; and splines. These have been integrated into the OPTARM II program. The results presented below use Bezier splines [10]. The matrix form of the Bezier Spline used here is:

$$\underline{P}(u) = \underline{U}(u) \, M_b \, R(\underline{a}) \qquad (2)$$

$\underline{P}(u)$ is the the 6 x 1 vector which defines the position and orientation of the manipulator's end effector on its path, u is a scalar variable related to the distance S, $\underline{U}(u)$ is the vector $[1\ u\ u^2\ u^3]$, and M_b is a constant matrix.

To optimize the velocity of the manipulator along a given path the partial derivatives of the path with respect to S are required [4]. These are obtained by differentiating the equation (2). For example:

$$\frac{\partial \underline{P}}{\partial S} = \underline{U}_s\ M_b\ R(\underline{a}) \qquad (3)$$

where:

$$\underline{U}_s = \frac{\partial \underline{U}(u)}{\partial u} \frac{\partial u}{\partial S} \qquad (4)$$

These matrix representations of the spline paths and their derivatives are easily integrated into the path optimization method described above, [9]. $R(\underline{a})$ is composed of coordinates of the control points and defines of the parameter vector, $\underline{a}$, in the path optimization.

A number of methods exist for the numerical solution of equation (1) and several were successfully used in this study. The Pattern Search method proved to be the most effective for the majority of the problems treated [11].

Path Optimization with Constraints Using Penalty Functions.

The presence of work space obstacles and joint motion limitations add additional constraints to the optimization problem described above. These are important because manipulator paths must not be too close to other objects, and because most manipulator joints have only a finite range of movement. These constraints can be written in the form:

$$\begin{array}{ll} g_j(\underline{a}) < 0 & j = 1,2,3,\ldots,q \\ \theta_i - \theta_{imin} > 0 & \\ \theta_i - \theta_{imax} < 0 & i = 1,2,3,\ldots,6 \end{array} \qquad (5)$$

where $g_j(\underline{a})$ are the obstacle constraint functions, q is the number of obstacles, and θ_{imin} and θ_{imax} are the joint limits on the ith joint.

In this study it was found that it was most effective to recast this constrained optimization problem into an unconstrained form using the penalty function optimization approach [12]. In this method weighted penalties for coming close to constraints are added to the cost function, $t_p(\underline{a})$. For the results presented below the cost function was:

$$J = t_p(\underline{a}) + \sum_{i=1}^{N} \frac{w_i}{d_i^m} + \sum_{i=1}^{M}\left[\frac{w_{\theta i}}{(\theta_i - \theta_{i\min})^m} + \frac{w_{\theta i}}{(\theta_{i\max} - \theta_i)^m}\right] \quad (6)$$

The method can be used easily for constraints expressed in different coordinate systems, eliminating the need for computationally intensive coordinate transformations. Also, the weighting factors, the w's, provide flexibility in selecting the importance of maintaining a distance from a specific constraint. The m's are positive integers, usually taken as 2, the d_i's are the minimal distance to the ith obstacle from the path, M is the number of joints of the manipulator, and N is the number of obstacles in the environment. OPTARM II uses a table of various general geometric shapes for computing the minimum distance to obstacles. More complex objects can be considered using solid modelling techniques.

EXAMPLES

Optimization of the Motion Along a Specified Path.

Figure 1 shows a manipulator moving from S_0, to S_f, along an intuitively chosen "good path", avoiding the obstacles. In typical conventional control, the tool point moves along most of its path from S_0 to S_f at a constant speed, with constant acceleration and deceleration at the end points. The values of the constant acceleration, deceleration and velocity must be selected so that the manipulator does not leave the required path at any point. The best conventional control trajectory in the phase plane, $S - \dot{S}$, is shown in Figure 2. The time required for this path with the conventional control is 1.23 seconds. Figure 2 also shows the optimal trajectory for this path.

The path was then optimized using OPTARM II, but without considering the obstacles. The end points were kept the same. The top view of the resulting optimal path is shown in Figure 3 along with the intuitive path. The time for this path is 0.37 seconds, compared to .82 seconds for the optimal motion along the intuitive path. This optimal path passes through obstacle B which is not acceptable. The phase plane diagram for this optimial path is shown in Figure 4 along with its limit curve. Also shown in this figure is the optimal trajectory for the intuitive path from Figure 1.

Next the distances from the obstacles were added to the penalty function. The resultant path is shown in Figure 5 along with the unconstrainted optimal path. The travel time has increased from the unconstrained case value of 0.37 seconds to 0.55 seconds. This is a relatively small increase, and the tool point no longer passes through the camera field of view. It should be noted that this path is further away from the camera than the original intuitive optimal path and has a much shorter move time. The minimum distance to the obstacle could be changed simply by changing the weighting factor in the penalty function. Of course, there would be a corresponding increase or decrease in the motion time.

The computation times for this method do not represent a severe burden; this off-line planning technique does not need to have real time computational speed. OPTARM II requires just minutes of computation time on a MicroVax II to optimize the motion for a given move. For many applications, the increase in productivity would most certainly make this a cost effective procedure.

This method has also been applied to optimizing more complex tasks [13]. It combines relatively simple optimal motions in an optimal manner to achieve time optimal tasks. This method can also be used to structure manipulator tasks and design their work environments so that practical tasks can be completed in mimimum time.

SUMMARY AND CONCLUSIONS

This paper presents a method for finding the minimum time motions for a manipulator between given end states. It considers the full non-linear dynamics of the manipulator and the saturation characteristics of its actuators. Using a penalty function approach, it accounts for the presence of obstacles in the work space and restrictions on the motions of the manipulator's joints. The method has proven to be computationally practical and has been implemented in a Computer Aided Design (CAD) software package, called OPTARM II. The results show that substantial improvements in system performance can be achieved with the technique.

ACKNOWLEGMENTS

The support of this research by the Automation Branch of NASA Langley Research Center under Grant NAG-1-489 is acknowledged.

REFERENCES

1. Kahn M.E. and Roth B., "The Near-Minimum-Time Control of Open-Loop Articulated Kinematic Chains," Journal of Dynamic Systems, Measurement and Control, Vol. 93, No. 3, Sept. 1971, pp. 164-172.

2. Niv, M. and Auslander, D.M., "Optimal Control of a Robot with Obstacles," 1984 American Control Conf., San Diego, CA., pp. 280-287, June 1984.

3. Sahar, G. and Hollerbach, J.M., "Planning of Minimum-Time Trajectories for Robot Arms," Proc. IEEE Intern. Conf. on Robotics and Automation, pp. 751-758, St. Louis, Mo., March, 1985.

4. Bobrow, J.E., Dubowsky, S., and Gibson, J.S., "On the Optimal Control of Robotic Manipulators with Actuators Constraints," Proc. of the 1983 American Control Conf., San Francisco, CA, pp.782-787, June 1983.

5. Bobrow, J.E., Dubowsky, S., and Gibson, J.S., "Time-Optimal Control of Robotic Manipulators," The Intern. J. of Robotics Res., Vol.4, No. 3, 1985

6. Dubowsky, S. and Shiller, Z., "Optimal Dynamic Trajectories for Robotic Manipulators," Proc. of the CISM-IFTOMM Symposium on the Theory and Practice of Robots and Manipulators, Udine, Italy, June 1984.

7. Shiller, Z. and Dubowsky, S., "On the Optimal Control of Robotic Manipulators with Actuator and End-Effector Constraints," Proc. IEEE Intern. Conf. on Robotics and Automation, pp. 614-620, St. Louis, Mo., March 1985.

8. Rajan, V.T., "Minimum Time Trajectory Planning", Proc. of the 1985 IEEE Conference on Robotics and Automation , pp. 759-764, St.Louis, Missouri, March 1985.

9. Dubowsky, S., Norris, M.A. and Shiller, Z., Time Optimal Trajectory Planning for Robotic Manipulators with Obtacle Avoidance: A CAD Approach Proc. IEEE Intern. Conf. on Robotics and Automation, San Fransico, CA, April 1986..

10. Faux, I.D. and Pratt, M.J., Computational Geometry for Design and Manufacture, Ellis Horwood Limited, New York, 1981.

11. Bronson, R., "A Modified Pattern Search,"IEEE Transaction on Automatic Control, Vol. 25, No. 2, April 1980.

12. Walsh, G.R., Methods of Optimization, Jonh Wiley & Sons, London, 1975.

13. Dubowsky, S. and Blubaugh, T.D., "Time Optimal Robotic Manipulator Motions and Work Places for Point to Point Tasks," <u>Proc. of the 24th IEEE Conf. on Decision and Control,</u> Fort Lauderdale, FL, Dec., 1985.

FIGURES

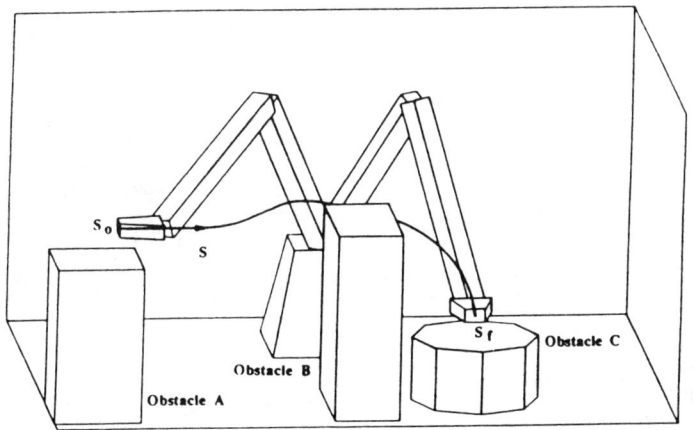

Fig. 1.   A Six Degree-of-Freedom Manipulator in its Environment

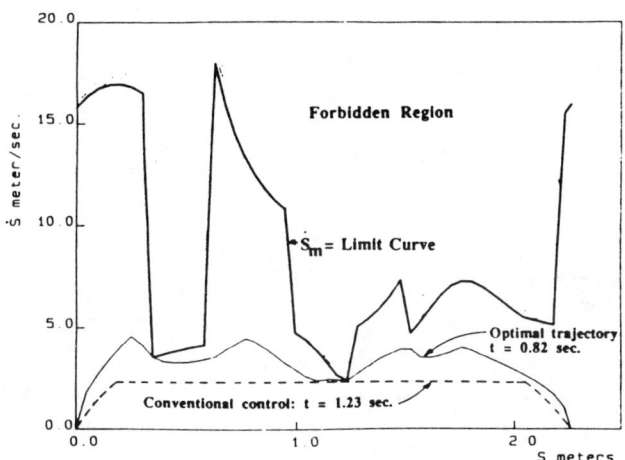

Fig 2.   Phase Plane Trajectories of Conventional and Optimal Motions for the Intuitive Path.

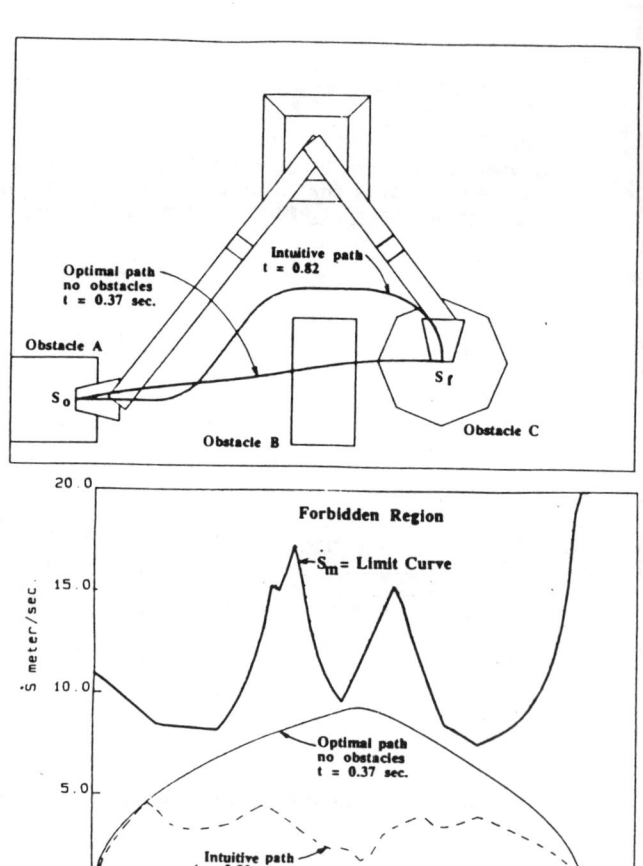

Figure 3.

Top View of Intuitive And Unconstrained Time Optimal Paths.

Figure 4.

Phase Plane Diagram for Unconstrained Optimal and Intuitive Paths.

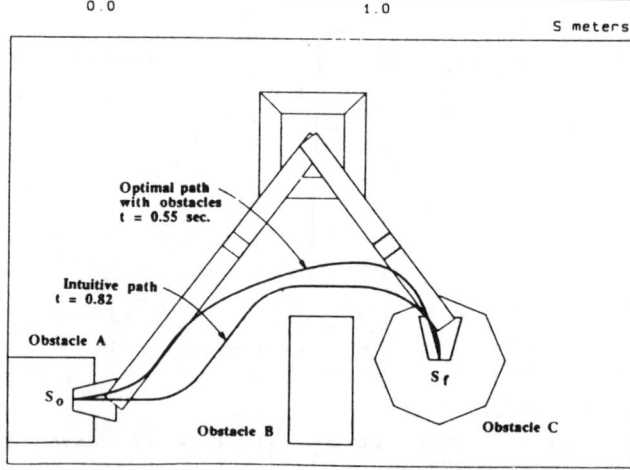

Figure 5.

Time Optimal Constrained and Intuitive Paths.

Time-Optimal Motions of Some Robotic Systems

L.D. Akulenko, N.N. Bolotnik,
F.L. Chernousko and V.G. Gradetsky

Institute for Problems in Mechanics, Academy of Sciences, Moscow, USSR

Summary

This paper is devoted to optimization of robotic motions with respect to operational time. Problems of time-optimal control of manipulation robots were earlier considered in a number of papers, see for example, [1 - 5]. In [1] the computer oriented procedure is given for obtaining open-loop optimal control of restricted joint torques for a six-degree-of-freedom manipulator. Here the path of the end effector of a robot is prescribed. Bang-bang time-optimal controls of a manipulator are considered in [2], provided all generalized coordinates change monotonically. The paper [3] is devoted to construction of suboptimal controls for multi-degree-of-freedom mechanisms; here a nonlinear system is replaced by a simplified linear one. In [4] a method has been proposed using constant maximum velocity and acceleration to minimize the traveling time on a path composed of straight lines connected by circular curves.

Following results are presented below. Optimal control problem for an articulated manipulation robot with two links is considered. A procedure for determining moments of switching bang-bang control is described. Controls calculated by means of this procedure are time-optimal in some cases. In general case they may be regarded as suboptimal. Feedback control of the electromechanical drive for the industrial robot is determined, which is time-optimal for the simplified model of the robot without interaction of different degrees of freedom. Mathematical simulation shows that obtained feedback control provides good performance quality for industrial robots with high gear ratios. Some results are also applicable to more complex electromechanical systems.

1. Open-loop control for a manipulator with two links

We consider here an articulated manipulator with two links (Fig. 1). Control is performed by two independent drives, located in the joints O_1, O_2 and producing torques M_1, M_2 about axes of the joints. The manipulator moves in a horizontal plane, therefore gravity is not taken into account. The system is governed by differential equations

$$(I_1 + ML^2)\ddot{\varphi}_1 + MaL\cos(\varphi_2 - \varphi_1)\ddot{\varphi}_2 - MaL\sin(\varphi_2 - \varphi_1)\dot{\varphi}_2^2 = M_1 - M_2 \tag{1}$$

$$I_2\ddot{\varphi}_2 + MaL\cos(\varphi_2-\varphi_1)\ddot{\varphi}_1 + MaL\sin(\varphi_2-\varphi_1)\dot{\varphi}_1^2 = M_2$$

Here φ_1, φ_2 are angles (generalized coordinates) introduced as shown in Fig. 1; L is the length of the link O_1O_2; a is a distance between the joint O_2 and the centre of mass of the link O_2O_3; I_1, I_2 are moments of inertia of links O_1O_2, O_2O_3 about axes of joints O_1, O_2 respectively.

We consider the following problem of optimal control for the system (1). It is necessary to determine open-loop control laws $M_1(t)$, $M_2(t)$ which transfer the manipulator from an initial state

$$\varphi_1(0) = 0, \quad \dot{\varphi}_1(0) = 0, \quad \varphi_2(0) = \varphi_2^0, \quad \dot{\varphi}_2(0) = 0 \quad (2)$$

to the given terminal state

$$\varphi_1(T) = \varphi_1^1, \quad \dot{\varphi}_1(T) = 0, \quad \varphi_2(T) = \varphi_2^1, \quad \dot{\varphi}_2(T) = 0 \quad (3)$$

in minimal time T, provided control torques M_1, M_2 are restricted

$$|M_1(t)| \leq M_0, \quad |M_2(t)| \leq M_0' \quad (4)$$

We use below the dimensionless variables and parameters:

$$t' = M_0^{1/2}(ML^2)^{-1/2}t, \quad a' = a/L, \quad I_i' = I_i(ML^2)^{-1}$$
$$M_i' = M_i/M_0', \quad i = 1, 2 \quad (5)$$

If we omit primes, then equations (1) - (4) still hold in dimensionless form with $M = 1$, $L = 1$, $M_0 = 1$.

Exact solution of optimal control problem for nonlinear system (1) is rather difficult and time consuming; it was obtained for some cases in [5, 8]. Here we obtain more simple bang-bang controls $M_1(t)$, $M_2(t)$ having minimal number of switchings (three):

$$M_1 = M_0 \mu_\alpha(t), \quad M_2 = \nu_\beta(t) \quad \text{(a)}$$
$$M_1 = M_0 \nu_\beta(t), \quad M_2 = \mu_\alpha(t) \quad \text{(b)}$$
$$\mu_\alpha = (-1)^\alpha \operatorname{sign}(\tau_0 - t)$$
$$\nu_\beta = (-1)^\beta, \quad t \in [0, \tau_1) \cup [\tau_2, T]$$
$$\nu_\beta = (-1)^{\beta+1}, \quad t \in [\tau_1, \tau_2)$$
$$0 \le \tau_0 \le T, \quad 0 \le \tau_1 \le \tau_2 \le T, \quad \alpha, \beta = 0, 1$$
(6)

Both torques (6) satisfy constraints (4) and have maximal admissible values. One of the torques M_1, M_2 has one switching instant τ_0, while the other has two switchings (at τ_1, τ_2). Formulas (6) describe eight types of control regimes, which differ from each other by the number of the torques with single switching (cases a, b) and by conbinations of values of $\alpha, \beta = 0, 1$ determining alternation of signs for controls. Only four of these regimes are considered below

1: (a), $\alpha = 0$, $\beta = 0$ 2: (a), $\alpha = 1$, $\beta = 1$
3: (b), $\alpha = 0$, $\beta = 0$ 4: (b), $\alpha = 1$, $\beta = 1$
(7)

Denote by $y_i^{(k)}(t, \tau_0, \tau_1, \tau_2)$, $i = 1, 2$ the solution of the initial value problem (1), (2) for controls of the type $k = 1, 2, 3, 4$, see (7). For each type parameters $\tau_0, \tau_1, \tau_2, T$ can be obtained from terminal conditions (3):

$$y_i^{(k)}(T, \tau_0, \tau_1, \tau_2) = y_i^1, \quad \dot{y}_i^{(k)}(T; \tau_0, \tau_1, \tau_2) = 0$$
$$i = 1, 2$$
(8)

We suggest a simple technique for calculating of control parameters using special diagrams. To plot the diagram for the fixed initial value ψ_2^0 it is necessary to carry out of the following operations for each type $k = 1,2,3,4$ of control regimes (7).

1. Fix two of the parameters τ_0, τ_1, τ_2 (e.g. τ_1, τ_2) and find (numerically) the third parameter (τ_0) and the time T from two velocity conditions (8).

2. Using coordinate conditions (8), determine the terminal position (ψ_1^1, ψ_2^1) corresponding to parameters τ_0, τ_1, τ_2.

3. Carry out stages 1 and 2 for all possible values τ_1, τ_2 changing them with some steps. Then we obtain the domain Ω_k of terminal positions (ψ_1^1, ψ_2^1) reachable from the fixed initial state using different controls of the type k.

4. Inside all domains Ω_k, $k = 1,2,3,4$ plot the lines of equal values for parameters τ_0, τ_1, τ_2, T.

Calculations show that domains Ω_k, $k = 1,2,3,4$ do not intersect each other and cover all of the plane (ψ_1^1, ψ_2^1).

The typical example of the diagram is presented in Fig. 2. Thick curves are bounds between domains Ω_k. Instead of lines of equal values for τ_2, lines of equal values for $\tilde{\tau}_* = \tau_2 - 2\tau_1$ are drawn in Fig. 2. Lines $\tau_1 = $ const and $\tilde{\tau}_* = $ const are plotted by solid curves. Dash and dotted curves correspond to parameters τ_0 and T respectively. The domains Ω_k are marked by the number $k = 1,2,3,4$. Fig. 2 corresponds to the following dimensionless parameters of the robot, see (5): $I_1 = I_2 = 1$, $a = 0.1$, $M_0 = 2$, $\psi_2^0 = 0.5$ rad.

Diagrams similar to Fig. 2 permit for each terminal position (ψ_1^1, ψ_2^1) to determine the type of the control regime, which depend on the domain Ω_k containing the point (ψ_1^1, ψ_2^1). Using lines of equal values we can obtain moments of switching τ_0, τ_1, τ_2 and the operational time T.

It is shown, see [6], that in some limit cases ($a = 0$, or $|\varphi_1^1| \ll 1$, $|\varphi_2^1 - \varphi_2^0| \ll 1$) control laws calculated by the above approach are time optimal. In general case they may be considered as suboptimal.

2. Feedback optimal control for an electromechanical robot

Consider now the manipulator, similar to the one shown in Fig.1, equipped with electromechanical drives D_1, D_2. The drives are located in joints O_1, O_2 and consist of direct-current motors and reduction gears. To describe the motion of this robot it is more conveient to use generalized coordinates $q_1 = \varphi_1$, $q_2 = \varphi_2 - \varphi_1$ rather than φ_1, φ_2. Then Lagrange equations are

$$(A_{11} + 2MaL \cos q_2)\ddot{q}_1 + (A_{12} + MaL \cos q_2)\ddot{q}_2 -$$
$$- 2MaL \sin q_2 \dot{q}_1 \dot{q}_2 - MaL \sin q_2 \dot{q}_2^2 = M_1 \qquad (9)$$

$$(A_{12} + MaL \cos q_2)\ddot{q}_1 + A_{22}\ddot{q}_2 + MaL \sin q_2 \dot{q}_1^2 = M_2$$

$$A_{11} = I_1 + I_2 + J_1 n_1^2 + (M+m)L^2, \qquad A_{12} = I_2 + J_2 n_2,$$

$$A_{22} = I_2 + J_2 n_2^2$$

where J_i, $i = 1, 2$ are moments of inertia of rotors of motors with respect to their axes of rotation; m is a mass of the rotor of the motor D_2; n_i, $i = 1, 2$ is a gear ratio for the drive D_i; parameters M, a, L, I_1, I_2 are defined in section 1. Inductances of armatures of motors are usually small, and transient times for electric currents in rotor curcuits are much less then time of robotic transport operations. Therefore we neglect electric transient processes.

To describe the dynamics of the robot with electromechanical drives it is necessary to supplement equations (9) with equations of voltage balance as well as with relationships between torques M_1, M_2 and currents j_1, j_2 in circuits of motors:

$$R_i j_i + k_i n_i \dot{q}_i = u_i, \qquad M_i = n_i k_i j_i, \qquad i = 1, 2 \qquad (10)$$

Here R_i is the resistance of the armature of the motor D_i; k_i is a constant, u_i is a control voltage.

Inertial characteristics and gear ratios for the majority of industrial robots are such that interaction of different degrees of freedom is weak, and the motion of the system (in dimensionless variables) is described with high accuracy by the following decoupled equations:

$$R_i \ddot{q}_i + k_i \dot{q}_i = u_i, \qquad i = 1, 2 \qquad (11)$$

We consider time-optimal control problem for the simplified system (11) under the constraints $|u_i| \leq u_i^0$, $i = 1, 2$. The terminal state is fixed ($q_1(T) = q_1^1$, $q_2(T) = q_2^1$, $\dot{q}_1(T) = \dot{q}_2(T) = 0$), while the initial state is arbitrary. The optimal feedback control for the system (11) is well known, see e.g. [7]. We apply this control to the original system (9), (10) and study its behaviour.

Figs. 3, 4 show phase trajectories describing two degrees of freedom for the robot Universal - 5.02. Figs. 3, 4 present rotations of the platform and of the arm (respectively) of the manipulator in a horizontal plane. Solid lines correspond to decoupled system (11). Dash lines refer to the original system (9), (10). Analysis of these trajectories shows the applicability of control laws $u_i(q_i, \dot{q}_i)$, $i = 1, 2$, based on a decoupled model (+) for control of the manipulator. It is advisable to use another control law in the neighbourhood of the terminal position, for instance linear feedback control.

Optimal times of transport operations calculated theoretically for the robot Universal-5.02 were compared with experimental data obtained for the robot with standard control unit. The comparison shows that optomization permits to obtain a considerable gain in productivity of the robot.

The problem of time-optimal control of the system similar to (9) was considered in [9], where the nonlinear system of equations of motion was also reduced to decoupled linear equations. However, only open-loop controls were considered in the mentioned paper.

References

1. Dubowsky S., Shiller Z. Optimal dynamic trajectories for robotic manipulators. - In: Fifth CISM-IFToMM Symposium on theory and practice of robots and manipulators. Preprints. Udine, Italy, 1984.
2. Kiriazov P., Marinov P. A method for time-optimal control of dynamically constrained manipulators. - In: Fifth CISM-IFToMM Symposium on theory and practice of robots and manipulators. Preprints. Udine, Italy, 1984.
3. Kahn M.E., Roth B. The near minimum-time control of open-loop articulated kinematic chains. - Journal of Dynamic Systems, Measurement and Control, 1971, Vol. 93, N 3.

4. Luh J.Y.S., Lin C.S Optimum path planning for mechanical manipulators. - Journal of Dynamic Systems, Measurement and Control, 1981, Vol. 102, N 2.
5. Akulenko L.D., Bolotnuk N.N, Chernousko F.L., Kaplunov A.A. Optimal controls of manipulation robots. - In: Preprints of the 9th World Congress of the International Federation of Automatic Control, Budapest, Hungary, July 2 - 6, 1984, Vol. 9, pp. 91 - 95.
6. Avetisyan V.V., Bolotnik N.N., Chernousko F.L. Optimal program motions for a manipilator with two links. - Engineering Cybernetics, 1985, N 3, pp. 123 - 131 (In Russian).
7. Smolnikov L.P. Synthesis of quasioptimal systems of automatic control. - Leningrad, Energiya, 1967 (In Russian).
8. Bolotnik N.N., Kaplunov A.A. Optimization of control and configurations for a manipulator with two links. - Engineering Cybernetics, 1983, No4, pp. 144 - 150 (In Russian).
9. Sato P., Shimojima H., Kitamura Y. Minimum-time control of a manipulator with two-degrees of freedom. - Bull. JSME, 1983. N 218.

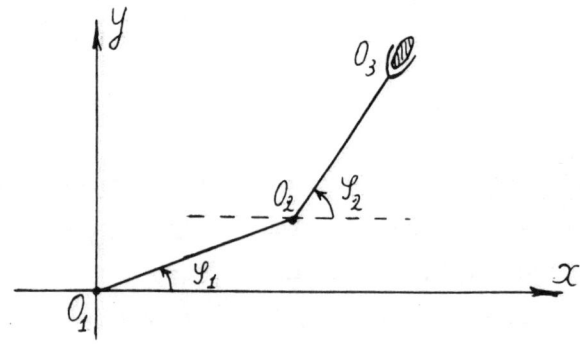

Fig. 1

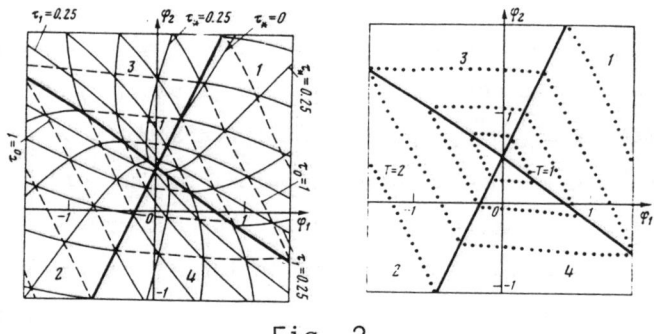

Fig. 2

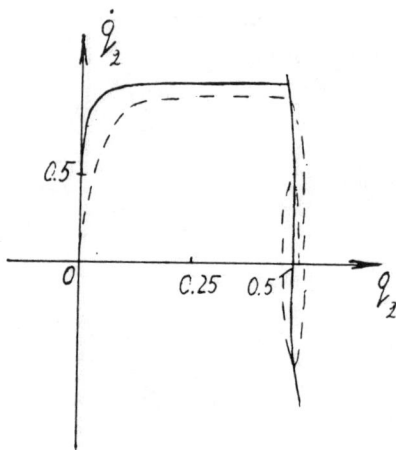

Fig. 3　　　　　　　　　　　　　Fig. 4

Frequency Space Synthesis of a Robust Dynamic Command

A. Oustaloup and B. Bergeon

E.N.S.E.R.B., Université de Bordeaux I, Talence, France

SUMMARY

Real time programming of manipulators implies two distinct, real time operations, namely the *generation* and the *checking* of trajectories defined by an operator. In another contribution, we discussed the real time generation of trajectories by a syntaxer, focussing on an optical, three dimensional syntaxer. Here, therefore, we shall study the real time control of the trajectory and the design of a high performance dynamic command for a manipulator.

Emphasis is placed on robustness, since this is of fundamental importance for dynamic command of articulated mechanical systems, where parameters of the process may vary widely and as fast as do the variables.

I - INTRODUCTION

Robustness is an idea describing the insensitiveness of the stability of a process, hence of its damping, to changes in the parameters. It is thus fundamental in the control of non-stationary processes, where changes in the parameters produce changes in the degree of stability.

An adaptive solution to the problem is to reduce variations of the stability by real time adaptation of the parameters of the regulator to those of the process. This implies either on line estimation of the parameters of the regulator, or on line estimation of those of the process, allowing calculation of estimated parameters of the regulator. The corresponding adjustable regulator is said to be self-adjusting.

The purpose of this paper is to present the principle of robustness in frequency space, and to synthesize in the same space a robust dynamic command based on a fundamental second order process, since this example is frequent in robot and thermal control. The method described here can be generalised to other processes, by adding an appropriate cascade of compensating filters, with the aim of reducing the problem to the same frequency configuration in an open circuit.

Study of this synthesis shows that robustness of such a command may be obtained only through a robust regulator, whose order must be non-integer, thus introducing the idea of a fractional derivative. The frequency synthesis of such a regulator is given.

Successive steps are the choice of the structure of the regulator, the determination of its parameters and finally the synthesis of its structure.

This fractional order regulator can in fact be adapted to some domain of variation of the parameters. When the parameters vary within such a domain, real time adaptation of the parameters of the regulator is unnecessary, contrary to the case of self-adaptive regulators.

The regulator proposed below is distinct from self-adapting regulators. Its obvious advantage is that it does not require on line identification of the process and on line estimation of its parameters. When the parameters vary within a domain compatible with the robustness of the regulator, off line identification of the process is sufficient to adapt the parameters of the regulator.

The present work is a quite new formulation of the remarkable properties of fractional orders which we exhibited about ten years ago.

True, we were at that time interested in the frequency response of fractional order systems, but temporal work being fashionable we also worked in the time domain. However, the time space is not the best one for understanding the remarkable way in which fractional orders do away with the classical dilemma stability-accuracy. It seems, in view of our experience, interesting to reconsider the frequency domain. The formulation is of course different, taking into account our detour in time space.

Indeed, a Black's diagram clearly reveals the relation between robustness and fractional order. This just shows that ideas as new as that of fractional orders must be integrated with the history of this field of work, i.e. Black's work in this case.

II - DEFINITION OF ROBUSTNESS

The term robustness describes the insensitiveness of the damping to changes in the parameters of a system. This insensitiveness can be obtained in practice only for some domain of the parameters. Robustness is greater when the domains are broader.

III - THE FUNDAMENTAL IDEAS OF ROBUSTNESS DESCRIBED IN FREQUENCY SPACE

A *robust dynamic command* is defined in time space by damping independent of changes of the parameters within some domain.

In frequency space this implies that the Black's locus of the open circuit frequency response should be a *segment of a vertical line*, lying between the abscisse $-\pi$ and $-\pi/2$, tangent to the contour of the Black's diagram corresponding to the desi-.ered damping for the process being controlled (figure 1).

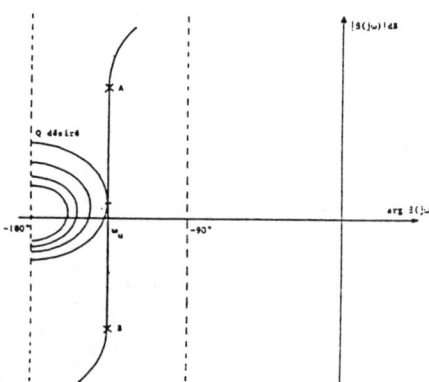

Figure 1 - Robustness illustrated by a Black's diagram. The problem is to synthesize segment AB.

When the parameters take new, constant values, the straight line representing the frequency response in open circuit about the unit gain frequency ω_u moves vertically along itself, remaining tangent to the original contour of damping amplitude. The resonance factor and thus the damping factor, remain unchanged. Clearly, the longer the straight portion AB, the greater the robustness, since the movement of this segment depends on changes of the parameters of the process.

IV - FREQUENCY SPACE SYNTHESIS OF A ROBUST, DYNAMIC COMMAND : INTRODUCTION OF A FRACTIONAL ORDER

The synthesis in frequency space of a robust dynamic command implies synthesis of a Black's locus of the frequency response in open circuit like the one in figure 1, namely an asymptotic frequency behaviour over a wide domain arround the unit gain frequency in open circuit, ω_u. The greater the desired robustness, the longer the asymptote must be.

Consider, for example, a second order, scalar, linear, stationary process, described by the transmittance :

$$G(p) = \frac{G_0}{1 + 2\zeta_0 \frac{p}{\omega_0} + \frac{p^2}{\omega_0^2}} \quad . \quad (1)$$

The modulus and the argument of the frequency response $G(j\omega)$ are given by :

$$|G(j\omega)| = G_0 \left\{ \left[1 - \left(\frac{\omega}{\omega_0}\right)^2\right]^2 + 4\zeta_0^2 \left(\frac{\omega}{\omega_0}\right)^2 \right\}^{1/2} \quad (2)$$

$$\text{Arg}(j\omega) = -\tan^{-1}\left[\frac{2\zeta_0(\omega/\omega_0)}{1 - (\omega/\omega_0)^2}\right] . \quad (3)$$

The regulator likely to provide a robust command (or robust regulator) is supposed to be a "high pass corrector" placed in series with the process to be controlled, in the feedback loop shown in figure 2.

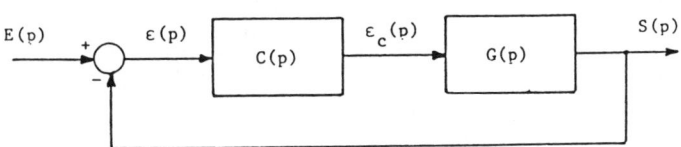

Figure 2 - C(p) is the transmittance of a high pass corrector placed in series in hope of creating a robust regulator : $\varepsilon(p)$ is the error signal and $\varepsilon_c(p)$ is the corrected error signal.

The open loop frequency response is then

$$\beta(j\omega) = C(j\omega) \, G(j\omega) \quad (4)$$

writing $\qquad C(j\omega) = C_0 \, C^*(j\omega) \quad (5)$

where C_0 is the static gain of the corrector ; (4) becomes :

$$\beta(j\omega) = C^*(j\omega) \, C_0 \, G(j\omega) . \quad (6)$$

Figure 3 shows the Black's locci of $G(j\omega)$, $C(j\omega)$, $C^*(j\omega)$, $C_0 \, G(j\omega)$ and $\beta(j\omega)$. C_0 is chosen so that the Black's locus of $C_0 \, G(j\omega)$ has an asymptotic behaviour over a wide range of frequencies about its unit gain frequency ω'_u.

$\beta(j\omega)$ will have an asymptotic behaviour over a wide range of frequencies about ω_u if points N_1, N_2, N_3, N_4, ... lie on the same vertical line between the abcissae $-180°$ and $-90°$. The figure shows that this requires that M_1, M_2, M_3, M_4, ... should also be on a straight line lying between the abcissae 0 and $\pi/2$.

Consequently, the locus of the frequency response of the corrector, $C(j\omega)$, must have an asymptote between 0 and $\pi/2$ (figure 4). Now control of a maximal phase variation between 0 and 90° *must* introduce a *fractional order*, so a truly robust regulator can be achieved only through use of a fractional order corrector.

V - FREQUENCY SYNTHESIS OF A ROBUST REGULATOR (OR A "FRACTIONAL ORDER REGULATOR")

V.1 - STRUCTURE

The structure must lead to a Black's locus of $C(j\omega)$ like that of figure 4, par-

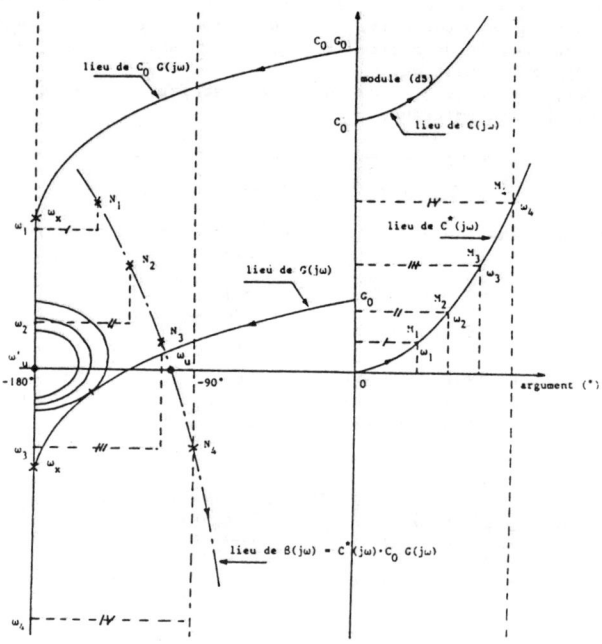

Figure 3 — Black's locci of $G(j\omega)$, $C(j\omega)$, $C^*(j\omega)$, $C_0\,G(j\omega)$ and $B(j\omega)$. ω_x is the frequency beyond which $G(j\omega)$ has asymptotic behaviour ; ω'_n is the unit gain frequency of $C_0\,G(j\omega)$ and ω_u that of $B(j\omega)$. C_0 is the static gain of $C(j\omega)$.

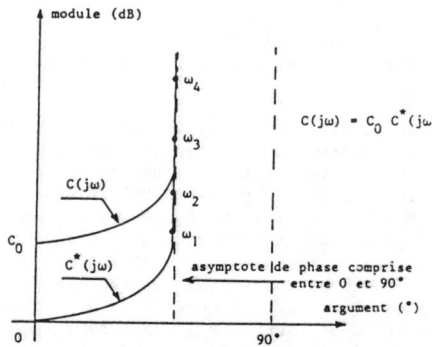

Figure 4 — The locci of both $C(j\omega)$ and $C^*(j\omega)$ must both have asymptotes at phases between 0 and $\pi/2$. Control of asymptotic phase variation on an asymptote between 0 and 90° implies automatically the introduction of a fractional order derivative.

ticularly the phase asymptote lying between 0 and $\pi/2$.
This may be achieved if the transmittance is of the form :

$$C(p) = C_0 \left[1 + \frac{p}{\omega_{n'}} \right]^{n'} ,$$

where $0 < n' < 1$.
The frequency response is then :

$$C(j\omega) = C_0 \left[1 + j \frac{\omega}{\omega_{n'}} \right]^{n'} \tag{7}$$

with modulus and argument :

$$|C(j\omega)| = C_0 \left[1 + \left(\frac{\omega}{\omega_{n'}}\right)^2 \right]^{n'/2} \tag{8}$$

and

$$\arg C(j\omega) = n' \arctan \frac{\omega}{\omega_{n'}} . \tag{9}$$

The corresponding Bode's diagrams are presented in figure 5, for $C_0 > 1$.

V.2 - PARAMETERS OF THE REGULATOR

The parameters C_0, $\omega_{n'}$ and n' of the regulator must be such that it is adapted to the process to be controlled. The parameters should be adapted to the parameters of the process, or in our case, to some domain of values compatible with the robustness of the regulator.

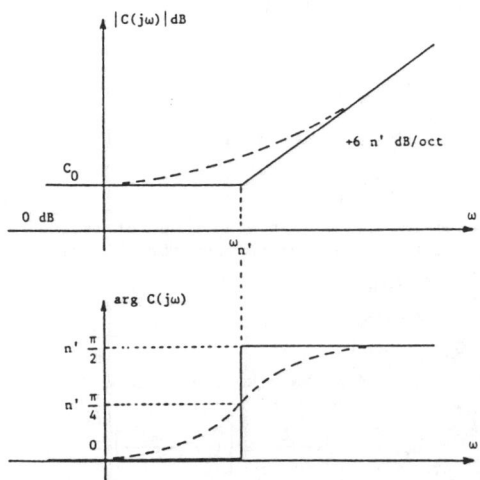

Figure 5 - Amplitude and phase characteristics of a fractional order, high pass regulator ; —— : phase and gain asymptotes ; --- : phase and gain diagrams ; $\omega_{n'}$: transition frequency ; C_0 : static gain ; n' : fractional order $(0 < n' < 1)$; $n'(\pi/2)$: phase asymptote $(0 < n'(\pi/2) < \pi/2)$.

V.2.1. - Choice of C_0

The frequency response $C_0 G(j\omega)$ for $\omega \gg \omega_0$,

$$C_0 G(j\omega) = -C_0 G_0 \left(\frac{\omega_0}{\omega}\right)^2 , \tag{10}$$

leads to
$$|C_0\, G(j\omega'_u)| = C_0\, G_0 \left(\frac{\omega_0}{\omega'_u}\right)^2 = 1, \qquad (11)$$

and finally to the ratio between ω'_u and ω_0 :
$$\frac{\omega'_u}{\omega_0} = (C_0\, G_0)^{1/2}. \qquad (12)$$

Let ω_x be once more the frequency beyond which $C_0\, G(j\omega)$ approximately touches its asymptote at phase $-\pi$. Writing $\omega_x = \mu\omega_0$, we have :
$$\arg C_0\, G(j\omega_x) = \operatorname{arctg} \frac{2\,\zeta_0\,\mu}{\mu^2 - 1} \qquad (13)$$

hence
$$\arg C_0\, G(j\omega_x) = -\pi + \operatorname{Arctg} \frac{2\,\zeta_0\,\mu}{\mu^2 - 1} \simeq -\pi + \frac{2\,\zeta_0\,\mu}{\mu^2 - 1} \qquad (14)$$

and
$$\arg C_0\, G(j\omega_x) + \pi = \frac{2\,\zeta_0\,\mu}{\mu^2 - 1} \qquad (15)$$

giving the distance from the phase asymptote $-\pi$. Write this as $\delta\pi$, where δ is the relative distance, so that
$$\frac{2\,\zeta_0\,\mu}{\mu^2 - 1} = \delta\pi. \qquad (16)$$

This gives a quadratic equation for μ :
$$\mu^2 - 2\,\frac{\zeta_0}{\delta\pi}\,\mu - 1 = 0, \qquad (17)$$

with determinant
$$\Delta' = \frac{\zeta_0^2}{\delta^2\,\pi^2} + 1 > 0 \qquad (18)$$

and two real roots of which only one has any physical meaning,
$$\mu = -\frac{\zeta_0}{\delta\pi} + \left[1 + \frac{\zeta_0^2}{\delta^2\,\pi^2}\right]^{1/2}. \qquad (19)$$

Robustness, or the existence of a wide range of frequencies about the unit gain frequency ω_u where the behaviour is asymptotic, implies :
$$\omega_x \ll \omega'_u \qquad (20)$$
or
$$\mu\omega_0 \ll \omega'_u \qquad (21)$$
or
$$\omega'_u/\omega_0 \gg \mu. \qquad (22)$$

Using (12) and (19), we see that the static gain C_0 of the regulator must satisfy
$$(C_0\, G_0)^{1/2} \gg -\frac{\zeta_0}{\delta\pi} + \left[1 + \frac{\zeta_0^2}{\delta^2\,\pi^2}\right]^{1/2}. \qquad (23)$$

V.2.2. - *Choice of* ω_n,

Let ω_y be the frequency beyond which $C(j\omega)$ is very close to its phase asymptote at $n'(\pi/2)$. Writing $\omega_y = \mu'\omega_{n'}$ the distance from the asymptote can be expressed as :
$$n'\frac{\pi}{2} - \operatorname{arctg} \frac{\omega_y}{\omega_{n'}} = \delta'\, n'\,\frac{\pi}{2}, \qquad (24)$$
or
$$n'\frac{\pi}{2} - \operatorname{arctg} \mu' = \delta'\, n'\,\frac{\pi}{2}, \qquad (25)$$

or $\quad\quad\quad\quad\quad \arctan \mu' = n' \dfrac{\pi}{2} (1 - \delta'),$ (26)

whence $\quad\quad\quad\quad\quad \mu' = \tan\left[(1 - \delta') \dfrac{\pi}{2} n'\right]$ (27)

where δ is the relative distance from the asymptote.
Robustness here implies :

$$\omega_y \ll \omega'_u,$$ (28)

or $\quad\quad\quad\quad\quad \mu' \omega_{n'} \ll \omega'_u$ (29)

or, using (12) : $\quad\quad \mu' \omega_{n'} \ll (C_0 \, G_0)^{1/2} \omega_0,$ (30)

hence $\quad\quad\quad\quad (C_0 \, G_0)^{1/2} \dfrac{\omega_0}{\omega_{n'}} \gg \mu',$ (31)

and finally, using the definition of μ' :

$$(C_0 \, G_0)^{1/2} \dfrac{\omega_0}{\omega_{n'}} \gg \tan\left[(1 - \delta') \dfrac{\pi}{2} n'\right].$$ (32)

The transition frequency ω'_n must satisfy this condition. ω_0 and G_0 are given. C_0 is determined by (23). δ' is fixed, of the order of a few percent. n' is determined by the method suggested below.

V.2.3. - *Choice of n'*
- *Approximate method*

We make the approximation by supposing that for a given damping factor ζ, the resonance factor Q is given by the relation applying to a fundamental, second order system

$$Q = \dfrac{1}{2\zeta \sqrt{1 - \zeta^2}}$$ (33)

so that $\quad\quad\quad\quad Q_{dB} = 20 \log Q.$ (34)

The value of n' then follows immediately from the distance $n'(\pi/2)$ (figure 6) between the critical point (0 dB, -180°) and the vertical tangent to the amplitude contour Q_{dB} in the Black's diagram.

- *Rigourous method*

We have shown elsewhere (1) that the damping factor ζ is related exactly to n', the order or the regulator, by

$$\zeta = -\cos \dfrac{\pi}{2 - n'},$$ (35)

hence $\quad\quad\quad\quad n' = 2 - \dfrac{\pi}{\arccos(-\zeta)}.$ (36)

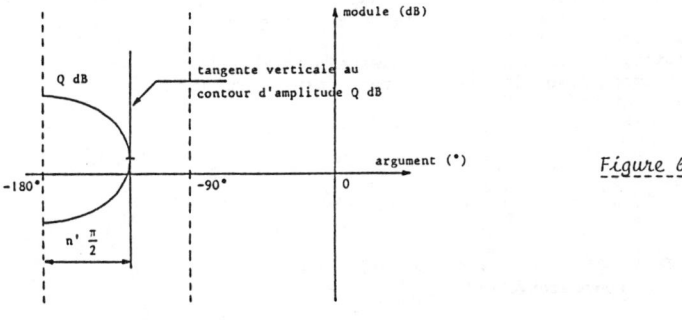

Figure 6

N.B. : It is also shown in (1) that the resonance factor is given by :

$$Q = \frac{1}{\sin (2 - n') \frac{\pi}{2}}, \qquad (37)$$

or

$$\frac{1}{Q} = \sin (2 - n') \frac{\pi}{2} \qquad (38)$$

Replacing $2 - n'$ by the expression derived from (35) gives the relation between Q and ζ :

$$\frac{1}{Q} = \sin \frac{\pi^2}{2 \arccos (-\zeta)}. \qquad (39)$$

The previous method may then be applied, with the exact form (39) replacing approximation (33).

V.3. - SYNTHESIS OF THE STRUCTURE OF THE REGULATOR

The synthesis of the transmittance

$$C(p) = C_0 (1 + \tau_n, p)^{n'}, \qquad (40)$$

is approached in an original way using a recursive distribution of localised elements, based on the idea of a fractal distribution. We outline this approach here, referring the reader to an exhaustive discussion to be published in a book we are writing.

The synthesis is based on a succession of operations on the functional representation of terms in the binomial expansion of $(1 + \tau_n, p)^{n'}$.

The first step is to group together large blocks of terms for which the inductive relation between terms may be taken to be independent of the rank in the block. Two cases occur, namely that when the boundary term is included in the block and that when it is excluded.

The next step is to represent each block by a transmittance defined by the sum of its member terms.

These steps lead to an infinite distribution of first order transmittances, both low and high pass. Extension to infinite distributions of impedances and admittances and introduction of a recursive distribution inspired by the concept of a fractal, lead to eight recursive arrangements of elementary RC and RL cells which meet the required synthesis.

- 1 parallel arrangement of RC series cells
- 1 series arrangement of RC parallel cells
- 1 cascade arrangement of RC gamma cells
- 1 cascade arrangement of CR gamma cells
- 1 parallel arrangement of RL series cells
- 1 series arrangement of RL parallel cells
- 1 cascade arrangement of RL gamma cells
- 1 cascade arrangement of LR gamma cells.

In practice, the number of cells is finite. The ratio between the transition frequencies of successice cells must make a compromise between conflicting requirements :

- It should be close enough to unity to limit modulations of gain and phase associated with the asymptotic frequency response, due to the connection of the cells.
- It must be sufficiently greater than unity to cover a wide range of frequencies with a reasonable number of cells.

This compromise is achieved by taking a ratio between 5 and 10 and a number of cells also between 5 and 10. Satisfactory results are obtained commonly with 8 cells and a ratio 6.

The non integer order n' is given by the relation (1) :

$$n' = \frac{1}{1 + \frac{\log \eta}{\log \alpha}}, \qquad (41)$$

where the parameters α and η are the ratios between the resistances and the capacitances of two consecutive cells.

CONCLUSION

Adaptive commands work on the principle of adapting the parameters of the regulator to those of the process to be controlled. This requires real time adaptation of the regulator's parameters to changes in those of the process. This in turn implies on line estimation of the regulator's parameters, or those of the process so that the regulator's parameters may be estimated.

Dynamic, fractional order control works on the same lines, namely adjusting the parameters of the regulator to those of the process. In this case however, adaptation is obtained with one set of control parameters not to one set of values of the process parameters (the desired result), but to a whole range of the process parameters. The regulator remains adapted during any change of the process parameters within this range. Real time adaptation of the regulator's parameters is therefore pointless, as is on line identification of the process : this is an obvious advantage.

In fact, if off line identification of the process shows that its parameters vary within a range compatible with the robustness of the regulator (i.e. adaptive ability), the regulator's parameters may be calculated and stored in the command once and for all. If, on the other hand, off line identification reveals that the parameters vary too widely, for a given robustness of the regulator, on line identification is necessary to allow real time adaptation of the regulator's parameters to those of the process. Such identification need not be very accurate, because the regulator is naturally adapted to a range of parameters.

Finally, if we grant that a regulator is robust when its parameters are adapted to a range of process' parameters and that the wider the range the greater the robustness, then we must admit that a fractional order regulator is not just robust, but *very robust*, judged by the very wide range of process' parameters to which any one setting of its own parameters is adapted.

AKNOWLEDGEMENT : We would like to thank E. IRVING for warmly encouraging us to revive and set forth in terms of robustness, one of the remarkable properties of fractional orders we found in 1975, when designing a 3/2 order feedback for a continuous wave dye laser (1).

BIBLIOGRAPHY

1 - A. OUSTALOUP - Systèmes asservis linéaires d'ordre fractionnaire, Théorie et pratique (Editions MASSON, Paris 1983).

2 - A. OUSTALOUP - Linear feedback control systems of fractional order between 1 and 2 (IEEE International Symposium on Circuits and Systems, Chicago, Illinois, April 27-29, 1981)

3 - A. OUSTALOUP et B. BERGEON - Commande d'ordre fractionnaire et introduction à la commande adaptative et d'ordre fractionnaire (Colloque "Commande adaptative", Grenoble, Novembre 1984)

4 - B. BERGEON et A. OUSTALOUP - Commande adaptative et d'ordre fractionnaire d'un manipulateur à deux degrés de liberté (Colloque "Commande Adaptative", Grenoble, Novembre 1984)

Structure Strategy Problem on a Redundant Manipulator

H. Asama, M. Onosato and H. Yoshikawa

Department of Precision Machinery Engineering, Faculty of Engineering, the University of Tokyo, Tokyo, Japan

SUMMARY

   Structure strategy is discussed concerning kinematic geometry of manipulator structure, which is a strategy to synthesize motion of a redundant manipulator developed for the purpose of automation of maintenance tasks in complicated and restricted environments. The motion synthesis process is considered as a planning flow in several subsystems. Subsystems for several phases of planning are presented, which synthesize manipulator motion according to the task commands from human operators to be fulfilled as a whole. Subsystems consist of an element task planning system, a trajectory planning system, a performance evaluation system, and an environment model system which provides the environment information. An integrated manipulator motion synthesis system is proposed for synthetic planning of the manipulator motion. It includes a knowledge base describing experience and knowhow, and is linked with CAD/CAM or sensors providing design information, manufacturing information, and information about the states of objects and obstacles. Examples of structure strategy to plan the manipulator's motion determining the redundant parameters' motion are shown to verify the effectiveness of motion inference applying knowledge engineering.

1. INTRODUCTION

   Maintenance automation is far behind design automation and manufacturing automation, while maintenance work including inspection, diagnosis, and repairing, have increasingly greater importance, and maintenance costs become rapidly expensive in contrast with lower manufacturing costs. We aim at automation of maintenance in accordance with new trends of maintenance technology. We developed a general purpose maintenance mobile robot, and an articulated manipulator with redundant degrees of freedom installed on the robot. However, special strategy in the motion synthesis planning process is required for the redundant manipulator operating in a complicated and restricted environment to fulfill various tasks. In the present paper, the integrated motion synthesis system for the manipulator is proposed, and the strategy with regard to the kinematic structure of the manipulator is discussed, which is a strategy to determine the redundant parameters' motion. Examples of trajectory and motion planning systems are shown, applying knowledge engineering to establish an efficient planning of the manipulator's motion.

2. MAINTENANCE MANIPULATOR

2.1 Manipulator Dexterity

Maintenance technology is characterised by such unique items as interdisciplinarity, diversity, obscurity, and non-repeatability (1), and that introduces the conclusion that a combination of single use machines cannot realise a flexible robot manipulating system. Therefore, new requirements should be given to the design process of a maintenance robot. Supposing that the environments where the robot runs and manipulates are such locations as containment vessels of nuclear power plants, a mobile robot system named "A MOOTY" (2) was developed. But dexterity is necessary not only for the vehicle but also for the manipulator installed on it. The manipulator's dexterity can be considered as versatility of orientation of the end effector and size of workspace, or ability of obstacle avoidance.

Extra degrees of freedom, representing redundancy, are required to be incorporated while constructing articulated manipulator structure, in order to realise arbitrary trajectory associated with position and orientation. Incorporation of too many degrees of freedom, however, causes a control problem. We decided to develop an articulated manipulator with nine degrees of freedom as the optimal number (3).

2.2 Requirements for Structure Strategy

Mastering structural dexterity, we should determine the strategy to control the manipulator. We define structure strategy as a strategy for planning motion, prescribing redundant parameters according to the environmental information and the task commands. The required specifications of the system are shown below.

(1) Task commands (final purposes of the task) can be instructed intentionally and intelligibly using natural language by human operator.

(2) Global planning is required to cope with a diverse and intricate environment. Such local motion planning methods as sensor driven or potential driven strategy are ineffective.

(3) A unique environment modelling system is required to plan trajectories or motions of the manipulator, detecting collisions with obstacles scattered in the working environment.

(4) Performance of the manipulator can be evaluated, and the function of the manipulator can be recognised explicitly.

And from these requirements the following items are introduced.

(5) Information about entities representing objects or obstacles is required from design information or manufacturing information.
(6) Information about states representing objects or obstacles is required from the environment.

We consider that the function of the manipulator can be recognised by operation experience, not deduced from common theories independent of different robot configurations. Therefore, we discuss only the function of the 9 DoF manipulator of original structure and geometry.

2.3 Motion Synthesis Process

We can obtain the image of the whole process of the manipulator's motion synthesis from the requirements. The process can be illustrated as in fig. 1.

First of all, an element task planning system is required as an interface between the system and the human operator, or an interpreter of the task commands issued by the operator. It comprehends the task purposes, and translates the commands into sets of elementary task sequences, each of which the manipulator should fulfill. Next, a trajectory planning system is required. Focusing on the motion of the objects restricted by the surrounding units (they can also be regarded as a part of the environment), and taking performance of the manipulator into account, we can generate the trajectories of the end effector. Moreover, a motion planning system is required, which generates the manipulator's motion referring to the obstacles restricting the manipulator's motion along the prescribed trajectory. It is also efficient to recognise the performance of the manipulator in motion planning. For the trajectory planning or motion planning system, an ad hoc modelling system is required which describes the environment including not only objects, tools and jigs, but also obstacles. The planned motion of the manipulator must be evaluated again to confirm the execution feasibility of the motion, compared with the manipulator performance kinematically, dynamically and structurally using various models.

3. SUBSYSTEMS FOR MOTION SYNTHESIS

3.1 Element Task Planning System

This is a subsystem intended to interpret the task commands as essential purposes of maintenance tasks, and devolve into the sequence of element tasks each of which corresponds to task level command of the manipulator, and which are usually given to maintenance operators in the form of check lists of predetermined periodic inspection, or repairing requests from users. Generally speaking, interdisciplinarity of the maintenance technology permits the calling up of various types over various levels of instruction maintenance tasks, and that makes it obscure to define the so-called maintenance task, and to find essential aims or intentions of the task we want to be achieved by the robot automatically. Without analysing maintenance works we can never recognise the purposes of tasks for such an intelligent robot, and can never describe the concrete requirements for the system which operates the manipulator to treat manufactured machines in the maintenance stage in their life cycles.

We analysed a task of disassembling RW resin transfer pumps as an example, which are at work in the nuclear power plant in large numbers, by MTM (Methods Time Measurement) which is a kind of motion study of PTS (Predetermined Time Standard) in the field of industrial engineering.

From the result we can abstract and represent knowledge which is supposed to be applied when we introduce motion sequences. We call a primitive unit of the sequence an element task corresponding to the one motion of the subject. Frame, which is one of the knowledge representation methods in knowledge engineering, is quite effective for representing such knowledge using a prominent method of default, class concepts, inheritance, and attached procedure. In order to generate the element task sequences according to the given task instruction from an operator, they should be inferred, applying internal knowhow (experiments and common sense when handling mechanical parts), referring external data or knowledge, such as design or manufacturing information (from CAD/CAM), operation manuals, and environment models (from sensors). We are implementing this system based on frame representation currently.

3.2 Trajectory Planning System

Obtaining the element task sequence mentioned above, we can know what the target object of the present element task is and in which direction it should be moved. If we have enough knowledge about both the target object and the end effector, the condition of the trajectory of the end effector, which guarantees the element task to be achieved, can be obtained. To discuss concretely how the condition could be derived, imagine a key which is used to fasten a gear along a shaft, and a typical end effector which has parallel faces for grasping, respectively as shown in fig. 2. First, each surface of the key is checked as to whether it is open or not. If there is any pair of open surfaces facing, that is chosen as a pair of grasping surfaces. And if there is none, it should be concluded that it is impossible to grasp the key with the specified end effector. Secondly, we have to decide what position and orientation of the end effector could be taken as the grasping state. In this case, the contact between the key and the end effector is a face-to-face one, and to keep this contact three of six parameters which specify the position and orientation of the end effector must be fixed relative to the key. Thus, we can automatically decide three of the six parameters which specify the position and orientation of the end effector at each state of the key when moving. The other three parameters - two are the translation along the grasped surface and the remaining one is the rotation around the normal to it - are not fixed and any value of parameters corresponding to the freedom of motion might be allowed so far as the relationship between the key and the end effector is concerned. These redundancies give flexibility to the trajectory planning. If we indicate all parameters of the end effector beforehand without consideration of the feasibility, it would often be impossible to grasp the key with specified position and orientation. Adopting some adequate evaluation or strategy which evaluates the manipulator's performance, control method, collision of the end effector, etc, we could choose candidates from a set of the free parameters and decide the trajectory of the end effector.

3.3 Environment Model

Particular inventions are necessary to construct an environment model system intending generation of the manipulator's trajectory and motion. It is possible for both the static and the dynamic environments to be modelled and referred to for collision detection and management of the environment transition. The former denotes a set of the task objects and the tools or jigs which will be modified as the task sequence is proceeded with. The latter denotes the set of obstacles scattered in the surrounding environment of the objects and subjects. Concerning the dynamic environment model, not only such geometries as shapes or dimensions but also structural relations between the parts presented by some method as a directed graph should be represented, and representation based on frames is quite qualified for such models. Concerning the static environment model, we propose a phased model system which provides a set of models depending on the obstacle in order to detect collisions effectively. The referred model is picked up from the set and modified according to phases of collision possibilities. Basically, the collisions are detected by intersection between primitive shaped models. We do not describe this system in detail here.

3.4 Motion Planning System

To solve the motion problem in the environment where arbitrary shaped obstacles are scattered in large numbers, intentional planning applied as above is required. We evaluated the static kinematic performance by simulation in terms of the geometric structure of the manipulator, and clarified and analysed the function of the manipulator, which is defined as the ability to achieve the given task in a restricted environment.

In this system the input is a trajectory of the end effector, and the output is the motion of the manipulator. The trajectory is a sequence (function of time) with respect to position and orientation of the end effector from the starting point to the destination. The motion is a sequence with respect to control variables (joint angles or displacements) of the manipulator. Assuming selection of a single solution from the possible symmetrical solutions in the synthesis problem, prescription of redundant parameters implies motion designation. In this paper we take only displacement (not velocity or acceleration) into account because we derive the motion from static environment information.

It is quite free to interpret parameters equivalent to redundancy. We assign original redundant parameters focusing on such characteristics of the manipulator's structure that joint 2-3-4 and 7-8-9 can be considered as spherical joints (fig.3) for successive RPR intersecting at one point, equivalent to a spherical pair with three degrees of freedom. The parameters are:

(1) joint displacement of 1st prismatic joint (s_1),
(2) rotating angle of the plane (α),
(3) rotating angle of 4-bar link on the plane (β),

where the plane denotes the particular plane on which links L2, L3, and L4 lie (fig. 4). Using the parameter links, movement is symbolised, and they are convenient to analyse the manipulator's function intuitively. Fixing two redundant parameters, we experimented to produce collision-noncollision maps on a simulator. The maps represent the relation between one redundant parameter and the position of the end effector on the way from the starting point to the destination. Fig. 5 shows an example considering α as the redundant parameter. If we find path-tracing non-collision points from top to bottom, it is a possible prescription candidate, namely a possible motion. Pigeonholing the maps obtained by altering the states of the task and environment, it is recognised that specific patterns appeared regularly, which are shown in fig. 6.

While it is very time-consuming to create the map according to the task and/or the environment modification, we can estimate the map pattern and infer simply from the pattern the possible motion, namely the optimal value of the redundant parameter by judging conditions concerning the states. The conditions are as follows:

(1) the obstacle becomes obstructive as the manipulator moves,
(2) the obstacle exists near the end effector's trajectory,
(3) the obstacle exists near the manipulator's base,
(4) the obstacle approaches as the manipulator moves.

We should prepare procedural knowledge which produces the conditions from the given states with task commands and environment restriction, and the declarative knowledge which produces the map pattern and the best motion from the conditions by the former inference. We constructed the former by frames and its control mechanisms, and the latter by production rules. Concerning the former knowledge, referring the task command and environment restriction both of which are prepared previously by instances of the class frames, the system infers the conditions and sets them in the condition frame using the mechanism of attached procedures (fig. 7). These frames are implemented in the frame operating system FRODIRD II (4) developed in our laboratory. Concerning the latter knowledge, the rules are prepared in the knowledge base, and are obtained as the result of pigeonholing the simulation results, and the motion is inferred as facts standing for the conditions which are added to the working memory of the production system. These rules are implemented in the production system PROSCODE-2 (5), also developed in our laboratory. Fig. 8 shows examples of the rules, and fig. 9 shows execution examples of FRODIRD II and PROSCODE-2.

Appending the frames or rules including meta-knowledge to resolve the conflicts when the number of obstacles becomes huge, can make the system extensible. Its total execution CPU time on SUN-2 workstation is 500msec at most until the system finds the motion, which is about 1/60 of CPU time taken to build a map for path searching.

4. INTEGRATED MANIPULATOR MOTION

From the process and subsystems discussed in the preceding section, integration with CAD/CAM and diverse sensors should be required to achieve the motion synthesis of an advanced manipulator. Fig. 10 shows the construction of the integrated manipulator motion synthesis system.

Each of the subsystems applies plentiful and extensive knowledge of experience or knowhow. Multiform knowledge is dependent on design information which offers attributive values of the objects, and sensed information which indicates the states of the environment including objects and obstacles. The subsystems specify not only the specific knowledge base peculiar to planning in respective systems, but also the common knowledge base typified by knowledge about the manipulator's performance. They also specify the environment model which will be constructed with design information and sensed information. The supervisor manages systematically the sequential planning through the several subsystems, and introduces the information flow.

The functional motion synthesis based on structure strategy of the redundant manipulator can be achieved using planning flow from task commands to motion sequences, and the systematic planning can be realised in the integrated manipulator motion synthesis system with knowledge bases and an advanced modelling system.

5. CONCLUSION

We propose the construction of the systematic trajectory and motion generation system for maintenance manipulator with redundancy. We implemented some parts of the system using such knowledge representation methods as frames or rules in knowledge engineering by analysing the manipulator's function with a simulator. And we corroborated the effectiveness of this approach in structure strategy problem with the result of inference time to find the manipulator's motion.

REFERENCES

[1] Yoshikawa, H.: "Necessity and Potential of Maintenance Technology — Automation of Maintenance —," 7th European Federation of National Maintenance Society, Venice, May 1984.
[2] Arai, T., et al.: "A Stair-Climbing Robot for Maintenance : 'A MOOTY'," Seminar on Remote Handling Equipment for Nuclear Fuel Cycle Facilities, International Atomic Energy Agency (IAEA), OECD Nuclear Energy Agency, Harwell, UK, Oct. 1984.
[3] Asama, H., Yoshikawa, H.: "Development of a Metamorphic Manipulator with 9 Degrees of Freedom," Proc. of 15th ISIR, pp. 415-422, Tokyo, Sep. 1985.
[4] Takeshige, A., Tomiyama, T., Yoshikawa, H.: "An application of Frame System to CAD," in V. Hubka & Programme Committee (ed.), Theory and Practice of Enginnering Design in International Comparison, Heurista, pp. 763-770, Zurich, 1985.
[5] Tomiyama, T., Yoshikawa, H.: "Requirements and Principles for Intelligent CAD Systems," in J. S. Gero (ed.), Knowledge Engineering in Computer-Aided Design, pp. 1-23, Amsterdam, 1985.

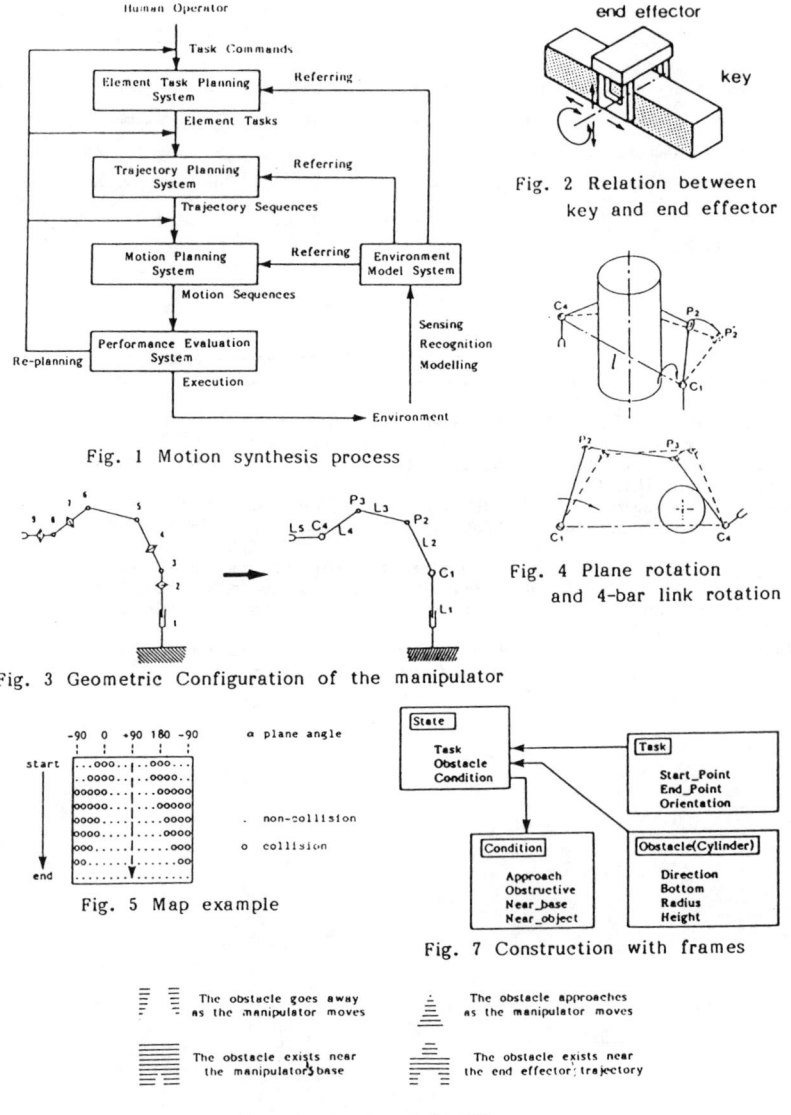

Fig. 1 Motion synthesis process

Fig. 2 Relation between key and end effector

Fig. 3 Geometric Configuration of the manipulator

Fig. 4 Plane rotation and 4-bar link rotation

Fig. 5 Map example

Fig. 6 Specific patterns of maps

Fig. 7 Construction with frames

Structure Strategy Problem on a Redundant Manipulator

```
<1:Obstacle becomes obstructive as the manipulator moves>
<2:Obstacle exists near the end effector trajectory>
<3:Obstacle exists near the manipulator base>
<4:Obstacle approaches as the manipulator moves>

if <1> and <3>
   then write (<Give up>)

if ~<1> and ~<2> and ~<3> and ~<4>
   then write (<Select plane angle alpha as J6 becomes farthest>)

if <1> and ~<2> and ~<3> and <4>
   then ask (<Find solution at the end point keeping J6 nearest>)
```

Fig. 8 Examples of production rules

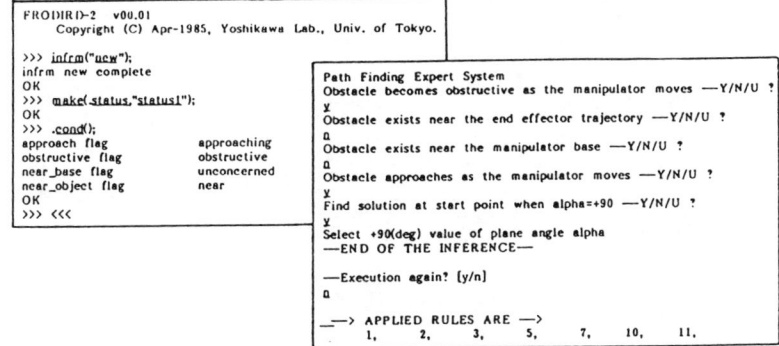

Fig. 9 Execution examples of FRODIRD II and PROSCODE-2

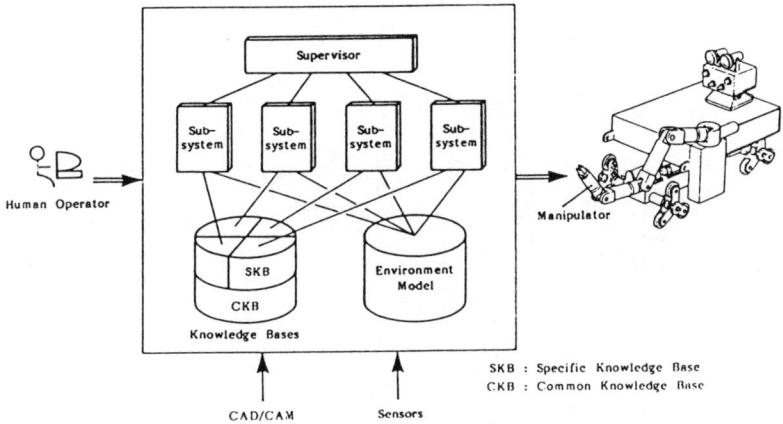

Fig. 10 Integrated Manipulator Motion Synthesis System

Participants

Australia

Samuel Emery, Department of Mechanical and Industrial Engineering, The University of Melbourne.

Bulgaria

Kiriazov P., Institute of Mechanics and Biomechanics, Bulgarian Academy of Sciences, Sofia.
Konstantinov Michael, Robotics Department, Higher Institute of Mechanical and Electrical Engineering, Sofia.
Lorer Maxim, Institute of Mechanics and Biomechanics, Bulgarian Academy of Sciences, Sofia.

France

André Guy, IRISA, Campus Universitaire de Beaulieu, Rennes.
Bidaut Philippe, Laboratoire de Mécanique et Robotique, Université Pierre et Marie Curie, Paris.
Fayet Michel, INSA de Lyon.
Fournier Raymond, CEA, Campus Universitaire de Beaulieu, Rennes.
Gaillard Jean, LRE, Campus Universitaire de Beaulieu, Rennes.
Gravez Philippe, CAL, Lille.
Guinot Jean-Claude, Laboratoire de Mécanique et Robotique, Université Pierre et Marie Curie, Paris.
Jutard Alain, INSA de Lyon.
Kessis Jean-Jacques, Laboratoire de Robotique et Intelligence Artificielle, Université de Paris 7.
Lallemand Jean-Paul, Laboratoire de Mécanique des Solides, Université de Poitiers.
Redarce T., Laboratoire Automatique Industriel, INSA, Lyon.

FRG

Rauch Jochen, Institute B of Mechanics, University of Stuttgart.

GDR

Barthel Helmuth, Astr. Ltr Projektirung, VEZB Werkzeug Maschinen Kombinat "7 Oktober", Berlin.
Bögelsack Gerhard, Section Gerätetechnik, Technische Hochschule Ilmenau.
Heiman Bodo, Institute of Mechanics, Berlin.
Luck Kurt, Technische Universität Dresden.
Palm Reiner, Central Institute of Cybernetics and Information Process, Academy of Sciences of the GDR.
Schultz Armin, VEB Kombinat Poligraph, Leipzig.

Hungary

Csaki Csaba, Technical University of Budapest.
Gabor Stepan, Department of Mechanical Engineering, Technical University of Budapest.
Filemon Elizabeth, Technical University of Budapest.

Israel

Shiller Zvi, Department of Mechanical Engineering, Massachusetts Institute of Technology, Cambridge, USA.

Italy

Bertozzi Armilla, CISM, Udine.
Bianchi Giovanni, Institute di Meccanica Applicata alle Machine, Politecnico di Milano.
Ceccarelli Marco.
Parenti Castelli.
Rovetta Alberto, Dipartamento di Meccanica, Politecnico di Milano.
Siciliano Bruno, Dipartamento di Informatica e Sistemistica, Universitá di Napoli.

Japan

Asada Haruhiko, Department of Applied Mathematics and Physics, Kyoto University.
Asama Hajime, Department of Precision Machinery Engineering, The University of Tokyo.
Fukuda Toshio, The Science University of Tokyo.
Hirose Shigeo, Tokyo Institute of Technology.
Matsushima Kozo, Institute of Engineering, University of Tsekuba.
Sugano Shigeki, Department of Mechanical Engineering, Waseda University, Tokyo.

Poland

Buśko Zbigniew, Warsaw University of Technology.
Frączek Janusz, Warsaw University of Technology.
Galicki Miroslaw, Department of Applied Mathematics, Higher College of Engineering, Zielona Góra.
Gielo-Perczak Krystyna, Warsaw University of Technology.
Golaś Andrzej, Academy of Mining and Metallurgy, Cracow.
Jarominek Wladyslaw, IBIP, Polish Academy of Sciences.
Jaworek Krzysztof, Warsaw University of Technology.
Kasińki Andrzej, Technical University of Poznań.
Kędzior Krzysztof, Warsaw University of Technology.
Knapczyk Józef, Technical University of Cracow.
Kosiński Witold, Institute of Fundamental Problem of Technic, PAS, Warsaw.
Kowalski Andrzej, Institute of Precision Mechanics, Warsaw.
Kozlowski Krzysztof, Technical University of Poznań.
Kuzan Pawel, Warsaw University of Technology.
Macukow Bohdan, Institute of Mathematics, Warsaw University of Technology.
Malczyk Grzegorz, Information Division, CBKO, Pruszków.
Marszalec Elzbieta, Polytechnic of Lublin.
Marszalec Janusz, Polytechnic of Lublin.
Morecki Adam, Warsaw University of Technology.
Olędzki Andrzej, Warsaw University of Technology.
Puchalka Tadeusz, Technical University of Poznań.
Samborska Malgorzata, Institute of Precision Mechanics Warsaw University of Technology.
Tomaszewski Karol, Academy of Mining and Metallurgy, Cracow.
Wojnarowski Józef, Technical University of Gliwice.
Zielińska Teresa, Warsaw University of Technology.

Romania

Duditza Florea, University of Brasov.

Switzerland

Jau Bruno, JPL, California Institute of Technology, Pasadena.

United Kingdom

Bicker Robert, University of Newcastle-upon-Tyne.
Maunder Leonard, University of Newcastle-upon-Tyne.
Prentis Jim, Engineering Department, University of Cambridge.
Vernon Geoffrey, Mechanisms and Machines Group, Liverpool Polytechnic.

USA

Dubowsky Steven, Department of Mechanical Engineering, M.I.T. Cambridge.
Duffy Joseph, Center for Intelligent Machines and Robotics, University of Florida, Gainesville.
Hollerbach John, MTI, Cambridge.
Khatib Oussama, Artificial Intelligence Lab., Stanford University.
Mason Matthew, Computer Science Department, Carnegie-Mellon University, Pittsburgh.
Michalowski Stefan, Department of Mechanical Engineering, Standford University.
Raibert Marc, Carnegie-Mellon University, Pittsburgh.
Salisbury Kenneth, M.I.T.-AI Laboratory, Cambridge.
Waldron Kenneth, Department of Mechanical Engineering, OSU, Columbs.
York Carl, System Development Fundation, Palo Alto.

USSR

Bolotnik Nikolai, Institute of Problems in Mechanics, USSR Academy of Science, Moscow.
Korenovsky Vladimir, Mechanical Engineering Research Institute, USSR Academy of Sciences, Moscow.
Nachapetjan Evgeny, Institute for the Study of Machines, USSR Academy of Sciences, Moscow.
Sergeev Vladimir, Mechanical Engineering Research Institute, USSR Academy of Sciences, Moscow.
Tzeitlin Yurij, The Institute of Geotechnical Mechanics, Ukrainian Academy of Sciences.

Yugoslavia

Kircansky N., Mihailo Pupin Institute, Beograd.
Leanarcić Jadran, Stephan Institute, Ljubljana.
Martinović Radwan, Musiuski Faoultet, Novi Sud.
Stokić Dragan, Mihailo Pupin Institute, Beograd.
Vukobratović Miomir, Mihailo Pupin Institute, Beograd.

Achevé d'imprimer en avril 1987
sur les presses de l'imprimerie Laballery
58500 Clamecy
Dépôt légal : avril 1987
Numéro d'impression : 703081

RAYMOND H. FOGLER LIBRARY
DATE DUE

BOOKS ARE SUBJECT TO
RECALL AFTER TWO WEEKS

OCT 1 9 1987